Quantitative Scanning Electron Microscopy

Quantitative Scanning Electron Microscopy

Edited by

D. B. HOLT, M. D. MUIR,

P. R. GRANT and I. M. BOSWARVA

Imperial College of Science and Technology,
London, England

1974

ACADEMIC PRESS

London New York San Francisco

A Subsidiary of Harcourt Brace Jovanovich, Publishers

ACADEMIC PRESS INC. (LONDON) LTD.
24/28 Oval Road,
London NW1

United States Edition published by
ACADEMIC PRESS INC.
111 Fifth Avenue
New York, New York 10003

Library of Congress Catalog Card Number: 745645
ISBN: 0–12–353850–5

PRINTED IN GREAT BRITAIN AT
THE UNIVERSITY PRESS, ABERDEEN

Contributors

J. A. BELK *Metallurgy Branch, The Royal Military College of Science, Shrivenham, Swindon, Wiltshire, England.* (p. 389)

H. E. BISHOP *Materials Development Division, Atomic Energy Research Establishment, Harwell, England.* (p. 41)

A. V. CREWE *The University of Chicago, Chicago, Illinois, U.S.A.* (p. 65)

D. J. DINGLEY, *H. H. Wills Physics Laboratory, University of Bristol, England.* (p. 487)

D. A. GEDCKE *ORTEC Inc., Oak Ridge, Tennessee, U.S.A.* (p. 403)

D. W. GIBBARD *Image Analysing Computers Ltd., Melbourn, Royston, Hertfordshire, England.* (p. 75)

A. GOPINATH *School of Electronic Engineering Science, University College of North Wales, Bangor, Caernarvonshire, North Wales.* (p. 95)

P. R. GRANT *Department of Geology, Imperial College of Science and Technology, Prince Consort Road, London S.W.7, England.* (p. 287)

D. B. HOLT *Department of Metallurgy, Imperial College of Science and Technology, Prince Consort Road, London S.W.7, England.* (pp. 213, 335)

A. HOWIE *Cavendish Laboratory, University of Cambridge, England.* (p. 183)

*B. JEFFERIES *Department of Mineralogy and Petrology, University of Cambridge, England.* (p. 517)

M. P. JONES *Department of Mineral Technology, Imperial College of Science and Technology, Prince Consort Road, London S.W.7, England.* (p. 531)

D. C. JOY *Department of Metallurgy, University of Oxford, England.* (p. 131)

J. V. P. LONG *Department of Mineralogy and Petrology, University of Cambridge, England.* (p. 517)

M. D. MUIR *Department of Geology, Imperial College of Science and Technology, Prince Consort Road, London S.W.7, England.* (pp. 3, 287)

J. W. STEEDS *H. H. Wills Physics Laboratory, University of Bristol, England.* (p. 487)

H. YAKOWITZ *Metallurgy Division, Institute for Materials Research, National Bureau of Standards, Washington, D.C., U.S.A.* (p. 451)

* Present address: *Department of Geology, University of Edinburgh, Scotland.*

Preface

In the decade since commercial scanning electron microscopes first became available, the modes of operation and the applications have expanded rapidly.

Initially, attention was confined almost exclusively to the exploitation of the special features of secondary emission. Micrographs of such emission compare with those obtained by reflected light microscopy, but have inherently a much greater depth of focus and even in the earliest instruments, a point to point resolution better than the best optical microscope. Among the earlier accessories were specimen current amplifiers and picture manipulating modules. Very rapidly, modes of operation such as selected area diffraction, cathodoluminescence, scanning transmission, voltage contrast and X-ray analysis became commonly ancillary, if not, in some cases, dominant.

These innovations were realized quite early in the history of the commercial scanning electron microscopes. Thus with the basic hardware available latterly, there has been a marked trend toward the development of empirical and theoretical methods for the quantitative interpretation of the wide range and vast amount of information available from multi-mode scanning electron microscopes.

This book collates accounts of the current state of development of quantitative scanning electron microscopy. Many leading authorities in this field have contributed not only statements of the present level of development but also their views on the directions and form of future advances.

A number of chapters arose from lectures given at an Advanced Summer School on Scanning Electron Microscopy held at Imperial College. Others were specifically requested contributions which we found necessary to round out certain aspects or to act as links between sections.

Although we have aimed at the adoption of an acceptable set of common symbols throughout, this has not been wholly achieved. However, lists of symbols and, for easy reference, of contents precede each chapter, and relevant references are listed at the end of each chapter.

In an attempt to ensure that this book will be useful for teaching and for study, though an acquaintance with the basic features of the scanning electron microscope is assumed, the treatment herein is relatively complete, consecutive and, we hope, comprehensible.

Special emphasis has been given to new developments and new instruments. Because of their recent development and potential, three chapters have been devoted to electron channelling and Kossel patterns. In contrast, electron probe analysis using crystal spectrometers has long been at a high level of development and has therefore only been summarized in a single chapter. However, references to the standard literature make this latter chapter complete.

We believe that future developments will increasingly be concerned with electronic signal processing and computer analysis of SEM data. Therefore, we have included chapters on image analysis by computer, on energy dispersive X-ray spectrometers and the employment of multichannel analysis in such instruments, and on automatic computer controlled point analysis, and stereological analysis in the X-ray mode. We hope that this book will be found valuable by expert and novice scanning electron microscopists alike.

Much thought and effort has been given to the preparation of this volume by the individual authors, whom we would like to thank for their prompt and conscientious work.

We would like to thank Professor J. G. Ball for his interest in this project in particular and together with Professor J. Sutton for support and encouragement for scanning electron microscopy at Imperial College.

June, 1974

D. B. Holt
M. D. Muir
P. R. Grant
I. M. Boswarva

Contents

PART I

Instrumental

Chapter 1

Recent Instrumental Developments

M. D. MUIR

Geology Department, Royal School of Mines, London, England

Symbols

J = the current density in the image
J_0 = the current density at the cathode surface
e = the electronic charge
V = potential difference between the cathode and the point where the image is formed
k = Boltzmann's constant
T = absolute temperature in the cathode
α = the semi-angle of the cone of rays which converge to form a point on the image
β = the useful current density in the crossover—the brightness of the source
V_p = the voltage at the crossover
d_0 = the Gaussian probe size, i.e. the total demagnification of the beam's first crossover
C_s = the spherical aberration coefficient of the final lens
α_1 = the beam semiangle of convergence at the first crossover
C_c = the chromatic aberration coefficient of the final lens
E_0 = the mean energy of the beam
δE_0 = the energy spread of the beam passing through the lens
δz = the spacing of the two stigmatic foci
λ = the de Broglie wavelength of the electrons
α_i = the half angle of convergence of the beam at the specimen
$C_s\alpha_1^3$ = the diameter of the disc of least confusion due to the use of lenses with finite numerical apertures
I_b = the incident probe current
d_{min} = the minimum probe diameter
A = a constant approximately equal to one
T_s = picture exposure time

N = number of lines in square picture (field of view)
c = fractional threshold contrast
R_V = radius of the virtual cathode source
R = actual radius of the tip
E_t = average transverse energy of electrons leaving the tip
V_1 = voltage between the tip and first anode
V_0 = the desired potential of the gun
E_s = field enhancement factor due to the effective lowering of the potential barrier—the Schottky effect
ϕ = the work function
F = electric field magnitude (V cm^{-1})
J_F = cathode current density for field emission
$J(T)$ = cathode current density at temperature T
ΔE_{tot} = total energy distribution in the final beam
ΔE_{th} = thermal energy spread
$\Delta E_{interact}$ = Boersch effect
subscript d = diffraction disc
subscript c = chromatic aberration
E = total energy
d_{CT} = limit of CTEM resolution
$\bar{d}(t)$ = limit of resolution in a specimen of thickness t_0
d_{eff} = effective beam diameter
d_s = multiple scattering
d_f = depth of focus
d_c = chromatic scatter
t = depth of structure of interest in specimen
M = mass of coating material of density ρ
T_f = film thickness in nm
R = distance in cm from evaporation source to sample
θ = angular deviation from the normal of the source to specimen surface
$\sigma(t)$ = surface charge density at time t
$\sigma(0)$ = surface charge density at time zero assumed = zero
σ_m = maximum surface charge density
η_r = specimen resistivity
ε = specimen permittivity
M_g = magnification
β_{SE} = secondary emission coefficient
A_v = area of visual display
η = backscattering coefficient
$\mathscr{V}$ = a hole created by electron impact
X = a higher level
Y = a lower energy
$E_{\mathscr{V}}$ = the energy level of the initial hole
$E_{\mathscr{X}}$ = the binding energy at level X
$E_{\mathscr{Y}}$ = the binding energy at level Y
$\bar{a}_{\mathscr{V}}$ = the mean Auger yield
$\bar{\omega}_{\mathscr{V}}$ = the X-ray fluorescence yield

I. Introduction

This book is intended as a summary of present understanding of quantitative scanning electron microscopy; for introductory accounts of scanning microscopes, their operation, and reviews of their applications in particular fields, reference must be made elsewhere.

An introductory account of the field is given by Thornton (1968), and Oatley (1972) describes the instrument, its design and the factors limiting its performance. The proceedings of the annual SEM Symposia held at the Illinois Institute of Technology (SEM 1968 through SEM 1974) provide a useful starting point for entry into the detailed SEM literature and the volumes contain bibliographies of the whole field of SEM investigations.

Here I have tried to summarize the rapid evolution of the instrument in recent years and to introduce the considerations involved in intercomparing the related techniques of scanning transmission electron microscopy (STEM) and conventional transmission electron microscopy (CTEM). These topics are further dealt with in Chapters 3 and 7 below.

II. The Scanning Electron Microscope

Although the scanning electron microscope (SEM) has as long a history as the transmission electron microscope (CTEM) both instruments being derived from the laboratories of Knolle and Ruske (Zworykin, Hillier and Snyder, 1942), its real development and exploitation began after the development of the electron probe family of instruments (Castaing, 1951). These are basically bipartite instruments: they comprise two systems with the specimen at the interface. The electron optical column provides the scanned electron beam "illumination" of the specimen. The second section of the SEM comprises at minimum a signal collection and display system. This section may be expanded for quantitative work into a signal detection, amplification, data processing and read-out system. Recent developments in instrumentation for the first section of the microscope are discussed in this chapter. The remainder of the book is devoted to the various quantitative aspects of the second section of the SEM.

In the earliest electron probe instruments, the electron beam did not scan. This was because the prime function of these instruments was the generation of X-rays for elemental analysis in metallurgical and mineralogical specimens. The X-rays were emitted from a volume near the specimen surface and could be diffracted by a crystal spectrometer on the Rowland Circle. Knowledge of the conditions limiting Bragg reflections of the X-rays from the diffracting crystal enables a quantitative determination to be made of the element present in the excited volume of the specimen. Moving the crystal round the Rowland circle enables X-rays of different wavelengths and hence different elements to be identified. The geometry of the take-off angle of the X-rays is critical and for quantitative electron probe microanalysis depends upon having the angle of incidence of the electron beam precisely normal to the specimen surface. Because of these geometric limitations, it was advantageous for Castaing (1951; also Castaing and Guinier, 1949) to use a static electron beam in his electron probe.

However useful point analyses may be, it is obviously faster to be able to look at relatively large areas of the specimen surface, and electron probes designed after 1959 (Cosslett and Duncumb, 1956; Duncumb and Melford, 1959) all possess scanning coils in the electron optical system. The beam scans in a raster or square, and by scanning a larger or smaller area, the ultimate signal from the specimen gives a lower or higher magnification when enlarged to a constant size in the display system. The later electron probes could also use the single element X-ray output to generate the display signal, thus enabling distribution maps of single elements to be produced.

The other elements of the electron optical column of the SEM are to be found in the CTEM as well as the electron probe. The electron source is still commonly a thermionic tungsten filament in an electron gun. In such a gun, the electrons are emitted from a cathode, and accelerated by a field produced by the anode, usually at a positive potential in the order of 1 kV with respect to the cathode. A third electrode lies between the anode and the cathode and is negative in potential with respect to the cathode (Fig. 1). This last electrode is generally described as the Wehnelt electrode (or grid, or modulator).

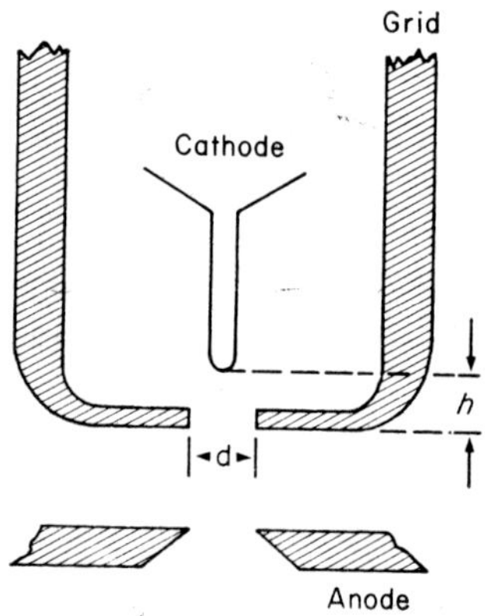

FIG. 1. Diagram of a triode gun with a tungsten hairpin filament. d = the aperture in the Wehnelt electrode (called the Grid in the diagram); h = filament height from the Wehnelt. (After Haine and Einstein, 1952.)

This gun, the triode gun, is assumed to have a planar filament, and electrons leaving different point sources along the filament have a Maxwellian distribution of initial velocities which causes them to crossover at a point beyond the anode. This crossover can be regarded as a small electron source, and the demagnified image of this crossover forms the electron probe impinging on the specimen. The useful current density in the crossover is given by β—the maximum permitted brightness according to Langmuir's Law (1937) which gives for the upper limit of the current density in the image:

$$J = J_0[(eV/kT)+1] \sin^2\alpha \tag{1}$$

where

J = the current density in the image
J_0 = the current density at the cathode surface
e = electronic charge
V = potential difference between the cathode and the point where the image is formed
k = Boltzmann's constant
T = absolute temperature of the cathode
α = the semi-angle of the cone of rays which converge to form a point on the image.

β is expressed in units of amps per square metre per steradian and can be calculated from the following equation (Oatley, 1972):

$$\beta = J_0 eV_p/\pi kT \tag{2}$$

where V_p is the voltage at the crossover.

The image of the crossover region is demagnified through (usually) three magnetic lenses, although electron probes and some commercial SEMs use two. As with any type of magnetic lens, those in the SEM are subject to spherical and chromatic aberrations and astigmatism. However, since all the lenses are demagnifying, only the aberrations of the final lens need be taken into account (Joy, 1973) and the probe size actually obtained, d, may be written as:

$$d^2 = d_0^2+(\tfrac{1}{2}C_s\alpha_1^3)^2+\left(C_c\alpha_1\frac{\delta E_0}{E_0}\right)^2+(\alpha_1\delta z)^2+\left(\frac{1{\cdot}22\lambda}{\alpha_i}\right)^2 \tag{3}$$

where

d_0 = the Gaussian probe size, i.e. the total demagnification of the beam's first crossover;
C_s = the spherical aberration coefficient of the final lens;
α_1 = the semiangle of convergence of the beam on the first crossover;
C_c = the chromatic aberration coefficient of the final lens;
E_0 = the mean energy of the beam;
δE_0 = the energy spread of the beam passing through the lens;
δz = the spacing of the two stigmatic foci;

λ = the de Broglie wavelength of the electrons which is inversely proportional to the square root of the energy of the beam;

α_i = the half angle of convergence of the beam at the specimen;

and

$C_s\alpha_1^3$ = the diameter of the disc of least confusion due to the use of lenses with finite numerical apertures;

$\frac{1{\cdot}22\lambda}{\alpha_i}$ = the diameter of the disc of least confusion as limited by diffraction considerations;

$\delta z\alpha_1$ = the diameter of the disc of least confusion when astigmatic factors are minimized.

$\frac{C_c\delta E_0\alpha}{E_0}$ = the diameter of the disc of least confusion when the high voltage supplies are not perfectly stable and lenses are not perfect thus causing a spread of energies of the beam about a mean value.

In a well set up microscope, the chromatic and astigmatic contributions are negligible compared with the other factors, and the equation can be simplified to:

$$d^2 = d_0^2+(\tfrac{1}{2}C_s\alpha_1^3)^2+\left(\frac{1{\cdot}22\lambda}{\alpha_i}\right)^2 \tag{4}$$

The incident probe current is given by:

$$I_b = 2{\cdot}5\beta\alpha_1^2d^2-(\tfrac{1}{2}C_s\alpha_1^3)^2-\left(\frac{1{\cdot}22\lambda}{\alpha_i}\right)^2 \tag{5}$$

and thus it is clear that use of too small a probe size may give a probe current insufficient for the purpose for which the SEM is to be used.

III. Electron Sources

At the time of writing, there are two basically different types of electron source. The first of these is the thermionic emitter, and the most commonly used version of thermionic cathode is the tungsten hairpin. This is operated by heating the filament directly by passing a current through it. This activates a few of the electrons near the Fermi surface to a sufficiently high energy to enable them to escape over the surface potential barrier. The thermionic gun is usually of a triode design as described above. However, in recent years, pointed filaments, i.e. solid rods of a suitable material, fabricated to a fine point have been used, with some advantages becoming apparent. Tungsten pointed filaments are available, but offer no very great advantage over the hair-pin type of filament because of their high work function. Plomp (1972) gives a table which shows the work function for various potential cathode materials (Table I) and it is clear that tungsten is very considerably worse then the other materials considered.

TABLE I.

Compound	Melting point (°C)	Work function (V)
YB_6	2300	2·2
LaB_6	2200	2·7
ZrC	3530	3·2
TaC	3800	3·2
W	3370	4·5

(After Plomp, 1972.)

This factor led several workers to consider some of these other emitters as possible cathodes, and Broers (1969) described the experimental and estimated characteristics of a lanthanum hexaboride rod type of electron gun. This gun gave an appreciable gain in brightness, and because it, therefore, gave a larger number of electrons in a small spot size, better resolving power became possible because of the improved signal to noise ratio.

Further developments of this type of gun have been reported by Ahmed (1972) who shows how not only gun brightness but rod lifetime have been greatly improved (Fig. 2). Ahmed and Nixon (1973) describe how different boride guns give better performances compared with the conventional tungsten hairpin gun.

An unusual development of the heated cathode is the indirectly heated cathode of le Poole *et al.* (1972). Their cathode is a thin, straight tungsten wire a small fraction of which is indirectly heated by an electron beam to a temperature just below the melting point. At this much higher temperature, 3500°K compared with the normal emission temperature of 2800°K, much more efficient emission can take place. They have contrived an ingenious mechanical device for moving the filament away from the heating electron beam after it has reached a critical thickness, and it is possible that this type of gun, which requires only normal vacuum and operating conditions may be used in the SEM of the future.

Thermionic cathodes can be operated in a vacuum of about 10^{-5} Torr, for the tungsten hairpin type, and a vacuum of 10^{-6}–10^{-7} Torr for the pointed rod LaB_6 type cathodes. The second important type of electron source, the field emission source, suffers from the disadvantage that it will only operate under ultra-high vacuum conditions, typically 5×10^{-10} Torr or better, and this leads inevitably to difficulties in microscope design and operation. However, this type of source has many advantages when compared with the thermionic cathode, and most manufacturers are pressing on with development of field emission guns for commercial SEMs.

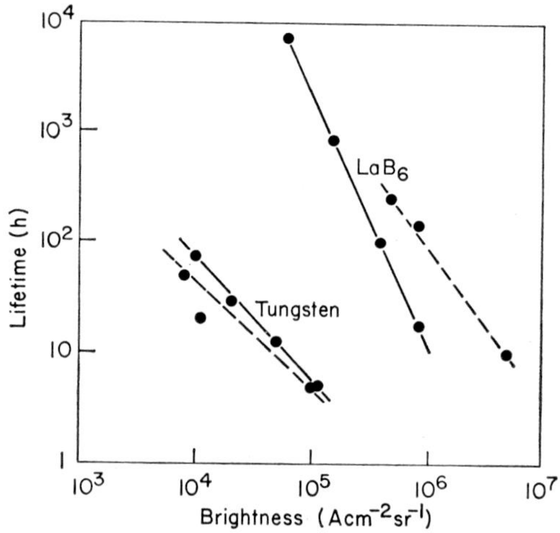

FIG. 2. Comparison of a tungsten hairpin filament with a LaB_6 rod cathode. Brightness is plotted against lifetime, and in the case of the LaB_6 cathode, the experimental results are better than those calculated theoretically. (After Ahmed, 1972.)

In its simplest form, the field emission gun is a two electrode or diode gun. Figure 3 (Smith, 1972) is a schematic diagram of such a gun. If a voltage of 1–3 kV is applied between the surface of the emitter, usually a crystallographically oriented tungsten rod with a tip radius about 200 Å, and the anode, a field is produced at the emitter surface which gives rise to emission by tunnelling through the potential barrier. The emission current density produced in this way is extremely high compared with that produced by thermionic emission. It is also non-uniform, being strongly influenced by the crystallographic orientation of the emitter at the point of emission, and certain orientations, e.g.

(111) and (310) (see Swann and Smith, 1973) are considerably more efficient than others, although emission efficiency is dependent both upon the accelerating potential and upon the vacuum conditions.

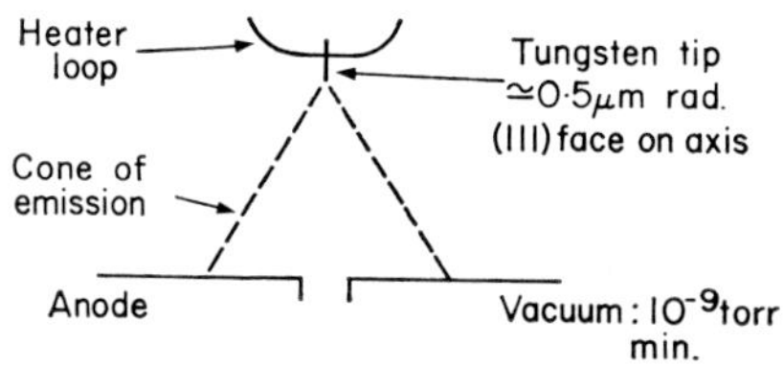

FIG. 3. Diagram of a diode field emission gun. (After Smith, 1972.)

In comparing various types of electron source, the most important parameters are minimum probe diameter, and the brightness or emission current density of the source. Other parameters such as the required field of view, picture recording time and contrast required have also to be taken into account. Smith (1972) gives the following relationship for the probe diameter:

$$d_{\min} = AC_s^{\frac{1}{4}}\lambda^{\frac{3}{4}}\left\{1{\cdot}52\,.\,10^{-6}\frac{N^2}{c^2T_s}\frac{T}{J_0}+1\right\}^{\frac{3}{8}} \quad \text{cm} \tag{6}$$

where

$d_{\min}$ = minimum probe diameter
A = a constant approximately equal to one
C_s = coefficient of spherical aberration of final probe forming lens (cm)
λ = wavelength of electrons
T_s = picture exposure time (sec)
N = no. of lines in square picture (field of view)
c = fractional threshold contrast
T = cathode temperature (°K)
J_0 = cathode emission current density (A cm^{-2}).

Substituting typical values, $N = 400$, $c = 0{\cdot}1$, $T_s = 100$ sec, and assuming $A = 1$:

$$d_{\min} = C_s^{\frac{1}{4}}\lambda^{\frac{3}{4}}\left(0{\cdot}24\frac{T}{J_0}+1\right)^{\frac{3}{8}} \quad \text{cm} \tag{7}$$

If $0{\cdot}24\ T/J_0$ is sufficiently small, then $d_{\min}$ approaches $C_s^{\frac{1}{4}}\lambda^{\frac{3}{4}}$ (the resolution parameter) which depends only on the gun accelerating voltage and the spherical aberration coefficient of the final probe forming lens.

T/J_0 for the tungsten hair-pin is about 2800/2, and $d_{\min} = 9\times$ resolution parameter, whereas for the pointed cathode (e.g. LaB_6) type of gun, $d_{\min} = 2{\cdot}4\times$ resolution parameter. If we assume that the resolution parameter is about 20 Å (C_s = 2 cm, V = 20 kV), then d_{min} (tungsten hairpin) = 180 Å, and for the LaB_6 = about 50 Å. Because the field emission gun is not (usually) heated—the so-called cold cathode—T/J_0 for the field emission source is about three orders of magnitude better than that for the thermionic cathode. $0{\cdot}24\,T/J_0+1$ consequently approaches unity and d_{min} becomes more or less equal to the resolution parameter (Crewe, 1970). Plainly, then for high resolution work, the cold cathode field emission source is likely to give rise to the smallest spot size available. However, because the effective diameter of the field emission source is small (typically of the order of 100 Å) the advantage of the field emission gun decreases with increasing probe diameter until at a probe diameter of 1000 Å, more current can be obtained from a thermionic emitter (Fig. 4, after Cleaver and Smith, 1973). Thus for X-ray microanalysis, conductive mode and cathodo-luminescence studies in the SEM, where high current density is a prime requirement, and resolution is restricted anyway because of electron spreading in the specimen, it would be advantageous to use a thermionic source, such as the LaB_6 rod type emitter.

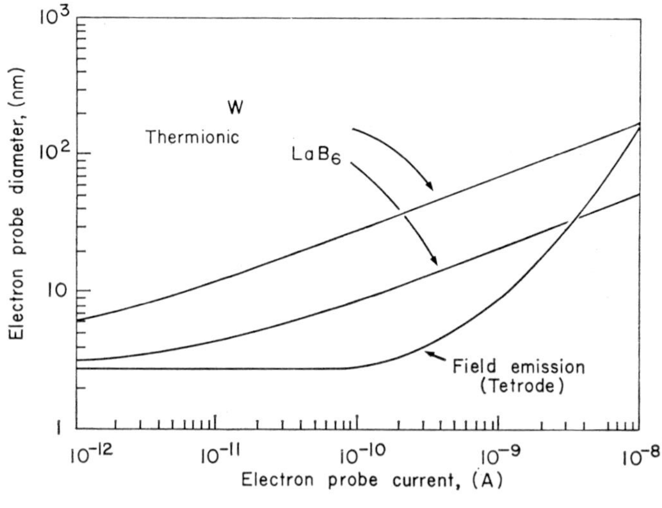

FIG. 4. Comparison of field emission and thermionic probe forming systems in terms of beam current, and probe diameter. (After Cleaver and Smith, 1973.)

Extensive research and development work is being carried out on improving field emission sources. An early improvement on the type of field emission source described above was that designed by Crewe *et al.* (1968) who were able to combine theories of field emission and electron optics with ultra-high vacuum technology, and construct a very satisfactory field emission gun using a two electrode system (Fig. 5). The brightness of the gun is proportional to the radius of the

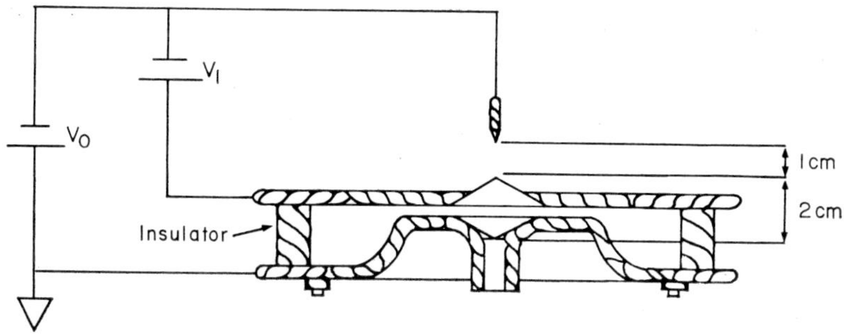

FIG. 5. Diagram of a triode electron gun for field emission. The cathode is shown as a small solid rod at the top of the gun. The distance between the Wehnelt and the anode is small. (After Crewe *et al.*, 1968.)

virtual cathode source. If we assume that the cathode is spherical and that are no aberrations then:

$$R_V = R\left(\frac{E_t}{V_1}\right)^{\frac{1}{2}} \tag{8}$$

where

R_V = radius of the virtual cathode source
R = actual radius of the tip
E_t = average transverse energy of electrons leaving the tip
V_1 = voltage between the tip and first anode.

For both a spherical and paraboloid tip, with an emission angle of less than 0·1 rad, aberrations caused by the field between the cathode and first anode are negligible compared with those in the later stages of the gun (Plomp, 1972). Spherical and chromatic aberrations in the gun, and diffraction at the limiting aperture increase R_V, and hence decrease the brightness. Using the parameters of Crewe *et al.*'s (1968) design, and the optical properties given in Fig. 6, with an emission angle of $1{\cdot}5 \times 10^{-3}$, then a probe diameter of 95 Å can be calculated. Using a similar two electrode lens with a different geometry, a twofold improvement in the coefficient of spherical aberration over Crewe's design is claimed (Munro, 1972). Further lens improvements have been suggested by Plomp *et al.* (1968), by Veneklasen and Siegel (1970) who make use of an electrostatic lens for the first anode; and by Septier (1966) who designed an asymmetric lens with a minimum coefficient of spherical aberration value of 4 f.

To obtain the inherent high brightness of the field emission source, the beam must have a crossover with a very small aperture angle and beyond the crossover the beam must be accelerated to a higher potential. In the two electrode system, V_1 is the voltage between the cathode and the first anode, and V_0 is the desired potential achieved by giving a post crossover acceleration using the second electrode. Combining the geometry of the system (see Fig. 6) with the ratio V_0/V_1, we can obtain an image of the source after the second electrode (Plomp, 1972). Hence if we require an image of the source at a certain position (i.e. in focus on the specimen) then the geometry controls V_0/V_1. Since the value of V_1 is determined by the required emission current, then V_0 is fixed. Thus the two electrode system provides less flexibility than is generally required in the SEM.

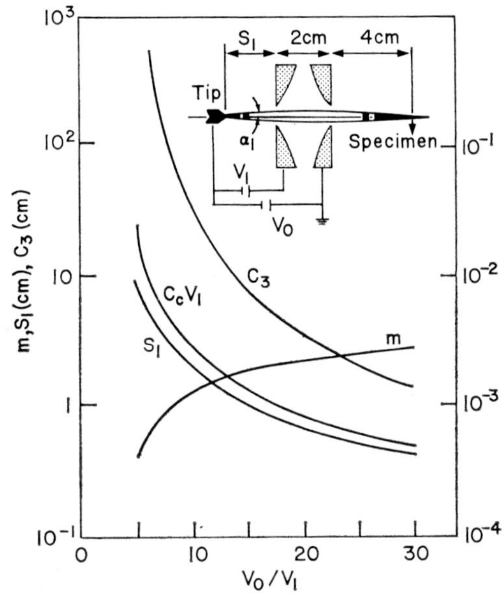

FIG. 6. Optical properties of the gun shown in Fig. 5. (After Crewe *et al.*, 1968.)

Cleaver and Smith (1973) have suggested improvements on the diode field emission gun namely the triode gun (Fig. 7) and the tetrode gun (Fig. 8). In the triode gun, the emission current is determined primarily by the potential of the Wehnelt and can be regarded as being independent of the accelerating potential. It appears to offer considerably greater flexibility than the diode gun, and Cleaver and Smith (1973) believe that it is likely to be most suitable for a general purpose instrument. The tetrode gun has optical properties significantly poorer than the triode gun (Fig. 9 and Fig. 10) and in this case, the probe brightness is reduced to about 2% of the source brightness compared with about 15% in the case of the triode gun. The principal limitation on the performance of the triode gun is the spherical aberration coefficient of the condenser lens, while with the tetrode gun, aberrations in the second accelerating system are of greater importance in controlling the final probe size. Figure 11 shows the comparative performance of the tetrode and triode guns. From this it is clear that the triode gun produces its highest brightness at relatively high probe currents (10^{-9}–10^{-8} A).

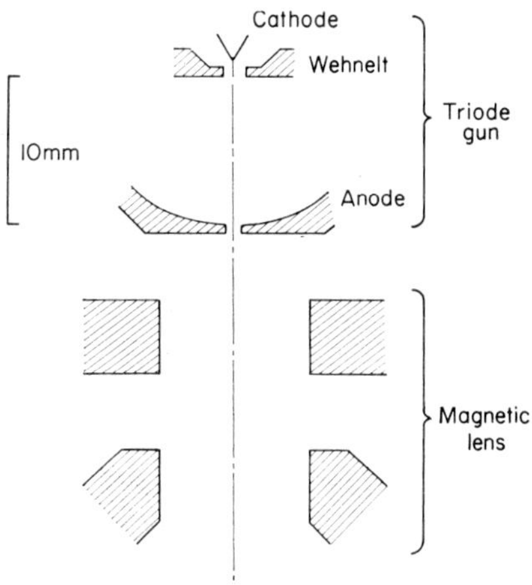

FIG. 7. Diagram of a triode field emission gun with different geometry from Crewe's gun. (After Cleaver and Smith, 1973.)

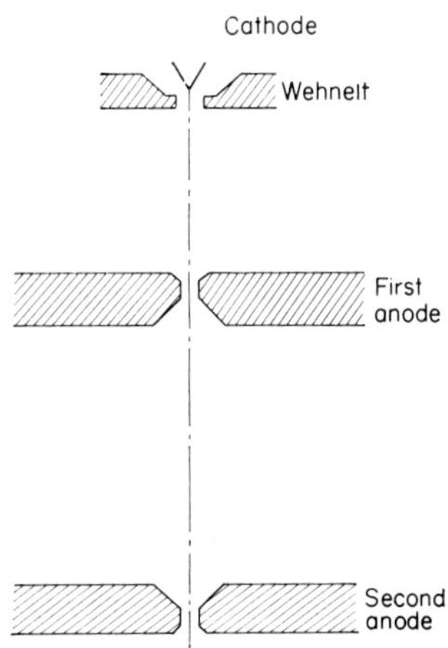

FIG. 8. Diagram of tetrode gun. (After Cleaver and Smith, 1973.)

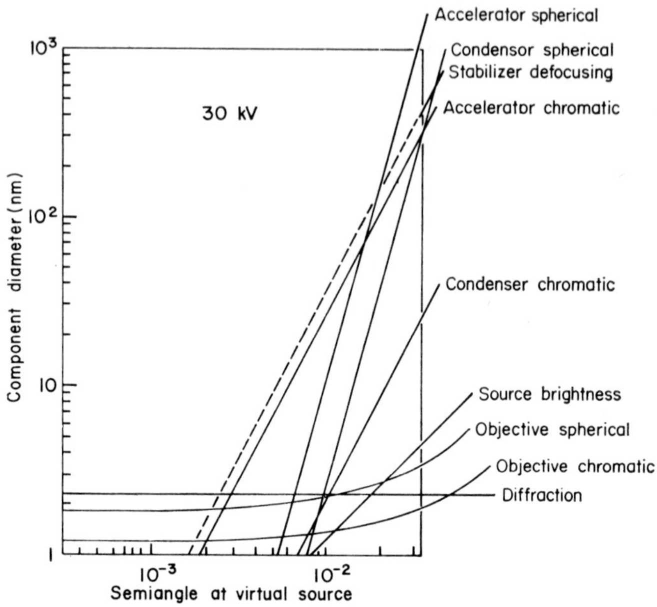

FIG. 9. Optical properties of the triode gun illustrated in Fig. 7. (After Cleaver and Smith, 1973.)

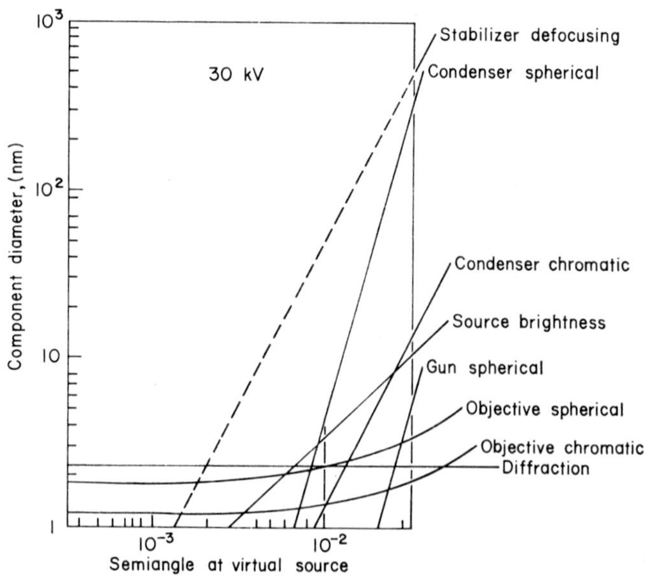

FIG. 10. Optical properties of the gun in Fig. 8. (After Cleaver and Smith, 1973.)

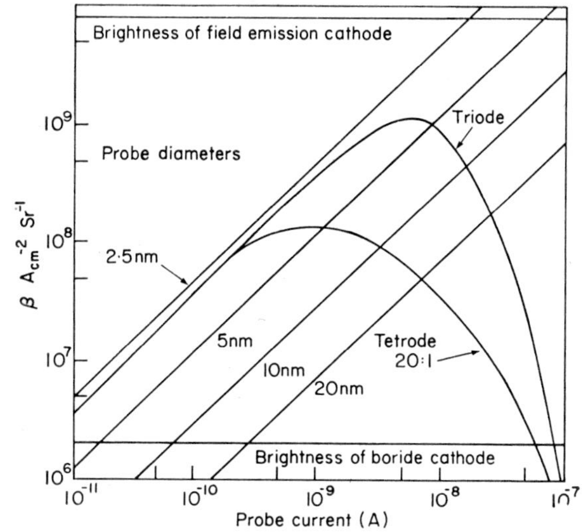

FIG. 11. Brightness of the probe-forming systems of the triode and tetrode field emission guns as a function of the probe current, and diameter. (After Cleaver and Smith, 1973.)

Other improvements in field emission sources will undoubtedly come from alternative materials being used in the fabrication of the emitters. The current density of a thermionic emission source at the cathode (J_0) is given by:

$$J_0 = 120 \,.\, T^2 \, E_s \,.\, (-\phi/kT) \tag{9}$$

where

T = temperature (°K)
E_s = field enhancement factor due to the effective lowering of the potential barrier—the Schottky effect
ϕ = the work function
k = the Boltzmann constant
$$E_s = \exp\left[\frac{(1{\cdot}441 \times 10^{-7}F)^{\frac{1}{2}}}{kT}\right]$$
with kT in eV,
F = electric field magnitude (V cm^{-1}).

The Fowler–Nordheim equation for cold cathode emission gives the cathode current density J_F (A cm^{-2})

$$J_F = 1.54 \times 10^{-6}(F^2/\phi t^2) \exp\left[-6{\cdot}83 \times 10^7 \,.\, \phi^{\frac{3}{2}} \,.\, v/F\right] \tag{10}$$

t and v are slowly varying functions of F and ϕ.

If ϕ lies between 2·0 and 7·0 eV, and J_F varies from 1·0 to 10^7 A cm^{-2}, then t will vary from 1·03 to 1·08, while v ranges from 0·3 to 0·8. For typical field emission, in tungsten, v is in the range of from 0·61 to 0·69 V. This equation only holds true at 0°K, and for higher temperatures, e.g. room temperature, the current density can be approximated as

$$J(T) = J_F \pi x / \sin(\pi x) \tag{11}$$

where $x = kT/d$. The approximation is true for $x \leqslant 0{\cdot}7$.

$$d = 9{\cdot}76 \times \frac{10^{-9}F}{\phi^{\frac{1}{2}}t} \tag{12}$$

and for field emission from tungsten $d = 0{\cdot}166$ eV, giving $j(T)$ for tungsten at room temperature $= 1{\cdot}04\, j_F$ (Wiesner, 1973). Lanthanum hexaboride gives an improvement of 10^2 in current density over tungsten because of its much lower work function (see Table I, after Plomp, 1972).

Instability problems associated with field emission sources are usually created by back-scattering of ions from the anode, which cause etching of areas of the tip. This can cause severe "migration" of the emission area, and has to be corrected by periodic cleaning of the tip, usually by electrical flash over. Fluctuations in the emission current can also be controlled by a feed-back system operating on the Wehnelt (Nomura *et al.*, 1973). A similar system has been proposed by Shimizu *et al.*

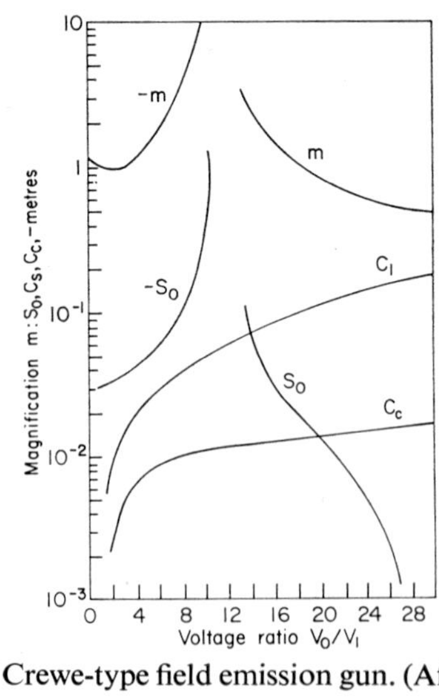

FIG. 12. Properties of Crewe-type field emission gun. (After Swann and Kynaston, 1973.)

(1973), who combine this with a gun using parallel plate electrodes for high resolution. A similar planar triode gun has been successfully used by Swann and Kynaston (1973). The optical properties of the planar gun are compared in Figures 12 and 13 (after Swann and Kynaston,

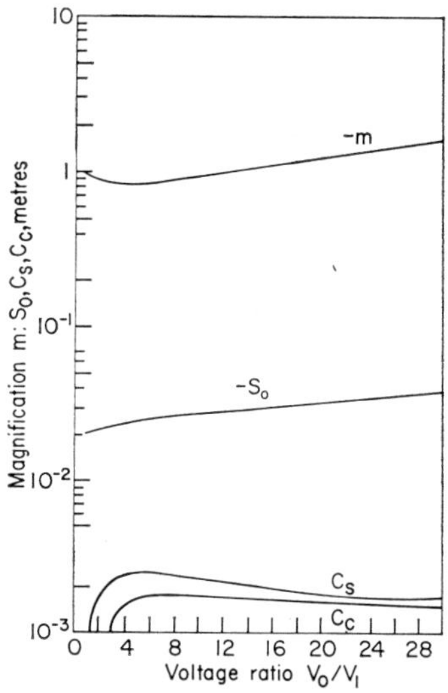

FIG. 13. Properties of planar field emission gun. For comparison with the gun whose properties are illustrated in Fig. 12. (After Swann and Kynaston, 1973.)

1973), and these authors have managed, by means of differential pumping, to attach a field emission source to a conventional SEM with some success. Other attempts to achieve the same results have been made by Joy *et al.* (1972) and by Cleaver and Smith (1972). Furthermore, Someya *et al.* (1972) have been developing a field emission electron gun for a high resolution 100 kV electron microscope. It thus seems likely that, in the near future, it will be possible to use field emission sources on most electron microscopes with the consequent increase in brightness and resolution.

IV. LENS SYSTEMS

Because of the recent publication of "Image Processing and Computer-aided Design in Electron Optics" (Hawkes, 1973) there is little point in repeating any of the chapters from that volume here. However, mention should be made of the contribution by Munro (1973, p. 284), who predicts the form of the magnetic field in a wide variety of lens designs used for different functions. Two further useful reviews come from the Proceedings of the Vth European Electron Microscope Congress (Munro, 1972; Mulvey, 1972), while Pfeiffer (1972) described some basic limitations on probe forming systems due to electron-electron interactions in the beam. According to Pfeiffer, the ultimate resolution of reflection (as opposed to transmission) SEMs is affected by energy broadening caused by electron–electron interaction, known as the Boersch effect.

$$(\Delta E_{tot})^2 = (\Delta E_{th})^2 + (\Delta E_{interact})^2 \tag{13}$$

where

ΔE_{tot} = total energy distribution in the final beam
ΔE_{th} = thermal energy spread
$\Delta E_{interact}$ = Boersch effect.

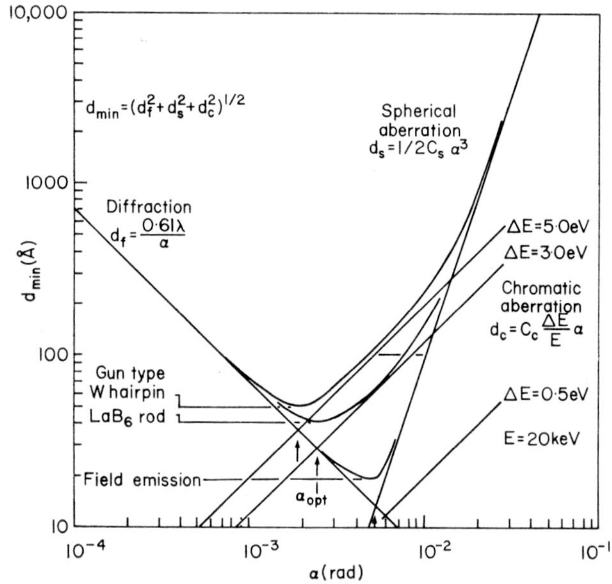

FIG. 14. Effects of beam spreading on the beam from different types of cathode. (After Pfeiffer, 1972.)

The major energy broadening is generated at the gun crossover, and for a tungsten hairpin can be as much as 4–5 eV at a brightness of 10^5 A cm^{-2} str; while for a lanthanum hexaboride gun it can be 3 eV at 5×10^5 A cm^{-2} str. Using field emission guns, the energy spread is extremely small due to their different geometrical configuration. The theoretical limit on minimum probe size is generally accepted as being affected by the spherical aberration coefficient and the diffraction disc only, but because of beam spreading effects, chromatic aberration should also be taken into account:

$$d_{\min,d,c} = 1{\cdot}2\left[\lambda C_c \frac{\Delta E}{E}\right]^{\frac{1}{2}} \tag{14}$$

where

$d_{\min}$ = final probe diameter
$_d$ = diffraction disc
$_c$ = chromatic aberration
λ = electron wavelength
C_c = coefficient of chromatic aberration
ΔE = total energy spread
E = total energy.

Figure 14 (after Pfeiffer, 1972) shows the result of taking account of both factors.

V. Scanning Electron Microscopy, Scanning Transmission Electron Microscopy, and Conventional Transmission Electron Microscopy

Probably the major difference between the conventional transmission microscope and the scanning transmission electron microscope is that in the former, the image is formed simultaneously over the observation area using image forming lenses, whereas in the latter, the image is formed in a time sequence of single picture points produced by the sweep of the electron probe over the specimen surface and the synchronous sweep of the electron beam on the display cathode ray tube (Kimoto, 1973).

The limit of CTEM resolution d_{CT} is determined by C_s, λ, and the apertural angle α. Assuming α to be optimum, then

$$d_{CT} = AC_s^{\frac{1}{2}}\lambda^{\frac{3}{4}} \tag{6}$$

The limit of SEM resolution is determined by the probe diameter, and has been given before as

$$d_{\min} = A'C_s^{\frac{1}{4}}\lambda^{\frac{3}{4}}$$

A and A' are constants nearly equal to unity, and therefore in both types of instrument the resolution parameter is more or less the same. Because of this, the same requirements must be met to improve resolution: C_s must be reduced as far as possible, and λ must also be reduced, i.e. the accelerating voltage of the electrons must be increased (Koike *et al.*, 1972) and Crewe (this volume, Chapter 3). According to Smith (1972) A' for tungsten hairpin cathodes is 9, and for a lanthanum hexaboride cathode, A' is 2·4, and very nearly the resolution parameter for field emission. It is clear, therefore, that for the highest resolution, field emission sources must be used.

Although there is a strong resemblance between the STEM and CTEM, in the information obtainable, and the resolution achievable (because of reciprocity considerations—see Howie, 1972, and this volume) where the STEM probe size is as small as or smaller than the size of structures in the specimen being examined, then the STEM will give information different from that obtained using CTEM which always has a large beam size (Kimoto, 1973). The present practical limit of resolution for both instruments is about 3 Å, but see Crewe (this volume) for the near future limit, 1 Å.

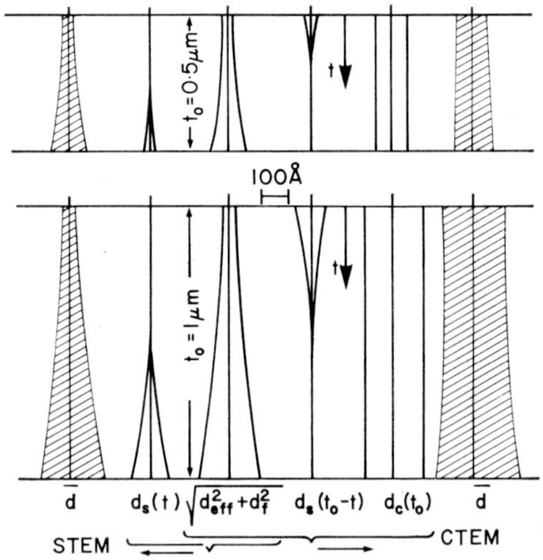

FIG. 15. Dependence of resolution on thickness (t) of specimens for scanning or conventional transmission electron microscopy. (After Reimer, 1972.)

For thicker specimens, the STEM gives better resolution than the CTEM because chromatic aberration is a negligible factor in the STEM where no image forming lenses are used. However, because a transmission specimen has a finite thickness a new parameter becomes important, namely, depth of focus d_f. The relationships between factors limiting resolution area:

$$\bar{d}(t) = \{d_{\text{eff}}^2 + d_f^2(t) + d_s^2(t)\}^{\frac{1}{2}} \quad \text{for STEM} \tag{15}$$

$$\bar{d}(t) = \{d_{\text{eff}}^2 + d_f^2(t) + d_s^2(t_0 - t) + d_c^2(t_0)\}^{\frac{1}{2}} \quad \text{for CTEM} \tag{16}$$

where

$\bar{d}$ = resolution
d_{eff} = effective beam diameter
d_s = multiple scattering
d_f = depth of focus
d_c = chromatic scatter
t = depth of structure of interest in the specimen
t_0 = thickness of specimen.

Thus STEM gives a much higher resolution at the top of the specimen (Reimer, 1972). This is shown diagrammatically in Fig. 15 (after Reimer, 1972).

TABLE II. Resolving Power—Practical Limitations at 1 MeV

	Conventional (CTEM)	Scanning (STEM)
Filament stability	satisfactory (1–10 sec exposure)	poor (100 sec or more exposure)
Accelerator stability	satisfactory	poor (vibrations, long exposures)
Column stability	good	good (?)
HV, lens stability required during	observation—100 sec exposure—1–10 sec	observation (min) exposure (min)
Focusing	satisfactory	poor—in sequence
Astigmatism correction (Fresnel fringes)	fair	very poor—in sequence small number of electrons
Contamination in moderate vacuum (10^{-5} Torr)	moderate	very high (reference 9) ($\propto$ 1/probe diameter^2)
Lens alignment	easy	difficult
Chromatic aberration as compared with 100 keV	substantial decrease	no lenses after the specimen
Energy analysis	difficulties	feasible
X-ray protection	expensive, bulky	negligible X-ray production
Photographic material (X-rays, vacuum trouble)	difficulties	pictures taken from a CRT
Specimen at atmospheric pressure	difficulties poor resolution	in principle easy no lenses after the specimen

Note: same gun brightness (hairpin or pointed filament) and same objective lens focal length and spherical aberration constant are assumed for both CEM and SEM. (After Strojnik, 1972a.)

Apart from this, because the final picture of the specimen is displayed on a CRT, using STEM, it is possible to use the contrast facilities of the CRT to enhance the specimen contrast in a way which cannot be applied to the CTEM. Careful exploitation of the contrast facilities available using a CRT display coupled with computational enhancement of contrast has enabled Crewe (1972a,b) to image single atoms supported on thin films of amorphous carbon using a field emission STEM.

However, it seems clear that in order to maximize the information obtained from specimens, a combination instrument such as that designed by Nauta and Thompson (1972) and Kuypers *et al.* (1973) will prove most satisfactory. In this ingenious design, they place the object in the lower part of the objective lens, and the upper part of the objective lens acts as a demagnifier and can produce a spot size of 10 Å with a thermionic source, and even smaller with a field emission source. The Arizona microscope with its 1 MeV gun and possible combination of STEM and CTEM (Strojnik, 1972a,b) also should be capable of giving high resolution in both modes, but see Table II (after Strojnik, 1972a).

VI. Interface between the Electron Optical System and the Collection and Display System: the Specimen

Almost any material can be examined in the SEM. Solids of all kinds can be examined at room temperature, and with suitable cooling facilities on the specimen stage, liquids, and even gases can be examined, frozen. Because bombardment of the specimen causes, among many other phenomena, the build-up of surface charge on all but perfectly conducting specimens, a few words may be in order at this point to discuss some of the ways that have been suggested for the removal of accumulated charge. General discussions on charging may be found in Pawley (1972) and Pfefferkorn *et al.* (1972).

The technique most commonly used to conduct away charge from insulating specimens is to coat them with carbon or a metal. A useful review of the rationale and mode of application of thin conducting films is given in Echlin and Hyde (1972; see Table III). Metals or carbon may be vacuum evaporated on to the specimen surface, or may be ion sputtered to give a thin coherent film. The thickness of the film should preferably be less than the resolution it is hoped to achieve in the SEM. Pfefferkorn (1973) quotes a formula for roughly estimating the thickness of metal to be laid down:

$$M = \frac{4}{3} \cdot \frac{T_f R^2 \rho 4\pi \,.\, 10^4}{\cos\theta} \tag{17}$$

where

M = mass of coating material of density ρ (g cm^{-3}) evaporated
T_f = film thickness in nm
R = distance in cm from evaporation source to sample
θ = angular deviation from the normal of the source to specimen surface.

However, in many cases, it is desirable not to have to coat specimens, and in recent years, various ingenious ideas have been put forward. Garner *et al.* (1973) suggested that charging could be controlled by passing a stream of ionized gas over the specimen surface, and gave a series of excellent micrographs to illustrate their method. Pachter *et al.* (1973) wished to examine uncoated thick sections of biological tissue and were able to do this by carbon coating a glass substrate and placing the specimen upon this. The results of their experiment were to give usable micrographs which had two components; the secondary electron distribution from the specimen surface, and a primary "transmitted" electron image produced by electrons back-scattering from the carbon-coated glass substrate and passing through the specimen before being collected. The combination of these two signals gives micrographs of good contrast and relatively high resolution.

Welter and McKee (1972) demonstrate how the use of fast scanning at television rates permitted the examination of samples that would otherwise charge at slow scan rates. They explain that the instantaneous charge on a given picture element for weakly conducting specimens can be given as:

$$\sigma(t) = \sigma_m(1-\exp(-t/\eta_r\varepsilon)) \tag{18}$$

where

$\sigma(t)$ = surface charge density at time t
$\sigma(0)$ = surface charge density at time zero, assumed = zero
σ_m = maximum surface charge density
η_r = specimen resistivity
ε = specimen permittivity
t = time

$\sigma(t)$ tends towards a maximum value independent of build up time of charge:

$$\sigma_m = M_g^2 I_b \eta\varepsilon(1-\beta_{SE})/A_v \tag{19}$$

where

M_g = magnification
I_b = beam current
β_{SE} = secondary emission coefficient
A_v = area of visual display.

TABLE III. Data and Techniques for the Evaporation of Thin Films

Element	Atomic number	Atomic weight	Density kg $m^{-3} \times 10^{-3}$	Thermal conductivity W cm^{-1} $°C^{-1}$ at 0°C	Resistivity $\mu\Omega$ cm^{-1} at 20°C	Melting point °C	Boiling point °C	Vapourization temp. at 1 N m^{-2}	Latent heat of vapourization at B.P. (kg J^{-1} g^{-1} atoms)	Evaporation technique
Beryllium	4	9·01	1·80	2·18	4·57	1284	2970	1246	309·35	Ta, W, Mo; Basket; Wets. Toxic
Carbon	6	12·01	2·26	0·04	35·00	3800	4200	2681	586·04	Pointed rods. Resitive/Evaporation
Magnesium	12	24·31	1·74	1·59	4·60	651	1107	443[a]	136·05	Ta, W. Mo; Basket, coat; Wets freely
Aluminium	13	26.98	2·70	2·37	2·80	659	2057	1000	284·23	Ta, W; Coil; Wets and alloys freely
Calcium	20	40·08	1·55	0·98	4·60	810	1240	605	153·63	W; Basket
Titanium	22	47·90	4·50	0·20	42·0	1727	3000	1546	445·81	Ta, W; Loop, Coil, Basket; Alloys
Chromium	24	51·99	7·10	0·87	13·0	1900	2480	1205	305·16	W; Basket
Manganese	25	54·94	7·20	—	5·0	1244	2087	980	224·79	Ta, W. Mo; Loop, Basket. Wets freely
Iron	26	55·85	7·86	0·80	10·0	1535	3000	1447	354·14	W; Loop, Coil; Alloys very readily
Nickel	28	58·71	8·90	0·89	6·10	1452	2900	1510	380·93	W; Coil (heavy). Alloys freely
Copper	29	63·54	8·92	3·98	1·67	1083	2336	1120	304·74	Ta, W, Mo; Basket boats, Not wet easily
Zinc	30	65·38	7·14	1·20	5·92	420	907	343[a]	114·82	Ta, W, Mo; Loop, Basket. Wets freely
Strontium	38	87·62	2·6	0·65	23·0	771	1384	549	141·49	Ta, W, Mo; Basket. Wets freely
Molybdenum	42	95·94	10·2	1·40	5·70	2620	3700	2533	537·48	Refractory

Rhodium	45	102·91	12·5	1·50	4·69	1967	2500	2149	531·62	Sublimation from W at very low pressure
Palladium	46	106·4	12·0	0·70	11·0	1550	3560	1566	372·55	W. Loop, coil
Silver	47	107·9	10·5	4·27	1·60	960	1950	1047	254·09	Ta, W, Mo; Coil, Basket, Boat; W. Not wet easily
Cadmium	48	112·40	8·65	0·93	7·50	321	767	264[a]	99·21	Ta, W, Mo; Basket, Boats; Wets freely Toxic
Tin	50	118·69	7·31	0·64	11·4	232	2337	1189	293·02	Ta, Mo; Basket, boat; Wets readily
Barium	56	137·34	3·50	—	50·0	717	1638	629	149·44	Ta, Mo, W; Basket, Boats; Wets freely
Tantalum	73	180·95	16·6	0·54	13·1	2996	4100	3000	753·48	Refractory
Tungsten	74	183·85	19·30	1·78	5·50	3396	5900	3309	772·74	Refractory
Platinum	78	195·09	21·5	0·73	10·0	1755	4300	2090	510·69	Stranded with W, Ta; Alloys
Gold	79	196·97	19·30	3·15	2·40	1063	2600	1465	342·41	Mo, W; Loop, Basket, Boats; Wets. Alloys Ta
Lead	82	207·21	11·34	0·35	22·0	328	1620	718	177·49	Fe; Basket, Boat; Does not wet W, Ta, Mo

[a] sublimes.

Compiled by P. Echlin, University of Cambridge. Some of data from "Vacuum Deposition of Thin Films", L. Holland, Chapman and Hall, London, 1958. (From Echlin and Hyde, 1972.)

The authors assume that adjacent scan lines on the specimen are overlapping or nearly so, and that the discharge time is long compared with the frame time (F). Table IV shows the difference between the line and frame times for a fast and slow scan SEM. Thus using fast scanning, a charge equilibrium results and non- or poorly-conducting specimens can be observed uncoated at relatively high accelerating voltages. At low accelerating voltages, below about 3 kV, charging generally is not a problem.

TABLE IV. Typical Values for Obtaining a Micrograph of an Uncoated, Poorly Conducting Specimen Using a Fast or Slow Scan Type Instrument. Assumptions: $M = 10^4$, $I = 10^{-10}$ A, $\beta = 0{\cdot}99$, A $= 10^4$ mm^2.

	Fast Scan Instrument	Slow Scan Instrument
Charge time constant ($\eta\varepsilon$)	1–5 sec	same
Line time l	60 μsec	60 msec
Frame time F	60 msec	60 sec

After Welter and McKee, 1972.

Discussion of methods for the prevention of build-up of charge can be found in most texts on the SEM (Hearle *et al.*, 1972).

I cannot begin to enumerate the many methods of specimen preparation applied to the SEM, although it may be said that biological materials have particular associated difficulties. Recent reviews cover freezing and freeze drying (Boyde and Echlin, 1973) and critical point drying (Lewis and Nemanic, 1973), and the reader who wishes to pursue problems of specimen preparation as applied to his material should consult one of the many excellent SEM bibliographies (Wells, 1972b; Carteaud, 1973; Young, 1973; Boyde *et al.*, 1973).

VII. Specimen Stages

The specimen to be examined is mounted on a suitable holder, usually metallic, and placed in the specimen chamber on or in the specimen stage. In most commercial instruments, the stage has external controls for manual or motor driven movements in the X and Y directions. Some have continuous movement in the Z position, whilst others have fixed Z positions. Most instruments provide tilting facilities and rotation, which may or may not be eucentric, is usually a standard feature.

In addition, many new stages have been developed over the last few years for special purposes. Heating and cooling stages are commercially available from liquid nitrogen temperatures to about 400°C, but Hörl (1972), Cort and Steeds (1972), Griffiths and Venables (1972) and Griffiths *et al.* (1972) have all independently constructed stages to operate at liquid helium temperatures. All have been able to reach temperatures of between 4·6°K and 5·6°K to examine condensed gases, to study magnetic contrast, and to observe martensitic transitions in metals.

Echlin *et al.* (1970) reported on the use of a liquid nitrogen cooled stage to examine biological materials, and this work has been continued (Echlin, 1971; Nei *et al.*, 1972), and appears to have great value in X-ray microanalytical studies of labile ions in biological materials.

Reports on high temperature stages are less frequent, but Fulrath (1972) described the use of a stage which contained heating elements capable of reaching temperatures of 1600°C. A stage which combines high temperatures with straining equipment was described by Attwood *et al.* (1972) who used this stage for the study of deformation processes in superplastic alloys, high temperature creep and stress-induced phase transformations. Thomas *et al.* (1972) and Jongebloed *et al.* (1972) have designed tensile/compression stages and Bywater *et al.* (1972) described a stage for testing frictional damage in the SEM. Frictional damage in keratin fibres has been examined by Brown and Swift (1972) who designed and built a precision stage which can be used for very accurate relocation of points of interest on specimen surfaces after various frictional tests.

Although it is not exactly a specimen stage, mention ought to be made of the various ion beam attachments available for most SEMs. In these, an ion gun is inserted into the specimen chamber, and a stream of ions strike the specimen surface. The interpretation of the etch structures produced remains difficult. Even with relatively simple systems such as glass or carbon fibres, structures are produced which cannot be interpreted in terms of our knowledge of the structure of the material (Goodhew and Hepburn, 1972). However, there remains a small but continuing amount of work, particularly in the biological field, where interpretation is particularly difficult (Fulker *et al.*, 1973; Hodges *et al.*, 1972).

Finally, various schools are investigating the possibilities of using SEMs as transmission scanning electron microscopes by the use of simple modifications to the stage. The system proposed by Nemanic and Everhart (1973) uses a metal foil placed more or less opposite to the signal collector to catch electrons transmitted through a thin specimen. The collected signal comprises secondaries from the foil, and has been able to give a resolution of 80 Å on mitochondria.

The second type of transmission stage has been developed at the Pennsylvania State University by Morgan *et al.* (1973) and comprises a small enclosed cell with a thin formvar window as the specimen support. Electron detection is by a small diode detector directly below the specimen, and provision is made to let gases into the cell for environmental, controlled atmosphere experiments. Up to the present these have been carried out mainly on biological systems, in the examination of wet tissues, but chemical and metallurgical experiments could also be carried out. Thirdly, Titchmarsh *et al.* (1972) designed and built a stage for transmission and reflection scanning of crystal defects.

Undoubtedly one of the major lines of technological development in the next few years will be the development of more and more specialized specimen stages for dynamic experimental work in the SEM.

VIII. Signal Detection and Collection

In the early SEM built by McMullan (1953), the specimen was tilted at about 30° to the incident beam, and the emission from the specimen was collected by an electron multiplier system. This was a very efficient collection system compared with earlier methods, but the collected signal was very largely a primary electron signal, and the ratio of signal to noise was poor. Furthermore, only electrons whose trajectory took them in a straight line to the collector could be detected, so that if the specimen surface was very irregular, holes appeared completely dark, and information could only be derived from surfaces directed towards the detector. However, back-scattered electrons can be used to give topographic detail, as well as information about the composition of the specimen. The back-scattering coefficient (η) is defined as the ratio of the total back-scattered current (excluding secondary electrons which by definition have a voltage of less than 50 eV) to the incident electron beam current. The back-scattering coefficient varies with Z, the atomic number of the material under bombardment, and thus back-scattered electrons can be used to produce images with atomic number contrast. Figure 16 shows the variation of η for accelerating voltages of 5 eV and 30 eV respectively. The angle to which the specimen surface is tilted with respect to the incident beam also affects the back-scattering coefficient (see Fig. 17). If the fraction of incident electrons scattered into a solid angle $\delta\Omega$ is $\eta(\theta)\delta\Omega$ making an angle θ with their direction of incidence then

$$\eta(\theta) = (\eta/\pi)|\cos\theta| \qquad (2)$$

and the emergent electrons have a more or less Lambert cosine distribution. For an incident electron beam normal to the specimen surface, electrons are elastically back-scattered in all directions. For this reason, and because by virtue of their high energy they can emerge from a large volume of the specimen, the resolution of back-scattered electron micrographs is not particularly high, say about 500 Å compared with the 100–200 Å for secondary emission.

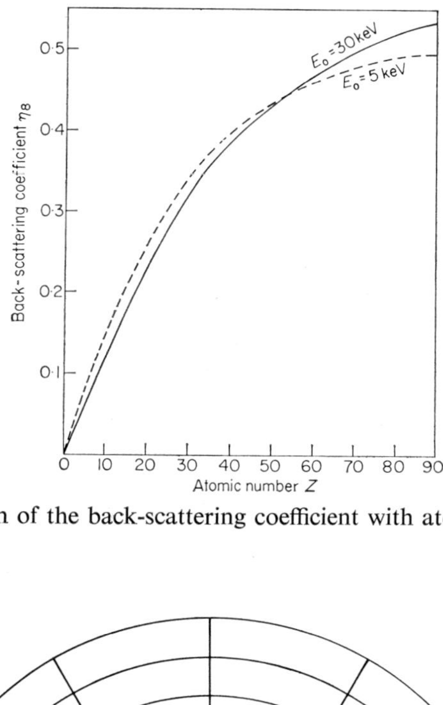

FIG. 16. Variation of the back-scattering coefficient with atomic number. (After Oatley, 1972.)

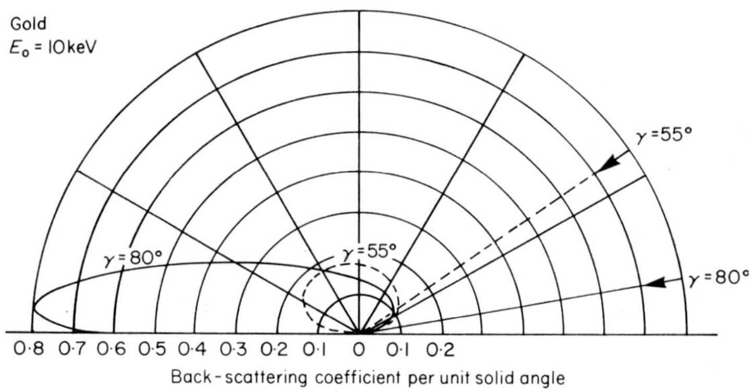

FIG. 17. Variation of the back-scattering coefficient with angle of incidence. (After Kanter, 1957.)

However, if the specimen surface is tilted with respect to the incident electron beam, then the total back-scattering coefficient tends towards unity as the angle of tilt of the specimen tends towards 90°. The scattered electrons leave the specimen surface preferentially in the direction of the incident beam, and this preference is more marked when the angle of incidence exceeds 55° (Oatley, 1972; Wells, 1972c; Fig. 18). Thus in more or less tangential incidence, penetration of the incident beam is reduced, improving the resolution. Wells (1972b,c) has used this effect, combined with an energy filter (which rejects electrons that have lost more than 400 eV) to collect his "low loss" electron signal. This is a primary electron signal collected from near the specimen surface which is capable of giving resolutions of 170 Å (for a detailed mathematical explanation of the low loss signal, see Wells, 1972b). Because of the shallow depth of collection of signal, using the low loss technique, it is possible to detect thin surface layers overlying subsurface structures which are undetectable by any other means (see also Wells, 1974).

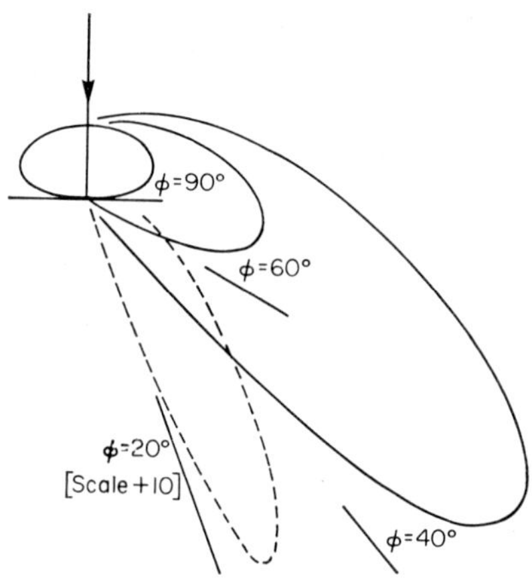

FIG. 18. Calculated angular distribution of low-loss electrons. (After Wells, 1972b.)

Using the type of experimental set up shown in Fig. 19 (after MacDonald, 1971), it is possible to detect Auger electrons, given good vacuum conditions, say 10^{-7} Torr. These are produced by electronic transitions in an atom that occur after excitation of atomic levels. Figure 20 illustrates schematically the Auger transitions for Fe. The energy for the Auger electron transitions is given by the relationship:

$$E(\mathcal{V}XY) = E_{\mathcal{V}} - E_{\mathcal{X}} - E_{\mathcal{Y}} \tag{21}$$

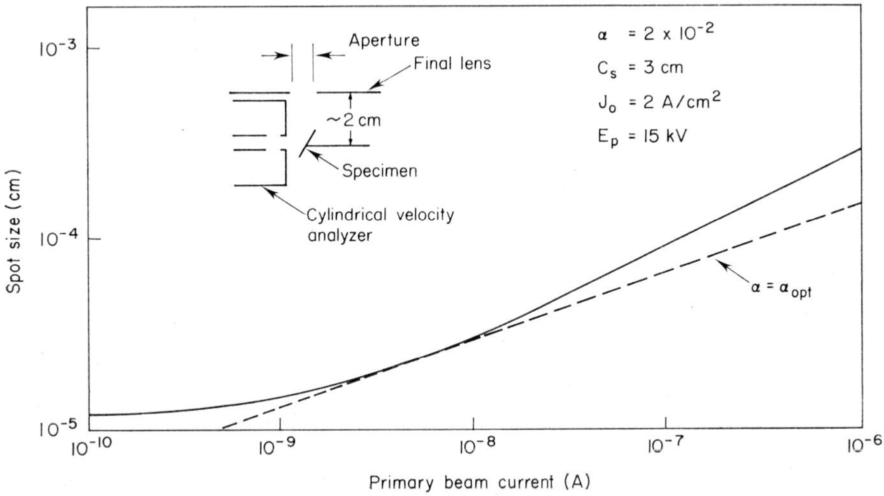

FIG. 19. Experimental arrangement for the detection of Auger electrons in the SEM. Also shown is the relationship between beam current and spot size. (After MacDonald, 1971).

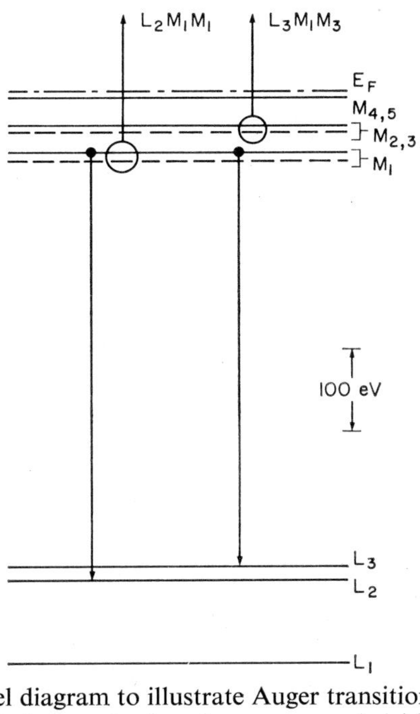

FIG. 20. Energy level diagram to illustrate Auger transitions for Fe. Solid line = energy level in unexcited atom, broken line = energy level in singly ionized atom. E_F = Fermi level. (After MacDonald, 1971.)

where

$\mathscr{V}$ = a hole (or electron vacancy) created by electron impact
$\mathscr{Y}$ = a higher level (e.g. M_1)
$\mathscr{X}$ = a lower level (e.g. M_3)
$E_{\mathscr{V}}$ = the energy level of the initial hole
$E_{\mathscr{X}}$ = the binding energy at level $\mathscr{X}$
$E_{\mathscr{Y}}$ = the binding energy at level $\mathscr{Y}$.

On excitation, the atom may emit an X-ray, or an Auger electron. For the Auger electron to be ejected from the specimen surface, it must be able to reach the specimen surface, and surmount the surface potential barrier. The escape depth for Auger electrons is limited by their low energy. If they are to escape from the specimen without significant energy loss (say 1 eV) the escape depth must be extremely small. Palmberg and Rhodin (1968) determined that the escape depth for 72 eV gold Auger electrons is 4 Å, while that for the silver 362 eV peak is 8 Å. Thus the emission of Auger electrons is essentially a surface phenomenon, and MacDonald (1971) quoted an average depth of 1–10 Å for the thickness of layer analysed in Auger spectroscopy. The Auger yield can be calculated using the formula:

$$\bar{a}_V + \bar{\omega}_V = 1 \tag{22}$$

where

$\left.\begin{array}{l}\bar{a}_V = \text{the mean Auger yield} \\ \bar{\omega}_V = \text{the X-ray fluorescence yield}\end{array}\right\}$ for a vacancy yield in the Vth shell.

Figure 21 shows the mean Auger yield plotted against the atomic number Z. For elements with Z greater than 20, the Auger yield is low for transitions from the K shell, and for the L shell, the Auger yield only falls off for elements above $Z = 45$. For light elements, therefore, the Auger yield is near to unity. This phenomenon is extremely useful because in this region, the X-ray yield is low, and the X-rays occur in an energy range where present day X-ray detectors are insensitive. It is possible, therefore, to use the Auger electron output to characterize materials chemically in a region where X-ray analysis is inefficient. The electron binding energy, generally, increases with Z for a given shell, and decreases as we pass from K to L to M etc. The absolute Auger yield increases with respect to primary (incident) beam energy until $E_{\text{incident beam}}$ approaches 3 E_V. Thus, to excite an Auger transition, the incident beam energy should be greater than 3 times E_V. Table V shows the Auger peak energies for some common surface impurities. It is clear that excitation energies of, say, 3 kV will produce Auger electrons from most elements.

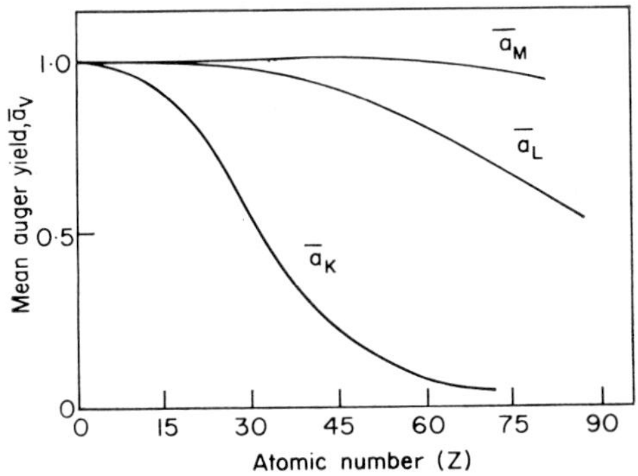

FIG. 21. Mean Auger yield for vacancies in the *K*, *L*, and *M* atomic levels. Plotted as a function of atomic number. (After MacDonald, 1971.)

TABLE V. Auger Peak Energies for Common Surface Impurities

Z	Element	Energy
15	P	118
16	S	150
17	Cl	181
18	A	215
19	K	254
6	C	273
20	Ca	290
7	N	390
8	O	515

After MacDonald, 1971.

MacDonald (1971) has used the SEM to record Auger spectra on areas of his specimen surface which he has selected by a topographic SEM analysis. It is possible to build up an Auger display by allowing the amplitude of the Auger signal to modulate the beam of a CRT. The resolution of this display, however, is comparable with that of back-scattered primary electron displays, but only represents information about a thin surface layer. The Auger electrons are collected by an energy analyser (see Fig. 19) interfaced with the specimen chamber. Changes in potential can also be detected using Auger analysis (MacDonald, 1970) and it seems likely that this technique will become increasingly used with the advent of the newer, ultra-high vacuum SEMs.

The means of collecting primary electron signals has greatly advanced since the electron multiplier system of McMullan (1953), and most instruments nowadays use a detector of the Everhart/Thornley type (1960). This is illustrated in Fig. 22, and consists of a scintillator which generates photons as a result of the impinging electrons. The photons are conducted along a light pipe to the photomultiplier tube, which converts the photons into an electron signal. The intensity of this signal is amplified and used to modulate the brightness of the display c.r.t. Using this simple system, the bulk of the signal will consist of primary electrons. If however, a biased grid at a positive potential is placed in front of the scintillator, then the low energy (less than 50 eV) electrons will be collected, and if these are then accelerated on to the scintillator by a high voltage of 10 kV or more, the primaries will form an insignificant part of the total detected signal. Collection of secondaries led Smith (1956) and Smith and Oatley (1955) to exploit this mode of emission.

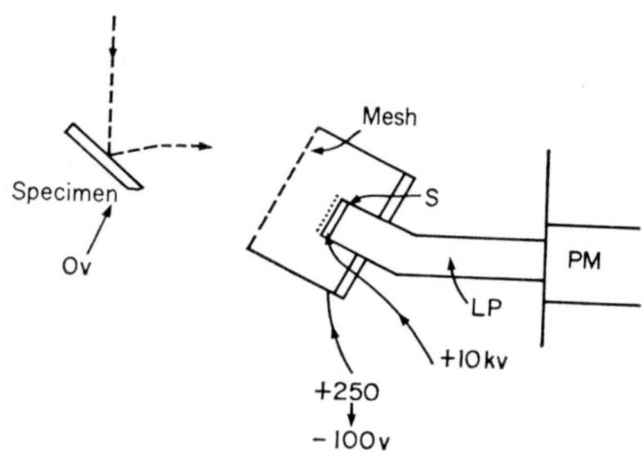

FIG. 22. The Everhart/Thornley detector. (After Joy, 1973.)

Improvements in the detection and processing of the signals employed have played a vital role in the development of each of the modes of operation of the SEM: emissive, luminescent, conductive, transmissive and X-ray and detailed accounts of these matters will be found in the later chapters dealing with each particular mode.

ACKNOWLEDGEMENTS

I wish to acknowledge the help and support of P. R. Grant, D. B. Holt and C. A. McGrath.

REFERENCES

AHMED, H. (1972). *Proc. Vth European Congr. Electron Microscopy*, 10–11.

AHMED, H. and NIXON, W. C. (1973). *Proc. Vth Ann. SEM Symposium, IITRI, Chicago*, 217–224.

ATTWOOD, D. G., HAZZLEDINE, P. M., JOY, D. C., NEWBURY, D. E. and PEASE, D. E. (1972). *Proc. Vth European Congr. Electron Microscopy*, 382–383.

BOYDE, A. and ECHLIN, P. (1973). *Proc. VIth Ann. SEM Symposium, IITRI, Chicago*, 759–766.

BOYDE, A., JONES, S. A. and BAILEY, E. (1973). *Proc. VIth Ann. SEM Symposium, IITRI, Chicago*, 697–734.

BROERS, A. N. (1969). *J. sci. Instrum.* **2,** 273–276.

BROWN, A. C. and SWIFT, J. A. (1972). *Proc. Vth European Congr. Electron Microscopy*, 386–387.

BYWATER, N. E., BUCKLEY, T., HEPWORTH, A. and SIKORSKI, J. (1972). *Proc. Vth European Congr. Electron Microscopy*, 384–385.

CARTEAUD, J.-P. (1973). *Proc. VIth Ann. SEM Symposium, IITRI, Chicago*, 629–636.

CASTAING, R. (1951). Publ. No. 55, O.N.E.R.A., Chatillon-sous-Bagneux (Seine).

CASTAING, R. and GUINIER, A. (1949). *Proc. Ist Internat. Conf. on Electron Microscopy, Delft, 1949*. Martin Nijhoff, Delft, 1950, 60–63.

CLEAVER, J. R. A. and SMITH, K. C. A. (1972). *Proc. Vth European Congr. Electron Microscopy*, 18–19.

CLEAVER, J. R. A. and SMITH, K. C. A. (1973). *Proc. Vth Ann. SEM Symposium, IITRI, Chicago*, 49–56.

CORT, D. M. and STEEDS, J. W. (1972). *Proc. Vth European Congr. Electron Microsc.* 376–377.

COSSLETT, V. E. and DUNCUMB, P. (1956). *Nature, NATUA* **177,** 1172.

CREWE, A. V. (1970). *Q. Rev. Biophys.* **3,** 137–175.

CREWE, A. V. (1972a). *Proc. Vth European Congr. Electron Microscopy*, 14.

CREWE, A. V. (1972b). *Proc. Vth European Congr. Electron Microscopy*, 640–643.

CREWE, A. V., WALL, J. and WELTER, L. M. (1968). *J. appl. Phys.* **39,** 5861–5868.

DUNCUMB, P. and MELFORD, D. A. (1959). "Proceedings of II International Symposium on X-ray Microscopy and X-ray Microanalysis, Stockholm, 1959." (A. Engström, V. E. Cosslett and H. Pattee, eds), p. 358. Elsevier, Amsterdam, 1960.

ECHLIN, P. (1971). *Proc. IVth Ann. SEM Symposium, IITRI, Chicago*, 225–232.

ECHLIN, P. and HYDE, P. J. W. (1972). *Proc. Vth Ann. SEM Symposium, IITRI, Chicago*, 137–143.

ECHLIN, P., PADEN, R., DRONZEK, B. E. and WAYTE, P. (1970). *Proc. IIIrd Ann. SEM Symposium, IITRI, Chicago*, 49–56.

EVERHART, T. E. and THORNLEY, R. F. M. (1960). *J. sci. Instrum.* **37,** 246.

FULKER, M. J., HOLLAND, L. and HURLEY, R. E. (1973). *Proc. VIth Ann SEM Symposium, IITRI, Chicago*, 379–386.

FULRATH, R. M. (1972). *Proc. Vth Ann. SEM Symposium, IITRI, Chicago*, 17–24.

GARNER, G. E., COHEN, A. L. and STEEVER, R. G. E., Jr. (1973). *Proc. VIth Ann. SEM Symposium, IITRI, Chicago*, 189–196.

GOODHEW, P. J. and HEPBURN, M. J. (1972). *Proc. Vth European Congr. Electron Microscopy*, 300–301.

GRIFFITHS, B. W. and VENABLES, J. A. (1972). *Proc. Vth Ann. SEM Symposium, IITRI, Chicago*, 9–16.

GRIFFITHS, B. W., MILLS, C. A. and VENABLES, J. A. (1972). *Proc. Vth European Congr. Electron Microscopy*, 372–373.

HAINE, M. E. and EINSTEIN, P. A. (1952). *Br. J. appl. Phys.* **4,** 40.

HAWKES, P. W. (Ed.) (1973). "Image Processing and Computer-aided Design in Electron Optics," pp. 1–441. Academic Press, London and New York.

HEARLE, J. W. S., SPARROW, J. T. and CROSS, P. M. (1972). "The Use of the Scanning Electron Microscope," pp. 1–278. Pergamon Press, Oxford and New York.

HODGES, G. M., MUIR, M. D., SELLA, C. and CARTEAUD, J.-P. (1972). *J. Microsc.* **95**/3, 445–451.

HÖRL, E. M. (1972). *Proc. Vth European Congr. Electron Microscopy*, 374–375.

HOWIE, A. (1972). *Proc. Vth European Congr. Electron Microscopy*, 408–413.

JONGEBLOED, W. L., ARENDS, J. and MOLENAAR, I. (1972). *Proc. Vth European Congr. Electron Microscopy*, 380–381.

JOY, D. C. (1973). *Proc. VIth Ann. SEM Symposium, IITRI, Chicago*, 734–750.

JOY, D. C., TITCHMARSH, J. M., WOOLF, R. J. and BOOKER, G. R. (1972). *Proc. Vth European Congr. Electron Microscopy*, 16–17.

KANTER, H. (1957). *Ann. Phys. Leipzig* **20,** 144.

KIMOTO, S. (1973). *Proc. VIth Ann SEM Symposium, IITRI, Chicago*, 1–8.

KOIKE, H., MATSUO, T., UENO, K. and SUZUKI, M. (1972). *Proc. Vth European Congr. Electron Microscopy*, 496–497.

KUYPERS, W., THOMPSON, M. N. and ANDERSEN, W. H. J. (1973). *Proc. VIth Ann. SEM Symposium, IITRI, Chicago*, 9–16.

LANGMUIR, D. B. (1937). *Proc. I. R. E.* **25,** 977.

LE POOLE, J. B., KRAMER, J. and VAN DER MAST, K. D. (1972). *Proc. Vth European Congr. Electron Microscopy*, 8–9.

LEWIS, E. R. and NEMANIC, M. K. (1973). *Proc. Vth. Ann. SEM Symposium, IITRI, Chicago*, 767–774.

MACDONALD, N. C. (1970). *Proc. IIIrd Ann. SEM Symposium, IITRI, Chicago*, 481–487.

MACDONALD, N. C. (1971). *Proc. IVth Ann. SEM Symposium, IITRI, Chicago*, 89–96.

MCMULLAN, D. (1953). *Proc. I.E.E. B* **100,** 245.

MORGAN, R. S., LEBIEDZIK, J. and WHITE, E. W. (1973). *Proc. VIth Ann. SEM Symposium, IITRI, Chicago*, 205–210.

MULVEY, T., 1972. *Proc. Vth European Congr. Electron Microscopy*, 64–69.

MUNRO, E., 1972. *Proc. Vth European Congr. Electron Microscopy*, 22–23.

MUNRO, E. (1973). "Image Processing and Computer-aided Design in Electron Optics" (P. W. Hawkes, ed.), pp. 284–323. Academic Press, London and New York.

NAUTA, R. and THOMPSON, M. N. (1972). *Proc. Vth European Congr. Electron Microscopy*, 46–47.

NEI, T., YOTSUMOTO, H., HASEGAWA, Y., NAGASAWA, Y. (1972). *Proc. Vth European Congr. Electron Microscopy*, 252–253.

NEMANIC, M. K. and EVERHART, T. E. (1973). *Proc. VIth Ann. SEM Symposium, IITRI, Chicago*, 25–32.

NOMURA, S., KOMODA, T., KAMIRYO, T. and NAKAIZUMI, Y. I. (1973). *Proc. VIth Ann. SEM Symposium, IITRI, Chicago*, 65–72.

OATLEY, C. W. (1972). "The Scanning Electron Microscope. Part I. The Instrument." pp. 1–194. Cambridge University Press.

PACHTER, B. R., PENHA, D., DAVIDOWITZ, J. and BREININ, G. M. (1973). *Proc. VIth Ann. SEM Symposium, IITRI, Chicago*, 387–394.

PALMBERG, P. W. and RHODIN, T. N. (1968). *J. appl. Phys.* **39,** 2425.

PAWLEY, J. B. (1972). *Proc. Vth Ann. SEM Symposium, IITRI, Chicago*, 153–160.

PFEFFERKORN, G. (1973). *Proc. VIth Ann. SEM Symposium, IITRI, Chicago*, 751–758.

PFEFFERKORN, G. E., GRUTER, H. and PFAUTSCH, M. (1972). *Proc. Vth Ann. SEM Symposium, IITRI, Chicago*, 147–152.

PFEIFFER, H. C. (1972). *Proc. Vth Ann. SEM Symposium, IITRI, Chicago*, 113–120.

PLOMP, F. H. (1972). *Proc. Vth European Congr. Electron Microscopy*, 2–7.

PLOMP, F. H., VENEKLASEN, L. H. and SIEGEL, B. M. (1968). *Proc. IVth European Reg. Conf. Electron Microscopy, Rome* 1968, **1,** 141.

REIMER, L. (1972). *Proc. Vth Ann. SEM Symposium, IITRI, Chicago*, 197–204.

SEPTIER, A. (1966). *In* "Advances in Optical and Electron Microscopy" (V. E. Cosslett and R. Barrer, eds), Vol. 1, p. 221. Academic Press, New York and London.

SHIMIZU, R., KURODA, K., SUZUKI, T., NAKAMURA, S., SUGANUMA, T. and HASHIMOTO, H. (1973). *Proc. VIth Ann. SEM Symposium, IITRI, Chicago*, 73–84.

SMITH, K. C. A. (1956). Ph.D. Dissertation, University of Cambridge.

SMITH, K. C. A. (1972). *Proc. Vth Ann. SEM Symposium, IITRI, Chicago*, 1–8.

SMITH, K. C. A. and OATLEY, C. W. (1955). *Br. J. appl. Phys.* **6,** 391.

SOMEYA, T., GOTO, T., HARADA, Y. and WATANABE, M. (1972). *Proc. Vth European Congr. Electron Microscopy*, 20–21.

STROJNIK, A. (1972a). *Proc. Vth Ann. SEM Symposium, IITRI, Chicago*, 215–224.

STROJNIK, A. (1972b). *Proc. Vth European Congr. Electron Microscopy*, 48–49.

SWANN, D. J. and KYNASTON, D. (1973). *Proc. VIth Ann. SEM Symposium, IITRI, Chicago*, 57–64.

SWANN, D. J. and SMITH, K. C. A. (1973). *Proc. VIth Ann. SEM Symposium, IITRI, Chicago*, 41–48.

THOMAS, K., BREAKWELL, P. R. and FURSE, J. E. (1972). *Proc. Vth European Congr. Electron Microscopy*, 378–379.

THORNTON, P. R. (1968). "Scanning Electron Microscopy," Chapman and Hall, London.

TITCHMARSH, J. M., JOY, D. C., WOOLF, R. J. and BOOKER, G. R. (1972). *Proc. Vth European Congr. Electron Microscopy*, 494–495.

VENEKLASEN, L. H. and SIEGEL, B. M. (1972). *J. appl. Phys.* **43,** 1600.

WELLS, O. C. (1972a). *Proc. Vth Ann. SEM Symposium, IITRI, Chicago*, 375–442.

WELLS, O. C. (1972b). *Proc. Vth Ann. SEM Symposium, IITRI, Chicago*, 169–176.

WELLS, O. C. (1972c). *Proc. Vth European Congr. Electron Microscopy*, 484–485.

WELLS, O. C. (1974). *Proc. VIth Ann. SEM Symposium, IITRI, Chicago*, 1–9.

WELTER, L. M. and MCKEE, A. N. (1972). *Proc. Vth Ann. SEM Symposium, IITRI, Chicago*, 161–168.

WIESNER, J. C. (1973). *Proc. VIth Ann. SEM Symposium, IITRI, Chicago*, 33–40.

YOUNG, J. (1973). *Proc. VIth Ann. SEM Symposium, IITRI, Chicago*, 775–780.

ZWORYKIN, V. K., HILLIER, J. and SNYDER, R. L. (1942). *ASTM Bull.* **117,** 15(4).

Chapter 2

Electron – Solid Interactions and Energy Dissipation

H. E. BISHOP

Materials Development Division, AERE, Harwell, England

Symbols

e = electronic charge
Z = atomic number of the nucleus
E = incident electron energy
θ = angular deflection of the electron
$F(\theta)$ = factor $F(\theta)$ allows for the screening effect of the atomic electrons and dominates at small angles
N = Avogadro's no.
A = atomic weight
ρ = the density
s = the actual distance travelled by the electron along its trajectory
J = the mean ionization potential
$\mathbf{r}$ = the position vector
$\mathbf{v}$ = a direction vector
$F(\mathbf{r},\mathbf{v},s)$ = the distribution function of electrons having travelled a distance s in the medium
$\eta(E)$ = the fraction of electrons with energy greater than E, back-scattered from the target
$\Phi(E)$ = the ionization cross section
E_c = the critical ionization potential
$\langle r^2(E) \rangle$ = the mean square radial spread for electrons with residual energy E
κ = a constant
R_{max} = maximum range of electrons
R_x = the extrapolated range of electrons
R_B = Bethe range of electrons

I. Introduction

Before discussing the contrast mechanisms available in the scanning electron microscope it is first desirable to have a general picture of how electrons interact with solids. The basic mechanisms of electron scattering are well understood but an electron suffers so many interactions before it either comes into thermal equilibrium with the target or escapes from its surface that no precise theory of electron scattering by solids is possible. In this chapter it is proposed to describe the basic scattering mechanisms, then to discuss briefly how these may be used to predict scattering in solid targets for primary beam energies in the range 5–40 keV and finally to look at experimental results for various phenomena such as electron penetration, back-scattering, secondary emission and energy dissipation in the light of the understanding obtained from the theory and to give some empirical relations for beam penetration in the target.

II. Basic Scattering Mechanisms

At the relatively low energies considered here the interaction between incident electrons and the solid are wholly electrostatic and may be divided into two types. The first is the scattering by the strong coulomb

Table I.

	Effect	Application
Elastic scattering	High energy back-scattered electrons	Atomic number contrast in the SEM Diffraction effects. e.g. channelling patterns
	Bremsstrahlung	
Inelastic scattering	Characteristic X-ray emissions	Electron probe microanalysis
	Auger electrons	Surface analysis
	Luminescence	SEM information Cathode ray tubes
	Secondary electron emission	Topographical, voltage and magnetic contrast in the SEM
	Radiation damage	Cathodochromic displays Exposure of photoresist
	Heating	Electron beam machining
	Electron hole pairs	SEM conductive mode charge collection effects

field of the nucleus which causes large angular deflections in the path of the incident electron. Because of the large discrepancy in the masses of the particles, there is essentially no energy transfer and the collision is elastic. At higher energies there can be a significant loss of energy during these large deflections due to the emission of X-radiation, *bremsstrahlung*, but at the energies of interest here the contribution to energy loss due to this process is negligible although it is responsible for the continuous background observed in X-ray spectra. The incident electrons lose their energy in electron–electron interactions when the atomic electrons are excited into vacant higher energy states leaving holes which are subsequently filled by electronic transitions. It is the various phenomena concerned with the filling of these holes that made the SEM such a powerful tool in the study of materials. Some of these effects are listed in Table I.

A. ELASTIC SCATTERING

Scattering cross sections of increasing complexity have been derived over the years to describe the scattering of an electron by the nucleus. In the context of scattering in solids which concerns us here, a simple screened Rutherford cross section is usually found to be adequate.

$$\frac{d\sigma(\theta)}{d\Omega} = \frac{e^4(Z-F(\theta))^2}{16E^2 \sin^4(\theta/2)} \text{cm}^2 \tag{1}$$

where e is the electronic charge, Z the atomic number of the nucleus, E (ergs) is the incident electron energy and θ is the angular deflection of the electron. The factor $F(\theta)$ allows for the screening effect of the atomic electrons and dominates at small angles but is negligible for large angles. This expression is only an approximation and becomes inadequate for energies below about 5 keV, particularly for heavy elements. However, calculation of electron scattering in bulk targets based on this cross section are found to give good agreement with experiment (Bishop, 1966, 1967). The two main points to notice about this cross section are that it is inversely proportional to the square of the incident energy and that for large angle scattering it is proportional to the square of the atomic number so that angular scattering is most significant in the heavy elements.

B. INELASTIC SCATTERING

Scattering cross sections similar to eqn (1) can be derived for electron–electron scattering. This cross section is a function of both the binding energy of the atomic electron and of the state into which it is excited. It is therefore not practicable to use individual scattering cross sections when considering the energy loss of a fast electron in a solid and it is customary to use an expression due originally to Bethe (1930) for the mean rate of energy loss

$$\frac{\mathrm{d}E}{\mathrm{d}s} = -\frac{2\pi e^4 NZ\rho}{EA} \ln(1{\cdot}166\, E/J) \tag{2}$$

where N is Avogadro's number, A the atomic weight, ρ the density, s is the actual distance travelled by the electron along its trajectory ignoring deviations caused by the elastic scattering (Fig. 1) and J is the mean ionization potential. Various theoretical estimates of J have been made but the value is found to depend critically on the atomic model used in

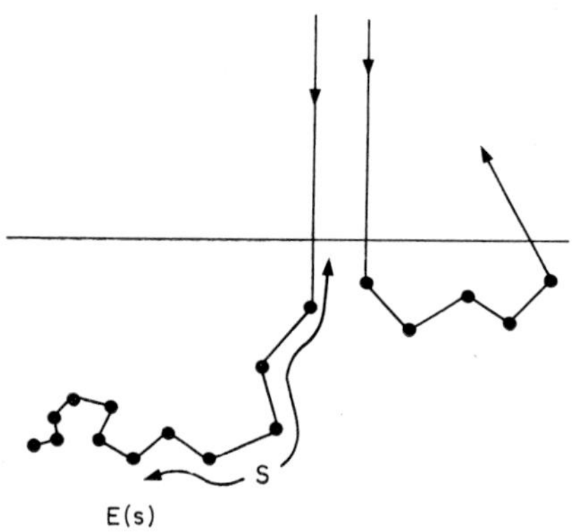

FIG. 1. Schematic representation of electron scattering in a target.

the calculation. The mean ionization potential is normally determined experimentally by measuring the ranges of high energy heavy particles such as protons for which angular scattering is negligible. There are several sets of results to choose from: for instance Everhart and Hoff (1971) have used an empirical formula attributed to Sternheimer.

$$J = (9{\cdot}76 + 58{\cdot}8Z^{-1{\cdot}19})Z \text{ eV}. \tag{3}$$

It is necessary to be a little cautious in relying on the experimental values of J obtained for high energy particles when considering relatively low energy electrons. Duncumb and Reed (1968) in a careful evaluation of data obtained in the electron probe microanalyser obtained a better fit to their results if they used a set of J values fitting the curve shown in Fig. 2. Also once the energy of the incident electrons drops below the binding energy of the inner atomic electrons these can no longer contribute to the energy loss and a reduced value for J should be used. However as J only appears in a logarithmic term its exact value is not very critical except where high accuracy is required. Indeed for many calculations in the literature a value of $J/Z = 11{\cdot}5$ eV has been used.

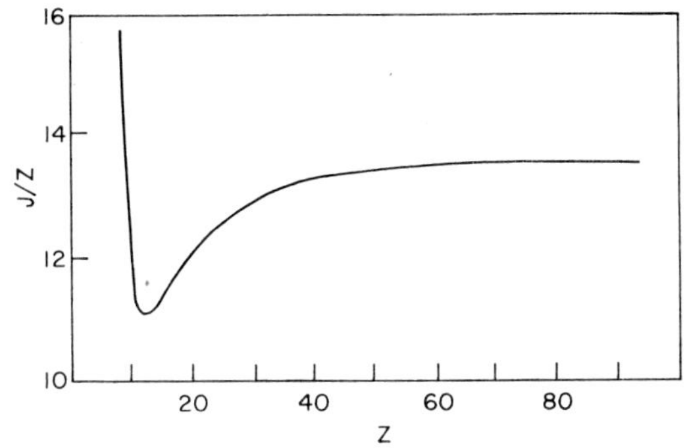

FIG. 2. Variation of mean ionization potential with atomic number determined from a series of microprobe analyses. (After Duncumb and Reed, 1968.)

Integration of eqn (2) allows us to determine two important relations. The first is an expression for the mean energy loss of electrons that have travelled a distance s in a specimen (Fig. 1)

$$s(E) = \int_{E}^{E_0} \left(\frac{dE}{ds}\right)^{-1} dE \tag{4}$$

where E_0 is the energy of the incident electrons. The other gives the mean range of electrons, called the *Bethe range*

$$R_B = \int_{0}^{E_0} \left(\frac{dE}{ds}\right)^{-1} dE. \tag{5}$$

If eqn (2) is substituted directly into (5) a singularity occurs when E approaches J. For high beam energies one can avoid this by choosing a lower value for the integral just greater than J, but for lower energies it is better to adopt the approach used by Everhart and Hoff (1971) and use a modified form of eqn (2) as E approaches J. They obtain an expression for the Bethe range

$$\rho R_B = K\left(\int_{\xi_1}^{\xi_0} \frac{\xi \, d\xi}{\ln \xi} + \frac{\xi_1^2}{(1+m)\ln \xi_1}\right) \tag{6}$$

where

$$K = 9{\cdot}40 \times 10^{-12} J^2 A/Z \text{ g cm}^{-2}$$

where the reduced energy is given by

$$\xi = 1{\cdot}166E/J$$

with

$$\xi_1 = 5 \text{ and } m = 0{\cdot}3787.$$

In Fig. 3 the curve of $R_B = \rho R_B/K$ as a function of ξ calculated by Everhart and Hoff (1971) are shown, together with power laws that fit various parts of the curve.

The Bethe range is a very important parameter in understanding electron scattering in matter because electron penetration and spread are found to be proportional to the Bethe range for a given material over a wide range of incident beam energies. Experimental measurements of electron ranges are usually found to obey power laws close to those indicated in Fig. 3. In order to understand the general dependence of the Bethe range on atomic number let us substitute for the

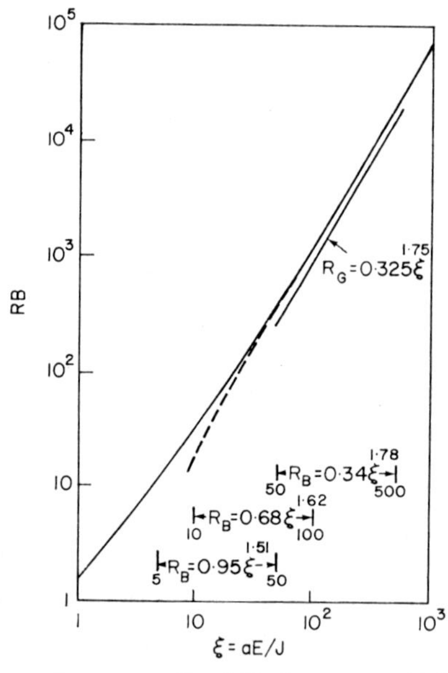

FIG. 3. Universal curve for normalized Bethe range. (After Everhart and Hoff, 1971.)

integral in eqn (6) the power law obtained for R_B for values of ξ between 10 and 100 and substitute for J with eqn (3) which gives an expression of the form

$$\rho R_B = \text{const}\,(9{\cdot}76+58{\cdot}8Z^{-1{\cdot}19})^{0{\cdot}38}A\,E^{1{\cdot}62}Z^{-0{\cdot}62}. \tag{7}$$

We note first that mass range ρR_B rather than absolute distance is significant. The small power of the first term confirms the earlier statement that the exact form of the expression for J is not very important. As the atomic number is proportional to atomic weight we see that the range increases with atomic number, but not very rapidly because of the small value of the exponent.

It is important to remember that eqn (2) only gives the mean energy loss and that the Bethe range only gives a mean range. Landau (1944) has derived an expression for the energy loss distribution of electrons which have travelled a small distance in a target. From this expression it is possible to calculate a distribution for electron ranges. The extent of this range straggling is shown in Fig. 4 which gives the distribution of actual electron path lengths S_R in copper calculated by a Monte Carlo technique from the Landau expression.

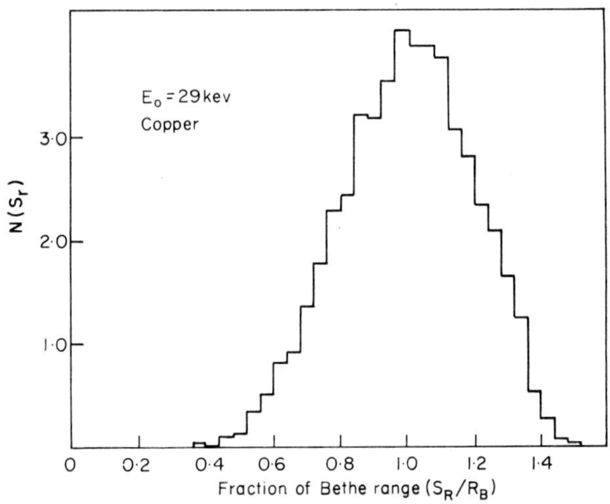

FIG. 4. Range straggling in a copper target. The figure shows the distribution of the actual distances travelled by different electrons, s_R, as a fraction of the Bethe range. (Bishop, 1966.)

C. DIFFRACTION EFFECTS

So far the discussion has assumed that the target is amorphous. The wave properties of electrons and the periodicity of the target material have been ignored. Diffraction and channelling effects are well known in transmission electron microscopy and one might expect to find similar effects in the scanning electron microscope. In the conventional scanning electron microscope where a bulk sample is used the incident electron beam suffers significant elastic and inelastic scattering events relatively early in its trajectory even when directed in preferred channelling directions with the results that any effects depending on the periodicity of the specimen usually appear as a small ($\sim$ a few per cent) modulation on a large amorphous background. A detailed discussion of these effects is given by Joy and Howie (Chapters 6 and 7, this volume).

III. Multiple Scattering Theory

There is only space here to give the most cursory discussion of this complex subject. The theory has to show how the elastic and inelastic scattering processes combine to give the scattering observed for a given target. In the previous section the first major approximation that is made in nearly all cases has been indicated. That is to use a continuous energy loss approximation so that eqn (4) may be used to relate the residual energy of an electron in the specimen with the distance it has travelled. Having made this approximation we have in principle to solve a Boltzmann transport equation of the type (Lewis, 1950):

$$\frac{\mathrm{d}f}{\mathrm{d}s}+\mathbf{v}\cdot\nabla f=\frac{\mathbf{N}}{A}\int[f(r,\mathbf{v}',s)-f(\mathbf{r},\mathbf{v},s)]6(\mathbf{v}-\mathbf{v}')\,\mathrm{d}v' \tag{8}$$

where $\mathbf{r}$ is a position vector, $\mathbf{v}$ a direction vector and $f(\mathbf{r},\mathbf{v},s)$ the distribution function of electrons having travelled a distance s in the medium.

No simple analytical solutions have been found for this equation and in particular no way has been found for including boundary conditions. Lewis has derived expressions for the moments of the electron scattering distribution from which Spencer (1955, 1959) has calculated the energy dissipation distribution (depth dose distribution) for a large variety of materials. These distributions are available in tabular form and some examples are shown in Fig. 14. Although they are only strictly valid for an infinite target,* it will be shown later that for many purposes they may be usefully employed for the practical situation of semi-infinite targets if some discrepancy near the surface is not too important.

Once boundary conditions are introduced it is necessary to make further approximations to simplify eqn (8), or to employ numerical methods. The simplest case to deal with is one often met in practice; that is, the scattering of electrons by thin films. Provided energy losses in the film are not great enough to change the elastic scattering cross section greatly, the angular scattering may be determined by a simple convolution technique. A detailed review of small angle angular scattering is given by Scott (1963). Many of the concepts introduced in small

* This represents the mathematically convenient situation where the incident beam originates at some point within a continuous scattering material.

angle scattering theory, in particular the transport mean free path and depth of diffusion have been carried over to help explain scattering in more complex situations where energy loss becomes significant. These ideas are useful in leading to a qualitative interpretation of scattering and may form the basis for semi-empirical models of electron scattering. However, if more precise scattering information is required it is necessary to adopt a purely numerical approach to the problem.

There are two main approaches to a numerical solution to eqn (8). The first is via finite difference equations, the other to use Monte Carlo simulation. The finite difference method has been used by Brown and Ogilvie (1966) for calculation of X-ray production in solid targets. It should be noted, however, that they have simplified eqn (8) by using a small angle approximation for the elastic scattering cross section; Dashen has used a more precise method to calculate the back-scattering from a semi-infinite target. Although his approach could be used in principle to calculate other aspects of electron scattering it is in practice rather too complex to be of much practical use.

Monte Carlo simulation is a method now widely used for solving complex mathematical problems. In the case of electron scattering the approach is to simulate a large number of electron trajectories in a target with the required boundary conditions by random sampling from the single scattering distributions. The use of Monte Carlo techniques in electron scattering problems is reviewed by Berger (1963) and examples of their application to the type of problems met in the SEM are given by Bishop (1966, 1967) and by Shinoda (1969). For calculations of electron scattering in bulk targets or thick foils where energy loss is insignificant the Monte Carlo approach is probably the most flexible and convenient to use in practice.

Before leaving multiple scattering theory it is worth mentioning a very simple scattering model due to Archard (1961). Although in practice it does not give a very accurate picture of the electron scattering it does give a clear qualitative picture of how the atomic number of a target influences the electron scattering. When an electron beam has only travelled a short distance in a target most electrons will be travelling in directions close to that of the incident beam. However, as the beam penetrates further into the target, elastic scattering deviates the electrons more and more from their initial direction until eventually there is no memory of the incident direction and the beam is said to be diffused. It is possible to define a depth of diffusion z_d where the average cosine of the angle between the incident beam and the actual direction is $1/e$, and it may be shown that z_d is approximately given by

$$z_d/R_B = 40/7Z.$$

Archard proposed that a simple picture of electron scattering is given by allowing the electron beam to penetrate undeviated down to a depth z_d and then to radiate out randomly up to their total range (Fig. 5). From this model it is possible to obtain an expression for the fraction of back-scattered electrons

$$\eta = (7Z-80)/(14Z-80).$$

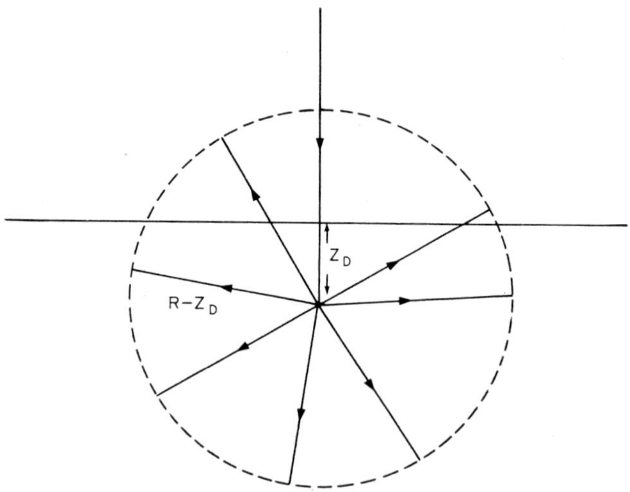

FIG. 5. Archard diffusion model.

For low atomic numbers the model clearly breaks down but for the heavier elements the model gives reasonable agreement with experiment. The value of this model is that it illustrates how the much stronger angular scattering in the heavy elements leads to many more electrons being scattered back out of the target and also that the mean penetration of electrons (as a fraction of the Bethe range) decreases with increasing atomic number.

An alternative empirical model has been proposed by Everhart (1960). which accounts for back-scattering in terms of large angle elastic scattering events. This model works well for the lighter elements but predicts too much back-scattering for atomic numbers greater than 30. More complex models combining these two approaches have been developed which can be made to fit experimental data by varying various parameters (Kanaya and Okayama, 1972). The expressions involved, however, are rather complex and this severely limits the usefulness of such models.

IV. Experimental Results

There is a large body of experimental data available on the transmission of electrons through thin films and on the back-scattering of electrons. There is only space here to discuss those results with most relevance to scanning electron microscopy using scattering theory to help understand the trends with atomic number and energy.

A. Electron Back-Scattering

The typical form for the energy spectrum of electrons emitted from a target is shown in Fig. 6. It has three distinct regions. At the high energy end there is an elastic peak together with some subsidiary maxima corresponding to discrete energy loss processes. At low incident beam

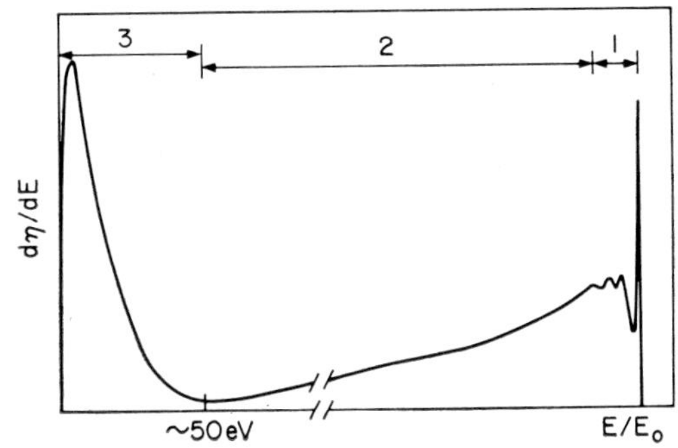

FIG. 6. Schematic energy spectrum for back-scattered electrons.

energies, say below 1 keV, quite a significant fraction of the back-scattered electrons fall in this region and show strong diffraction effects which are exploited in the LEED (low energy electron diffraction) technique. However, as the beam energy increases the proportion of elastically back-scattered electrons decreases so that in the energy range used in the SEMo ne has to go to glancing incidence to obtain sufficient elastically back-scattered electrons to give useful diffraction effects. The second region, ranging from just below the primary beam energy shown to about 50 eV, mainly represents electrons that have been scattered back through the surface of the specimen although small peaks are present, due to Auger electrons which may be used to identify elements present near the surface. The peak in the low energy region is due to secondary electrons excited by the primary beam which have escaped through the specimen surface. In this section we are concerned with the back-scattered electrons which for experimental convenience are defined as all those with energy greater than 50 eV. Auger electrons which occur at higher energies are also secondary electrons but do not contribute very much to the back-scattered current.

Electron back-scattering yields for normal beam incidence from a series of elements at different incident beam energies are given in Table II and Fig. 7. The two main points to note are that the back-scattering coefficient is almost independent of beam energy in the energy range 5–40 keV and that the yield increases monotonically with atomic number, the change being most rapid at low Z. The angular distribution of the back-scattered electrons is found to approximate to a Lambertian cosine distribution as shown in Fig. 8.

TABLE II. Measured Back-scattering Coefficients

		Incident energy (keV)		
	Z	5	10	30
Carbon	6	0·085	0·072	0·060
Aluminium	13	0·186	0·177	0·155
Silicon	14	0·197	0·186	0·162
Titanium	22	0·270	0·268	0·254
Chromium	24	0·285	0·283	0·270
Iron	26	0·300	0·296	0·288
Nickel	28	0·333	0·323	0·308
Copper	29	0·352	0·339	0·319
Zinc	30	0·352	0·342	0·330
Germanium	32	0·362	0·349	0·334
Molybdenum	42	0·367	0·381	0·385
Silver	47	0·418	0·420	0·420
Tungsten	74	0·472	0·483	0·501
Platinum	78	0·486	0·503	0·516
Gold	79	0·489	0·501	0·521
Uranium	92	0·495	0·513	0·534

The energy spectrum of back-scattered electrons, again for normal beam incidence is shown in Fig. 9. These spectra illustrate an important property of electron scattering in this energy range, that is, if the spectrum is plotted with reduced energy co-ordinates, the shape of the spectrum is almost independent of incident beam energy. From a practical point of view it is interesting to note that the main change in the spectra as the atomic number of the target increases is at high energies and as a result improved atomic number contrast in the SEM may be achieved by using an energy sensitive electron detector and also by discriminating against all those electrons with energy less than say, $0{\cdot}8E_0$.

So far all the results presented have been for normal beam incidence. At oblique incidence the back-scattering yield increases for reasons that can readily be appreciated from the diffusion model. In addition more energetic electrons will be able to escape from the specimen although qualitatively the same trends with beam energy and atomic

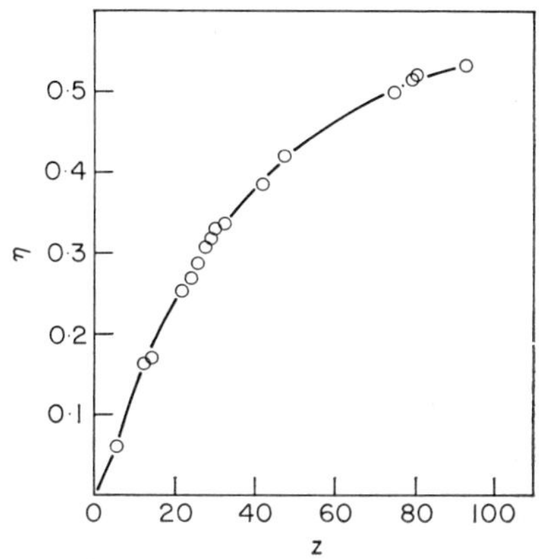

FIG. 7. Fraction of electrons back-scattered as a function of atomic number for normal beam incidence and an incident energy of 30 keV. (From Bishop, 1966, 1967.)

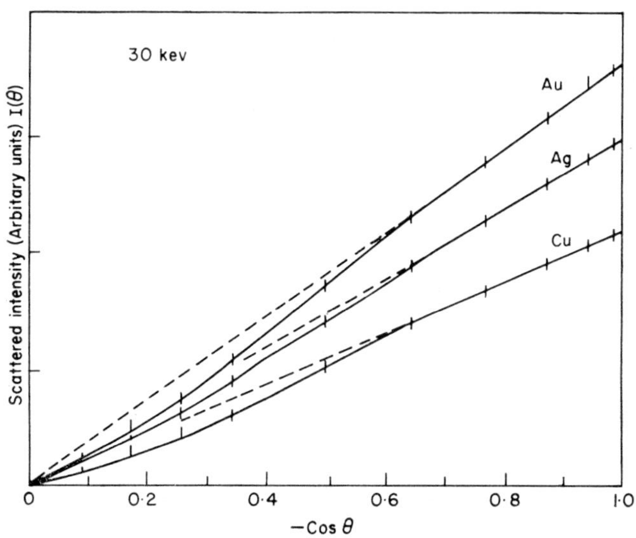

FIG. 8. Angular distribution of back-scattered electrons for normal beam incidence. (From Bishop, 1966, 1967.)

number are observed. Figure 10 gives a series of measurements by Kanter (1957) on the variations of η with the angle of incidence of the primary electrons.

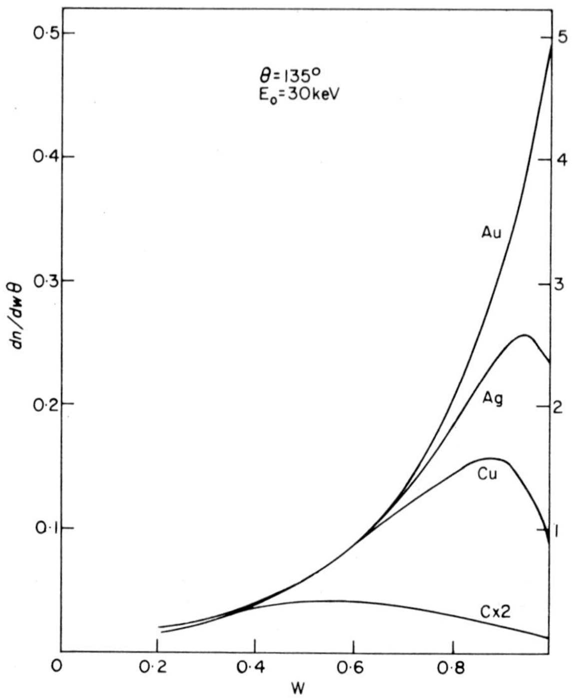

FIG. 9. Energy spectrum of back-scattered electrons. The reduced energy, $W = E/E_0$. (From Bishop, 1966, 1967.)

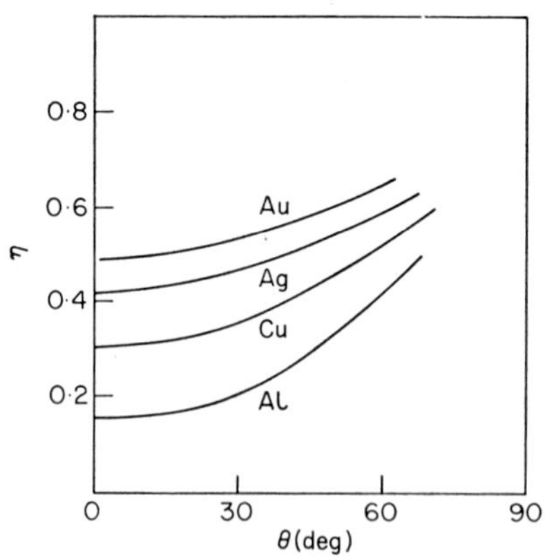

FIG. 10. Variation of back-scattering coefficient with angle of incidence of primary beam. (After Kanter, 1957.)

B. RANGE ENERGY RELATIONSHIPS

Measurements of electron penetrations are mostly made by measuring electron transmission through thin films. A comprehensive discussion of this type of measurement has been given in a series of papers by Cosslett and Thomas (1964). Experimentally the transmission through a film is measured as a function of incident beam energy giving a result of the type shown in Fig. 11. In practice it is found difficult to determine the cut off energy below which no electrons are transmitted, i.e. where the film thickness corresponds to the maximum range of electrons R_{max}. More reproducible results are obtained by extrapolating the transmission curve to cut the energy axis as shown in Fig. 11 defining a critical energy E_c corresponding to the extrapolated range R_x. Figure 12 shows values for R_x and R_{max} taken from Cosslett and Thomas (1964) which show the following features

(i) for a given element R_x and R_{max} are approximately proportional to R_B above 5 eV.
(ii) R_{max} is almost equal to R_B for lighter elements but is somewhat less than R_B for the heavy elements.
(iii) The mass penetration of the beam R_x is almost independent of atomic number.

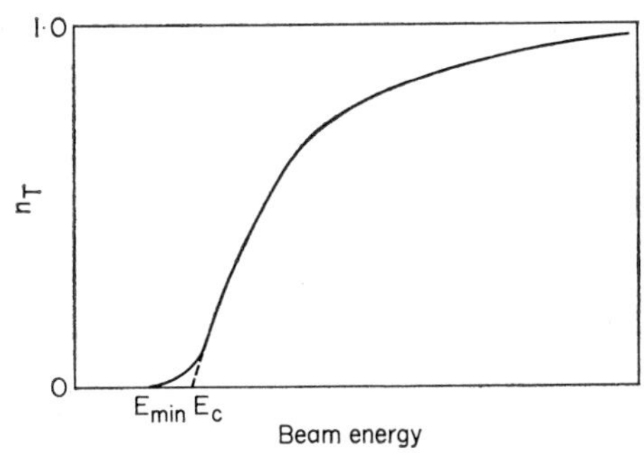

FIG. 11. Typical transmission curve for a thin film.

The first of these points illustrates a second important scaling property of electron scattering, that the linear dimensions of electron penetration may be scaled as the Bethe range. This is a very powerful property when combined with the energy scaling property noted for electron backscattering because if we know a set of scattering parameters such as penetration and energy loss at one energy we may with reasonable accuracy calculate the equivalent distributions at other energies for incident energies above 5 keV. Below 5 keV the situation is more complex and these simple scaling laws no longer apply.

The insensitivity of R_x to the atomic number of the target indicates that the increase in Bethe range with increasing atomic number is almost exactly balanced by the much higher angular scattering. Cosslett and

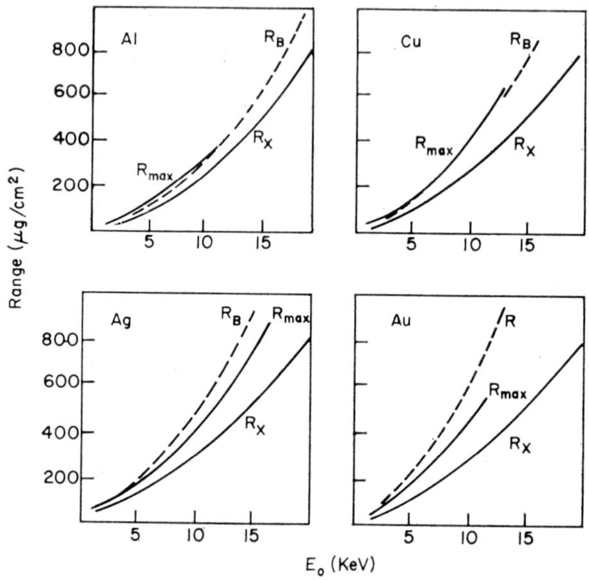

FIG. 12. Ranges of electrons in various elements. (After Cosslett and Thomas, 1964.)

Thomas have found that in the energy range 5–15 keV their experimental results for the elements aluminium, copper, silver and gold may be fitted within ±6% to an expression,

$$R_x = kE^{1\cdot 5}$$

with k = 7·4, 9·7, 8·6, 8·7 respectively for these elements where R_x is in $\mu g\,cm^{-2}$ and E is in keV. For order of magnitude calculations of electron penetration into a target it is reasonable to use this expression with a mean value of k = 8·6.

Cosslett and Thomas also discussed the energy loss of electrons transmitted through thin films. They found that the Thompson–Whiddington law which related the most probable energy E_p of transmitted electrons to the film thickness was only obeyed strictly if the mean energy loss E_m was used rather than E_p giving an expression

$$E_0^2 - E_m^2 = bx.$$

Experimental values of b/ρ obtained by Cosslett and Thomas are given in Table III. It is important to notice that b/ρ is a function of both atomic number and incident beam energy. This fact is sometimes ignored in empirical models of electron scattering.

TABLE III. Values of $b/\rho \times 10^{-11}$ (from Cosslett and Thomas, 1964).

E_0(keV)	9	15	18
Al	4·0	4·6	5·2
Cu	3·4	3·9	4·9
Ag	3·0	3·8	4·6
Au	2·7	3·1	3·8

C. ENERGY LOSS DISTRIBUTION

The major part of the energy lost by the incident electrons is soon degraded to electron hole pairs with a few eV energy close to the site of the primary energy loss. These ultimately lose their energy to the lattice as heat or in a phosphor a significant proportion produce photons. The electron hole pairs are responsible for a variety of phenomena, for example secondary electron emission, carrier production in semiconductors, cathodoluminescence and in some materials radiation damage, as well as heating of the specimen.

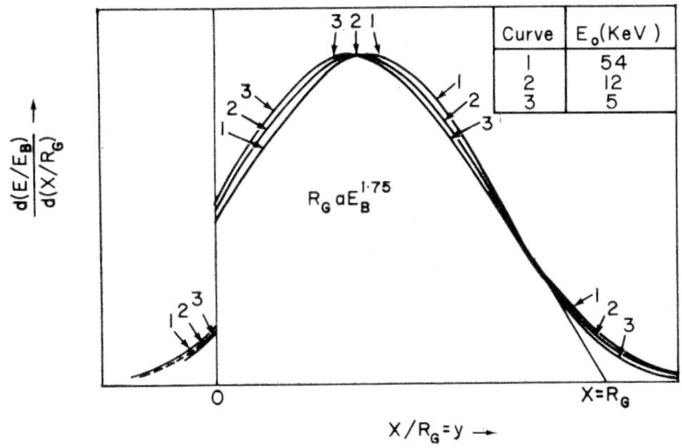

FIG. 13. Energy dissipation in penetration. (After Gruen, 1957.)

Gruen (1957) has measured the variation of energy dissipation with penetration distance (the depth dose distribution) for air and found that the shape of the distribution was almost independent of beam energy when plotted as a function of an extrapolated range R_G where

$$R_G = 4{\cdot}57\, E_0^{1{\cdot}75}\ \mu\text{g cm}^2$$

over the energy range 5–25 keV. As might be predicted from Fig. 13 the exponent in the power law here is higher than that obtained by Cosslett and Thomas who were working at lower energies with heavier elements.

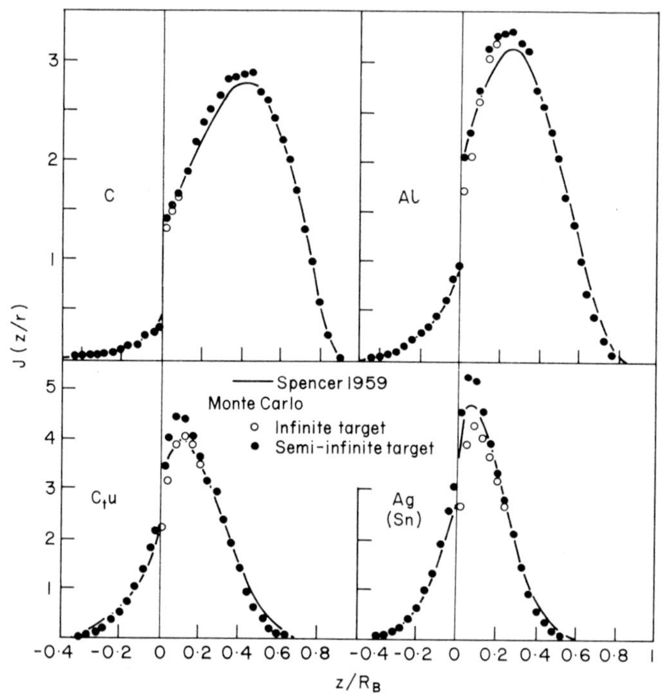

FIG. 14. Energy loss distributions normalized to the initial rate of energy loss of the incident beam. The negative values of z/R_B corresponding to scattering in the back hemisphere which is present for the infinite target.

Measurements have been made by Ehrenberg and King (1963) of energy dissipation profiles in phosphors which agree qualitatively with Gruen and more recently Everhart and Hoff (1971) have determined energy dissipation profiles from the beam induced conductivity in thin silicon dioxide layers. They found that for silicon

$$R_G = 3{\cdot}98 E_0^{1{\cdot}75}$$

and that the distribution function can be fitted to a polynomial

$$J(y) = 0{\cdot}60 + 6{\cdot}21 - 12{\cdot}4y^2 + 5{\cdot}69y^3$$

where y is the depth divided by R_G and that their results agree well with the equivalent distribution obtained from Monte Carlo calculations. Cosslett and Thomas also deduced energy dissipation profiles from their transmission measurements which agree well with Monte Carlo calculations for copper and gold. This general agreement between calculations and experiment means we can use the calculated distributions with some confidence. In Fig. 14 energy loss distributions for four elements are shown, comparing Spencer's (1955, 1959) results from fitting moments for an infinite target with Monte Carlo calculations for both infinite and semi-infinite targets. One should note how back-scattering in the semi-infinite targets reduced the energy dissipated near the surface particularly for heavy elements. However apart from a discrepancy at the surface the depth dose distributions tabulated by Spencer may be used at the beam energies used in the SEM.

V. Secondary Electron Emission

The third part of the energy spectrum of back-scattered electrons, the low energy secondaries, is the most widely exploited feature of electron scattering in the SEM. There is not space here to give a comprehensive discussion of secondary emission so we will confine ourselves to the major physical process involved. The origin of secondary electrons is the large number of low energy electrons generated in the energy dissipation process. Any of these electrons reaching the surface with an energy greater than the work function has a finite chance of escape. The secondary electron yield will be related to the energy dissipated in the surface layer, the range of the low energy secondary electrons in the target and the escape probability from the surface. The dominant factor is the range of the secondaries in the target. In conducting materials the secondary electrons can lose energy to conduction electrons and have relatively short ranges (a few tens of Å) with the result that yields are small, normally of the order of unity or less. For pure insulating materials on the other hand once the electron energy is less than approximately twice the band gap energy it can no longer lose energy to other electrons and can only give up its energy in relatively infrequent interactions with the lattice to produce phonons. The range of secondary electrons in insulators is thus much larger, resulting in very high secondary emission and yields as high as 20 have been observed. The presence of impurities leading to electronic states within the band gap reduces the range of secondary electrons and thus the secondary emission. Similarly an increase in temperature increases the probability of lattice scattering and again reduces the secondary yield for insulators whilst the secondary yield for conducting materials is independent of temperature.

The variation of total back-scattering yield with beam energy is illustrated schematically in Fig. 15. It rises to a maximum E_0^{max} between 100 and 800 eV for metals but the maximum may occur as high as 2000 eV for some insulators. The points where δ is unity are called the crossovers. At primary energies below E_0^{max} incident electrons lose their energy in a distance less than equal to the range of secondary electrons so the yield increases with beam energy, however above E_0^{max} the penetration of the primary electrons is significantly greater than the range of the secondaries and as the primary energy is increased the energy dissipated near the surface drops, leading to a reduction in secondary emission.

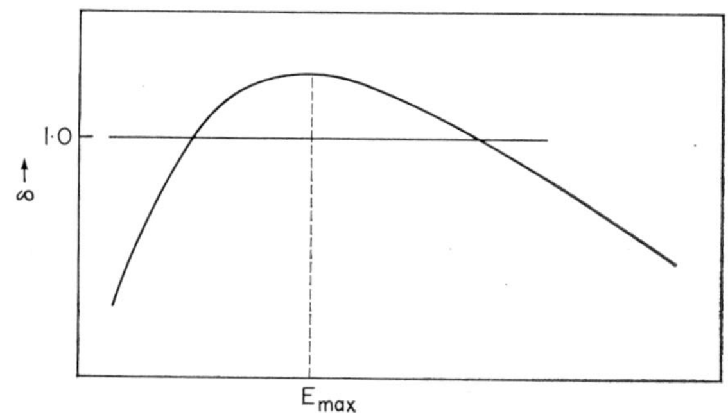

FIG. 15. Variation of secondary electron yield with incident beam energy.

Above E_0^{max} the secondary yield increases as the incident beam is moved away from normal incidence because the energy dissipated in the surface layer is increased. Bruining (1938) found an empirical expression for this variation

$$\ln(\delta(0)/\delta(\theta)) = k(1-\cos\theta)$$

where θ is the angle between the incident beam and the normal to the surface and k is a constant. Below E_0^{max} very little dependence of δ on the incident angle is found.

VI. Radial Spread of Electrons

The resolution that can be achieved in the SEM is ultimately limited by electron scattering effects. Rather less attention has been paid to this aspect of electron scattering in solids and our discussion will therefore be qualitative. The general shape of the electron energy loss distribution in copper is shown in Fig. 16. The spread of the scattering at the surface is rather less than the electron range. If we are using back-scattered electrons to produce contrast the resolution will therefore be related to the range of the electrons in the target. However, the actual resolution achieved will also depend to some extent on the detection system used. If the detector is energy sensitive better resolution will be achieved than with a simple current detector as the more energetic electrons on average leave the target closer to the incident beam than do less energetic ones. The situation for secondary electron detection is rather different. The radial origin of secondaries is illustrated schematically in Fig. 17, assuming an infinitely small incident beam. A large proportion of secondaries is produced by the incident beam and they are thus confined to a small peak whose width is determined by the range of secondary electrons in the target; the rest are produced by back-scattered primaries in a broad distribution whose size is of the order of the primary electron range. It is the sharp peak produced by the primary electrons that gives the SEM its high resolution the broad distribution just contributes a background level to the picture.

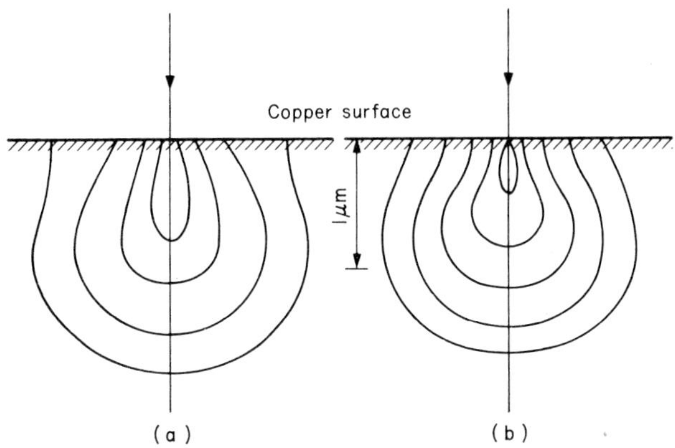

FIG. 16. Contours of equal energy loss (a) and characteristic K ionization concentrations (b) from a Monte Carlo calculation for a 29 keV electron beam incident normally on a copper target. (After Bishop, 1965.)

In the X-ray and cathodoluminescence modes of operation the resolution is in general poorer than in the corresponding electron modes. From Fig. 16 it may be seen that the radial spread is of the same order as the electron penetration. If a more precise value is required for the resolution it may be obtained experimentally, which is rather tedious, or may be calculated. Monte Carlo calculations have shown that the

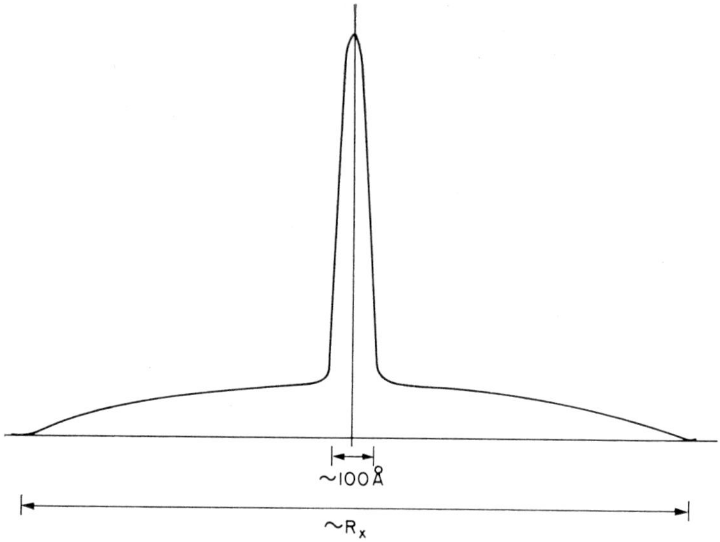

FIG. 17. Radial spread of secondary emission.

radial moments for electron scattering distributions do not differ significantly (Bishop, 1966, 1967) between infinite and semi-infinite targets. It is therefore relatively easy to calculate the mean square spread $\langle r^2 \rangle$ of X-ray and cathodoluminescent sources using the Lewis expressions for the moments of electron scattering distributions. For energy dissipation or cathodoluminescence which is proportional to energy dissipation

$$\langle r^2 \rangle = \frac{\int_0^{E_0} (1-\eta(E))\langle r^2(E)\rangle \, dE}{\int_0^{E_0} (1-\eta(E)) \, dE}$$

and for a source of characteristic X-rays

$$\langle r^2 \rangle = \frac{\int_{E_c}^{E_0} (1-\eta(E))\langle r^2(E)\rangle \Phi(E)\left(\frac{ds}{dE}\right) dE}{\int_{E_c}^{E_0} (1-\eta(E))\Phi(E)\left(\frac{ds}{dE}\right) dE}$$

where $\eta(E)$ is the fraction of electrons with energy greater than E, back-scattered from the target, $\Phi(E)$ is the ionization cross section, E_c

is the critical ionization potential and $\langle r^2(E)\rangle$ is the mean square radial spread for electrons with residual energy E. An expression for $\langle r^2(E)\rangle$ may be found in the paper by Bishop (1965).

VII. Concluding Remarks

It is hoped that this chapter has demonstrated that although direct calculations of electron scattering are relatively difficult it is possible to make reasonable estimates of electron beam scattering effects from an understanding of the basic mechanisms producing scattering, simple expressions derived from experimental results and the scaling properties of electron scattering distribution. The discussion here has necessarily been brief. A more comprehensive review of electron scattering in solids with many references is given by Thornton (1968) in his book on scanning electron microscopy.

References

Archard, G. D. (1961). *J. appl. Phys.* **32,** 1505.
Berger, M. J. (1963). *In* "Methods in Computational Physics", Vol. 1 (B. Alder, S. Fernbach and M. Rotenberg, eds). Academic Press, New York and London.
Bethe, H. A. (1930). *Ann. phys., Leipzig,* **5,** 325.
Bethe, H. A., Rose, M. E. and Smith, L. P. (1938). *Proc. Am. phil. Soc.* **78,** 573.
Bishop, H. E. (1965). *Proc. phys. Soc.* **85,** 855–865.
Bishop, H. E. (1966). Ph.D. Thesis, University of Cambridge.
Bishop, H. E. (1967). *Br. J. appl. Phys.* **18,** 703–715.
Brown, D. B. and Ogilvie, R. E. (1966). *J. appl. Phys.* **37,** 4429.
Bruining, H. (1938). *Physika,* **5,** 901.
Cosslett, V. E. and Thomas, R. N. (1964). *Br. J. appl. Phys.* **15,** 883–907; 1283–1300; and **16,** 779–795.
Dashen, R. F. (1964). *Phys. Rev.* **134A,** 1025.
Duncumb, P. and Reed, S. J. B. (1968). *In* "Quantitative Electron Probe Microanalysis", p. 133. Special Publication 298, National Bureau of Standards, Washington, D.C.
Ehrenberg, W. and King, D. E. N. (1963). *Proc. phys. Soc.* **81,** 751.
Everhart, T. E. (1960). *J. appl. Phys.* **31,** 1483.
Everhart, T. E. and Hoff, P. H. (1971). *J. appl. Phys.* **42,** 5837–5846.
Gruen, A. E. (1957). *Z. Naturforsch.* **12a,** 89.
Kanaya, K. and Okayama, S. (1972). *J. Phys.* **D. 5,** 43–58.
Kanter, H. (1957). *Ann. phys. Leipzig,* **20,** 144.
Landau, L. (1944). *J. Phys. USSR,* **8,** 201.
Lewis, H. W. (1950). *Phys. Rev.* **78,** 526–529.

SCOTT, W. T. (1963). *Rev. Mod. Phys.* **35**, 231.

SHINODA, G. (1969). *In* "Electron Probe Microanalysis", Advances in Electronics and Electron Physics, Supplement 6, pp. 15–44. Academic Press, New York and London.

SPENCER, L. V. (1955). *Phys. Rev.* **98**, 1597–1615.

SPENCER, L. V. (1959). National Bureau of Standards Monograph No. 1.

THORNTON, P. R. (1968). "Scanning Electron Microscopy." Chapman and Hall, London.

Chapter 3

Scanning Microscopes: The Approach to 1Å*

ALBERT V. CREWE

The University of Chicago, Chicago, Illinois, U.S.A.

Symbols

I_0 = intensity of electron beam at source
I_2 = intensity of electron beam at anode
β = source brightness, measured at some voltage V_1
R = ratio of final accelerating voltage V_0 to V_1 (the potential at the source)
δ_0 = radius of Gaussian image of the source
α_0 = convergence angle of the probe
r = radius of electron trajectory in lens
δ_s = radius of aberration disc of lens
A = a constant
S = distance of the focus from the lens
C_s = spherical aberration coefficient of the lens
α = semi-angle of the lens
L = $S+d$, where S = distance of focus from lens and d = diffraction of lens
δ_c = chromatic aberration coefficient of the lens
B = a constant
λ = de Broglie electron wavelength
M_L = lens magnification
f = focal length of lens

I. Introduction

The scanning transmission microscope has many advantages over the conventional transmission microscope. Among these are the fact that the various scattering events which take place in the specimen can

* Work supported by the U.S. Atomic Energy Commission.

readily be separated. Inelastically scattered electrons can be separated from the elastically scattered electrons (Crewe *et al.*, 1970), and the information from the microscope comes out in a well defined time sequence which can be readily handled by an on-line computer. The major disadvantage of the scanning microscope is that at the present time the number of image points is restricted by the limitations of the display systems available. This, however, can be overcome, at least in principle.

A scanning microscope can be used to provide all the types of micrographs that can be obtained with a conventional microscope such as bright field, dark field and phase contrast of lattice images (Crewe and Wall, 1970a). With the exception of the last mode the images are obtained at the expense of far less specimen damage and since the output of the microscope is an electrical signal it can be amplified and be suitably biased with d.c. to produce higher contrast.

In addition to the normal contrast modes a variety of others are available which can be of significant value. For example, energy loss electrons can be used to produce contrast either alone or in conjunction with elastically scattered electrons (Crewe *et al.*, 1970). The reduced specimen damage deserves further explanation since confusion seems to exist on this point in the literature (Scherzer, 1971). We give an explanation of this subject from an information theoretic point of view.

An object is visible in any microscope only if it scatters electrons. Scattering will be both elastic and inelastic. Either or both types of event can be used to provide contrast and either or both can cause specimen damage. A micrograph of given quality requires a given number of electrons incident upon the emulsion or detector. We will assume that the efficiency of detection is unity. That is, one electron will produce a detectable blacking of the emulsion or an electrical signal.

The least possible specimen damage for a given micrograph will occur when every electron which is scattered is used in the micrograph. More damage per picture will occur when the transmission of the optical system has an efficiency less than unity.

It is beyond the scope of this chapter to calculate transmission efficiencies for all forms of contrast and all microscope configurations, but we will indicate the argument for dark-field contrast.

For high resolution micrographs the objective aperture of a conventional microscope is approximately 10 mrad and this is the beam convergence angle which would be used in the equivalent scanning microscope. The mean angle for elastic scattering is several times this value and, in fact, the ratio between these two angles is very closely equal to the resolution of the microscope in Ångstroms.

Dark-field contrast in a scanning microscope can be achieved by surrounding the cone of unscattered electrons with an annular detector. In this way an efficiency of 0·8 can easily be reached, for a mercury specimen at 100 kV with a 2 Å resolution.

In a conventional microscope dark-field contrast is obtained either by the use of tilted illumination or by means of a central stop in the aperture. In either case the requirements of imaging mean that a defining aperture of about 10 mrad must be used. This is considerably smaller than the angular distribution of scattered electrons so that the maximum efficiencies which can be obtained are 0·09 and 0·15 for the two cases (again for mercury).

Consequently it is clear that the scanning microscope produces at least 5 times less damage than a conventional microscope. (For a light element such as carbon the figure is 2 times less damage.)

If one includes consideration of inelastic scattering processes the argument in favor of the scanning microscope is even better.

Complete calculations of this kind will be given in a forthcoming paper.

Because of the advantages of the scanning microscope principle, it is obviously advantageous to attempt to improve its performance as much as possible. This performance is determined by the quality of its electron optical components of which there are three—namely, the electron source, the electron gun and a system of lenses.

II. Electron Source

The intensity of the electron probe can be shown to be

$$I_0 = I_2 = \beta\pi^2 R \,.\, \delta_0^2 \alpha_0^2$$

(Crewe, 1972), where β is the brightness of the source measured at some voltage V_1, R is the ratio of the accelerating voltage V_0 to V_1, δ_0 the radius of a Gaussian image of the source, and α_0 the convergence angle of the probe. If we are to attain high resolution, then δ_0 must be small, namely smaller than the desired radius of the probe. The values of α_0 and R are determined by the resolution which is required in the microscope. The brightness β is therefore a fairly well defined quantity, and in order to have sufficient intensity in the probe, the brightness should exceed about 10^8 A cm^{-2} sr^{-1} at 3 kV. This requirement can only be met by a field emission source, and it is for this reason that we have concentrated on developing this kind of electron source (Crewe *et al.*, 1968).

For the present purposes we only point out the following facts.

Field emission sources must be operated in an ultra high vacuum environment and at pressures approaching 10^{-10} Torr. They can easily be operated at a brightness of 10^9 A cm^{-2} sr^{-1} with a total emission count of 10–100 μA and an effective source radius of 10 Å. Lifetimes vary from a few hundred hours to a few thousand. In recent years our tips have had an average life of 3 or 4 months with the record standing at 10½ months. The energy spread of electrons from the source is 0·2 V when, as in our case, no heating current is used.

A. THE ELECTRON GUN

The function of the electron gun is to accelerate electrons from a potential V_1 to a final voltage V_0. In so doing, any electron gun will act as a lens and produce an image of the electron source at some point in space. This image can be either real or virtual. Therefore, the electron gun introduces magnification and, unfortunately, it also introduces aberrations. In this context it is useful to consider the electron gun as a lens and to compare its performance with what one would expect of a thin lens.

The electron gun which we used has often been described in our publications (Crewe *et al.*, 1968, 1970), and consists of two anodes at potentials V_1 and V_0. These anodes are carefully shaped in order to reduce aberrations. We know of no other configuration with smaller spherical aberration and therefore this is the only gun we shall consider. We have previously published data on the various aberration coefficients and other properties of our electron gun, but we have recently found it useful to recast this information by comparing the gun with a lens.

1. *Magnification*

Data on magnification of the gun has already been published, and it will not be repeated here.

2. *Spherical Aberration*

We shall first review what we would expect from a thin lens. A perfect thin lens is one in which the angle of deflection in passing through the lens is proportional to the radius of the electron trajectory in the lens. The spherical aberration effect is due to imperfections in the lens which add a term proportional to r^1, which is the lowest order term which can occur.

The radius of the aberration disc is therefore

$$\delta_s \simeq Ar^3 . S$$

where A is some constant and S is the distance of the focus from the lens.

Normally we define

$$\delta_s = C_s \alpha^3$$

(Haine, 1961) where the spherical aberration coefficient is given by:

$$C_s = \frac{5f^3}{(L)^2}$$

for a thin lens where $L \sim (S+d)$, the dimensions, and f the focal length of the lens. This is defined for an incoming beam which is parallel to the axis so that

$$\delta_s = \frac{5r^3}{(L)^2}.$$

where r is the radius of the source. Therefore we can write

$$\delta_s = A' \frac{r^3 S}{f}$$

where A' is determined by the dimensions of the lens.

In the case of our electron gun we find that we can also write

$$\delta_s = A r^3 S$$

where A is a constant for any given voltage ratio. Roughly speaking $A \sim 8 \cdot 10^{-3}\ R^2\ \mathrm{cm}^{-3}$ for a 3 cm gun.

This is a somewhat surprising result since the gun is not small compared to other dimensions of the system. However, most of the focusing takes place in a small region near the first anode. We have calculated the value of A for distances varying from 10^4 cm to 1 cm outside the gun. The variation in A is less than 20% and may reasonably be taken to be constant (Crewe, 1972).

The value of A is given such that when all quantities on the right are in cm, the value of δ_s is in Ångstroms. (The same convention is used for chromatic aberration.)

B. CHROMATIC ABERRATION

For a thin lens the effect of using electrons of energy $V+\Delta V$ instead of V is to introduce an extra deflection in the lens such that

$$\frac{d\alpha}{\alpha} = \frac{\Delta V}{V}.$$

Therefore the radius of the aberration disc is

$$\delta_c = \alpha S \,.\, \frac{\Delta V}{V}$$

$$= \frac{rS}{f} \cdot \frac{\Delta V}{V}.$$

In the case of the electron gun we find that we can write

$$\delta_c = BrS\frac{\Delta V}{V}$$

and roughly speaking $B \sim 2\times 10^{-2}\ R^2\ \text{cm}^{-1}$.

Surprisingly, then, the electron gun behaves very much like a thin lens in that the parameters A and B are almost independent of S and are constant for any given operating condition of the lens. The properties of the electron gun can therefore be defined by two universal curves, one for A and one for B, which are accurate to about 20% for any real image produced by the gun (Crewe, 1972).

C. Lens

The final probe-forming lens can produce a probe of a radius given by the standard equation

$$\delta = 0{\cdot}43\ C_s^{\frac{1}{4}}\lambda^{\frac{3}{4}}.$$

However, we must take into account any contributions due to the electron gun and the probe must have adequate intensity.

The lens system will change the size of the gun image by virtue of the lens magnification M_L. It is clear, then, that all the effects of the gun can be characterized by the product SM_L.

Our calculations show that at 100 kV, for instance, if $SM_L < 0{\cdot}2$ the effects of the gun can be neglected (Crewe, 1972).

As an example, we take $\alpha_0 = 12$ mrad, $R = 10$ ($V_1 = 3$ kV, $V_0 = 30$ kV) and β we take to be $5\times 10^8\ \text{A cm}^{-2}\ \text{sr}^{-1}$ which is our normal operating condition. Then we have

$$I_0 = 3{\cdot}7\times 10^{-10}\ \text{A}$$

where δ_0 has a value of 0·7 Å.

III. 30 kV Microscope

Descriptions of our 30 kV microscope have been published elsewhere (Crewe and Wall, 1970b). It was previously characterized as a 5 Å microscope, but recent improvements in the alignment and a more careful optimization of the aperture have improved the resolution to approximately 3 Å.

The theoretical performance can readily be calculated. The expected value of C_s of the lens is 0.4 mm. At 35 kV we have

$$\delta = 0{\cdot}43\ C_s^{\frac{1}{4}}\lambda^{\frac{3}{4}} = 2{\cdot}52\ \text{Å}.$$

The contribution of source size is approximately 0·7 Å and chromatic aberration 0·8 Å. These three should probably be taken in quadrature giving a theoretical resolution of 2·74 Å.

IV. 100 kV MICROSCOPE

We have also designed and built a 100 kV microscope which should have a resolution of about 2 Å (Crewe and Retsky, 1972).

The source and gun are arranged to produce a focus close to the exit of the gun. A condenser lens of 5 cm focal length will be used to produce a parallel beam which will then enter the objective lens which has a focal length of 1 mm.

The scanning system, stigmator and defining aperture are placed in the parallel beam section.

As in our other microscope, we shall use an annular detector for the elastically scattered electrons and a spectrometer to separate the inelastically scattered electrons.

Two vacuum pumps each of 150 l sec^{-1} are used, one for the gun region above the aperture and one for the specimen region below.

This microscope represents about the best we can do by using conventional techniques. It would be difficult to improve on its theoretical performance since the focal length is already so small.

The gap between the pole pieces is now 2·4 mm and reducing this any further would not allow specimen tilt.

This microscope is now being put into operation. The high voltage has been applied and field emission obtained. All power supplies are stable to 1 ppm h^{-1} (Retsky and Wall, 1972). Attaining the design value of resolution will require some modification of the electronic system and careful attention to stray magnetic fields and vibrations.

V. FUTURE MICROSCOPES

There are two possible avenues to improvement of resolution, reducing C_s by correcting it or reducing λ.

Correction of C_s. In order to correct spherical aberration it is necessary to introduce a deflection of the beam proportional to r^3. This cannot be done in cylindrical symmetry. However, it can be done in one dimension using an octupole lens. Such a lens produces a deflection proportional to r^3 in one plane of symmetry and a deflection proportional to $-r^3$ in the other plane. The two planes are at an angle of 45°. Quadrupole lenses must be used to produce the necessary line foci for the octupole lenses. Scherzer (1949) has shown how this can be done, and Deltrap (1966) has shown that such a system can be made to work.

We have chosen to use a system similar to the one of Deltrap. Four quadrupole lenses are used in such a way that a parallel incident beam produces a parallel outgoing beam. Two intermediate line foci occur inside two octupole lenses and a third octupole is placed at the centre of the system.

Computer simulation studies have shown that such a system can be made to work and normal machining tolerances are adequate.

This system is now being constructed.

A. HIGH VOLTAGE MICROSCOPE

We are currently designing a 1 MeV scanning microscope. With a short-focus lens this microscope should have a resolution of ≈ 1 Å.

The requirements for a scanning microscope at 1 MeV are more stringent than for a conventional microscope. In particular the vibration problem can be severe. For this reason we must make the instrument as compact as possible. At the moment it looks possible to reduce the length of the acceleration column to 24 in. by using a "three stack" system—one column to generate the voltage, one for the measuring resistor, and one for the column.

The objective lens will be made as small as possible using high current density coils. One condenser lens will be used, together with a spectrometer.

At the moment it appears possible to build a system with an overall height of 9 ft.

VI. SUMMARY

We have demonstrated a resolution of approximately 3 Å with an accelerating voltage of 35 kV. A resolution of 2 Å can be expected from our new 100 kV instrument and perhaps 1 Å at 1 MeV. A resolution of 1 Å may also be possible with our 100 kV instrument if the spherical aberration of the lens can be successfully corrected.

REFERENCES

CREWE, A. V. (1970). *Q. Rev. Biophys.* **3,** 137.

CREWE, A. V. (1972). *In* "Progress In Optics", in press.

CREWE, A. V. and RETSKY, M. (1972). *Proceedings, 30th Annual Meeting Electron Microscopy Society of America* (in press).

CREWE, A. V. and WALL, J. (1970a). *Optik*, **30,** 461.

CREWE, A. V. and WALL, J. (1970b). *J. molec. Biol.* **48,** 375.

CREWE, A. V., EGGENBERGER, D., WALL, J. and WELTER, L. (1968). *Rev. scient. Instrum.* **39,** 576.
CREWE, A. V., LANGMORE, J. and WALL, J. (1970). *Science, N.Y.* **168,** 3937.
DELTRAP, M. (1966). Ph.D. Thesis, University of Cambridge.
HAINE, M. E. (1961). "The Electron Microscope." E. and N. F. Spon, Ltd. London.
RETSKY, M. and WALL, J. (1972). *Rev. scient. Instrum.* **43,** 384.
SCHERZER, O. (1949). *Optik*, **5,** 497.
SCHERZER, O. (1971). *Optik*, **27,** 221.

Chapter 4

The Application of Image Analysis Techniques to Scanning Electron Microscopy and Microanalysis

D. W. GIBBARD

Image Analysing Computers Ltd., Melbourn, Royston, Herts, England

Symbols

P_e = perimeter of object
V = volume of object
A = area of object
P_h = horizontal projection of object
P_v = vertical projection of object
F_h = horizontal Feret diameter of object
F_v = vertical Feret diameter of object

I. What is Image Analysis?

Image analysis is the process of extracting information from a "picture". It is most often performed by human beings making some sort of judgement about photographs, about what they see down a microscope, on the phosphor screen of an electron microscope, on the CRT of an SEM, etc. Human beings are generally good at making qualitative judgements, recognizing patterns and trends, and the thought process by which decisions are arrived at is not well understood.

However, human beings are generally bad at making quantitative measurements, counting objects and estimating size. They do this slowly and make errors through getting tired and distracted. They can also be misled or unintentionally select a non-representative population of objects for consideration (Fig. 1). This clearly defines occasions when machines should be considered. They *could* give:

(a) Objectivity and unbiased selection.
(b) Quantitative measurements of numbers and sizes of objects.
(c) High speed measurements and freedom from fatigue.

The questions that must be asked are:

(a) Feasibility—does the machine have the power to recognize what it is to measure and then the ability to measure it?
(b) Economics—most things usually boil down to how much money is available for the job.

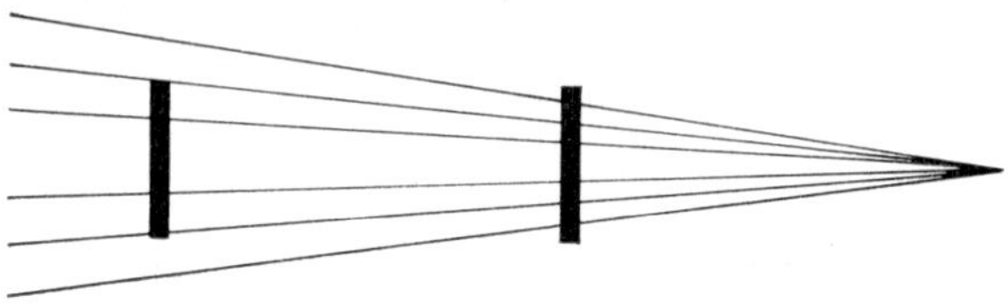

FIG. 1. An example of how a human observer can misinterpret picture information. The 2 vertical lines are actually the same size, although the right hand one appears larger.

II. Development of Image Analysis

In the 1950s Roberts and Young attempted applications (biological) too difficult for then current technology, and this pioneering work met with comparatively little success (Roberts and Young, 1956). Commercially successful equipment did not arrive until the 1960s with IMANCO'S early "Quantimet" machine (Fisher, 1966; Fisher and Cole, 1968). They were soon joined by Leitz, Zeiss, and Bausch and Lomb all offering analogue T.V., fixed raster scanning systems for specimens viewed through light microscopes. Applications were consequently limited until 1970 when IMANCO introduced the digital, modular "Quantimet" 720 (Fisher, 1971) with image scanning system designed for image analysis which allowed images of finer detail to be analysed, and had a low noise level and corrections for image shading which allowed images with many grey levels to be analysed. The later addition of pattern recognition facilities added computing power (Gibbard *et al.*, 1972). Applications of image analysis were developed, for example to epidiascopes—for analysing photo-micrographs (Fig. 2) etc., to film and slide scanners and to transmission electron microscopes (Strang, 1969; Crawley and Gardner, 1972), and scanning electron microscopes (Braggins *et al.*, 1971; Figs 2 and 4). At the same time, other workers have been using conventional digital computers for image analysis (White *et al.*, 1970; Ingram and Preston, 1970).

Such systems have not generally been available commercially, the high cost and limited operating speed restricting the potential market.

It is true, however, that such systems can be very powerful in terms of object discrimination and pattern recognition, whilst fixed raster scanning systems with special hardware begin to get expensive as their computing power is increased. The future probably lies with hybrid systems offering fast operation and vast computing power, whilst there will always be a market for the lower cost limited performance systems.

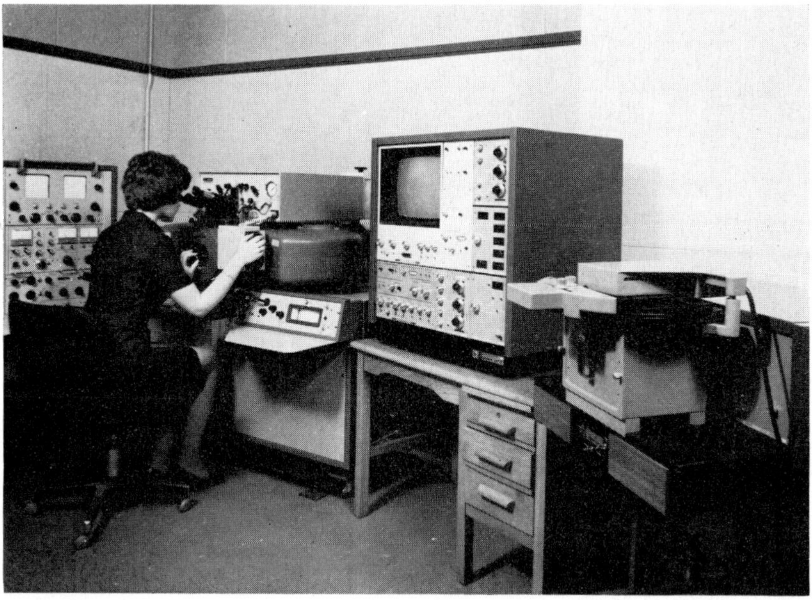

FIG. 2. An image analysis installation coupled to a Cambridge Instruments "Microscan" for direct image processing. Note also there is an alternative input device, an epidiascope, suitable for analysing photo-micrographs etc.

III. IMAGE ANALYSIS SYSTEMS

A. GENERAL SYSTEM

Figure 3 shows a block diagram of a typical image analysis system. The important parts are:

1. *Image Producing and Scanning*

The system will be no better than the image it is working on. The electronic signal produced by the scanner must be an adequate description of the specimen for subsequent computation. This means:

(a) high resolution, for accurate definition of shapes and ability to handle a large range of particle sizes;

(b) low distortion and faithful geometrical reproduction;
(c) low noise and correction for image "shading" for good grey level discrimination.

For example the Quantimet 720 has a specification which includes an image definition of over 6×10^5 "picture points", total geometrical distortion less than 1% and grey level discrimination up to 30 levels.

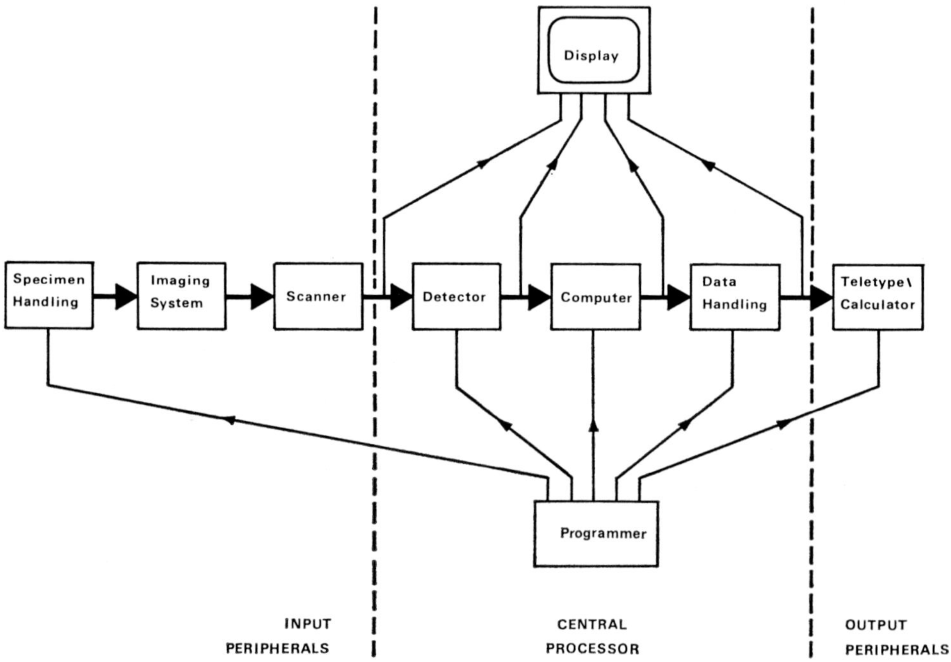

The Quantimet 720 Block Diagram.

FIG. 3. Block diagram of an Image Analysis System.

2. *Detection*

In image analysis terms this is the process of extracting regions of interest from the image for computation. It is most often simply the application of a grey level threshold or thresholds. For a description of the techniques and some interesting new developments, see Wadlow, 1972.

3. *Computation*

This is the conversion of the digitized signal into "data". These data can range from a simple measurement of the area of a grey "phase", through to a particle count, taking into account size and shape factors (see section IIIC.)

4. *Display*

A display is essential in image analysis because it enables the machine to be set up initially (or programmed) and provides a constantly available check on the operation. As well as the image, detected and computed functions may be displayed, superimposed on the scanned image.

5. *Automation*

In order to take full advantage of image analysis machines it is desirable to be able to run a program of measurements automatically, including automatic selection of many fields on the specimen and further data processing, e.g. statistical analysis of specimen data obtained from many fields by use of an on-line digital computer.

B. THE IMAGE ANALYSIS SYSTEM APPLIED TO THE SEM

The SEM takes the place of both the image producing device and the scanner. Figure 4 shows how the SEM signals are processed

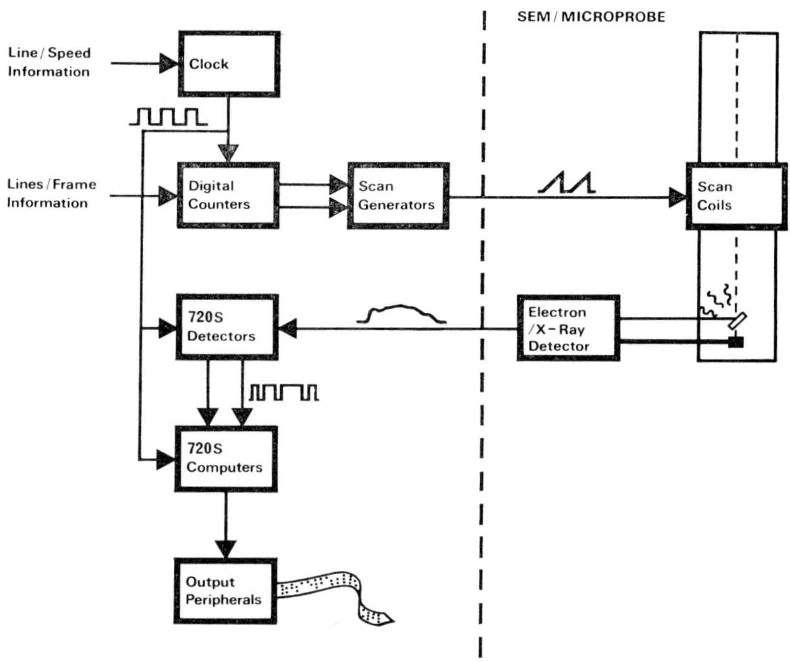

FIG. 4. Block diagram of an SEM/Image Analysing computer link.

directly by the computer. Since several signals are available simultaneously, this enables the computer to analyse objects using these different descriptions, e.g. the objects to be counted may be identified by one signal but the criteria used to classify them defined by another—see later section. In this arrangement, the SEM is controlled by scans generated in the computer from a very stable crystal-controlled digital "clock". The same "clock" also controls the detection and computation, giving precise geometrical fidelity. The SEM display is used to display computer functions as well as the images.

Two examples of the simpler type of applications are shown in Figs 5 and 6. Figure 6 in particular shows the limitation of simple computation techniques and the necessity for having what is loosely termed "pattern recognition".

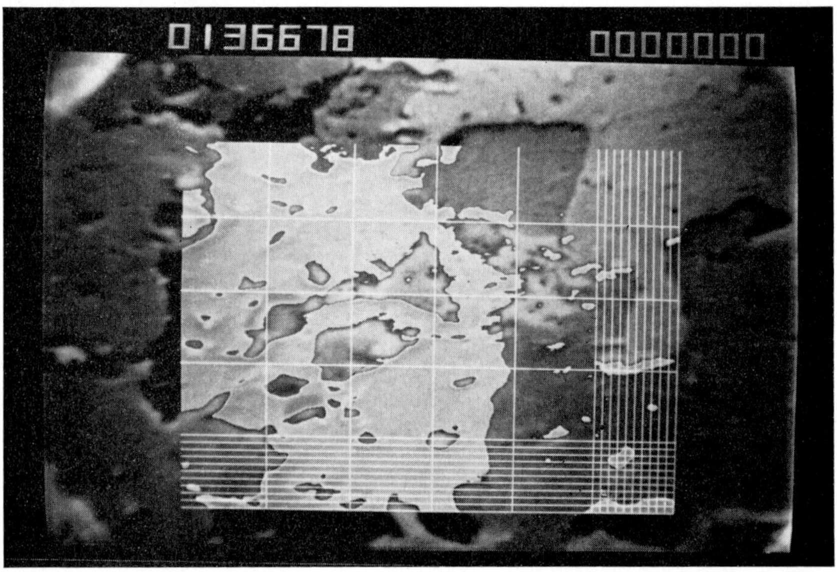

FIG. 5. Mineral section, specimen current image. Darkest "phase" is detected and its area measured within the region with the electronic graticule. The number displayed is the area in picture points.

C. PATTERN RECOGNITION

1. *The Need for Pattern Recognition*

Particle counting techniques have, until recently suffered from a limited range of acceptance criteria, usually taking the form of discrimination on the grounds of "*maximum chord*", or Martin's diameter.

There are many applications where this is an unsuitable parameter and more comprehensive descriptors are needed, e.g. area, density, perimeter (e.g. Fig. 6).

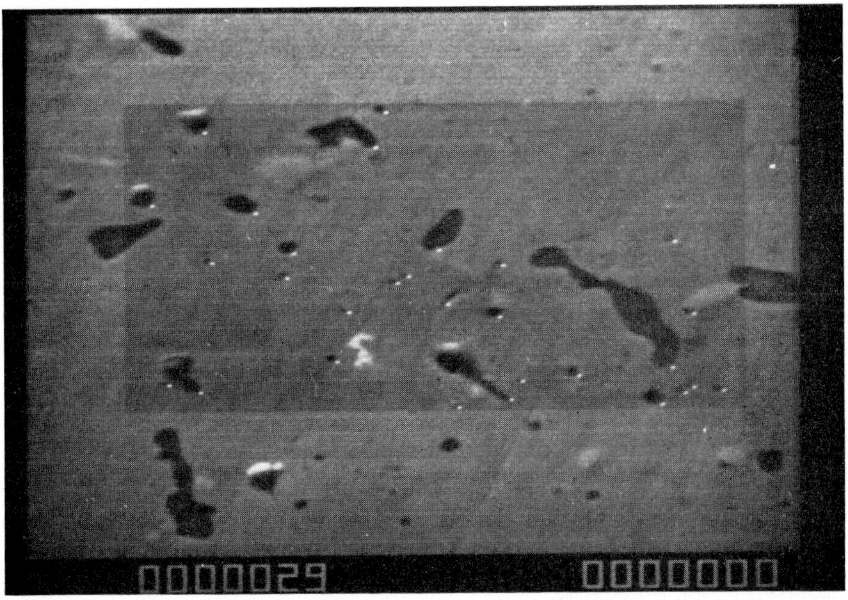

FIG. 6. Manganese sulphide inclusions in steel, secondary electron image. Inclusions in the central region are being counted and identified by the bright "count flags". Size classification of these inclusions by any descriptor other than their individual areas would have little meaning, due to their random shapes.

These descriptions can then also be combined to enable shapes to be separated, as illustrated in Fig. 7. This is important for two reasons:

(i) to separate an image of mixed groups of objects into groups for independent size distribution analysis, e.g. the flaky and globular groups of inclusion in cast iron (Fig. 8). Another application of this is where objects occur singly, or in agglomerated groups. The groups may be eliminated by virtue of shape to allow only the single particles to be analysed e.g. powder sample (Fig. 9).

(ii) To separate objects from unwanted background it is often found, particularly in biological applications, that it is impossible to "detect" the object to be analysed without also detecting unwanted areas, e.g. damaged tissue (Fig. 10). One way of overcoming such problems is by laborious light pen methods (Fig. 11) using the human operator as a pattern-recognition extension to the machine. This, however, removes the speed advantage of automatic analysis and is difficult to use on the SEM, making automatic methods of pattern recognition clearly desirable.

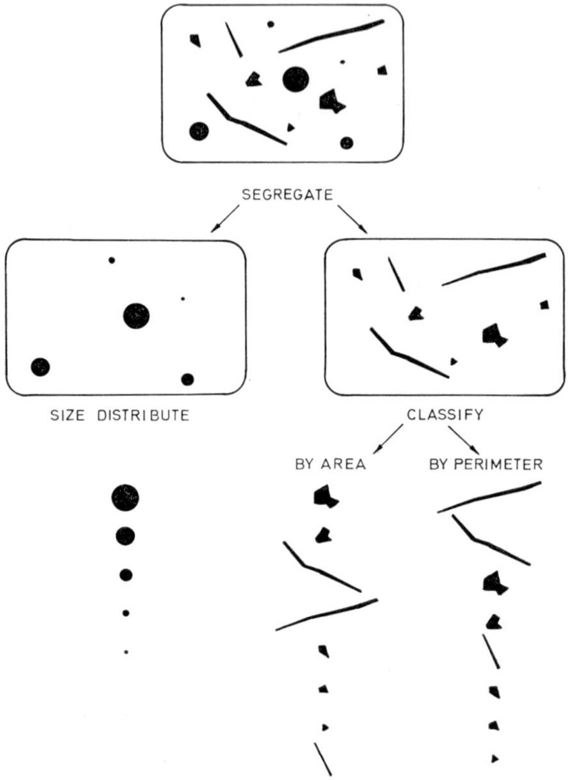

FIG. 7. The basic pattern recognition problem is separating mixed populations into their different sets. Full geometrical and densitometric descriptions of every object in the field allows this to be done in real time while retaining the data on the wanted set for subsequent classification and analysis.

There is a whole set of problems which may be called *topological* where an object is analysed in relation to its surrounding background or neighbouring objects. To be included for analysis it may be necessary for an object to surround, or touch or be enclosed by another type of object or phase, for example Fig. 12 shows the selection of only those cells surrounded by cytoplasm by a Quantimet "Cytoscreen" (Sturgess, 1971). Another example is shown in Figs 13 and 14 where Quantimet 720 is recognizing only the fish surrounded by the net.

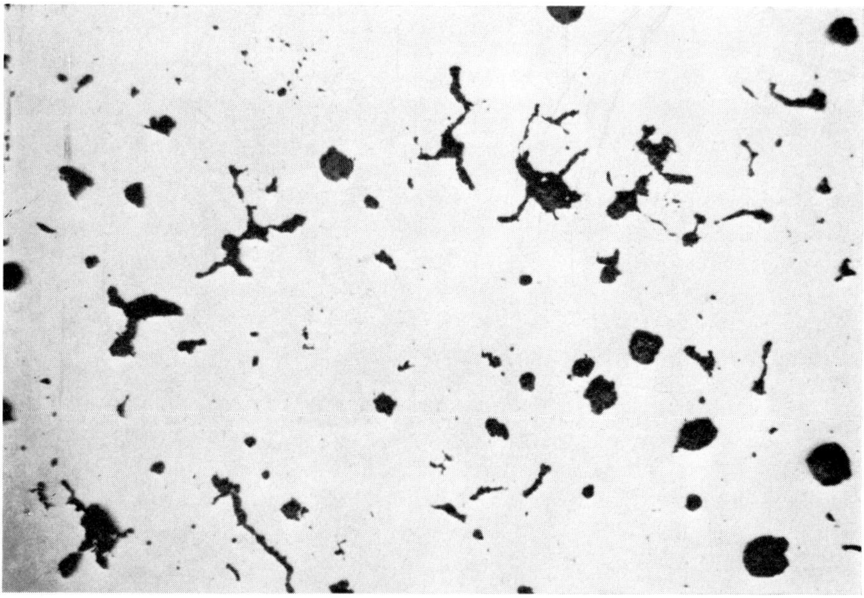

FIG. 8. Graphite inclusions in cast iron consist of 2 types, flaky and globular, and these must be analysed separately.

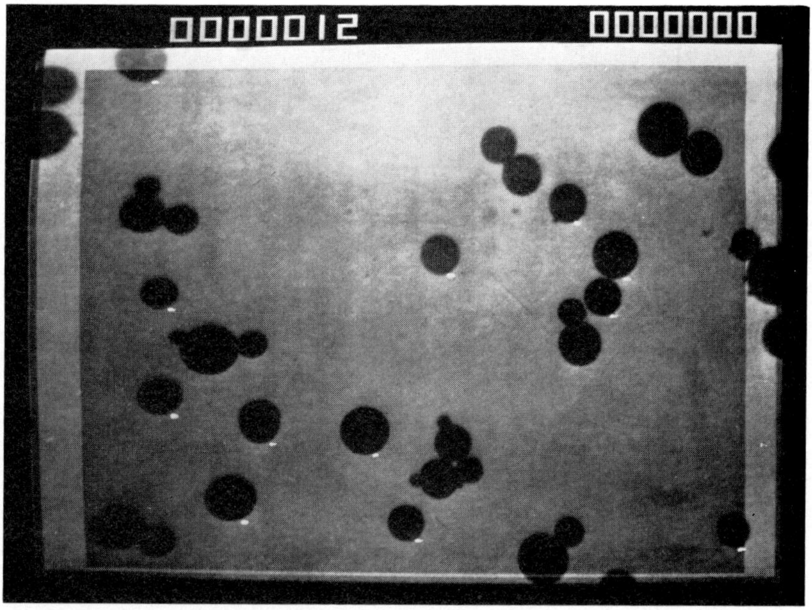

FIG. 9. Agglomeration is a common problem in particle sizing and errors are made without pattern recognition. Using the dimension less form factor, Area/Perimeter^2, the 720 Pattern recognition systems can avoid counting the agglomerates.

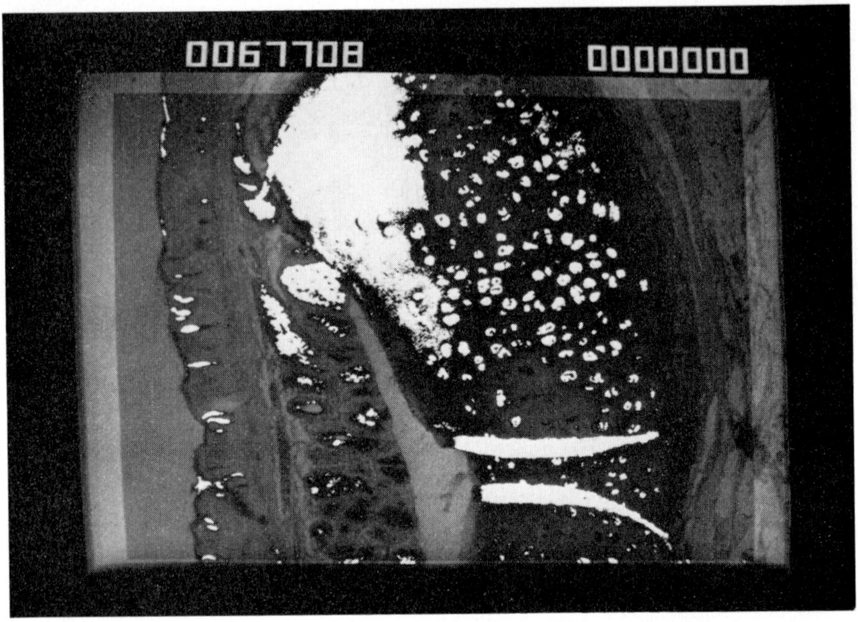

FIG. 10. Dark unwanted tissue confuses a simple detector which cannot select the individual glands in this stained section of cat trachea.

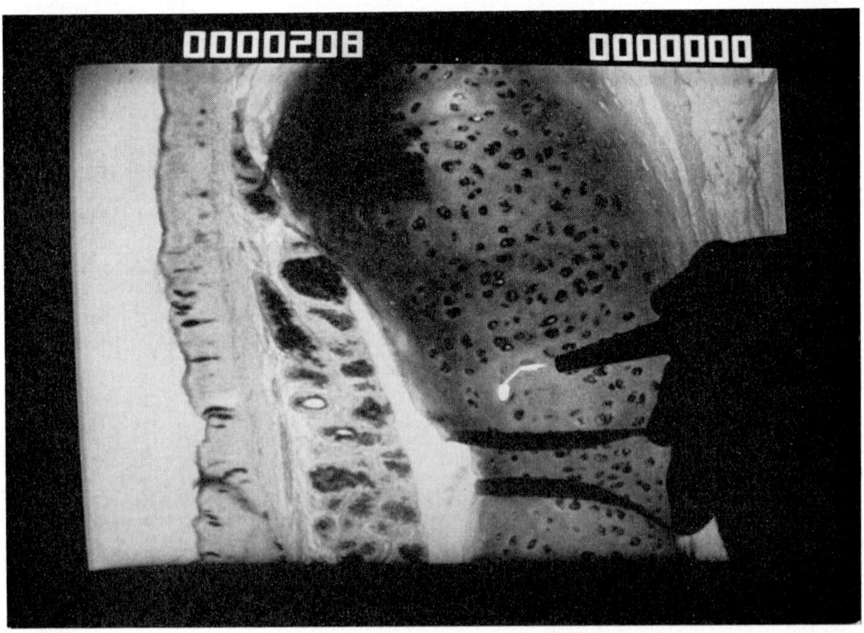

FIG. 11. An operator performing the pattern recognition operation using a light pen.

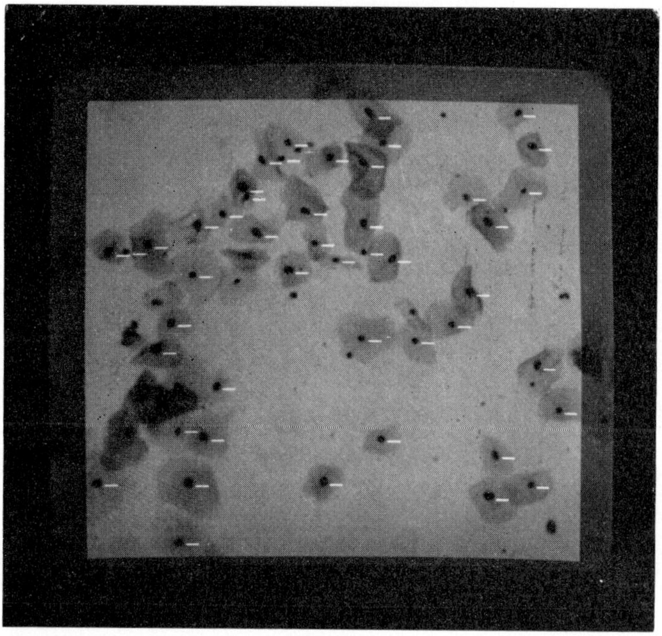

FIG. 12. Quantimet Cytoscreen identifies cell nuclei (marker pulses). One of many pattern recognition techniques used is that cell nuclei must have cytoplasm surrounding them.

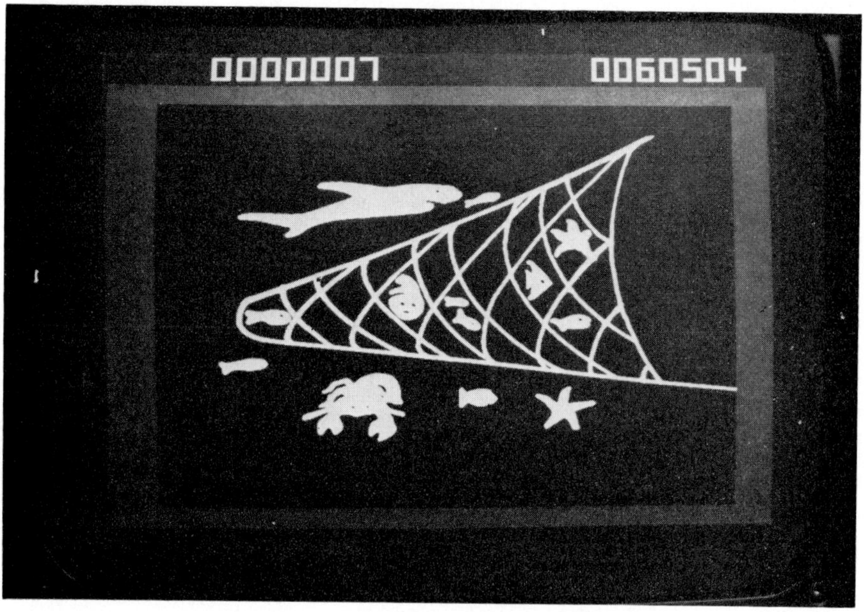

FIG. 13. Fish and net are detected all together.

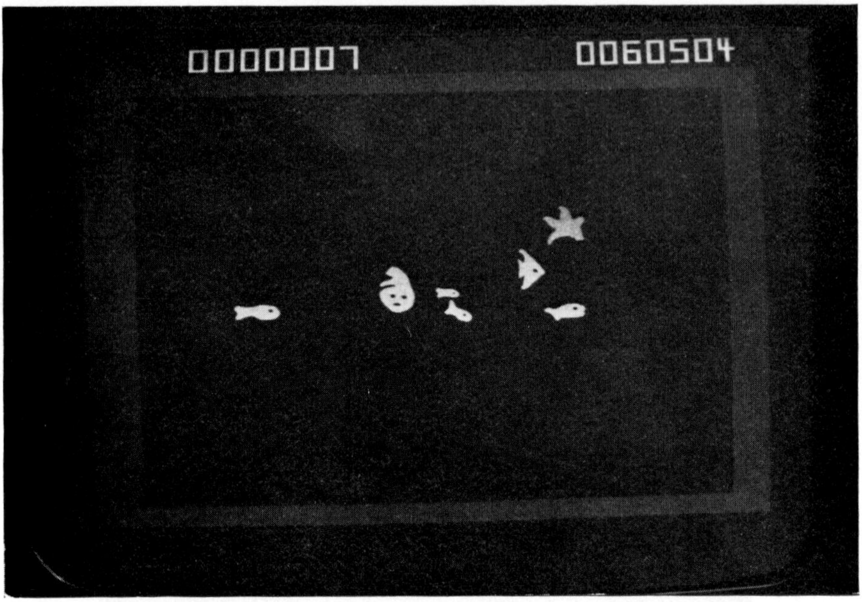

FIG. 14. Only those fish surrounded by the net are being counted.

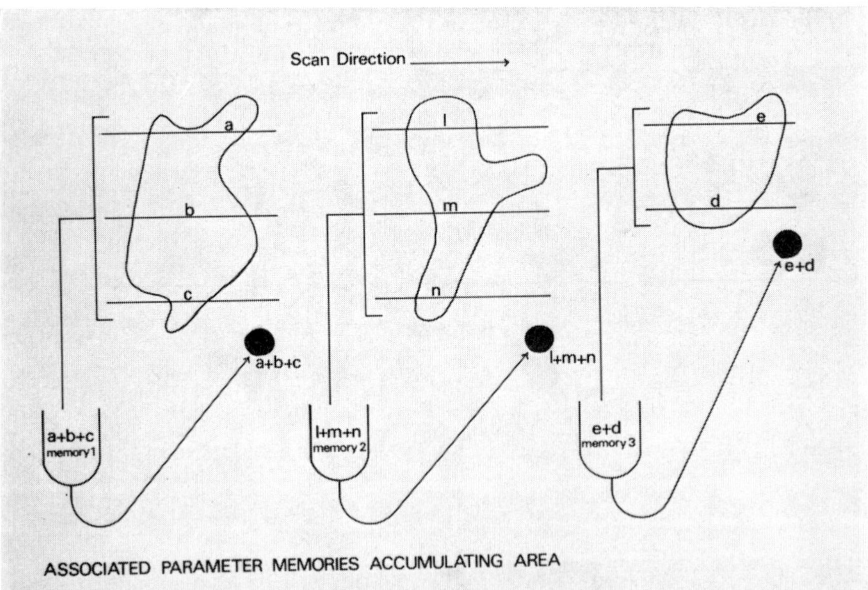

FIG. 15. A field of view containing three separate objects has all the chord information, allocated to three separate memories.

2. *The Method*

The basis of the method is shown diagrammatically in Fig. 15. Each object has a high-speed store associated with it that may be accessed in real time and descriptors of each object are put into store as the object is scanned and accumulated in each line, so that at the feature end (anti-coincidence point ACP) a complete description is fed out. At this point, the empty store is then available for processing another feature. This reduces the storage needed and makes it possible to process thousands of objects in a single scan.

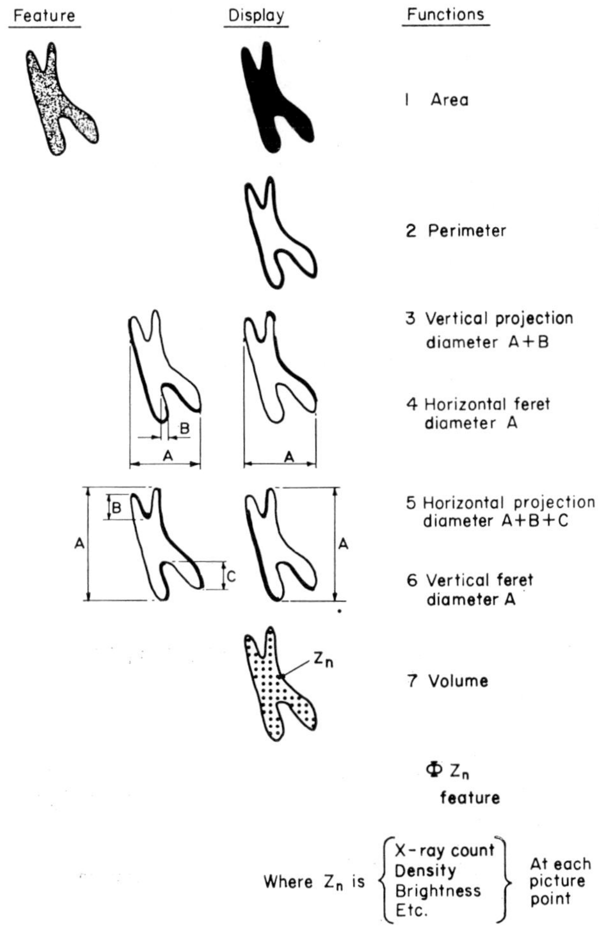

FIG. 16. The seven descriptor functions available for all objects in the field of view. Area—A, Perimeter—P_e, Horizontal and Vertical Projections—P_h and P_v, Horizontal and vertical Feret diameters F_h and F_v.

The descriptions used at present are of two types, comparative functions and accumulative functions. The former compares the description to date with that of the new chord and updates the data as required, the latter simply adds the contributions from all chords of the object.

A "primary set" of functions that describe objects is shown in Fig. 16. There are seven functions. Area, perimeter, horizontal and vertical projections, horizontal and vertical Feret diameters and volume. The Feret diameters and projections are sensitive to object orientation while area, perimeter and volume are independent of orientation. Volume is a loose term referring to the sum of an instantaneous parameter at each picture point over the whole area of the object. In conjunction with a densitometer the parameter is instantaneous object density and the computed function is integrated density for each object. On a scanning electron microscope system the parameter could be the X-ray count from a spectrometer and the "volume" function would then be the total characteristic X-ray emission over each object. The objects themselves would be defined by a simultaneously available electron signal enabling them to be classified by elemental composition at electron signal processing speeds (Fig. 17).

Displays characteristic of these functions may be superimposed on the image. The display is very important in image analysis, both from

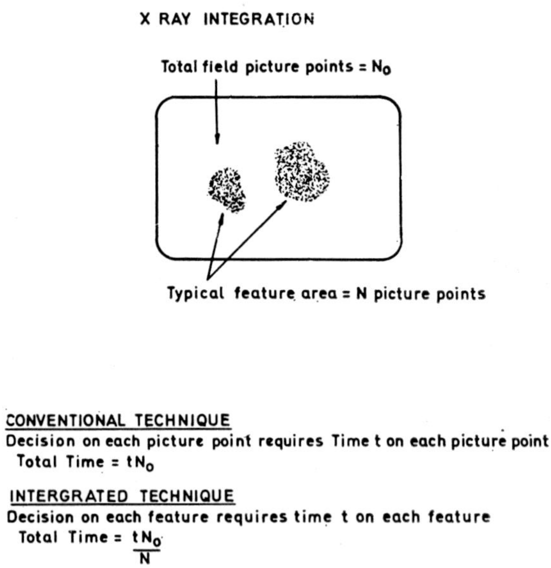

FIG. 17. Characterization of objects in an image by the corresponding X-ray emissions. Use of object specific analysis allows a significant increase in speed to be realized.

the point of view of setting up the instrument and also as an indication of exactly where errors are being made, if any. With computation in real time, display facilities are easily achieved. From the primary set of functions, a "secondary" set may be generated. These functions are sometimes general and sometimes applicable to particular object types.

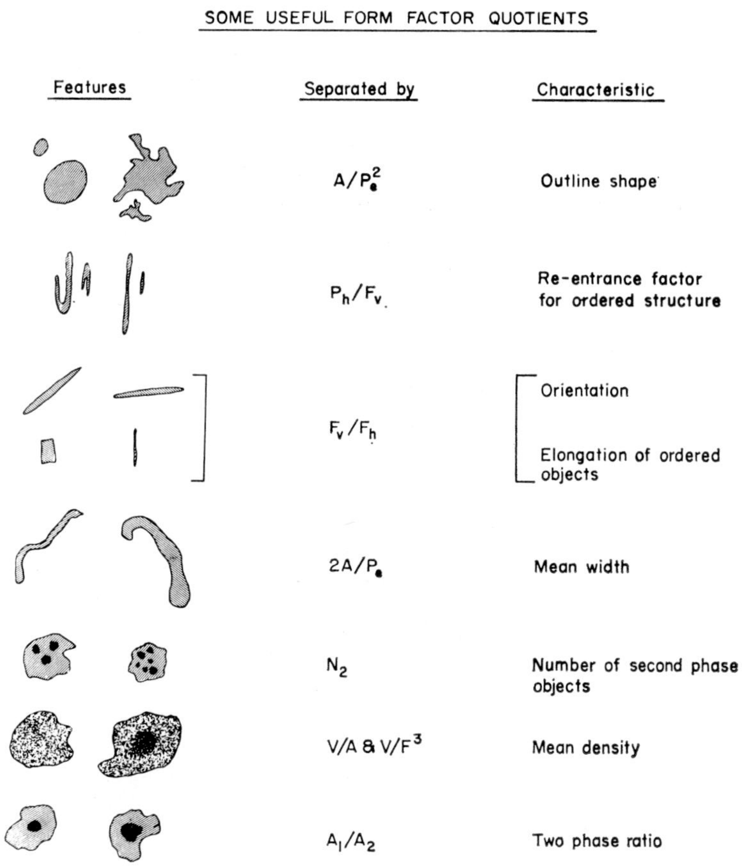

FIG. 18. Some useful form factor quotients, including secondary functions. See Fig. 16 for definitions. $F = F_v$ or F_h, whichever is appropriate.

Some examples of a secondary set are shown in Fig. 18 and include such functions as:

$$\text{Mean width} = \frac{2 \times \text{area}}{\text{perimeter}} = \frac{2A}{P_e}$$

This applies to long thin features of any shape or orientation (from straight fibres to spirals)

$$\text{Orientation} = \frac{\text{volume}}{\text{area}} = \frac{V}{A} \qquad \text{Generally applicable.}$$

Longest dimension = greatest of 4 Feret Diameters; horizontal, vertical $+45°$ and $-45°$.

Functions may also be combined to calculate more complex information about particles. Figure 18 shows some examples of "*form factor quotients*" which include the "secondary set "of functions described earlier as well as shape descriptions and quotients involving a second phase. A "Form Separator" may be used to enable these quotients to be calculated, by combining functions in the formula $A \times B/C^n$ where A, B, C are descriptive functions and $n = 1$, 2 or 3.

When Form Separation is being used the operator may forget which function he has chosen to feed into the $A \times B/C^n$ formula.

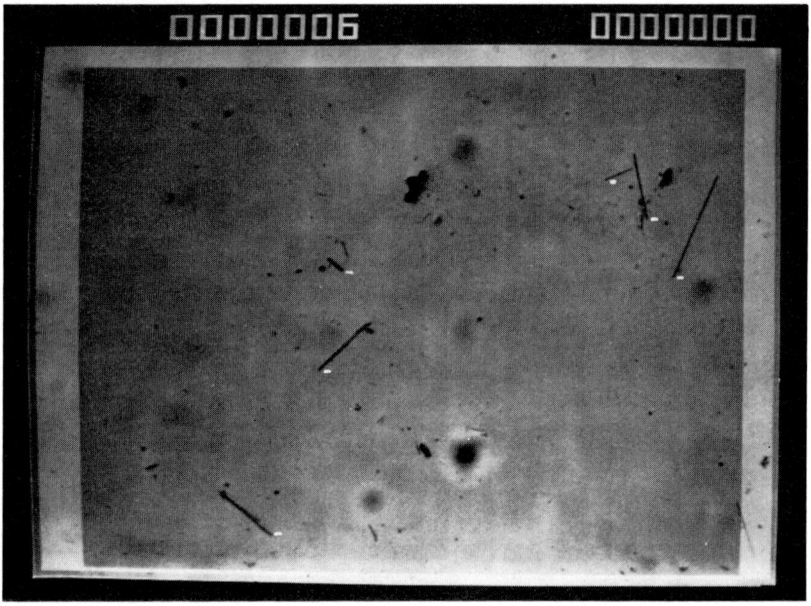

FIG. 19. Pattern recognition being used to separate the asbestos fibres from other dust in this sample of airborne dust. Length discrimination is also being used simultaneously to select only the longer fibres.

This problem has been recognized in the design and there are "Display" buttons on the Collector/Classifier and Form Separator modules to identify which function computer is measuring the relevant parameter and to display what function it is computing. Pressing

"Display *A*" on the Separator will cause the display lamp on the appropriate function computer to come on and also put the corresponding characteristics display on the monitor. Pressing "Display *B*" and "Display *C*" leads to similar displays which remind the operator instantly what he has programmed the equipment to do. Figure 19 shows long thin asbestos fibres being selected from other dust.

IV. Conclusions

Image analysis is becoming a technology in its own right, being applied to almost any image-producing device, including SEMs with their particular advantages of multiple signal descriptions, high magnification and large depth of focus.

There is no longer a rigid distinction between special purpose hardware and general digital software methods and consequently hybrid systems are being developed for many applications. This enables the advantages of each method to be combined in one instrument. Such instruments will have the computing potential of a conventional digital computer, combined with the speed of operation of specially designed "hardware".

Automatic image analysis has been proving its worth in hundreds of laboratory installations. Some applications are of such general interest that special machines can be built to cater for them. The Quantimet 360 is an example of this for the steel industry, with just enough complex computation to do the required analysis and a minimum of operator controls. Such quality control applications will now increase in number, and there is also considerable interest in mass medical screening applications. Already an experimental machine, the Cytoscreen, is undergoing trials in an attempt to detect pre-cancerous conditions in cervical smears. Slightly different machines are being considered for routine screening of sputum and blood. The problems involved in these applications are usually formidable but the potential rewards in terms of saving human lives and saving on hospital services by early detection of disease are enormous.

References

Braggins, D. W., Gibbard, D. W. and Gardner, G. M. (1971). *SEM/1971 Proc. 4th Ann. SEM Symposium IITRI*, 393–400.

Crawley, J. A. and Gardner, G. M. (1972). *EMCON 72 Proc. Vth Ann. Congr. Electron Microscopy Manchester* (*Sept.*).

Fisher, C. (1966). *Proc. Particle Size Analysis Conference, Loughborough.*

Fisher, C. and Cole, M. (1968). *Microscope*, **16,** 2, 81–84.

FISHER, C. (1971). *Microscope*, **19,** 1, 1–20.
GIBBARD, D. W., SMITH, D. T. and WELLS, R. A. (1972). *Microscope*, **20,** 1, 183–202.
INGRAM, M. and PRESTON, K., Jnr. (1970). *Scient. Am.* (*Nov.*), 72–82.
ROBERTS, F. and YOUNG, J. Z. (1956). *Instn. elect. Eng.* **99,** IIIA, 747.
STRANG, A. (1969). *J. Scient. Instrum.* **2,** 45–47.
STURGESS, G. L. (1971). *Laboratory Equipment Digest*, **9,** 9, 66–71.
WADLOW, D. E. (1972). *Microscope*, **20,** 183–202.
WHITE, E. W., GORZ, H., JOHNSON, G. E. and MCMILLAN, R. E. (1970). *SEM/1970 Proc. 3rd Ann. SEM Symposium IITRI*, 57–64.

PART II

Quantitative Interpretation of Contrast in the SEM Modes

Chapter 5

The Emissive Mode

A. GOPINATH

University College of North Wales,
School of Electronic Engineering Science, Bangor, North Wales

Symbols

N = number of secondary electrons that can escape surface barrier/unit area

N_0 = number of secondary electrons that can escape surface barrier/unit area at the creation point $x = 0$

x = distance from creation point

S = sharpness of image, i.e. distance over which emulsion changes by 90% of the total variation

d = source diameter

M = magnification

$\bar{n}$ = number of electrons/sec arriving at specimen, on average

$\bar{\delta}$ = mean value of secondary emission coefficient

δ = value of secondary emission coefficient at any instant

$b = 1/\bar{\delta}$

σ = yield coefficient-emitted current/incident current

I_0 = output of either half of a split annular ring Si-diode detector when beam hitting perfectly flat section of specimen

I_{d1}, I_{d2} = output of halves 1, 2 when beam on any section

Δ_1,Δ_2 = increase on half 1, and decrease on half 2 w.r.t. I_0
E = potential difference across a p–n junction
ϕ = potential at any point (x,y) in region of junction
x = the co-ordinate along specimen surface, y along beam direction
F = field in volts/cm along y-direction
x_0 = width of junction

I. Introduction

The emissive mode of operation of the scanning electron microscope is probably the one most widely used. This mode maps the electron emissive property of the specimen under beam excitation and the micrographs thus obtained are geometrically similar to the optical pictures from reflection light microscopes. This chapter provides a brief introduction to the mode, the collection and detection systems used, contrasts available, and its resolution capability. Selected topics on specific contrasts are also discussed in detail in subsequent sections, and finally we outline the progress made in the study of fast periodic events in the microscope.

The scanning electron microscope scans a fine electron-beam probe over the surface of the specimen, and, in synchronism with this, the electron beam of a cathode-ray tube is scanned over its whole display face. The specimen, under bombardment of the electron probe, emits electrons, whose number varies according to the local emissive properties of the specimen. This emissive current is collected and amplified to intensity modulate the display tube beam, and the resultant picture is the emissive micrograph of the specimen. If the area scanned by the probe over the specimen is the same as that scanned by the display tube beam, the magnification is unity. By reducing the specimen scan area of the probe, while retaining the scan in the display tube over its whole face, magnification of the image is obtained. The magnification thus obtained is limited by the size of electron probe, since the instrument cannot resolve details of the specimen smaller than the probe cross-section. The purpose of the electron optical column is therefore to reduce the size of the electron probe spot. With tungsten filament gun sources spot sizes of 50 Å (Nixon and Pease, 1965) have been obtained for probe currents of 8×10^{-13} Å. Guns with other emitters (Broers, 1970a; Crewe *et al.*, 1970) such as field emission tips have provided smaller spots for comparable currents. (Additional details will be found in Chapter 1.)

The emissive micrograph is geometrically similar to the light reflection picture, and the reason for this is illustrated in Fig. 1. In (a) of this figure the eye or, for example, a series of lenses could focus the

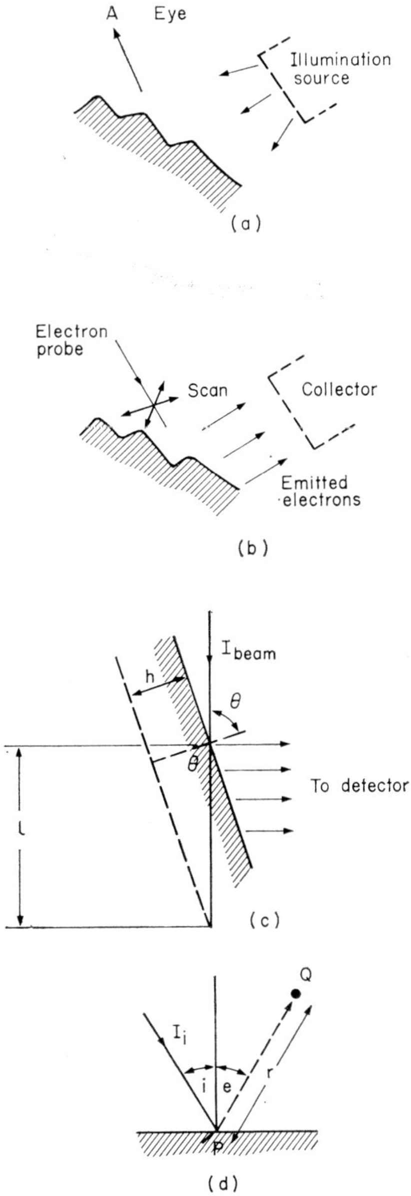

FIG. 1. Analogy between: (a) the reflection light image and (b) scanning electron microscope image production. (c) High electron emission occurs for an inclined surface as secondaries generated escape from a depth *h*. (d) Lambert's Cosine Law for diffuse reflection of light. (After Houston, 1949.)

reflected light from the specimen to form an image of the object. In the scanning microscope (b), the probe is from the former position of the eye and the collector of the emissive current is at the position of the illumination source of the previous case. The scanning, however, causes the picture on the display tube to form sequentially. A surface facing the illuminating source in (a) and facing the collector in (b), appears brighter in both cases and a shadowed region in (a) and a similar shadowed region with respect to the collector in (b) also appears dark.

In the SEM, it is found that the number of emitted electrons depends upon the orientation of the beam to the specimen surface, and the collected fraction depends on the position of the collector. For a smooth surface inclined to the beam as shown in Fig. 1(c) nearly all the secondaries that escape through the surface originate within a small distance from the surface h of about 50–100 Å. The number of secondary electrons emitted increases with the increasing angle of inclination simply because the length of the beam path, l, within a distance h of the surface increases and $l/h = \sec\theta$. Hence the number of secondaries emitted N and the detector current are given by

$$I = C_1 N = C_1 C_2 l = C_1 C_2 h \sec\theta = K \sec\theta, \ (C_1,\ C_2,\ K \text{ are constants}).$$

Experimental agreement has been found for θ in the range 50–70° by Everhart *et al.* (1959), when the specimen faces the collector.

The SEM micrograph resemblance to the optical picture of the same object arises through the similarity of the above relation to that derived from Lambert's Cosine Law for the diffuse reflection. For diffuse reflection the intensity received at Q is given by (Houston, 1949)

$$\frac{CI_i \, dS \cos i \cos e}{r^2} \quad \text{(Lambert's Law)}$$

where C is a constant, I_i is the incident intensity, dS is an element of diffusing surface area at P, and e and r are defined in Fig. 1(d).

Let $i = 90°$, then

$$\text{signal} = \frac{CI_i \, ds}{r^2} \cos e.$$

Therefore

$$I_i \propto \frac{\text{signal}}{\cos e} = \text{signal} \times \sec e.$$

This is the same as the situation for the secondary electron in the SEM (the electron detector being assumed vertical above the surface) but *with source and detector reversed.* Even if the secondary detector is not vertically above the specimen, most of these electrons are collected because of the positive bias on the collector. However, in this case even when the surface does not face the collector, some of the secondaries reach the collector. Thus, the geometric contrasts are similar in a large measure to its light reflecting property.

A. COLLECTION AND DETECTION SYSTEMS

The emissive current has to be collected and amplified to provide the video signal, and the electrode systems used to facilitate the collection process are called collection systems. In general these currents are small, of the order of 10^{-10} A or less, and therefore it is usual to amplify the current almost immediately by means of a detector.

The electrons emitted from the specimen surface may be divided into two classes: secondary electrons which leave the specimen at energies of 50 eV and less and back-scattered or reflected electrons which include primary (i.e. beam-probe) electrons which are both elastically and inelastically scattered, possessing energies greater than 50 eV.

It is also necessary to distinguish between two classes of back-scattered primaries (Wells, 1971), those which undergo plural scattering, and have one or two scattering events before being emitted, and the others which undergo multiple scattering and participate in several events before emerging. However, the class of primary back-scattered electrons may be regarded as including both categories. The division of secondaries at 50 eV, however, is arbitrary.

The basic collection system due to Everhart and Thornley (1960) is sketched in Fig. 2. It consists of a scintillator at the end of a perspex light guide, surrounded by a shielding cage covered at one end by a metal gauze. The scintillator is covered with a very thin layer (typically 500 Å) of aluminium and held at 10–12 kV. The gauze and cage are directed towards the specimen and held at some potential, typically 250 V. The perspex light pipe terminates at a glass vacuum window, pointing to a photomultiplier tube. To collect the emitted secondary electrons, the specimen is tilted towards the cage and the secondaries are drawn towards the collector. Some of them travel through the gauze, and are subsequently accelerated to the coated scintillator, pass through the aluminium film and create photons, about 50 per incident electron. This light is conveyed by the light guide to the photomultiplier and converted into electrical signals. Some of the back-scattered primary electrons also find their way to the scintillator and provide a further component to the signal. Other back-scattered electrons strike the final lens, chamber and other objects and produce further low energy electrons, often called tertiaries, some of which reach the scintillator. An investigation by Everhart *et al.*, (1959) has shown that about 5% of the signal is due to reflected primaries, 65% secondaries and about 30% due to tertiaries. This type of detector with the photomultiplier is found suitable for currents in the range 10^{-6}–10^{-13} A, provided that the current density at the scintillator does not exceed 1 μA cm^{-2}. Typical loss in the light pipe bends and reflections at the ends reduce the signal to 35% of the input photons. Additional loss occurs at the junction of the light pipe and the vacuum window of the order of another 10% at worst. The decay time for the scintillators presently in use in the system is of the order of 1 nsec and the response time of the photomultiplier is of the order of 10 nsec.

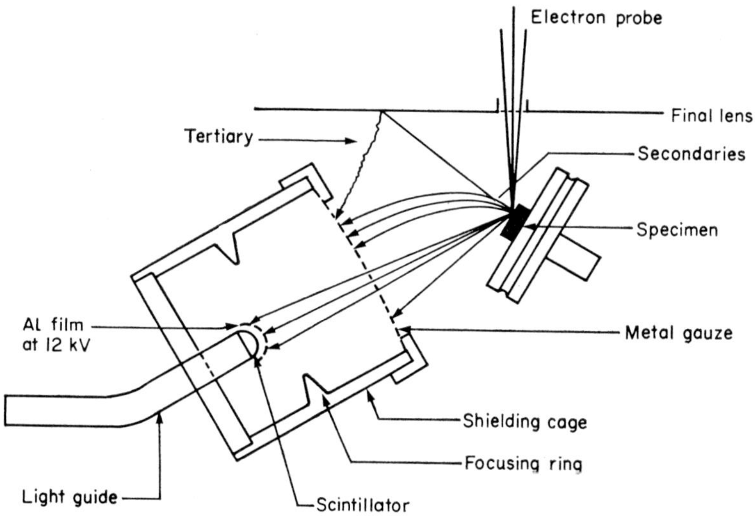

FIG. 2. Everhart-Thornley collection–detection system showing shielding cage, scintillator and light guide arrangement.

The scintillator at the end of the light pipe is usually hemispherical, and commercial light pipes arrange to glue this wet-machined scintillator (usually a plastic) to a flat light-pipe face. However, losses occur at this interface, and it is found that the scintillator also degrades with machining. Hatzakis (1970) has suggested that if the perspex rod is machined with a hemispherical end, the scintillator plastic, dissolved in ultrapure toluene, can be used to coat this end. Coated with an aluminium film, the scintillator light pipe is found to be much better, by almost a factor of 2.

Alternative detectors sometimes used in the microscope are channel multipliers in the usual form of a spiral or plates. Thornton (1968) has discussed the use of these in the SEM in some detail. The difficulty with these is that the output floats at, say, 2 kV, and therefore coupling of the signal into a fast circuit such as a head amplifier is not easy. One scheme is to use GaAs photodiodes and detectors to decouple optically. The supply to the channel multiplier may now take the form of a battery, since the current drain will be fairly low. The advantage of the multiplier is the small physical size of the device which may be used in confined regions. It is also quite fast, about the same as a photomultiplier tube.

Semiconductor diode detectors are also suitable for electron detection. Typically these are lithium drifted Ge or Si diodes and are cooled to liquid nitrogen temperatures to increase sensitivity and reduce noise. A much cheaper alternative is to use Si solar cells which work on the same principle. The fast back-scattered primary electrons create hole-electron pairs, of the order of 10^3 per incident electron (depending on the beam energy), and this is sufficient to be used as the signal for amplification. (A more detailed account of charge collection current signals is given in Chapter 8.) Detection of currents of the order of 0·1 nA is possible, but for lower currents this detector is not satisfactory—unless sophisticated diodes and circuitry are used. The advantage of the diode detector is that its gain is energy dependent, unlike the biased scintillator–photomultiplier or the channel multiplier. With diode detectors, it should be possible to energy sort the incident electrons using a pulse–height sensitive system. To detect the secondaries the diode could be floated up to, say, 5 kV, when sufficient multiplication exists. Again the difficulty is the coupling out of the signal from the diode which is at a high voltage.

The collection systems vary with the type of detector used. With the scintillator–light-pipe system, the pipe must terminate at a vacuum glass window in the front of the photomultiplier or the light pipe may itself form the vacuum seal. The geometry of the system is fixed within this constraint. Thus, the simple cage shielding the high voltage on the scintillator from the main beam can take a variety of shapes, provided of course that there is a gauze or mesh region through which the emitted electrons can reach the scintillator. The types of collectors range from hemispherical and cylindrical cages to simple flat pieces of mesh, portions of which may be held at different potential.

With other detectors the cage may be dispensed with unless the electrode (or detector) close to the specimen and primary beam-probe is at a high potential. If the detector is capable of detecting slow secondaries then the collector cage with a potential on it may serve the purpose of attracting these slow electrons. Special purpose collection systems are used to enhance particular types of contrasts and these are discussed in detail in the relevant sections.

B. OUTLINE OF CONTRASTS

Several different types of contrast occur in the emissive mode. We may differentiate between the signals obtained from reflected or back-scattered primaries, and those obtained with secondaries. In what follows the basic cage–scintillator–light-pipe collector detector is assumed to be used. When the cage is biased to some negative voltage, say −50 V, most of the slow secondaries will not reach the scintillator.

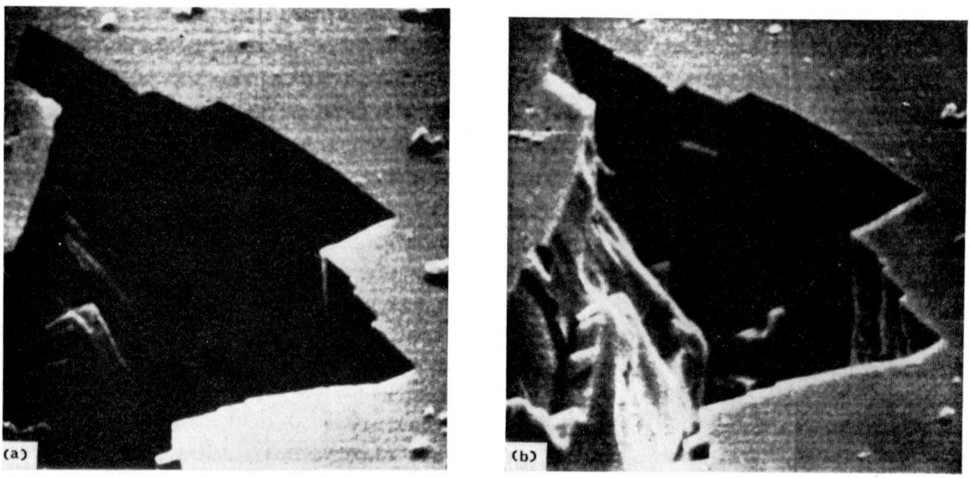

FIG. 3. Micrographs showing (a) signal from back-scattered electrons only, (b) signal from secondaries combined with back-scattered electrons. (After Oatley *et al.*, 1965.)

Thus the signals under this condition are mainly due to back-scattered primaries. Since these are fairly energetic electrons, they travel in straight lines (for all practical purposes) from the specimen to the scintillator. Thus, these cannot reveal details of any part of the object which is not in line of sight to the scintillator. Secondaries are not limited by this consideration and therefore may reveal details not seen with the reflected primary signals. When the cage is biased to say 250 V, the signals are now a combination of the primary and secondary electrons. A typical set of micrographs showing the topographical detail obtained with the two cage bias conditions, i.e. (a) primary signals only and (b) primary with secondary signals superimposed, are shown in Fig. 3. Thus for topographic detail the combined signals are by far the best and these are normally used.

The video signals obtained from back-scattered electrons are from either plurally or multiply scattered primaries. In general, the plurally scattered electrons suffer small angle scattering, less than 90°, and predominate at primary beam glancing angles, and have small penetration depths, of the order of the mean free path. Multiply scattered primaries undergo a larger number of scattering events and therefore may emerge with random scattering angles and from greater specimen depths. Wells (1971) has performed some interesting experiments in which he has examined the plurally scattered signals and compared these with the combined secondary and back-scattered signals. Figure 4 shows an aluminium film with a hole which has a thin oxide layer covering it. The secondary micrograph shows the hole, but the plurally scattered image (he calls these low-loss electrons) shows no hole. The estimated thickness of films that can be seen with the plurally-scattered primaries is the order of the mean free path (in aluminium with 15 kV beam, 75 Å). To obtain the low-loss image a retarding field analyser was used.

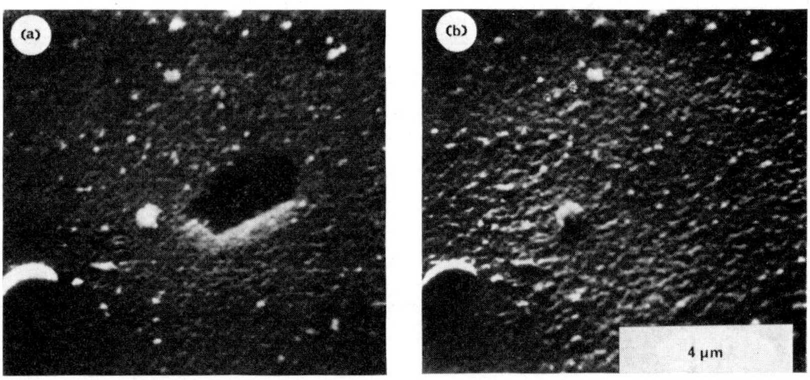

FIG. 4. Micrographs of aluminium surface (a) secondary image showing hole (b) plurally-back-scattered image showing thin film covering hole. (Courtesy O. C. Wells.)

The number of reflected primaries is a function of the specimen atomic number (Bishop, 1966—see Chapter 2). Since most of the information in the SEM is on an intensity basis, this provides a means of obtaining compositional contrast. The detection of the primaries could be by any of the detectors discussed above. The limitation is that elemental distributions in a uniform compound or alloy cannot uniquely be determined, unless segregation occurs. The contrast is used in qualitative interpretation of micrographs. The other form of compositional contrast arises from Auger electrons which are the electrons of characteristic energy peaks. These require special vacuum conditions (10^{-8} Torr or better) and essentially provide a surface examination technique (McDonald, 1971).

Potentials on the surface of a specimen affect the number of electrons collected, giving rise to voltage contrast. In general the more negative regions become brighter, and the positive regions become darker. It thus becomes possible to see voltages and potential distributions on the surfaces of semiconductor devices. The technique is used in device laboratories and production lines for trouble-shooting, and also on a qualitative basis for specific studies (Thornton, 1968). Contrast on magnetic and ferro-electric materials due to local domains has also been observed (Joy and Jakubovics, 1968). In all these cases the contrast information is contained in the secondary electrons.

C. RESOLUTION CAPABILITY OF MODE

The resolution of the instrument is largely dependent on the specimen being examined, and the spot size of the probe at the specimen. Additional problems frequently affecting the ultimate resolution of an instrument are due to mechanical vibrations, stray field interference, and surface contamination, but as these may be overcome with careful design they will not be considered here.

Probe spot sizes obtained to date (Broers, 1970a) are:

1. Tungsten filament source: 50 Å, with 8×10^{-13} A.
2. LaB_6 Schottky Emission Tip: 25 Å with 2×10^{-12} A.
3. Tungsten field emission tip: 5 Å with 10^{-10} A.

These probe sizes can only be obtained with meticulous design and construction of the electron optical column. In theory it should be possible to resolve two points or edges, spaced a little more than the spot diameter apart. In practice the specimen will not emit uniformly over the same spot area—in fact the emission is over a much larger area, and depends on the type of specimen. The main source of secondaries appears to be the area under the probe spot as the impacting electron flux is greatest. The reflected and back-scattered electrons escape from a much wider area, and in turn also create additional secondaries which are also emitted over their whole escape area. This larger area can be anything up to the order of one tenth of the primary beam range but fortunately does not degrade the resolution significantly, since the number of secondaries thus emitted is comparatively small.

Everhart *et al.* (1959) have postulated, and this has since been supported by several others, that the primary beam loses energy in a pear-shaped region as shown in Fig. 5. In the process of losing energy the primaries excite secondary electrons which may be assumed to travel isotropically outwards. Bruining (1954) suggested that the number of secondary electrons which can escape the surface barrier varies as

$$N = N_0 \exp(-\alpha x)$$

where x is the distance from the beam impact point, N_0 is the total created at this point, α is a constant, with suggested values for $1/\alpha$ in the range of 100 Å. The primary beam, however, loses a small fraction of its total energy in this distance $1/\alpha$, and also does not spread significantly within this distance. Therefore the majority of the emitted secondaries are from an area of the size of the spot, and from a maximum distance of the order of $1/\alpha$.

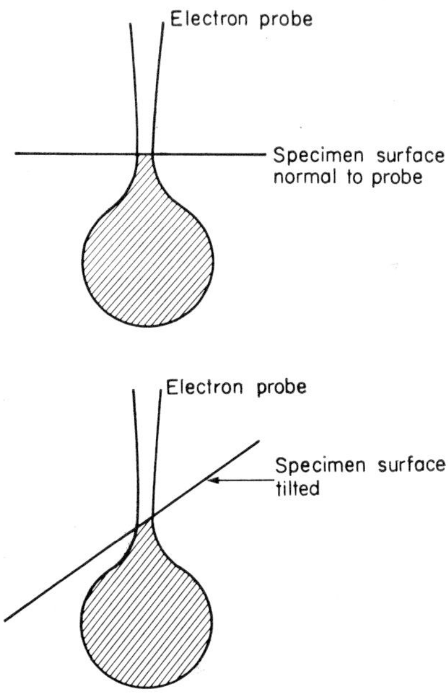

FIG. 5. Primary beam energy loss region in specimen.

When the specimen is tilted to the primary beam the primary energy-loss region is closer to the surface and results in a larger number of secondaries being emitted. However, their emission area is now much larger than the probe spot size and the resolution degrades (see Fig. 5).

Pease (1965) carried out tests to examine the resolution obtained with back-scattered primary electrons. In these he examined the source size (diameter d) of these primaries obtained by a fine probe spot by examining the projection image of a fine grid on a photographic plate. Defining S the sharpness of the image as the distance over which the emulsion changes by 90% of the total variation, it can be shown that

$$d = \frac{S}{M}$$

where M is the magnification obtained. With a probe size of about 1000 Å, he estimated the typical escape diameter for the back-scattered electrons was about 4500 Å, though there may be some error in this value. This is about $\frac{1}{10}$ of the range of the primaries in the target material.

More recently, however, Wells (1971) has shown that for a glancing angle of 30°, the resolution with the secondaries combined with back-scattered electrons, all back-scattered electrons, and with the plurally scattered component of the back-scattered electrons, was substantially the same. He has resolved 170 Å steps in all three modes, which is in contrast with the results of Pease (1965). It is apparent that the resolution capability attributed to the back-scattered electrons has therefore not been clearly resolved.*

Thus, the typical specimen-determined resolution in the emissive mode is of the order of the range of the low energy secondaries in the material (typically 100 Å), assuming that the spot is small compared to this range. However, the electron optical limitations of the instrument are also of importance, but as these have been discussed in Chapter 1 they will not be repeated here.

Noise considerations are of importance in the emissive signal processing. The primary beam may be assumed to have n electrons per sec on average at the specimen. If the source is random the mean deviation for this is $\sqrt{n}$ and the signal to noise ratio which is the mean value divided by the deviation is also $\sqrt{n}$. However, the fluctuations in the secondary emission coefficient also degrade this ratio. Everhart *et al.* (1959) have shown that this is worsened by a factor $|1+\overline{(\delta-\bar{\delta})^2}/\bar{\delta}^2|^{\frac{1}{2}}$ where $\bar{\delta}$ is the mean value of the secondary electron emission coefficient. If δ is a random variable then we have:

$$\overline{(\delta-\bar{\delta})^2} = \bar{\delta}$$

and the factor above may be written as $(1+b)^{\frac{1}{2}}$, where $b = 1/\bar{\delta}$. From the above calculation it seems that randomness in the emission process

* More recent results (Well, 1974) suggest that the plurally scattered primaries provide the best possible resolution in this mode.

may degrade the signal to noise capability by as much as a factor of $\frac{1}{4}$ or so in a specific case, depending on the value of b.

The noise due to the scintillator–photon conversion, attenuation in the light-guide and vacuum window, amplification in the photomultiplier and following amplifiers all add noise to the signal. In considering the noise contributions of the primary beam and emission process, we have ignored the noise additions of these components of the video chain and assumed that this is small. Practical experience with a poor scintillator or a noisy photomultiplier shows that this is not always the case, and therefore their contributions should be accounted for in any complete analysis.

If this secondary emission factor is set to unity, it becomes possible to estimate the beam current required for a particular scan time. Thus, Oatley (1972) maintains that the eye requires a signal-to-noise ratio of at least 5 to distinguish an area of brightness B from an adjacent area of brightness ΔB. Thus, we have:

$$\sqrt{n} \geqslant N(B/\Delta B)$$

and if we require N to be set to 10, i.e. ten levels of brightness, then for every picture point $n \simeq 2500$ electrons. With say 10^6 picture points in a frame, the requirement is that $2{\cdot}5 \times 10^9$ electrons must fall on the specimen during the micrograph record time. Thus, a 100 sec frame would require $2{\cdot}5 \times 10^7$ electrons per sec or approximately 4×10^{-12} A beam current. For any given gun emitter with known brightness, the minimum usable spot size can then be estimated for the particular value of current, and convergence angle.

D. SPECIMEN EXAMINATION CONDITIONS

The optimum examination conditions vary immensely with the specimen and the type of information required. As indicated earlier, the secondaries are suitable for topographic information and the backscattered primaries provide contrast in regard to the composition of the sample. We have further seen that the resolution is better with secondaries, and therefore where possible this advantage of the mode should be utilized.

Figure 6 shows the variation of yield coefficient σ against beam voltage, and the shape is found to be very similar for a large variety of materials (Hachenberg and Braner, 1959). The yield coefficient σ is defined as the ratio of the emitted current to the incident. For metals σ_{max} varies from 0·6 to 1·6 for a range of beam voltages from 100 to 800 V and for semiconductors and insulators σ_{max} is 1–20.

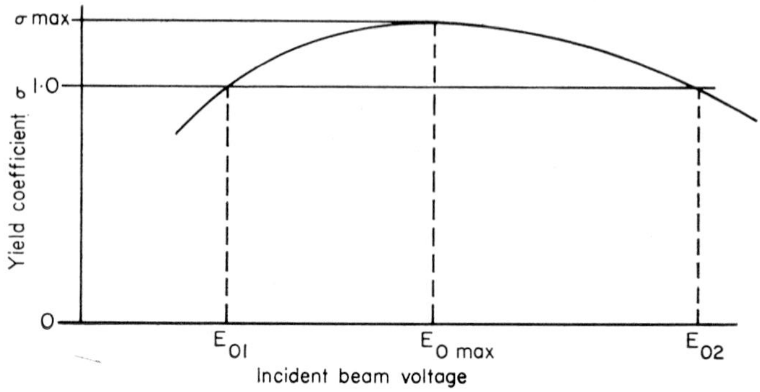

FIG. 6. Emission yield coefficient as a function of beam voltage. This curve is similar for a large number of materials.

1. *Insulators*

To examine uncoated insulators it is preferable to work at the point where σ is unity, but often this is rather low (say 2 kV) and does not provide sufficient surface detail. When the yield coefficient is greater than unity the specimen charges positively, which is preferable to negative charging and consequent probe repulsion. The spot size of the instrument may be large at low beam voltages but it is the penalty for not coating the specimen. Conductive coating of the specimen, to obviate charging, is by a thin deposit of carbon followed by a film of metal, typically gold–palladium, or an anti-static spray. The coating should be such that no artifacts are created nor should the details sought be obscured. Since a large variety of specimens are examined in the SEM, no specific rules can be given for the optimum examination conditions.

2. *Biological Specimens*

These can be subdivided into two broad classes, hard tissue structures whose water content is not significant to its topography, and soft tissues which collapse when their water content is removed abruptly. The hard tissue structure studies in the SEM often do not require special specimen preparation apart from coating, and possibly may yield both topographic and compositional information. Soft tissue specimens require very careful preparation so that the water content is removed without structural modification. The techniques presently used are reviewed by Boyde (1972) and several topical papers have been devoted to this subject. Since soft tissue specimens are largely composed of light elements, compositional contrast is poor. Considering the enormous number of specimens that have been examined through the SEM, no optimum examination conditions can be specified. It is, however, preferable to use as low a beam voltage as tolerable and as small a probe current as acceptable for a reasonable signal.

3. *Metallurgical Samples*

These also vary considerably from fracture surfaces showing considerable topological detail to flat polished specimen with little in the way of topography. The fracture surfaces are examined with topographically favourable signals, depending on the kind of detail that is required. Flat samples with little detail are best examined with the back-scattered signals.

II. Specimen Topography and Associated Measurements

The advantage of the scanning microscope is that a large depth of focus is retained at relatively high magnifications and this makes it eminently suitable for the study of specimen topography. The depth of focus F can be defined in this context as the distance measured parallel to the optical axis for which the image still appears in focus, and occurs due to the small angular beam divergence α, which is set by the final lens aperture and specimen to lens distance in the SEM. F is given by the equation (Booker, 1970):

$$F = d/\tan \alpha \simeq d/\alpha, \quad \text{for } \alpha \ll 1$$

where d is the resolution required. Thus, for $\alpha = 10^{-3}$ rad and $d =$ 500 Å, the depth of focus is 50 μm which is very large. The alternative definition for F includes the magnification setting M on the instrument, and d', the smallest distance that can be resolved on the display tube (say 0·1 mm), so that (Oatley, 1972):

$$F \simeq d'/(M\alpha).$$

We have seen in the introduction that the micrographs in the SEM show the same geometric contrast as a reflected-light picture, but the topographic detail shown in a two-dimensional picture is certainly not unambiguous. To resolve specific details a commonly used method is by examining the specimen at different tilts and rotations. However, such simple techniques are not suitable for topographically involved structures, and stereoscopy is the obvious solution. As this has been treated in detail by Boyde (1973) this technique will not be discussed here.

Corvin and Johari (1970) use the tilt and rotate technique for the identification of topography. In Fig. 7 we consider a triangular elevation and depression with the probe illumination direction and the position of the collector shown. The analogy between the reflected-light image and the SEM image shown in Fig. 1 should be borne in mind with these tests. Thus in Fig. 7, surfaces 3 and 5 should be very bright, 4 moderately so, 2 and 6 will be very dark, and surface 1 a light grey. Tilting and rotating the specimen enables one to identify a depression from an elevation, with little difficulty.

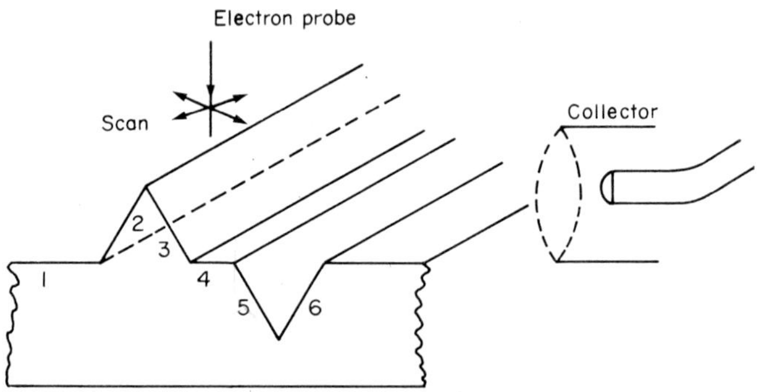

FIG. 7. Geometry of specimen for topographic identification tests.

Apart from studying topographical detail, it is possible to measure linear distances on the specimen provided the linear magnification is available. However, this may only be known approximately, since the working distance, i.e. the distance from specimen to the final lens, may vary from one specimen to the next and the beam voltage may also be changed. The simplest method of calibrating the magnification is to use a calibrated metal grid, for example the type obtained from the Scientific Instrument Research Association, Chiselhurst, Kent. Laying this alongside the specimen being examined, one may obtain a ready guide as to the true magnification. Calibrated latex spheres were used by Broers (1969) and also by Osborne (1971) to test the resolution of the instrument. Broers (1969) has also used a test sample of epitaxially grown layers alternately of GaAs and GaAsP, which is cleaved normal to the layers and preferentially etched to show parallel ridges separated by 1000, 500, 250 and 170 Å (Blakeslee, 1970). Other calibrated samples have been listed by Ballard (1972), and it appears that no single specimen is suitable under all conditions. In addition there are distortions in the display of the signal on the CRT screen so that for some purposes it is necessary to apply corrections to linear measurements (Murray *et al.*, 1973).

When the projection height or the depression depth on a flat sample is to be measured Hoover (1971) and Brandis *et al.* (1971) have suggested a simple technique. If the depth of a groove in a flat specimen shown in Fig. 8 is to be measured, then the specimen is mounted flat and a contamination line marked on it by letting the beam run on it for a short time. The specimen is then tilted, i.e. rotated, about this line to a known inclination angle. Knowing the magnification along the line, it is now possible to measure the depth or height required. However, the magnification along the line has to be known accurately and also the angle of tilt about the contamination line. The latter can be obtained from an accurately calibrated goniometer stage and the former by, say, a calibrated specimen.

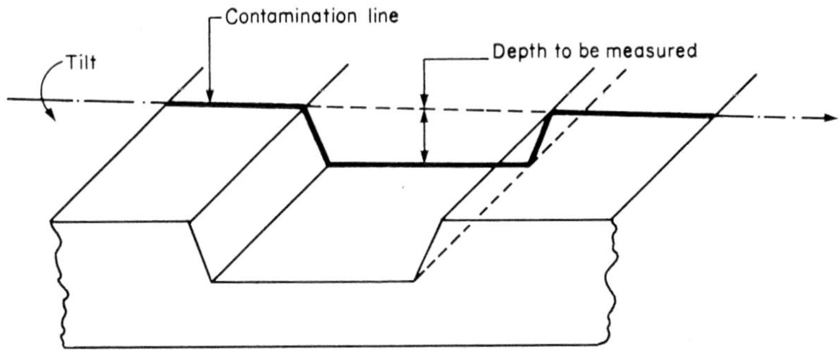

FIG. 8. Depth of channel in sample to be measured in SEM (see Hoover, 1971).

Brandis (1972) has shown that projections on a ground plane can be examined using a shadow casting technique. In this technique, gold is evaporated on the specimen from a known but shallow angle, and the projections cast a shadow over the plane which is not covered by the gold. When examined in the SEM the shadow is darker than gold, which appears very bright owing to its high emissivity. If the linear magnification is known the size of the shadow can be measured and the evaporation source angle to the plane then provides a measure of the specimen height. The disadvantage of the method is that the evaporation source size is finite and therefore the definition of the shadow edge will be poor.

While several problems on spatial measurement in the SEM have been discussed in the literature, we consider one of some interest: the measurement of the profile of a razor blade cutting edge (Secker *et al.*, 1970). As the cutting face tapers very quickly to a sharp edge, one method is to examine it end-on with the blade mounted in the vertical plane, the edge set at an angle to the beam. The result on a commercial razor blade is shown in the micrograph on Fig. 9, and it is clear that measurements can be made if the magnification in the plane normal to the edge and the tilt of the edge are known, but these will be inaccurate as no datum exists from which the measurement can be made. Estimates by the technique of Hoover (1971), discussed above, could also be used if the blade is placed flat and then tilted about a contamination line. Or alternatively one of the unsharpened edges of another identical blade is placed alongside the edge to be measured, and the process repeated. Similar measurement problems are encountered in estimating tip radii of field emitters. These are also primarily due to the beam penetration and resultant degradation in the spatial resolution.

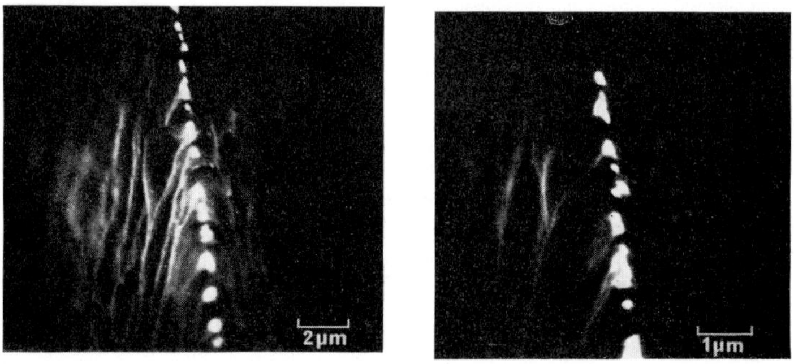

FIG. 9. Micrograph showing edge of commercial razor blade. Note flaring at some of the points. (Courtesy P. E. Secker)

III. COMPOSITION DETECTION

There are two forms of compositional contrast in the SEM, the one depends on the variation of the number of back-scattered electrons with specimen atomic number. The other examines the energy distribution of the secondary electrons for element-dependent characteristic peaks, called Auger electrons. Both these are discussed in some detail below.

A. BACK-SCATTERED COMPOSITIONAL CONTRAST

The back-scattering coefficient is a function of the specimen atomic number and Fig. 10 shows such a curve given by Bishop (1966) for 10 kV beam voltage. The compositional contrast in the SEM is attributed to this phenomenon, and it is apparent that any composition detector requires to differentiate between the relative numbers of back-scattered electrons. In practice the presence of the secondaries tends to obscure this type of contrast. Any of the detectors discussed in Section I.A, set to examine reflected or back-scattered primaries, would be adequate. If the specimen has topographic detail, the variation in number of back-scattered electrons reaching the detector due to changes

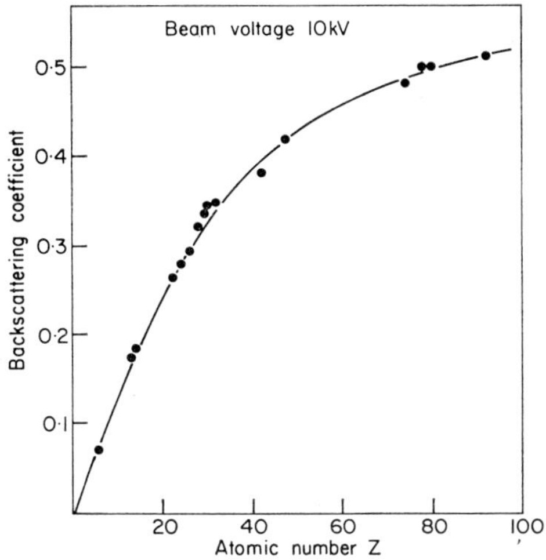

FIG. 10. Back-scattered coefficient as a function of target atomic number (Bishop, 1966).

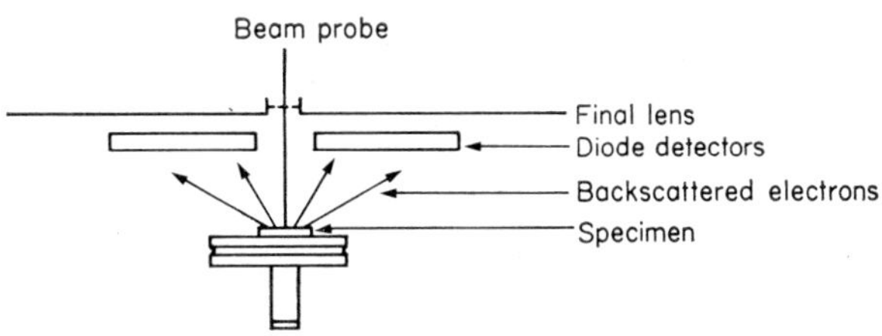

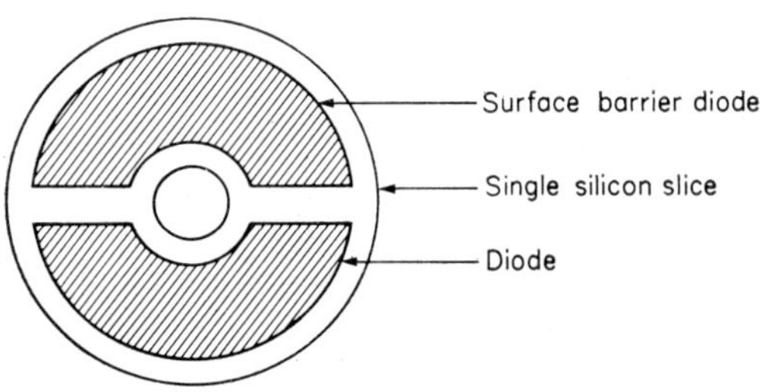

FIG. 11. Compositional contrast detector with surface barrier diodes (Shaw, 1970).

in the specimen inclination, would also be present in the signal. A composition detector should preferably be independent of specimen topography.

A simple scheme to use this phenomenon put forward by Kimoto and Hashimoto (1960) using two surface-barrier Si diodes has been very effective. A variation of the scheme used by Shaw (1970) used surface–barrier Si diodes made from *n*-type Si, 1000 Ω cm, with each of the two diodes in the form of half annuli as shown schematically in Fig. 11. The diodes are held in zero bias, depletion width ~ 10 μm, and are connected so that their outputs can be added or subtracted. A thin coating of gold on the barrier ensures that no infrared photons from the specimen reach the collector. On a perfectly flat specimen the back-scattered component reaching both diodes is identical and therefore their output summed is a measure of the specimen composition, and their difference, provided the diodes are well matched, will be negligible. On a specimen with topography present, say an inclination, the output of one diode will be enhanced by say Δ_1 and that of the other reduced by Δ_2:

i.e.

$$I_{d1} = I_0 + \Delta_1$$
$$I_{d2} = I_0 - \Delta_2.$$

Therefore

$$I_+ = 2I_0 + \Delta_1 - \Delta_2$$
$$I_- = \Delta_1 + \Delta_2.$$

The outputs summed is therefore a current that gives compositional contrast with a reduced topography signal, and the outputs subtracted gives essentially a topographic signal.

Shaw (1970) has carried out numerous experiments to show that adjacent elements in the Periodic Table can be distinguished in flat samples. He has examined thin films of various elements on Si substrates and shown the following:

1. Compositional signals with elemental films of atomic numbers greater than Si decrease with increasing beam voltage, from 5 to 20 kV. Thus, as the beam penetration increases, a greater fraction of the back-scattering could arise from the substrate. The exception to this rule was Zn on Si and this remains unexplained.
2. Compositional signals with elemental films of atomic numbers less than Si increase with increasing beam voltage, although this change is small. Note that in this case the back-scattering coefficient is higher for the substrate, and hence increased scattering from the substrate is an acceptable explanation.

From Bishop's results (1966), there is some evidence that the back-scattered electron angular distribution also changes with atomic number. Thus, with a fixed and rather small solid angle detector, this effect was also noticed by Shaw (1970). Other studies by Shaw have used two channel multipliers insead of diodes and the resultant compositional contrast is just as good.

In practice it should be possible to use any of the available SEM emissive detector systems to examine the compositional variation. However, a homogeneous compound or alloy without any sègregates will not be detected by this technique. Spatial resolution available with this technique depends on beam current and voltage and has been estimated to be about 4000 Å at worst. Sensitivity is determined by the slope of the curve in Fig. 10. It is apparent however, that no effort has been made to use this contrast for quantitative work.

B. AUGER–ELECTRON COMPOSITION DETECTION

Auger–electron analysis is a part of secondary electron spectroscopy and, in particular, examines the energy levels of the ionized atom leading to emission of electrons with characteristic energy levels. These characteristic energies of the electrons are unique for a particular element

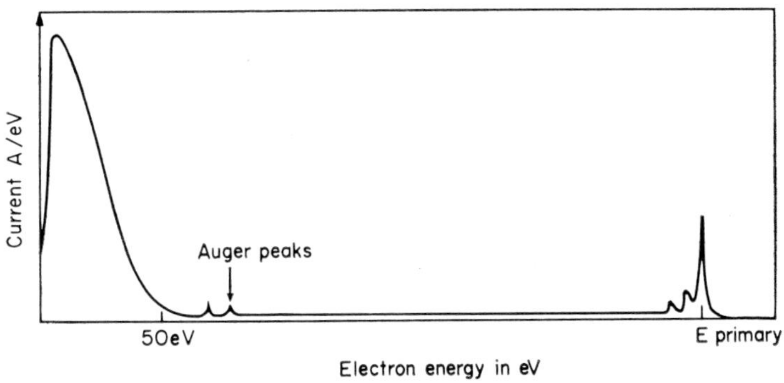

FIG. 12. Electron energy distribution curve showing position of Auger peaks.

and can therefore provide detailed compositional detection. Secondary electrons emitted within the first five or so atomic layers may possess energies belonging to one of these characteristic peaks. Energetic secondaries produced by the primary beam at greater depths may themselves create further secondaries, initiating a cascade of secondaries with decreasing average energies. These fast secondaries therefore lose their characteristic peak identity. A typical emission energy distribution curve is shown in Fig. 12.

Auger electron analysis has revealed traces of elements of bulk concentration of 10^{18} per cc and it is estimated that as little as concentrations of 10^{16} per cc may be detected by the techinque. Auger currents are about 10^{-5} times the primary current. The method is suitable for examining material surfaces which are atomically clean or those that can be cleaned *in situ*. Auger analysis is more favourable for the detection of light elements where X-ray compositional analysis is poor. Vacuum pressure at the specimen must be of the order of 5×10^{-8} Torr or lower for detailed examination of the specimen. For further details of the mechanism, the review article by Chang (1971) should be consulted.

The energy analysis of the emitted electrons for Auger electrons has used the three-grid retarding field analyser and the coaxial cylindrical mirror analyser. In the former an integrated energy distribution is obtained normally, and to obtain the differentiated curve, the energy distribution, it is usual to superpose an a.c. signal of say 70 Hz on the retarding field ramp. Using a phase-sensitive detector with the a.c. signal as reference at the output of the collector, the energy distribution curve is obtained. To enhance the Auger peak it is usual to use the phase-sensitive detector with the second harmonic (say 140 Hz) of the excitation (at 70 Hz) as the reference. The disadvantage of the technique is that the total collected current has shot noise on it, which is wide band. The phase-sensitive detector has a narrow pass band centred at the second harmonic, but the shot noise that passes through is proportional to the integrated collected current at that point and therefore is high. The other disadvantage is that the field penetration through the meshes of the grid causes degradation in the analyser resolution capability.

The coaxial cylindrical mirror analyser has excellent focusing properties for three-dimensional finite sources and also gives an output which is the energy distribution function. Thus, the noise in the signal is the shot noise of the collected current at any particular energy only. Thus, it is a better analyser for Auger electrons and therefore widely used (see Chang, 1971, for further details).

Commercial SEMs have vacuum pressures around 10^{-6} Torr at best, and ion pumps improve this to 10^{-7} Torr. A few special purpose instruments, however, have reached better pressures and it is likely that commercial instruments may offer higher vacuum options in the future to enable Auger work to be carried out. The disadvantage of using a higher pressure instrument is that the surface contamination builds up very quickly, and then the Auger peaks that are observed are those of the contaminant, in particular carbon.

MacDonald (1971) has carried out some pioneer Auger work in the SEM with a cylindrical analyser, and a beam probe current of around 10^{-7} A. The pressure in the chamber was at the high side of 10^{-8} Torr, and therefore the results had to be obtained rapidly to overcome the problem of contamination. It is apparent that the technique (i.e. Auger spectroscopy) has not been exploited extensively and is likely to be a growth area in the future. Its main disadvantage is that only surface studies can be performed, but in combination with ion etching *in situ*, it should be possible to examine the composition of specimens layer by layer.

IV. Voltage Contrast

Voltage contrast has been the subject of several papers over the years, and in this section we discuss the basic mechanism, or as much as is known about it, and then enumerate the means of using the contrast for measurement of voltages.

A. Basic Mechanism of Contrast

The contrast is dependent on the type of secondary electron collector used and we assume that the cage–scintillator–photomultiplier system is used unless otherwise stated. For a uniform specimen held at different potentials with respect to ground, it is found that there is a variation in the number of secondaries collected, for a fixed collector cage potential. Everhart (1968) has shown that change in the potential on the specimen results in a change in the trajectories of the slow secondary electrons. Since the number of slow secondaries in the electron energy distribution is large this results in a change of signal. We term this type of contrast as trajectory-contrast.

The second cause of contrast is due to local fields. For a composite specimen with two electrodes at different potentials, an electric field is present at the specimen surface and this affects the trajectories of the slow secondaries ejected at these regions and hence the number collected. We term this contribution as local field contrast. Wells (1971) and Plows (1969) have both theoretically considered the case of two half planes with origin as junction, one extending from $x = 0$ to $x = +\infty$ and the other from $x = 0$ to $x = -\infty$, as shown in Fig. 13. The collector is assumed to be infinitely wide extending from $x = -\infty$ to $x = +\infty$ and produces a field given by, say, F V cm^{-1} in the y direction. Solution of Laplace's equation with plane one at $\emptyset = +E/2$ and plane two at $\phi = -E/2$ by conformal transformation results in an equipotential distribution as shown in Fig. 13 (after Wells). Note that near the origin ϕ is a multi-valued function and therefore electrons emitted in the region $x = 0$ to $x = x_0$ have to pass through a retarding region, before reaching the field accelerating them to the collector. Thus, slow electrons of energies less than $E/2$ eV will not reach the collector. This corresponds to the case of a p–n junction of negligible depletion width.

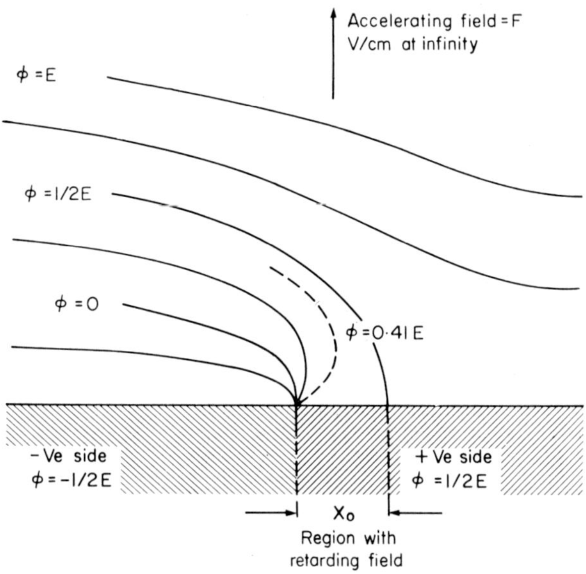

FIG. 13. Equipotentials for half planes at $\phi = -\frac{1}{2}E$ and $\phi = +\frac{1}{2}E$ with a junction at origin. Collector field is F V cm^{-1} in the y-direction. (After Wells, 1971.)

It can be shown that (see Wells, 1971)

$$\phi = (E/\pi)\tan^{-1}(x/y) + F$$

and

$$x_0 = E/(\pi F)\text{ cm}$$

where

$$E = \text{potential across the junction.}$$

Increasing the collector field will not cause x_0 to become zero, only make it smaller. Electrons with energy less than $0{\cdot}5E$ eV emitted within $0 < x \leqslant x_0$ will not reach the collector. Thus, the local field contrast results in a change of number of secondaries which reach the collector. The fields considered are not high enough to cause a variation in the number of electrons emitted.

Since the geometry of the specimen to collector cage and their surrounds is three dimensional and subject to significant variation, it is not possible to use the system for quantitative work. The contrast is also non-linear and requires some linearization before it can be used for measurements. However, a great amount of qualitative work uses the standard collector system. Contrast is best seen with the beam voltage set around 10 kV or lower, and with the cage potential slightly backed-off, ie. below the usual 250 V. A variety of specimens has been examined with this system, and the literature abounds with examples.

B. POTENTIAL MEASUREMENT WITH ENERGY ANALYSERS

If the specimen emission point is at a particular potential, it follows that thc secondary energy distribution function with respect to ground (zero) potential is shifted linearly. Thus, if the emission point is at +3 V, the electrons with energy less than 3 eV will not reach any earthed electrode, and similarly the emission point of −3 V results in a positive shift, as shown in Fig. 14. Thus, secondary electron energy analysis could provide information regarding the emission point potential. If the voltage variation of the emission point is small an analyser with a fixed window could be used, assuming that the change in signal is linear with potential (see Fig. 14). However, if the variation is large this technique will give erroneous values, and therefore the position of the energy function peak would provide the information required. Wells and Bremer (1968) have used this technique to measure the voltage across a *p–n* junction.

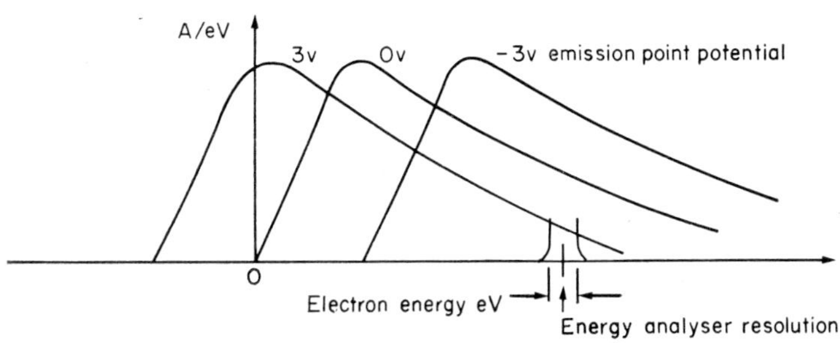

FIG. 14. Shift in secondary electron energy distribution curves with emission point potential.

Since the peak of the distribution function is at about 3 eV (typical half-width is 5 eV), it is apparent that the local fields will affect entry of these low energy secondaries into the analyser, and thus results may be in error. The alternative of using higher energy electrons was suggested by MacDonald (1970), who in fact used Auger electrons of the carbon contamination layer at 114 eV to identify emission point potential with a cylindrical coaxial analyser. Unfortunately, MacDonald's method, although extremely powerful, requires special vacuum conditions and is unlikely to be applied by users of commercial instruments. A third alternative uses a retarding field analyser (Gopinath and Sanger, 1971) and is discussed in the next section.

Wells and Bremer (1968) discussed sources of error in voltage measurement in the SEM with an energy analyser. He suggests that 70 mV may be the measurable limit as spurious signals of this order arise from contamination layers.

C. VOLTAGE CONTRAST LINEARIZATION AND ASSOCIATED MEASUREMENT TECHNIQUES

Banbury and Nixon (1970) enclose the specimen in a cylindrical cage which is part gauze to allow the secondaries to exit, and by imposing suitable potentials on parts of this cage linearize the voltage contrast. The detection is by means of a bare scintillator, held at 12 kV, with a light pipe and photomultiplier. Figure 15 shows two curves

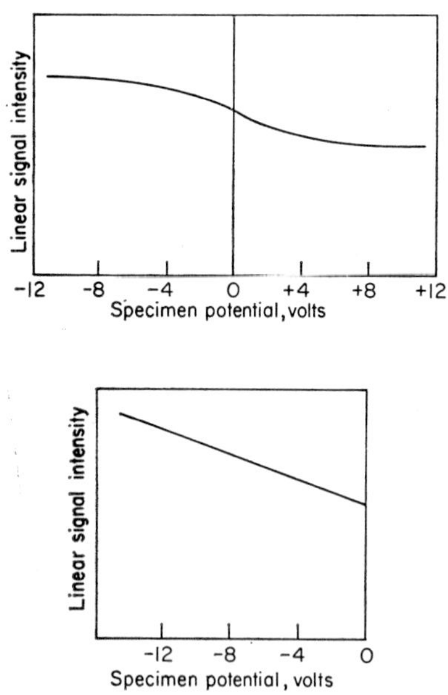

FIG. 15. Linearization of voltage contrast by Banbury's cylindrical cage enclosure. (Banbury, 1970.)

where the linearization over a range of voltages is indicated. It appears that, using this type of enclosure, it is possible to collect a certain fraction of emitted secondaries and tailor the collection to linearize the contrast. Further details of this work are given in Banbury's Ph.D. Thesis (1970).

A sophisticated form of linearization is due to Fleming (1969) who argued that if the local emission point potential affects the number of secondaries collected, a feedback loop could be used to keep the collection constant by varying the (floating) potential of the whole specimen. Thus the (floating) potential of the specimen, which is the output of the feedback-loop is a measure of the potential (plus or minus a constant) of the emission point (see Fig. 16). A large number of examples was used to illustrate the technique. However, it appears that the presence of local fields discussed in Section IV.A affects the total number of electrons collected and this is regardless of the collector to specimen potential. A grid held at 1 kV was placed near the specimen to overcome this effect, with considerable success. The limitation of the method is that the specimen is required to float, and that the high-voltage collection grid can cause destructive flash-overs. It is also not clear whether the local field effects can be completely overcome.

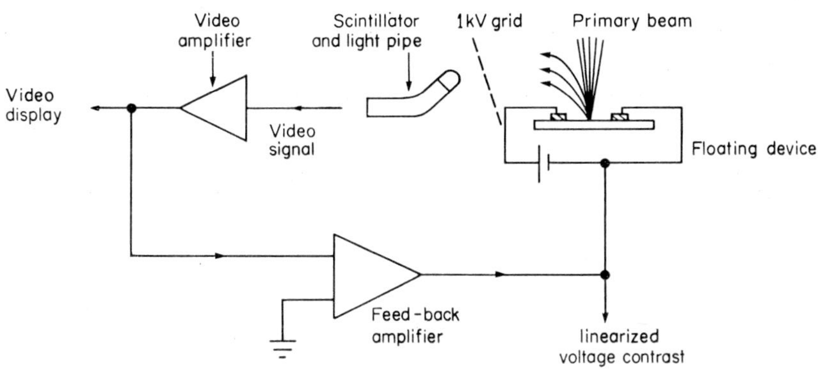

FIG. 16. Schematic diagram of feedback system for linearizing voltage contrast. (After Fleming, 1969.)

The alternative linearization technique due to Gopinath and Sanger (1971) is a compromise between the above two methods. It was suggested that the change in emission point potential results in a linear shift in secondary electron distribution and therefore a linear shift in the integrated curve is seen in a retarding field analyser as in Fig. 17. Thus, the linearity in shift could be employed by using a feedback loop to the retarding grid, to keep the output constant. Then the potential

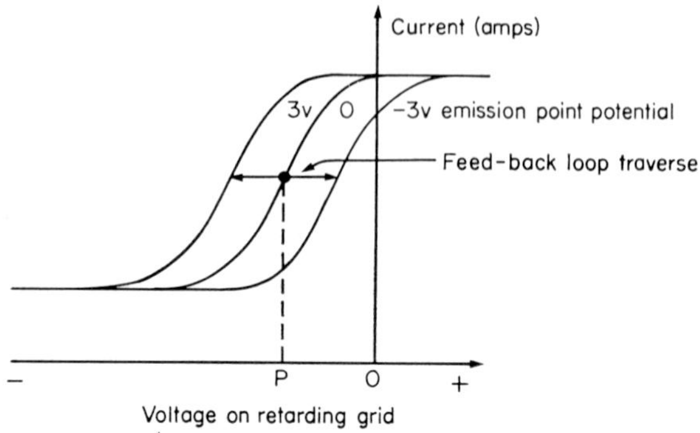

FIG. 17. Retarding-field integrated energy distribution curves, and shift due to emission point potential.

of the retarding grid plus or minus a constant is the potential of the emission point.

From Fig. 17, we note that the difference in the two feedback systems is that Fleming (1969) causes the whole retarding field curve for say −3 V for the particular emission point to shift entirely to coincide with say the zero curve which is assumed to be the initial collector setting, to maintain the output constant for a constant potential *P* on the retarding grid. This is accomplished by moving the specimen floating potential by the feedback loop to maintain the potential difference between the emission point and the retarding grid to be a constant. The second method, however, changes the retarding grid potential to maintain the video output to be a constant, and is dependent on the collection of secondaries being a constant. In practice this is not the

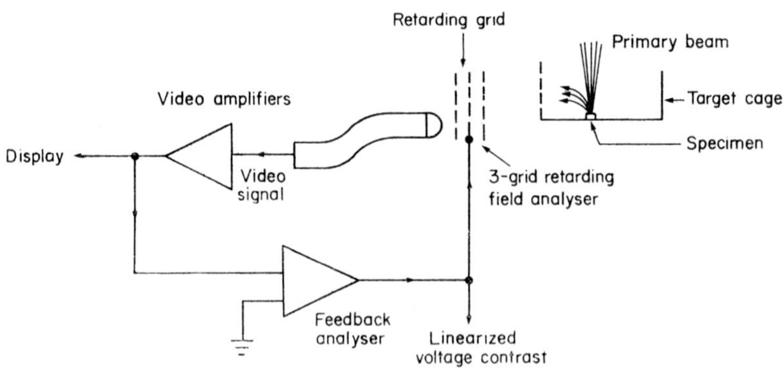

FIG. 18. Schematic diagram of retarding-field analyser type of feedback system for linearization of voltage contrast.

case and local fields affect the collection number. To overcome this to a limited extent, Gopinath and Sanger used a target cage in conjunction with a three-grid retarding field analyser at the scintillator as in Fig. 18. By empirical adjustment of the potentials on the cage, it is possible to obtain a set of retarding field curves open-loop such that their displacement is linear. With the potential of the retarding field set to give this constant output, the loop is closed and the potential or potential distribution measured. The method is exceedingly cumbersome but voltage distributions on several Gunn devices have been obtained. A modification using a hemispherical collector has been described (Sanger and Gopinath, 1972) which would simplify the setting up procedure very considerably.

V. Magnetic Contrast

As local (surface) magnetic fields also affect the slow secondary electron trajectories, magnetic contrast can be seen in the SEM. The first observation of magnetic contrast in uniaxial material was by Banbury and Nixon (1967) who showed that too high a voltage on the scintillator cage results in no contrast, while too little masks it since the signal is now predominantly back-scattered electrons. The contrast is best seen by placing apertures in front of the collector cage. It is also necessary to shield the specimen and the secondaries from the lens fields, and therefore a long working distance of about 2 cm was used. The contrast arises because of the Lorentz force acting on the slow secondaries affects the total number collected.

According to Joy and Jakubovics (1969), the magnetic field fringing on the surface of the specimen results in a net inclination of the axis of the cosine law distribution from the surface normal, of emitted secondaries. Thus, this change in inclination given by the angle β can be evaluated from the equation:

$$\beta \simeq \frac{|l|}{m\bar{v}} \int_0^{z_0} B_n d_z$$

where $\bar{v}$ is the average electron emission velocity, B_n is the transverse field fringing from the surface at $z = 0$ to some maximum value of z. From this the change in signal and hence the contrast can be estimated. However, these calculations are only approximate and indicate the order of magnitude of the effect and contrast. Further details of this calculation are included in Joy and Jakubovic's paper (1969).

Banbury (1970) after the earlier observations, has used the cylindrical cage discussed in section IV.C with the specimen in the centre to energy select secondaries within a certain range, and this was found to give excellent contrast. The cage is now available commercially as a magnetic contrast detector. Other observations on magnetic contrast have been reported by Joy and Jakubovics (1968) who used apertures in the cage.

A second type of magnetic contrast, first reported by Philibert and Tixier (1969), occurs in cubic materials which have no fringing fields on the surface and further results have been presented by Fathers *et al.* (1973). The specimen was examined at high beam voltages, upwards of 20 kV, and tilted towards the collector, and was seen in the emissive, secondary, back-scattered and specimen current modes. Rotation of the specimen about a normal to the tilted plane resulted in reversal of contrast at 180° rotation and extinction at 90° and 270° rotation. Fathers *et al.* (1973) suggest that Lorenz forces due to the domains affect the motion of the primary beam within the specimen, and this results in the contrast.

VI. Study of Fast Events in the SEM

To examine a periodic phenomenon in the SEM, it is necessary to operate the instrument as a stroboscope. Slow events, even if they are not periodic, may be observed by fast scanning in which case the physical movement or change in potential on an electronic device can be followed. For fast periodic events, the video signal has to be sampled synchronously for a time small compared to the cycle time and the display derived from this sample is then a two-dimensional picture of the dynamic state of the specimen. Moving the synchronous sampling pulse phase relative to the oscillation shows the dynamic state through the whole cycle.

The point at which the sampling is carried out depends on the frequency of the periodic event of interest, and the response time or bandwidth of the emission and collection process, detector and the associated electronic circuitry. The scintillator typical pulse decay time is of the order of a nanosecond, that of photomultiplier is about 10 nsec, and the head amplifier has a bandwidth of the order of say 10 MHz. It should be possible to examine oscillations up to say 100 kHz, by sampling at the output of the head amplifier. Thus, a typical system would pass the signal through a gate which has a sampling width of one-hundredth (100 nsec, say) of the cycle time, which is operated synchronously with the oscillator. The video amplifier bandwidth is generally smaller than 100 kHz and therefore the signal which is only present during the sampling pulse width will be integrated and the display will be the dynamic state of the specimen. The signal strength, however, is reduced considerably in the sampling process and therefore a slower scan time will be necessary for adequate noise integration. This technique assumes that the emission time is small, and that the times of flight of the emitted electrons with different energies do not smear out the details of the dynamic event excessively. The technique can be used for slow events (up to 100 kHz at least) and should be easily implemented.

When frequencies are higher, the electronic gating system above is not suitable, since the response times of the various components of the video chain are inadequate. It then becomes necessary to sample by pulsing the illumination which is the primary beam, in synchronism with the oscillator. Plows and Nixon (1968) have pulsed the beam by deflecting it past an aperture using a set of deflection plates, at 7 mHz repetition rate to obtain 10 nsec pulse widths. To ensure that only one beam pulse per deflection cycle is obtained it is necessary to use 3·5 mHz excition. They examined the voltage distribution over a MOST ladder network excited at 7 mHz using voltage contrast. It was necessary to use a retarding field and analyser to exclude the contrast due to slow secondaries whose information content was distorted by the repetitive local fields.

Macdonald *et al.* (1969) have examined acoustic domains in CdS crystals oscillating at a maximum frequency of 0·5 mHz using a continuous beam and fast detection and display. In this case, the response time of the photomultiplier, the decay time of the scintillator, and the display of the linescan on an oscilloscope (video signal as y modulation) were fast enough to display the signal directly. The curves obtained showing voltage distributions across the device are similar to those seen by capacitive probes with a sampling oscilloscope. They have also examined the motion of high-field domains in a Gunn device, 1 mm long, oscillating at 85 mHz using the stroboscopic technique. To obtain the relevant primary beam samples the Wehnelt cylinder on the electron gun was pulsed. Since the 1 mm long Gunn device used has to be examined under pulsed conditions, a delay was introduced between the device bias pulse, which was about 100 nsec long, and the electron-gun pulse. In practice it is found that the transit times of the secondaries from the sample to the collector cage requires a further sample of the signal to be taken as shown in Fig. 19, using a sampling oscilloscope, and the output linescan is then plotted on an XY recorder. The results show the propagation of a disturbance through the diode, though the voltage distributions are not in agreement with those obtained by capacitive probe measurements.

Robinson (1971) has extended the technique of stroboscopic studies to 2 GHz, and presented results of voltage distribution across a resistor. In this case the pulsing of the beam was by means of a travelling-wave deflection line, and degradation in the instrument resolution was found to be severe. Gopinath (1970) has also considered chopping at a 9·6 GHz repetition frequency by travelling-wave deflection lines and has also examined the spot degradation.

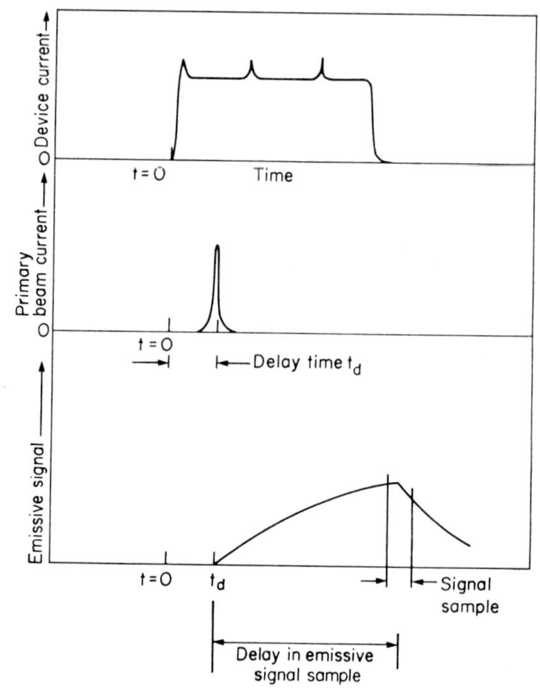

FIG. 19. Waveforms in the stroboscopic examination of Gunn devices. (After MacDonald *et al.*, 1969.)

Further work on stroboscopic operation of the SEM has been carried out by Gopinath and Hill (1973), who have operated the instrument at 9·1 GHz. They have obtained linescans across Gunn devices 8 μm in width, of voltage distribution curves with the high-field domain in mid-flight. These show agreement with capacitive probe results obtained with much larger devices (100 μm) oscillating at about 1·2 GHz. From these results it appears that stroboscopy of very fast periodic events is feasible and can be used for dynamic studies.

VII. Summary

This chapter has introduced the emissive mode of the scanning microscope and discussed in brief various collector and detector systems, the contrasts available, the resolution capability and the specimen examination conditions. The specimen topography section discusses the non-stereoscopic method of identification of specimen geometry, and the measurement of linear distances including depths and heights. The compositional contrast available is discussed and the techniques available for composition identification are outlined. The voltage contrast phenomenon is discussed together with the mechanism, linearization and voltage measurement. Magnetic contrast is also briefly considered. The chapter concludes with a review of work on the examination of fast events in the SEM.

References

Ballard, D. B. (1972). *Proc. Vth SEM Symposium, IITRI, Chicago*, 121–128.

Banbury, J. R. (1970). Ph.D. Thesis, Cambridge University.

Banbury, J. R. and Nixon, W. C. (1967). *J. scient. Instrum.* **44,** 889–892.

Banbury, J. R. and Nixon, W. C. (1970). *Proc. IIIrd SEM Symposium, IITRI, Chicago*, 475–480.

Bishop, H. E. (1966). Ph.D. Thesis, Cambridge University.

Blakeslee, A. E. (1970). *Proc. IIIrd Int. Symposium on GaAs, Aachen*, 285–287.

Booker, G. R. (1970). "Modern Diffraction and Imaging Techniques in Materials Science," pp. 613–653. North-Holland, Amsterdam.

Boyde, A. (1972). *Proc. Vth SEM Symposium, IITRI, Chicago*, 257–264.

Boyde, A. (1973). *J. microsc.* **98,** 3, 452–472.

Brandis, E. K. (1972). *Proc. Vth SEM Symposium, IITRI, Chicago*, 241–248.

Brandis, E. K., Anderson, F. W. and Hoover, R. (1971). *Proc. IVth SEM Symposium, IITRI, Chicago*, 505–510.

Broers, A. N. (1969). *Rev. scient. Instrum.* **40,** 1040–1045.

Broers, A. N. (1970a). *Proc. IIIrd SEM Symposium, IITRI, Chicago*, 3–8.

Broers, A. N. (1970b). *Proc. VIIth Int. Congr. Electron Microscopy, Grenoble*, **1,** 239–240.

Bruining, H. (1954). "Physics and Applications of Secondary Emission." Pergamon Press, London.

Chang, C. C. (1971). *Surface Sci.* **25,** 53–79.

Corvin, I. and Johari, O. M. (1970). *Proc. IIIrd SEM Symposium, IITRI, Chicago*, 73–80.

Crewe, A. V., Eggenberger, D. N., Wall, J. and Welter, L. M. (1970). *Rev. scient. Instrum.* **39,** 576–579.

Everhart, T. E. (1968). *Proc. Ist SEM Symposium, IITRI, Chicago*, 3–13.

EVERHART, T. E. and THORNLEY, R. F. M. (1960). *J. scient. Instrum.* **37,** 246–248.

EVERHART, T. E., WELLS, O. C. and OATLEY, C. W. (1959). *J. Electronics and Control,* **7,** 97–111.

FATHERS, D. J., JOY, D. C. and JAKUBOVICS, J. P. (1973). *Proc. VIth SEM Symposium, IITRI, Chicago,* 259–266.

FLEMING, J. P. (1969). *Electronics Letters,* **4,** 435–436.

GOPINATH, A. (1970). *Proc. VIIth Int. Congr. Electron Microscopy, Grenoble,* **1,** 203–204.

GOPINATH, A. and HILL, M. S. (1973). *I.E.E.E. Transaction on Electron Devices,* **ED-20,** 609–611.

GOPINATH, A. and SANGER, C. C. (1971). *J. Phys. E. scient. Instrum.* **4,** 334–336.

HACHENBERG, O. and BRAUER, W. (1959). *In* "Advances in Electronics and Electron Physics" (L. Morton, ed.), Vol. 11, pp. 413–499. Academic Press, London and New York.

HATZAKIS, M. (1970). *Rev. scient. Instrum.* **41,** 128.

HOUSTON, R. A. (1949). "A Treatise on Light," pp. 368–372. Longmans Green, London.

HOOVER, R. (1971). *J. Phys. E. Scient. Instrum.* **4,** 747–749.

JOY, D. C. and JAKUBOVICS, J. P. (1968). *Phil. Mag.* **17,** 61–69.

JOY, D. C. and JAKUBOVICS, J. P. (1969). *J. Phys. D. Appl. Phys.* **2,**1367–1372.

KIMOTO, S. and HASHIMOTO, H. (1960). "The Electron Microprobe" (McKinley, Heinrich, Wittry, eds). John Wiley, New York.

MACDONALD, N. C. (1970). *Proc. IIIrd SEM Symposium, IITRI, Chicago,* 483–485.

MACDONALD, N. C. (1971). *Proc. IVth SEM Symposium, IITRI, Chicago,* 89–96.

MACDONALD, N. C., ROBINSON, G. V. and WHITE, R. M. (1969). *J. appl. Phys.* **40,** 4516–4528.

MURRAY, S., GILLHAM, C. and WINDLE, A. H. (1973). *J. Phys. E. Scient. Instrum.* **6,** 381–384.

NIXON, W. C., PEASE, R. F. W. (1965). *J. scient. Instrum.* **42,** 81–85.

OATLEY, C. W. (1972). "The Scanning Electron Microscope." Cambridge University Press.

OATLEY, C. W., NIXON, W. C. and PEASE, R. F. W. (1965). *In* "Advances in Electronics and Electron Physics" (L. Morton, ed.), Vol. 21, pp. 181–247. Academic Press, London and New York.

OSBORNE, J. S. (1971). *J. Phys. E, Scient. Instrum.* **4,** 958–960.

PEASE, R. F. W. (1965). *J. scient. Instrum.* **42,** 158–159.

PHILIBERT, J. and TIXIER, R. (1969). *Micron,* **1,** 174–186.

PLOWS, G. S. (1969). Ph.D. Thesis, Cambridge University.

PLOWS, G. S., NIXON, W. C. (1968). *J. Phys. E, Scient. Instrum.* **1,** 595–600.

ROBINSON, G. Y. (1971). *Rev. scient. Instrum.* **41,** 251–255.

SANGER, C. C. and GOPINATH, A. (1972). *Proc. European Microscopy Congress, Manchester,* 482–483.

SECKER, P. E., BARDEN, H. G. and CLEEVE, J. W. (1970). *J. Phys. E*, **3**, 319–320.

SHAW, D. A. (1970). Ph.D. Thesis, University College of North Wales, Bangor.

THORNTON, P. R. (1968). "Scanning Electron Microscopy." Chapman and Hall, London.

WELLS, O. C. (1970a). *Proc. IIIrd SEM Symposium, IITRI, Chicago*, 486–487.

WELLS, O. C. (1970b). *Appl. Phys. Letters*, **16**, 151–153.

WELLS, O. C. (1971). *Appl. Phys. Letters*, **19**, 232–235.

WELLS, O. C. (1974). *Proc. VIIth SEM Symposium, IITRI, Chicago, 1–8.*

WELLS, O. C. and BREMER, C. G. (1968.) *J. Phys. E, Scient. Instrum.* **1**, 902–906.

Chapter 6

Electron Channelling Patterns in the SEM

D. C. JOY

Department of Metallurgy, University of Oxford, England

Symbols

$\mu^{(0)}$ = mean absorption coefficient for electrons

$p^{(0)}$ = mean back-scattering coefficient for electrons

$\mu^{(j)}$ = absorption coefficient associated with jth Bloch wave

$I_0^{(j)}$ = intensity of Bloch wave (j) at crystal entrance surface

$p^{(j)}\,dx$ = fraction of Bloch wave (j) back-scattered by slice dx

$I_F(x)$ = intensity of forward scattered beam at depth x

$I_B(x)$ = intensity of back-scattered beam at depth x

t = specimen thickness

t_b = limiting thickness above which all crystals can be treated as bulk crystals

θ = angle (of incidence) between the surface normal and the beam

d_{hkl} = spacing of Bragg reflecting planes

θ_B = Bragg angle

x = depth in the crystal

I_b = incident beam current

α = electron beam divergence angle

B = electron gun brightness

d = electron beam spot size

A = linear dimension of display screen
W = distance from the point about which the beam scanning action occurs to the specimen
θ' = semi-angle scanned
V_b = beam accelerating voltage
D = diameter of the final aperture
ϕ = contribution to the beam divergence at the specimen due to the operation of the final lens
ψ = contribution to the beam divergence at the specimen due to the finite source size of the electron gun
L = diameter of the filament cross-over image before the final lens
u = distance of the filament cross-over image from the final lens
s = distance of the specimen from the final lens
f = focal length of the final lens
λ = wavelength of the beam electrons
a = lattice parameter of the crystal specimen
UVW = Miller indices of a zone
hkl = Miller indices of a plane
C_s = coefficient of spherical aberration
R = radius of stereogram
X, Y = cartesian coordinates of a pole HKL referred to (001) as origin
Θ = average angular misorientation of the subgrains in the selected area
S = selected area
s = average subgrain size

I. Introduction

Until recently the scanning electron microscope (SEM) has treated all specimens as if they were amorphous. The information supplied by the SEM related only to the topography and chemical composition of the sample. This was a serious deficiency in the usefulness of the instrument when compared with the conventional transmission electron microscope in which the crystallography of the specimen can be observed by means of the diffraction and Kikuchi patterns produced and in which variations in the local crystalline orientation, due to defects of various types, can be made to yield visible contrast. Now, however, the electron channelling pattern (ECP) technique (Coates, 1967; Booker *et al.*, 1967) remedies this drawback and exploits the crystallinity of the specimen to provide detailed quantitative information about the symmetry, lattice parameter, orientation and perfection of the sample. This information is a necessary "by-product" of the scanning action of the microscope and the interaction of the incident electron beam with the specimen. To obtain this data it is therefore only necessary to set up the appropriate electron optical conditions and use a suitably prepared specimen. In the following pages the theoretical origin of the ECP contrast is briefly described and the electron–optical conditions needed to produce it are derived. The quantitative applications of the technique are then discussed in detail.

II. ECP Contrast

A. The Theory

If an electron beam is incident upon an amorphous specimen the electron intensity back-scattered can be described in terms of a constant back-scattering fraction. If now the beam is incident upon a single crystal of a material the electron intensity back-scattered from the crystal is found to be a function of the orientation of the beam with respect to the crystal lattice. This effect is the ECP contrast and is quite distinct from changes in the back-scattered signal due to topographical features. That is, the ECP contrast will always be present even under conditions where there is zero topographical contrast as shown in Fig. 1. The appearance of an ECP is shown in Fig. 1(a). A semi-quantitative explanation of the channelling effect was given by Booker *et al.* (1967) and quantitative theories have been given by several authors (Hirsch and Humphreys, 1970; Vicario *et al.*, 1970; Reimer *et al.*, 1971). The treatment presented here is that due to Spencer *et al.* (1972).

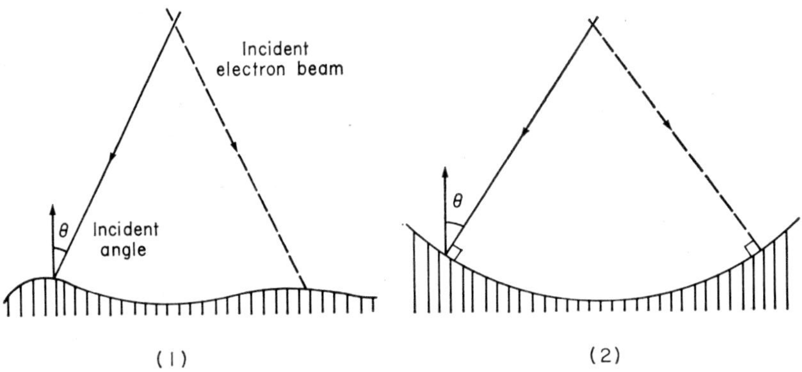

FIG. 1(a). (1) Topographic contrast arises from changes in angle of incidence relative to surface normal. ECP contrast arises from changes in angle of incidence relative to lattice planes. (2) No topographic contrast as beam is always normal to surface. Still ECP contrast as angle relative to lattice planes changes.

An electron inside a crystal lattice can be represented as a superposition of two or more Bloch waves. These Bloch waves are the solutions of the Schrodinger equation for a periodic lattice and in physical terms they represent the current flows inside the lattice.

If we consider the simplest case only two Bloch waves are excited. Their positions relative to the atoms in the crystal lattice are shown in Fig. 2 for the case where the beam is near a Bragg position. Obviously

Bloch wave type 1, which has its maxima along the line of atom centres, will be much more likely to be scattered than the type 2 Bloch wave which has its maxima between the atom planes. Consider now what will happen in a slice of the crystal of thickness dx at a distance x from the top surface of the crystal (Fig. 3) which is of total thickness t. We shall treat the electrons inside the crystal as being of two classes, those in the

FIG. 1(b) ECP from (110) orientation silicon specimen. 30 kV, specimen current.

Bloch wave field and those which have been inelastically scattered either forwards (i.e. scattered through angles less than 90°), or backwards (i.e. scattered through angles greater than 90°). As these scattered electrons are travelling in all directions they will experience a mean absorption coefficient $\mu^{(0)}$ and a mean back-scattering coefficient $p^{(0)}$. Since the slice scatters electrons out of Bloch wave (j) we also associate an absorption coefficient $\mu^{(j)}$ with this Bloch wave due to this process. Let $I_0^{(j)}$ be the intensity of Bloch wave (j) at the crystal entrance surface and $p^{(j)}$ dx be the fraction of the wave back-scattered by the slice.

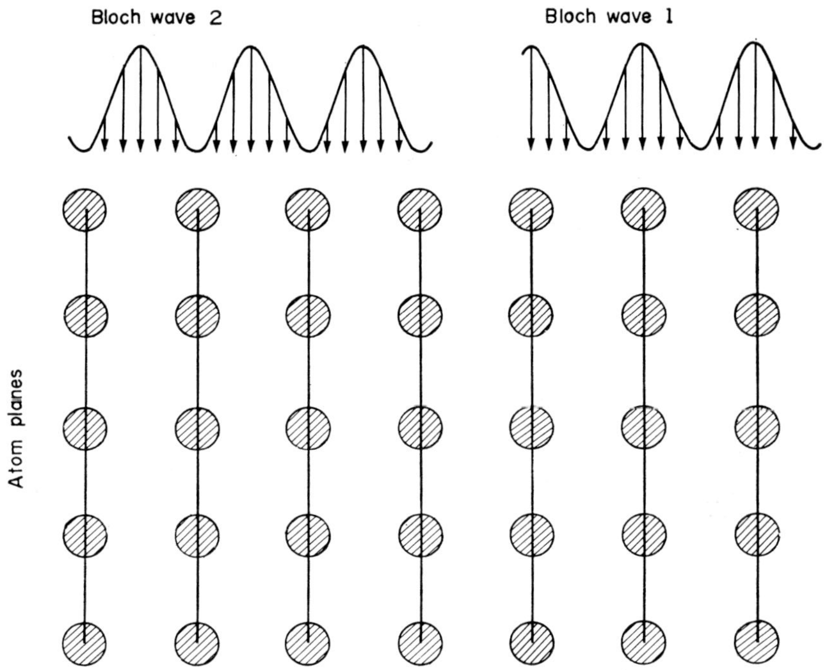

FIG. 2. Position of Bloch waves 1 and 2 relative to lattice planes.

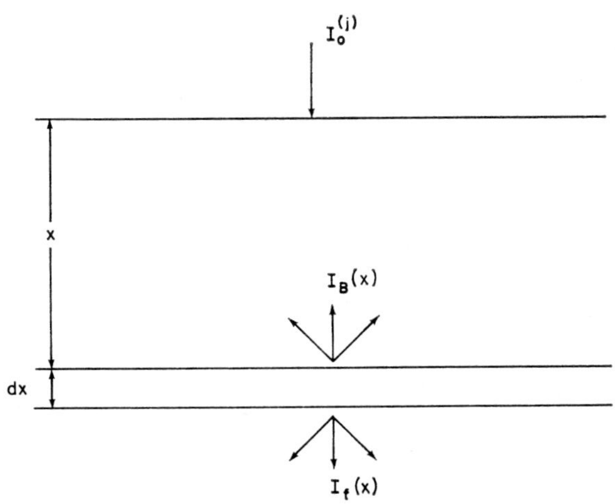

FIG. 3. Notation used in theory.

Call $I_F(x)$ and $I_B(x)$ the intensities of the forward and back-scattered beams at depth x, and $p^{(0)}\,\mathrm{d}x$ the fraction back-scattered from these beams by the slice. Then applying "electron conservation" to the slice we get

$$\mathrm{d}I_F = I_0^{(j)}\exp(-\mu^{(j)}x)(\mu^{(j)}-p^{(j)})\,\mathrm{d}x - I_F p^{(0)}\,\mathrm{d}x + I_B p^{(0)}\,\mathrm{d}x \qquad (1)$$

$$-\mathrm{d}I_B = I_0^{(j)}\exp(-\mu^{(j)}x)p^{(j)}\,\mathrm{d}x + I_F p^{(0)}\,\mathrm{d}x - I_B p^{(0)}\,\mathrm{d}x. \qquad (2)$$

Uncoupling these simultaneous first-order equations gives

$$\frac{\mathrm{d}^2 I_F}{\mathrm{d}x^2} = -I_0^{(j)}\exp(-\mu^{(j)}x)\{\mu^{(j)}(\mu^{(j)}+p^{(0)}-p^{(j)})\} \qquad (3)$$

$$\frac{\mathrm{d}^2 I_B}{\mathrm{d}x^2} = -I_0^{(j)}\exp(-\mu^{(j)}x)\{\mu^{(j)}(p^{(j)}-p^{(0)})\}. \qquad (4)$$

Applying the boundary conditions $I_B(x) = 0$ at $x = t$; $I_F(x) = 0$ at $x = 0$ we find that the total back-scattered intensity at the top surface from the jth Bloch wave is

$$I_B(0) = \frac{I_0^{(j)}}{1+p^{(0)}t}\left\{p^{(0)}t + \frac{(p^{(j)}-p^{(0)})}{\mu^{(j)}}(1-\exp(-\mu^{(j)}t))\right\}. \qquad (5)$$

The total intensity back-scattered is just the above expression summed over all Bloch waves (i.e. over all j). If we normalize the input beams so that

$$\sum_j I_0^{(j)} = 1.$$

Then

$$I_B^{(0)} = \frac{p^{(0)}}{1+p^{(0)}t} + \frac{1}{1+p^{(0)}t}\sum_j\left\{I_0^{(j)}\,\frac{(p^{(j)}-p^{(0)})}{\mu^{(j)}}[1-\exp(-\mu^{(j)}t)]\right\}.$$

In this expression the term $(1+p^{(0)}t)^{-1}.\ p^{(0)}t$ is essentially a background term whilst the terms summed over j (and multiplied by $(1+p^{(0)}t)^{-1}$) are the desired signal contrast. Note that as t goes to infinity the contrast from this expression appears to go to zero. In fact this is not true in practice and only arises in this expression because we have neglected the effect of energy losses in the incident beam. This can be taken care of by treating all crystals above a certain limiting value of thickness as bulk crystals. For silicon crystals at 20 kV this limiting thickness t_b is about 2 μm. Hence for bulk crystals the theory is applied with $t = t_b$, whilst crystals with $t < t_b$ are given the correct value of t in the expression.

To evaluate the intensity back-scattered the quantities $p^{(0)}$ and $p^{(j)}$ must be known and the quantities $\mu^{(j)}$ and $I_0^{(j)}$ must also be calculated. The way in which these vary with the angle of incidence of the beam determine the shape of the intensity profile calculated from the above expression.

Figure 4 shows the geometry of the beam scan, with the angle of incidence θ defined as being that between the surface normal and the beam. Figure 5 then shows the variation of $I_0^{(j)}$ with θ (calculated from the data of Hirsch *et al.*, 1965). Between the zero angle of incidence and the first point at which the Bragg equation for the lattice planes normal to the surface

$$n\lambda = 2d_{hkl} \sin \theta_B$$

is satisfied we see that Bloch wave 1 is strongly excited, but Bloch wave 2 and all the higher Bloch waves are only weakly excited. Around θ_B the excitation of the waves changes and Bloch wave 2 is now predominant until θ reaches the second order Bragg position, when Bloch wave 3 takes over, and so on.

The variation of the $p^{(j)}$ for the first four Bloch waves is shown in Fig. 6. The variation of the $\mu^{(j)}$ coefficients is similar in form but the magnitude of variation is less. The expected variation of the back-scattered signal with the orientation of the beam can now be calculated and Fig. 7 shows the intensity profile out to the third Bragg position with the individual contributions from each Bloch wave indicated. It can be seen that between the symmetry position and the first Bragg position Bloch wave 1 is the most strongly excited wave. As this wave is strongly back-scattered (i.e. $p^{(1)}$ is high) the back-scattered signal comes mainly from wave 1 and is high. Around the first Bragg position the excitation of waves 1 and 2 are approximately symmetrical but wave 1 has a higher back-scattering coefficient than wave 2 (see Fig. 6). Therefore just negative of the Bragg position the back-scatter will be stronger than for a position just positive of the Bragg position.

A similar change-over occurs at around the second order Bragg position as the highly excited Bloch wave changes from wave 2 to 3. Around this position wave 2 is more back-scattered than wave 3 so as the waves change over the back-scattered intensity will go from bright to dark. In general the back-scattered intensity is seen to be controlled by the values of $p^{(j)}\ I_0^{(j)}$ rather than by the $\mu^{(j)}$ term.

For a complete range of scan angles from, say, $+3\theta_B$ to $-3\theta_B$ the back-scattered signal will be in the form of a bright central band of width $2\theta_B$ with sharp bright–dark higher order lines spaced at θ_B on either side.

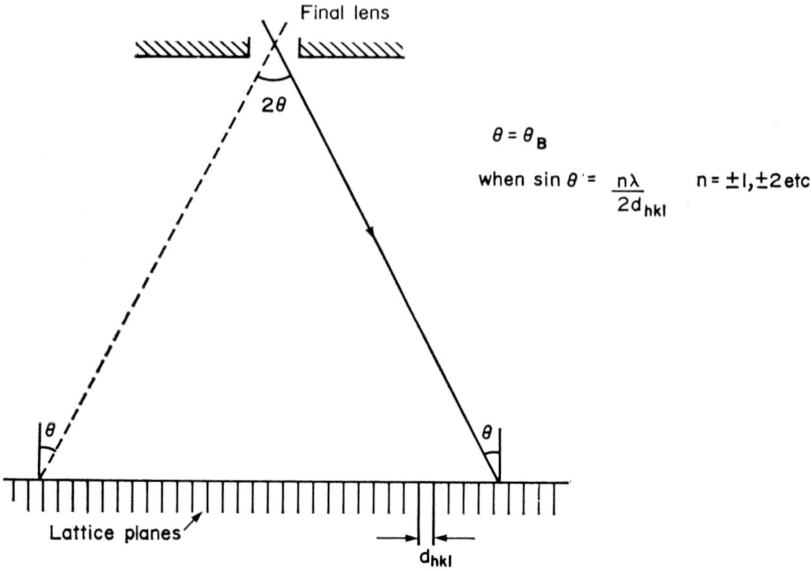

FIG. 4. Geometry of scan and symbols used.

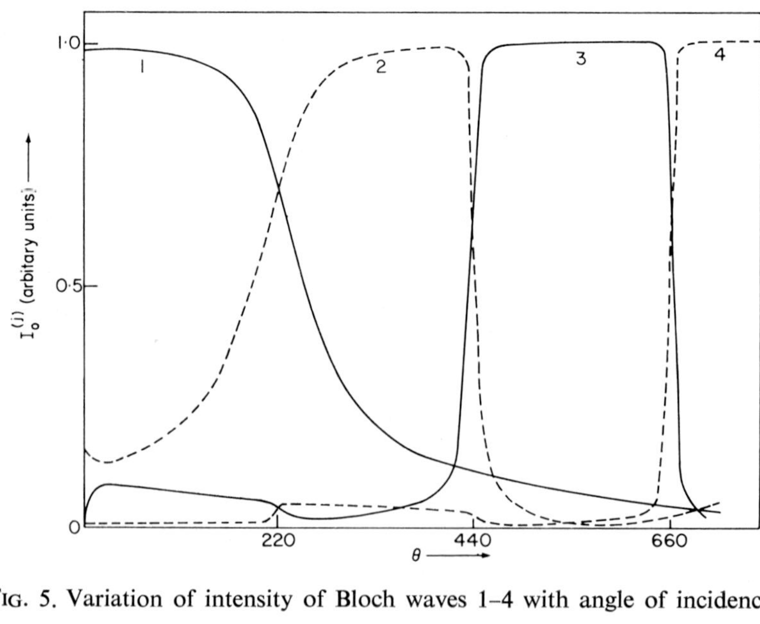

FIG. 5. Variation of intensity of Bloch waves 1–4 with angle of incidence.

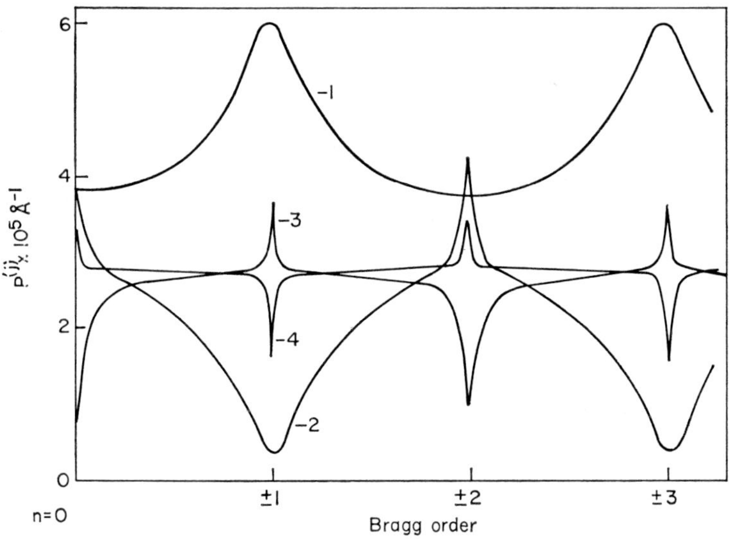

FIG. 6. Variation of back-scattering coefficients for first 4 Bloch waves.

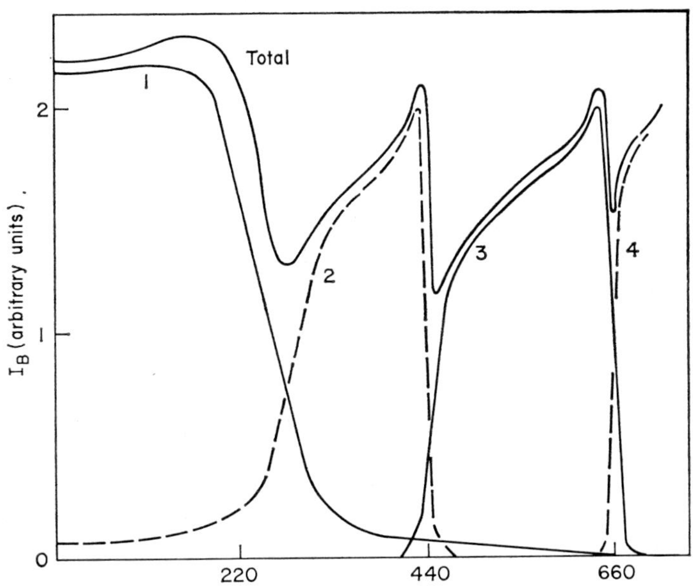

FIG. 7. Profile of ECP calculated from theory showing contributions of each of the first four Bloch waves.

Figure 8 shows the measured profile together with the profile calculated from the above theory (Spencer *et al.*, 1972) and it can be seen that very good agreement is obtained. The t_b limiting value described above has been set at 2 μm. In practice, of course, more than one set of planes is able to take part in the scattering because of the symmetry of the crystal and when the incident beam is scanned in two dimensions rather than one a complex pattern of bands will be built up when the back-scattered signal is displayed on the SEM viewing screen.

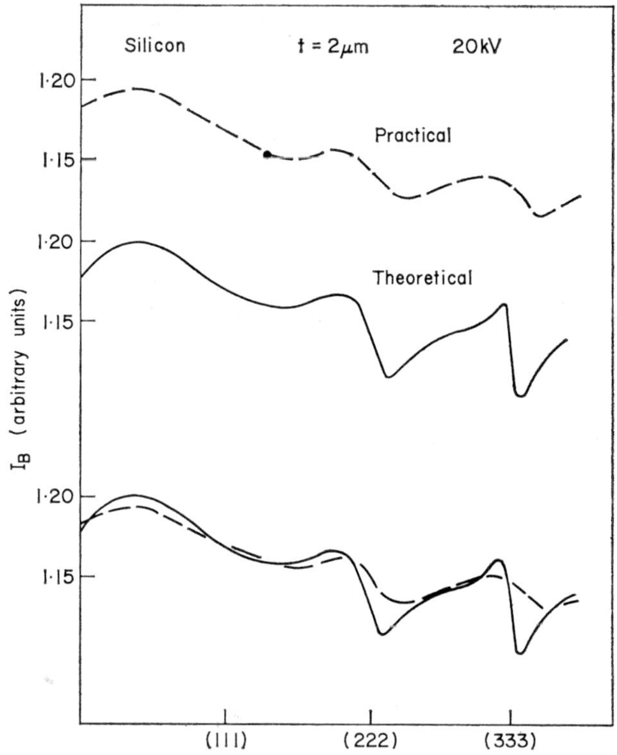

FIG. 8. Comparison of theoretical and practical profiles for silicon (111) band.

Figure 9 shows such a pattern taken from a (111) face of silicon showing the (220) bands together with the parallel (440), (660) etc. higher order lines. The three bands arise because of the three-fold symmetry of the (111) pole in the cubic system. The other lines in the pattern come from bands not visible on the image because of the restricted angle of scan used. This pattern is a function of the crystallography only, so lateral movements of the specimen will not change the pattern, as long as the crystal is of one orientation. If, however, the specimen

FIG. 9. (111) pole in silicon at 20 kV. Pattern differentiated to show fine line structure visible in the pattern.

is tilted then the pattern will move as if rigidly fixed to the lattice and new bands and lines will become visible as different sets of lattice planes are placed so as to diffract the beam.

The bands and poles appearing on the pattern in most cases can be predicted by the "allowed reflection" rules used in X-ray analysis. That is to say for an FCC crystal the only reflections visible will be those for which (h, k, l) are all odd or all even, and for a BCC those for which $(h+k+l)$ is even, and the similar relationships for lower symmetry crystals. However, certain lines which should be present are either of very low intensity or missing altogether—for example (111) in silicon, $(10\bar{1}0)$ in cobalt and (222) in gold. These effects can be understood in terms of the theory presented above and arise because of the relative numerical values of the different contribution to the back-scattered signal.

When, on average, the beam is normal to the specimen during the scan raster the ECP contrast is symmetrical about the centre position $\theta = 0$. If the specimen is tilted then the contrast becomes asymmetrical and for very high angles of tilt (of the order of 60°) the contrast across a band may even reverse in sign. These effects are similar to those observed in reflection Kikuchi patterns.

The contrast of the pattern is determined by the magnitude of

$$\frac{I_0^{(j)}(p^{(j)} - p^{(0)})}{\mu^{(j)}}$$

and for a bulk material this usually turns out to be about 5% contrast at 20 kV. The magnitude of the contrast will be higher for a thinner material (i.e. $t < t_b$) as the "background" term $p^{(0)}t(1+p^{(0)}t)^{-1}$ will be less and contrast values of 25% or more have been found on foils 1000 Å or less in thickness.

Although all the above theory has been for the back-scattered signal it is clear that there will also be contrast on the specimen current signal and in the secondary electron signal. The contrast in the specimen current signal arises because of the complementary nature of this signal to the back-scattered signal (Thornton, 1968)—but see Chapter 8 for conditions in which these signals are not complementary, and the magnitude of the contrast is of the same order of magnitude as that in the back-scattered image although the sign of the contrast is reversed so that the main band will be dark rather than bright. The contrast in the secondary electron signal is the same sign as that in the back-scattered signal and of a similar magnitude although changes in detector geometry very markedly affect the value obtained. Any of the standard detector systems in the SEM can thus be used to collect the signal. Although the magnitude and sign of the contrast may be different in each case the geometrical form will always be the same.

B. THE ELECTRON–OPTICAL CONDITIONS

We have seen from the above theory that ECP contrast comes about from changes in the angle of incidence of the electron beam with the crystal lattice. Since the contrast is purely a function of the angle of incidence of the beam it is important that the beam itself should have a clearly defined angle of incidence, that is to say the beam must be collimated. The more the beam is collimated the higher the angular resolution of the pattern and the finer the detail that can be observed. The maximum acceptable beam divergence is about one-fifth of the Bragg angle for the reflection to be observed.

For operation at 30 kV this implies the need for a beam divergence of

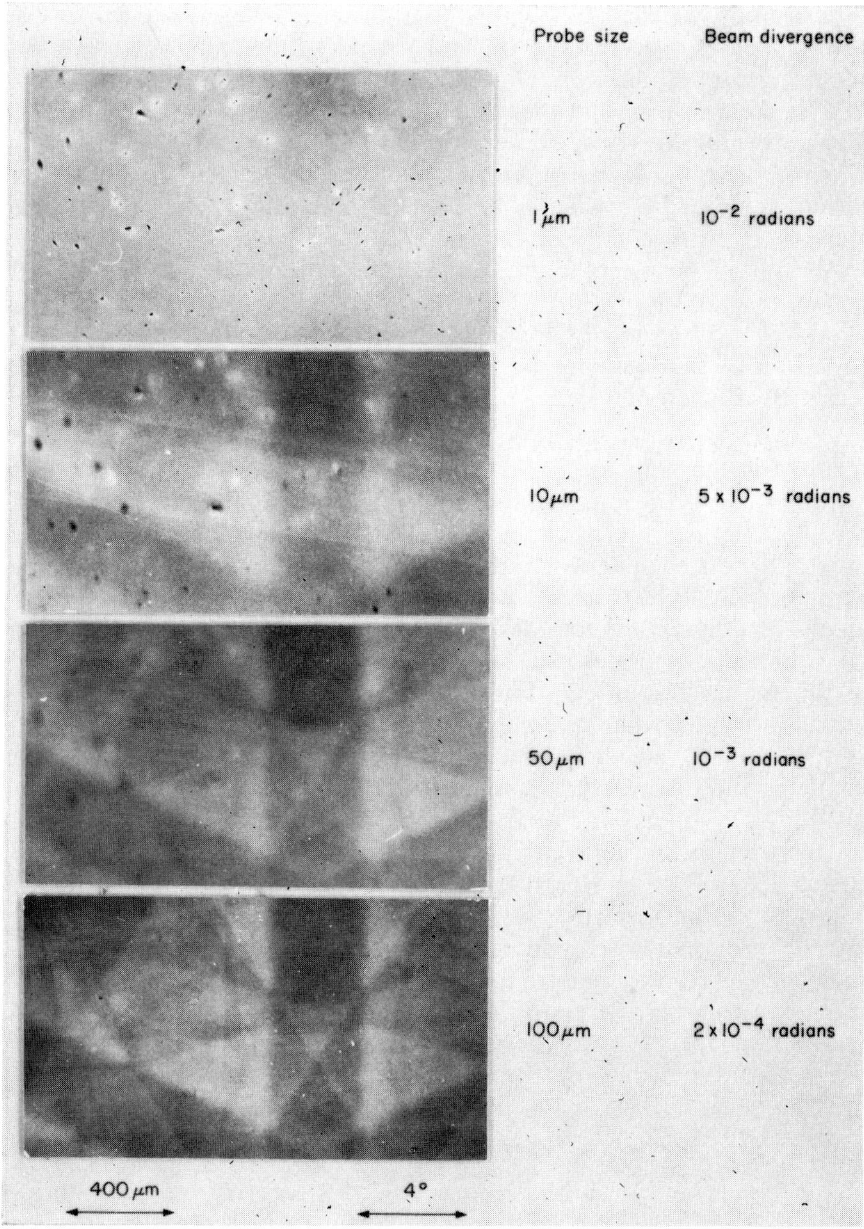

FIG. 10. Variation of ECP quality with beam divergence.

10^{-2} radians or less. At this divergence only the main bands on the ECP will be visible. As the divergence is decreased the quality of the pattern improves and the narrow, high-order lines become visible. For most purposes a divergence of 3×10^{-3} rad is adequate but for really good angular resolution a value of 5×10^{-4} rad is desirable.

We have also seen that the contrast of the ECPs is low, typically 5%. This means that it is necessary to use a fairly high incident beam current if this contrast is to be observed above the noise. The lowest usable current is about 1 nA when a 40 sec scan time is used. For operation at visual scan rates a current of around 10 nA is desirable. These two conditions of beam collimation and beam current also determine the spatial resolution of the microscope when operated in a condition suitable for observing ECP contrast. This is because of the brightness limitation of the electron source used. The incident current on to the specimen is given by

$$I_b = 2{\cdot}5\, d^2 \alpha^2 B \text{ Amps} \tag{6}$$

where d is the spot size in centimetres, α is the beam divergence in radians and B is the gun brightness in A cm^{-2} sr^{-1}. Since the gun brightness B is a constant of the gun for any voltage, the quantities I_b d and α cannot be varied independently. If both I_b and α are fixed then d is automatically fixed from the above relation. In particular once a minimum acceptable value of I_b is chosen it can be seen that the angular resolution of the pattern (i.e. α) can only be improved at the expense of the spatial resolution of the electron beam (i.e. α). This effect is shown in Fig. 10 where the microscope has been adjusted in such a way as to keep the incident beam current fixed at 10 nA while the beam diameter and divergence are varied. As one progresses from Fig. 10(a) to 10(d), 2α (the beam divergence) is progressively decreased so in order to retain the same current I, d has to be progressively increased. Figure 10(a) is therefore a standard micrograph showing topographical contrast. Going towards Fig. 10(d) the micrograph resolution (determined by d) progressively deteriorates until no topographical contrast remains, while the pattern angular resolution (determined by 2α) progressively improves until the lines are sharp.

III. The Standard Mode of Electron Channelling Pattern Operation

A. Operating Conditions

The conditions discussed above fix the necessary electron–optical conditions in detail. We shall now see how to adjust the SEM to meet these requirements.

1. *Scan Angle*

In order that the ECP contrast be easily recognizable it is necessary to scan the beam through at least $\pm 2\theta_B$ angles, this angular range being sufficient to display the band together with the first high order lines. From the table (Fig. 11) giving a selection of Bragg angles for common materials and low index planes at 15, 20 and 30 kV, it can be seen that this condition implies a scanned angle of at least 8° about the normal.

				θ_B (Degrees)							
Element	a (Å)	Type	Voltage kV	⟨110⟩	⟨111⟩	⟨200⟩	⟨211⟩	⟨220⟩	⟨301⟩	⟨311⟩	⟨222⟩
			15	1·40	—	2·00	2·43	2·80	3·13	—	3·45
Fe	2·86	BCC	20	1·23	—	1·71	2·10	2·43	2·71	—	2·99
			30	0·99	—	1·40	1·71	1·99	2·21	—	2·43
			15	1·24	—	1·72	2·11	2·44	2·73	—	3·00
Nb	3·29	BCC	20	1·06	—	1·50	1·87	2·20	2·37	—	2·60
			30	0·86	—	1·22	1·50	1·76	1·97	—	2·11
			15	—	1·40	1·61	—	2·29	—	2·68	2·79
Ni	3·52	FCC	20	—	1·21	1·40	—	1·98	—	2·33	2·43
			30	—	0·99	1·15	—	1·61	—	1·90	1·97
			15	—	1·37	1·58	—	2·21	—	2·60	2·71
Cu	3·61	FCC	20	—	1·19	1·36	—	1·94	—	2·26	2·36
			30	—	0·96	1·10	—	1·57	—	1·85	1·92
			15	—	1·21	1·40	—	1·99	—	2·32	2·43
Al	4·04	FCC	20	—	1·06	1·22	—	1·71	—	2·02	2·10
			30	—	0·85	1·00	—	1·40	—	1·65	1·71
			15	—	1·21	1·39	—	1·97	—	2·30	2·41
Ag, Au	4·075	FCC	20	—	1·05	1·21	—	1·70	—	2·01	2·09
			30	—	0·84	0·99	—	1·38	—	1·63	1·69
			15	—	(0·91)	—	—	1·49	—	1·74	1·82
Si	5·42	Diamond	20	—	(0·80)	—	—	1·29	—	1·52	1·60
			30	—	(0·64)	—	—	1·05	—	1·23	1·28
			15	—	0·87	—	—	1·43	—	1·68	1·74
Ge	5·65	Diamond	20	—	0·75	—	—	1·25	—	1·45	1·51
			30	—	0·63	—	—	1·03	—	1·20	1·25

FIG. 11. Table of Bragg angles for common cubic elements.

For a scanning microscope the magnification is given by

$$A/(W.2 \tan \theta')$$

where A is the linear dimension of the display screen, W is the distance from the point about which all the scanning action occurs to the specimen and θ' is the scanned semi-angle. In microscopes with double-deflection scan systems W is approximately the indicated working-distance of the final lens. For instruments with single deflection scan systems after the probe lens, e.g. the JEOL U3, the distance will depend on the exact positioning of the scan coil assembly. When these factors are determined the scan angle can be calculated from the relation above. For a Stereoscan with an indicated magnification of $\times 20$ (indicated wd 10 mm) the scan angle is $\pm 8°$, for a JEOL U3 at the 32 mm working distance condition the angle at $\times 20$ is about 2·5°. As the magnification is increased the scanned angle will fall and less of the pattern will be visible on the display screen. In general therefore the SEM will be set to the lowest possible magnification when used for ECP work. On most SEMs the magnification is kept constant when the instrument is switched between different accelerating voltages so the scanned angle remains the same. Since however the Bragg angle varies as the electron wavelength, the width of the bands on the screen will fall as $(V_b^{\frac{1}{2}})^{-1}$ where V_b is the beam accelerating voltage. A necessary consequence of operating at such a low magnification is that a large area of the specimen is scanned and the whole area observed must be a single crystal.

2. *Beam Collimation and Beam Current*

The way in which the required degrees of beam collimation and the necessary level of beam current are achieved depends in detail on the exact layout of the microscope being used. The most convenient case to consider is the general one of a three-lens instrument, as in this case there is independent control of the parameters. The two-lens microscope will be less flexible in operation but the principles will be the same.

The beam divergence value has been considered in detail by Schulson and van Essen (1969). The value at the specimen surface is the sum of two contributions, one due to the operation of the final lens and depending on the size of the final aperture and the other due to the finite source size of the electron gun. Figure 12 shows the variation of these two quantities for three different lens conditions: (a) the normal micrograph case where the beam is focused on to the specimen surface by the final probe forming lens; (b) the case where the beam is not focused because the final lens is switched off; and (c) the case where the beam is collimated by setting the focal length of the final lens equal to its object distance. Calling the contribution from the lens ϕ and from the finite source size ψ we get

Case A	$\phi = D/s$
	$\psi = L/u$
Case B	$\phi = D/u$
	$\psi = L/(u+s)$
Case C	$\phi = O$
	$\psi = L/u = (f)^{-1} . L$

where D is the diameter of the final aperture, L is the diameter of the filament crossover image before the final lens, u is the distance of the image from the final lens, s is the distance of the specimen from the final lens, and f is the focal length of the final lens.

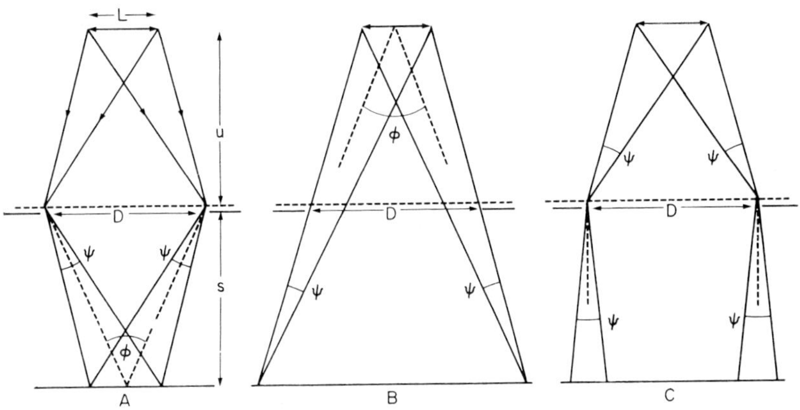

FIG. 12. Calculation of beam divergence under various operating conditions.

In case A the microscope is set up for normal high resolution microscopy. The source image L is therefore small after demagnification by the previous lenses so ψ is small, typically a few microradians. However, the contribution from the lens will be large as the conventional 100 μm final aperture will be used for microscopy, hence the total beam divergence for a 1 cm working distance will be about 10^{-2} rad. Under such conditions only a very poor pattern would be achieved. It would be possible to improve the pattern by reducing the size of the final aperture to say 30 μm, but under this condition the probe would contain little current if both the beam divergence and probe size were limited.

In case B the final lens has been switched off. The beam divergence will now be considerably less than in case A since $u \gg s$ and typically a value of 10^{-3} rad will be achieved. The contribution from the finite source size will still be negligible in comparison with the term due to the lens action. This technique of producing ECPs is a useful one as it is simple to apply and gives patterns of good quality. The probe size obtained in this way will be approximately the same as the final aperture size, that is of the order of hundreds of microns. No surface detail will therefore be imaged. Because of the large probe size this technique also produces a large beam current at the specimen (tens of nA).

The optimum technique is that of figure C where the beam is collimated by the final lens. The limit on the minimum beam divergence obtainable is now set by the source size only. A value of 10^{-4} rad is easily obtained in this way and smaller values can be achieved but at the expense of beam current. The probe size is once more approximately equal to the final aperture.

In a three lens microscope the most convenient way of obtaining high quality ECP without surface detail is technique C with the final lens weakly excited. If the final lens is switched off the divergence will increase but there will also be an increase in probe current. This condition (case *B*) is obviously applicable to a two-lens instrument for which the optimum condition is one in which the final lens is run as weakly as possible. The difficulty with this mode of operation is that slight misalignments in the column tend to give uneven illumination of the specimen.

3. *Signal Collection*

Although the channelling effect is primarily related to the back-scattered electrons, contrast is also present on the secondary electron and specimen current images. Any convenient detection system can therefore be used. On most SEMs the detector is the conventional scintillator/light pipe/photomultiplier assembly which responds to both primary and secondary electrons. This can be used to give good ECPs but there are two possible drawbacks with this arrangement. Firstly, the high beam currents produced by the electron optical system used cause the scintillator to deteriorate very rapidly and secondly, because the detector is assymetrically placed with respect to the specimen, the illumination will tend to be uneven.

For these reasons it is often more convenient to use the specimen current mode of operation. This works efficiently at high beam currents and does not suffer from uneven signal collection effects. The contrast will, however, be reversed relative to the back-scattered case, although the magnitude of contrast will be similar. Another approach is to use an annular plate or semi-conductor diode detector mounted parallel to the surface of the specimen and concentric with the incident beam. This arrangement gives efficient collection and eliminates uneven contrast arising from geometrical collection effects. In addition certain diode detectors have been found to enhance considerably the contrast obtained and improve the visibility of high order lines (Wolf and Everhart, 1969). For most purposes the accelerating voltage used should be the highest available. Firstly this gives the highest gun brightness and hence the maximum current, secondly the contrast appears to be greater at higher voltages (probably because the effect of surface contamination is less significant) and thirdly the bands are narrower and easier to see.

B. APPLICATIONS

The discussion above has shown that, with suitable electron–optical conditions, an electron channelling pattern is produced when a large single crystal is observed at low magnification in the SEM. A considerable variety of information can be derived from this pattern and we shall now discuss the ways in which this can be done. As a knowledge of basic crystallography (e.g. lattice symmetry, Miller indices, stereographic projections) is assumed, readers unfamiliar with the subject might consult a suitable reference text, such as Phillips (1956).

1. *Orientation*

The most basic piece of information available is the orientation of the crystal under examination. The way in which this is done will depend on the amount of data already available for the crystal. Initially let us consider the simple case of a material known to be of cubic symmetry and of known lattice parameter.

(a) *Cubic materials.* The ECP produced has an angular width of perhaps 10°. For an arbitrarily chosen orientation the pattern on the screen may have no prominent symmetry feature or even a clear band. If, however, the crystal is tilted the pattern will change and by appropriate adjustment of the tilt controls for the specimen it will be possible to follow a chosen band or line. If a large number of patterns are recorded as the specimen is successively tilted the patterns can be built up into a map. Starting from a major symmetry pole, say (001), it will be found possible to obtain patterns from a triangle bounded by the (011) and (111) poles before any patterns are repeated.

Any pattern obtained from this crystal at the same operating voltage can be matched to a pattern within this triangle. The triangle is known as the "Unit triangle" and comprises 1/48 of the stereogram of the crystal. All the symmetry elements of the crystal appear once inside the unit triangle which is therefore characteristic of the crystal. The interplanar angles and hence the angles between poles are the same for all cubic crystals so the dimensions of the unit triangle are constant. However, the poles which are prominent inside the triangle and the major bands visible will depend on whether the crystal is simple cubic, FCC, BCC or diamond cubic. Figure 13 shows channelling maps, together with the corresponding part of the stereogram for the two most common cubic systems. Although the angles between poles are constant the detailed appearance of the map will be a function both of the accelerating voltage and the lattice parameter of the crystal from which the map was made. An increase in either the accelerating voltage or the lattice parameter will make all the bands narrower and this has the effect of altering the appearance of the map at all the places where lines and bands overlap. The maps of two crystals of the same lattice type but different lattice parameter can be made identical by choosing an accelerating voltage such that λ/a is the same in both cases. For example iron (a = 2·867 Å) at 20 kV has a map identical to that obtained from niobium (a = 3·33 Å) at 15 kV. Since small changes in lattice parameter do not markedly alter the map this correspondence makes it possible to cover most common lattice parameters by a limited number of maps and an appropriate choice of accelerating voltage.

The ECP maps can be drawn by a simple construction from a stereogram. The unit triangle on the stereogram is taken and the major low index poles are marked in. Next the important zones are drawn in—the zones chosen are those with indices satisfying the "allowed reflections" rule for the lattice type. For example in BCC the low index zones for which $h+k+l = 2n$ will be (110), (200), (220), (211), (301), etc. The zone (UVW) passes through the pole (hkl) if $hU+kl+1W = 0$. Each zone trace is then turned in to a band by drawing parallel to it a pair of lines, one on each side of the trace, spaced by the angle $2\theta_B$ where θ_B is the Bragg angle calculated for a plane of the same indices as the zone (UVW). So for example the zone $(01\bar{1})$ which passes through the poles (111) and (011) is turned into a band by drawing about it a pair of lines spaced by $2\theta_{B(01\bar{1})}$.

In this way the major bands and the overall appearance of the map can be quickly determined. The resultant drawing is accurate enough for most purposes. Computer techniques have also been applied for this construction and programs have been described elsewhere.

When the ECP map and the constructed map appropriate to the crystal are both available the orientation of a crystal can be rapidly determined from the ECP. The pattern obtained is matched up with the

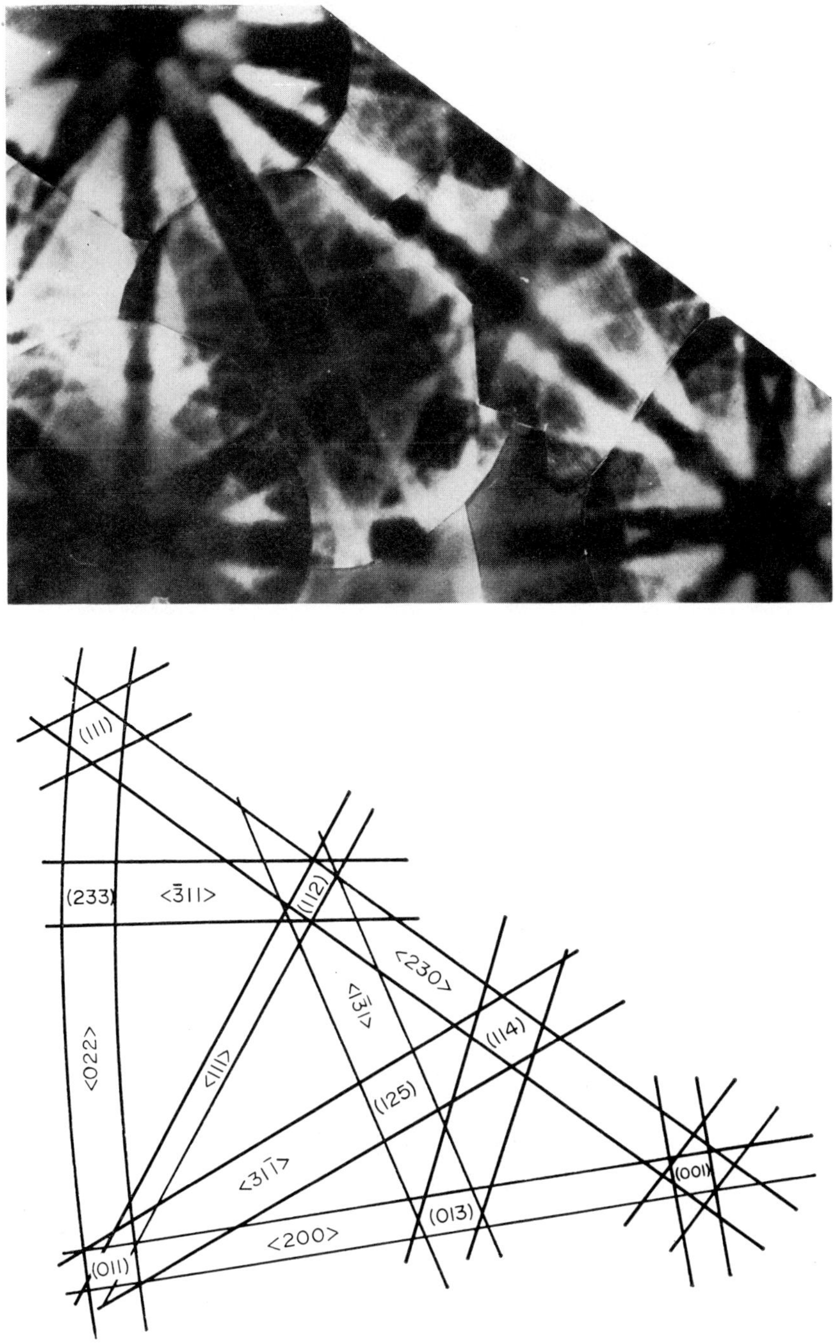

FIG. 13. (a) Indexed channelling map for BCC materials generated from single crystal of niobium at 20 kV by specimen current signal. Courtesy of C. van Essen.

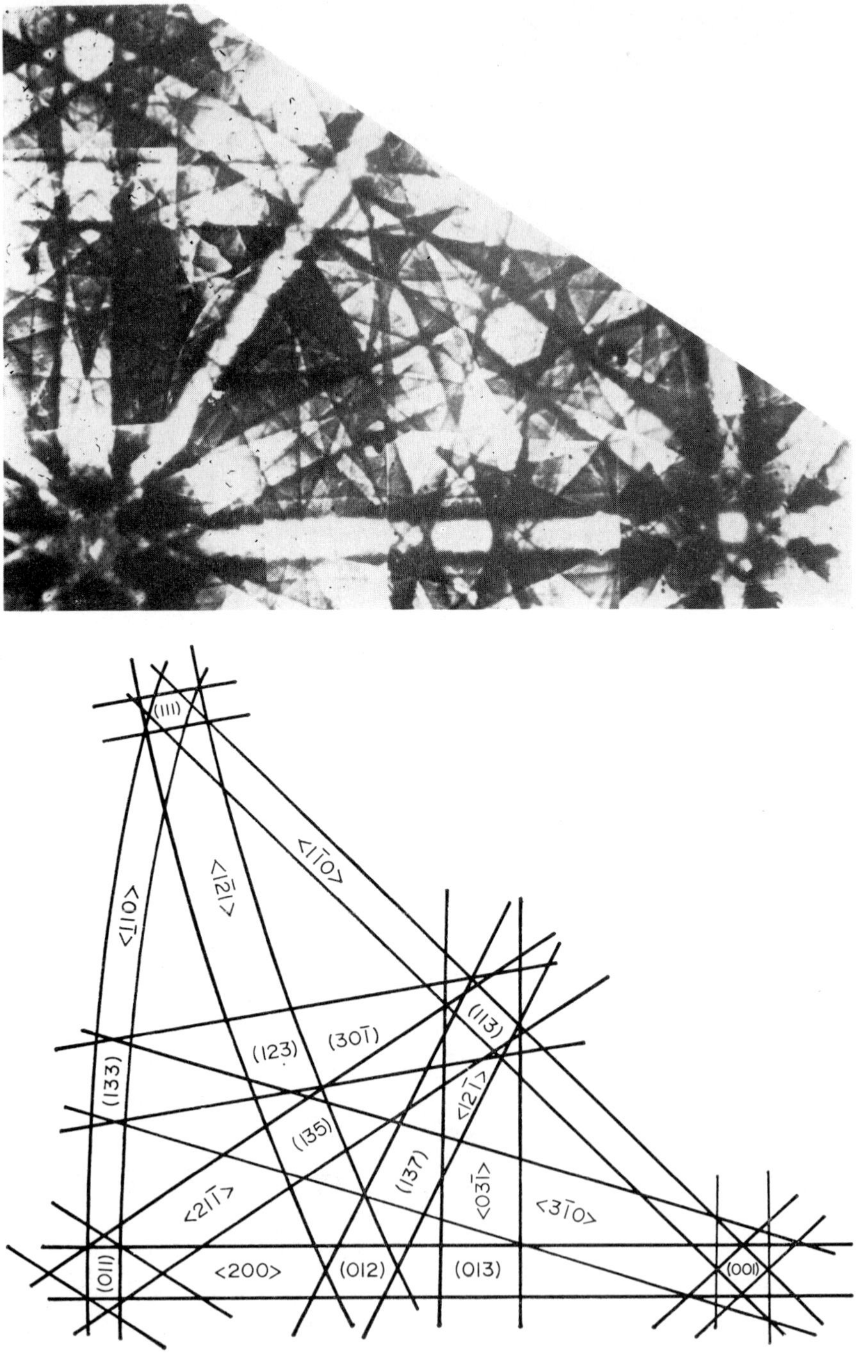

(b) Indexed channelling map for FCC materials generated from single crystal of copper at 20 kV by back-scattered signal. Courtesy C. van Essen.

map until the identical area is found. The centre of the unknown pattern (representing the direction of the surface normal) can then be marked on to the map and transferred to the constructed map by noting the position of the point relative to the neighbouring bands and the major poles. In practice it is very difficult to make an experimental ECP which is free from distortions of various kinds (due to scan and display errors) and for the highest accuracy it is therefore necessary to measure the position of the point relative to three or four bands and poles in order to reduce the effect of these errors. The point on the constructed map then represents the orientation of the crystal examined. In a later section a more convenient method of representing the orientation of a crystal will be described. The accuracy of the ECP technique for crystal orientation has been tested by comparison with the well-established Kossel technique and has been shown to be accurate to better than 1° for an absolute orientation (Joy *et al.*, 1971).

(b) *Non-cubic systems.* For materials known to be outside the cubic system a similar technique can be employed but the map will now have to cover a larger portion of the stereogram because of the lower system symmetry. For example, for a tetragonal material it will be necessary to generate a map covering the quadrant bounded by (001), (010), (100). Any map of this size constructed by assembling a large number of ECPs will be seriously distorted and in practice it is therefore better to make several smaller maps centred about the major poles. The difficulty of this procedure obviously increases greatly as the symmetry decreases and for non-orthogonal lattice systems the work involved may be too great. In addition it may not be possible to generate the whole map required because of the practical problems involved in tilting the specimen. In such a case it may be preferable to rely on a hand- or computer-constructed ECP map only or use the computer indexing technique described later.

(c) *Unknown materials.* If neither the crystal system nor lattice parameter of the specimen is known the problem is very difficult. In principle, however, it should be possible to deduce the symmetry by generating a large enough area of the ECP map and deduce the lattice parameter(s) from the measured width of indexed bands.

In this mode of ECP operation a large area of the specimen is scanned during the raster which gives rise to the pattern. If the whole area is of the same orientation then a single pattern will be produced, and the ECP technique therefore enables a very rapid visual check to be made of the perfection of large single crystals. Since there is a one to one correspondence between the beam position on the specimen, the beam on the display screen and the angle of incidence, local changes in the crystallographic orientation will show up as local changes in the pattern on the screen. The size and position of the regions of different orientation can be directly measured from the known magnification of the SEM although the accuracy will depend on the probe size of the instrument under the conditions being used for ECP generation. A high angle boundary, such as grain boundary, will be obvious and the relative orientations on either side of the boundary can be measured using the techniques described above. The misorientation across a low angle boundary, such as a sub-grain boundary, can be very accurately measured by noting the offset of lines and bands as they pass across the boundary. The accuracy of this procedure is limited only by the beam divergence and misorientations as low as 0·1° are measurable.

These methods of orientation analysis have been proved useful in many applications. One notable example has been in the study of epitaxial film growth. As shown in section II.A the contrast from a thin film is very high and in practice the majority of channelling contrast comes from the top few hundred ångströms of the specimen. This makes it possible to observe the orientation and perfection of very thin epitaxial films over large areas without any interference from the substrate (Vicario, 1970). The technique has also been used to study semiconductor materials and devices.

2. *Lattice Parameter Determinations*

It is obviously possible to use an ECP to deduce the lattice parameter(s) of the specimen. Since the angular width of the bands is $2\theta_B$ a knowledge of the accelerating voltage will enable a set of d_{hkl} values to be calculated and if the bands are self-consistently indexed (that is to say assigned indices consistent with the choice of poles in the unit triangle) the lattice parameters can be deduced. The accuracy with which this can be done is limited by several factors.

(1) The width of the bands displayed on the screens has to be converted to an angular width by a suitable calibration. This can be done by using a crystal of known lattice parameter to probably better than 0·5%.

(2) The display screens and their associated scan circuitry are not very linear and point-to-point nonlinearities of 5% have been measured (see Boyde, 1970). Thus the width of a band may apparently change by several percent depending on its position on the screen. This can only be overcome by careful calibration of the screen (Boyde, 1970) or the use of linearized display systems.

(3) The accelerating voltage must be known accurately. In practice it is commonly found that the indicated voltage may be in error by

several per cent and the value obtained for a given setting on the power supply may not be reproducible. Also, on some instruments the kV meter monitors the grid rather than the filament voltage of the gun so the indicated voltage is lower by the bias applied. The true voltage must therefore be measured with great care. The simplest way is that due to Russ (1972) of measuring the X-ray emission of a selected element as a function of beam voltage. This should extrapolate to zero at the absorption edge of the element. For example $Mo(K_\alpha = 17{\cdot}4$ kV) has an absorption edge of 20·00 kV. A more elegant technique is to observe the very high order lines found in the centre of low index poles (Fig. 14). Because the lines have very high indices, for example 11, 5, 1, their position is very dependent on beam voltage. For certain values of accelerating voltage the position of these lines becomes simply symmetrical (Vicario and Pitaval, 1969), and the voltage at which this occurs can be accurately calculated. These "spot" values can then be used to calibrate the meter.

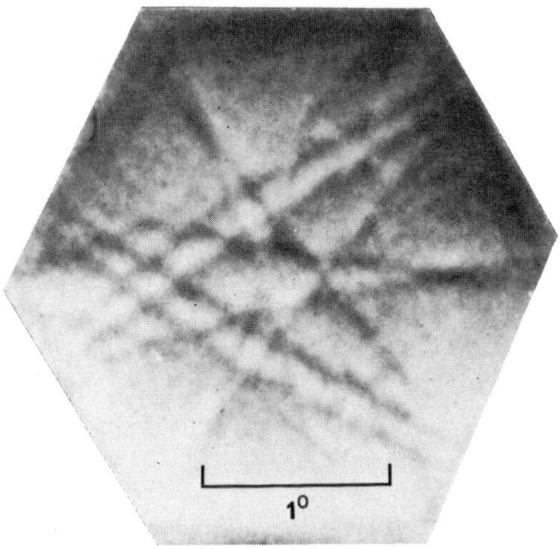

FIG. 14. Fine structure in centre of silicon (111) pole at 20 kV.

When all reasonable precautions have been taken it is still unlikely whether an accuracy of better than 1% can be achieved. For most purposes therefore it would be better to use an X-ray technique such as Kossel patterns for accurate lattice determinations (see Yakowitz, Chapter 13).

3. *Other Uses*

The applications dealt with above have depended on the geometrical form of the patterns. The quality of the pattern (that is its sharpness and contrast) also provides information about the specimen. Although this technique is of most use when applied to small selected areas of the specimen and will be considered more fully later it has been used for studies of bulk single crystals. The pattern quality depends on the degree of perfection of the lattice and the quality can be conveniently characterized in terms of the number of high order lines or the pattern contrast visible.

It has been shown (Davidson, 1970; Wolf and Hunsperger, 1970) that radiation damage to a crystal produces marked changes in the quality of an ECP. The observed quality can be calibrated in terms of the radiation dose received by a crystal and this calibration can then be used to study the variation of dose with depth or position in other crystals. In this respect it is useful that the ECP contrast information comes from only a thin layer of the specimen surface since by sectioning through, the depth distribution of damage can be easily determined unambiguously.

Again because the contrast does come from the direct surface region only, the pattern quality is greatly dependent on the surface preparation of the specimen. It has been shown (Davidson and Booker, 1970) that an amorphous surface layer of only 200 Å is sufficient to remove ECP contrast almost completely, and hence a clean specimen surface, produced by chemical or electrochemical polishing, is essential for good ECP production. Equally the ECP quality will be a sensitive test for surface cleanliness. Mechanical surface damage is also very deleterious to the pattern and even the lightest mechanical polish (for example $\frac{1}{4}$ μm diamond paste or even rough tissue paper) can produce obvious pattern damage. The ECP technique is therefore a sensitive test for the surface state of a specimen.

IV. Channelling Micrographs

When a large single crystal is viewed at low magnification we have seen that an ECP is produced because of the large angle of scan. As the magnification is increased the angle of scan falls proportionally and the pattern will spread out on the screen about the centre. At a high magnification therefore the screen will appear to be uniformly grey with a contrast level the same as that found at the centre of the original ECP. If the crystal were now tilted so as to change the pattern then at high magnification the contrast level on the screen would also change to a value appropriate to the new level at the centre of the new pattern. Therefore if the beam is allowed to scan across not a single crystal but a polycrystalline material containing grains of different orientation each grain in the micrograph image will show a uniform grey contrast with a level appropriate to the value that would be found at the centre of the ECP obtained from that grain. Each grain will therefore have a different grey level relative to its neighbours. Similarly any change in crystallography inside a grain, such as a twin or a sub-grain boundary will also show up as a difference in brightness.

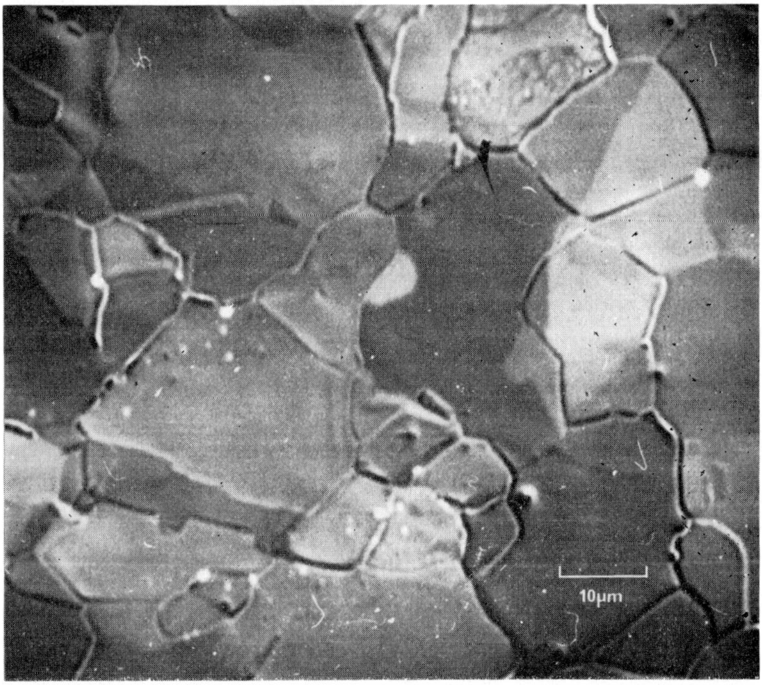

FIG. 15. Channelling contrast micrograph from Pb–1.5% Sn sample. Note that the channelling contrast shows that the grain boundaries have moved since the specimen was etched.

Figure 15 shows a micrograph taken in this mode of a polycrystalline sample of copper. The specimen was prepared by electropolishing and as a result the grain boundaries were slightly etched. The apparent grain boundaries are therefore displayed by normal topographic contrast. The variations in brightness show the actual regions of different crystallographic orientation and it can be seen that in fact the grains do not follow the apparent grain boundaries. This is because the material re-annealed between the time when the sample was polished and the time when it was examined in the microscope. Without the use of channelling contrast this change would not be apparent.

Because this form of contrast depends on channelling effects it is clear that the beam conditions necessary will be those for producing a normal ECP (that is a collimated beam and a high current) but in addition a small probe size will be needed to give adequate micrograph resolution. The brightness eqn (6) will obviously limit the performance in the usual way and the optimum conditions are those where the beam current is high (10^{-8} A) and the probe size just small enough to give the desired resolution—a value of 1 μm is typically employed. A beam divergence of around 3×10^{-3} rad is then attainable. These conditions can be most easily obtained using the conditions of (a) in Fig. 12. The final aperture is used to set the beam divergence to the desired value as in this case the ψ term can be ignored since the filament crossover size will be very small. The other lenses are set so as to produce a spot of the required size and current.

Once the correct electron–optical conditions have been set up the micrographs are obtained in the same way as usual. Because the channelling contrast is only of the usual 5% level it will be necessary to use a fair amount of black level on the video amplifier chains. Since this will also have the effect of increasing the visibility of the topographic contrast it is desirable to make this as small as possible by using specimens that are fairly smooth and flat. This is best achieved by chemical or electrochemical polishing of the specimen and these techniques also produce a surface sufficiently free from mechanical damage to make channelling contrast possible. The absence of crystallographic contrast, under suitable electron–optical conditions, from a specimen is a good indication of residual surface damage or contamination.

Channelling contrast can be distinguished from topographical contrast by several simple tests relying on the properties of crystallographic contrast. The simplest is to tilt the specimen by a few degrees. This will completely change the crystallographic contrast but leave any topographic contrast substantially unaltered. Alternatively the accelerating voltage can be changed. This changes the Bragg angle and hence the observed crystallographic contrast but again leaves the topographic contrast much the same.

The channelling mode of micrograph examination adds a new dimension to the information supplied by the SEM. The technique has been used to study grain boundary sliding and related effects (Joy and Newbury, 1971), residual strains after metal forming processes (Joy and Newbury, 1972; Stickler, R., personal communication) and the crystallographic changes occurring during phase transformations (Pease, D., personal communication). With the electron–optical conditions suggested, changes in crystal orientation of as little as 0·1°

can be detected and it will be shown later that this effect, combined with the selected area channelling pattern technique, enable the SEM to be used as a sensitive measure of local strain distributions.

V. The Selected Area Channelling Pattern

A. TECHNIQUE FOR PRODUCTION

Although the channelling micrograph mode of operation is a valuable one in the information it supplies about local changes in crystallography, the results are not capable of quantitative analysis. This is only possible where an actual pattern is produced since the detailed data about the crystal orientation and state is deduced from the geometrical form and exact state of the pattern. With the ECP technique described above, however, patterns can only be produced from large single crystals because the beam must scan through a large angle to produce the pattern. Since the contrast arises solely from changes in angle of incidence of the beam it is clear that the same pattern would be produced if no scanning action occurred but instead the beam "rocked" about a point on the specimen surface. In this case the full range of incident angles would still be scanned but because no lateral displacement of the beam would occur the pattern would come from an area of the specimen ideally equal to the diameter of the incident beam. This is the desired selected area channelling pattern (SACP) technique.

There are at least three ways of achieving the rocking action of the beam and each of these is considered briefly below. In every case the beam conditions will, of course, be the same, and similar to those described above for the production of channelling micrographs or ECPs with a high spatial resolution. That is to say we require a beam of adequate collimation (3×10^{-3} rad or better) containing at least 1 nA current and of a diameter less than the smallest area from which we wish to obtain the SACP. The principles of obtaining these conditions have already been considered above; practical considerations will be discussed below.

1. *Mechanical Method*

The simplest way of achieving a "rocking" beam would be to maintain the beam stationary but rock the specimen, about the two axes perpendicular to the beam, mechanically in such a way that the point of the specimen surface under the beam remains stationary. If the display CRTs are scanned synchronously with the specimen a selected area channelling pattern will be produced. Such a stage has been constructed by Coates (1967). This type of device offers advantages in that the electron optics are simple and a large angle of rock can be obtained (up to 90°). The drawbacks are that it is difficult to maintain the specimen point at rest during the mechanical operation because of tolerances in the components. The stage was designed to give a performance of 1 μm per 1 degree angle of rock giving a minimum area, for an identifiable pattern, of about 10 μm. The maximum scan speed is however limited (and careful optical setting up of the stage outside the SEM is required). No results have been reported with this device.

2. *Double Deflection Systems*

The most common form of scan system is the double deflection method. In this we have two sets of scan coils. The first deflects the beam off axis by some amount θ, and the second set then deflects the beam in the opposite sense so that it once more crosses the optic axis. Hence all the scan rays pass through one point on the optic axis. Under normal scan conditions this point will be well above the specimen surface so as to produce a large raster square for any given scan angle. However, it is clear that by appropriately adjusting the strength of the second set of scan coils the common scan crossover could be brought on to the specimen surface. The incident beam would then rock about this point and produce the SACP.

This method of SACP operation has been used on several types of instrument (van Essen and Schulson, 1969; Schulson *et al.*, 1969; commercially by JEOL and ETEC). It is simple to apply as either the normal scan coils can be used and the lower coils shunted to reduce their strength, or an auxiliary set of coils beneath the final lens can be employed. If the normal set of scan coils is used the final aperture of the microscope must be large enough to accept the full scan angle used and this means a size of 3 mm or larger depending on the working distance. In this case the beam divergence will have to be set to the desired value by limiting aperture further up the column. In the case of auxiliary coils the beam divergence can still be set with the final aperture. Normal micrograph operation can be obtained by restoring the scan coils to their normal strengths.

This method has the drawback that the minimum area attainable is not very small, a value of 50 μm diameter for a scan angle of 10° being typical. The reason is as follows. In essence the double deflection scan system is acting like a lens with the point from which the scan action starts and the final crossover being like the conjugate object and image points of the lens; the difference is only that not all the ray bundles are passing through the arrangement at the same time. Such a system thus displays all the normal aberrations of a lens and in particular that of spherical aberration. Spherical aberration arises because rays at a large angle to the optic axis are deflected more strongly than par-axial rays and consequently come to a focus above the other rays. The crossover point is therefore extended into a disc of least confusion. The size of this disc is given by $\frac{1}{2}C_s\theta'^3$ where θ' is the scan semi-angle and C_s the coefficient of spherical aberration of the "lens". Since a scan coils assembly is not a high precision device it turns out that C_s is typically many centimetres in magnitude and with a θ' of about 0·1 rad the minimum disc is 50–100 μm in diameter. Although this is considerably better than the area needed for simple ECPs it is still too large for most purposes and the method has not found wide application.

3. *Single Deflection Systems*

The defects of the double deflection system are overcome by the method due to van Essen *et al.* (1970), (1971) and applied to the Cambridge "Stereoscan" (and latterly the Cameca SEM). Figure 16(a) shows the arrangement of scan coils and final lens for a Stereoscan for normal micrograph operation. The double deflection scan is so set that

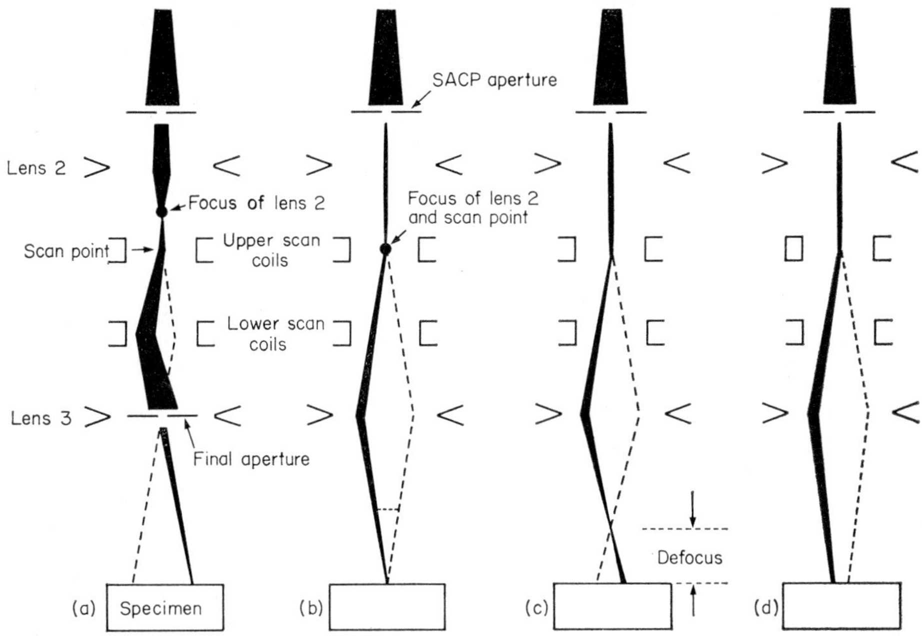

FIG. 16. Ray diagrams illustrating single deflection technique for SACP work.

all the scanned rays pass through the centre of the final lens bore and the beam is limited at this point by the final aperture. The scan action starts at the point labelled "Scan point" in the middle of the upper scan coil and the preceding lens is focused to a point some way above the scan point. Figures 16(b)–(d) show the ways in which rocking beam operation is achieved. The lower set of scan coils is switched off, the ray bundle deflected by the upper set from the scan point therefore continues to diverge from the optic axis until it enters the lens field. The focusing action of the lens will bring the bundle of rays back across the optic axis (Fig. 16(c) or (d)) at some point depending upon the strength of the lens. By correctly adjusting the lens excitation this point can be made to be the surface of the specimen (16(b)) and the beam will then rock about the surface normal. The lens has thus been used to make the scan point and the specimen surface conjugate.

As the strength of the lens is varied, the crossover point of the scan rays moves relative to the specimen surface and consequently on the display screen a variation in the effective magnification of the image will be seen. As the crossover point moves on to the surface the magnification will increase and ideally reach infinity at the exact rocking conditions, leaving only the required SACP contrast. Further adjustment of the lens will produce a micrograph again but rotated through 180° relative to the previous case. Normal micrograph operation is restored simply by switching on the lower scan coils. If the system was set up in this way it would, however, be necessary to refocus the final lens each time the system was changed from micrograph to SACP. This is because for micrograph operation the final lens is adjusted to produce a focused probe and this means that the focus point of lens 2 and the specimen surface are made conjugate. In SACP operation the scan point and the specimen surface are required to be conjugate. This difficulty can be simply overcome by adjusting lens 2 until it focuses at the scan point. On changing from micrograph to SACP the beam will therefore stay focused but rock rather than scan.

The required electron optical conditions are obtained in the usual way. Because of the large scan angle through the final lens a large aperture must be used as shown in Fig. 16(b). The beam divergence is therefore set by an aperture above the final lens. This can be done by replacing the 800 μm spray aperture with one of about 100 μm diameter.

In practice the procedure is as follows (Fig. 17). A suitable feature is selected (at a magnification of about $\times 500$) and brought to the centre of the screen and focused with lens 3. Go to $\times 20$ and switch off the lower set of scan coils. A high magnification micrograph of the previous area will be seen (Fig. 17(b)). Adjust the lens to make the

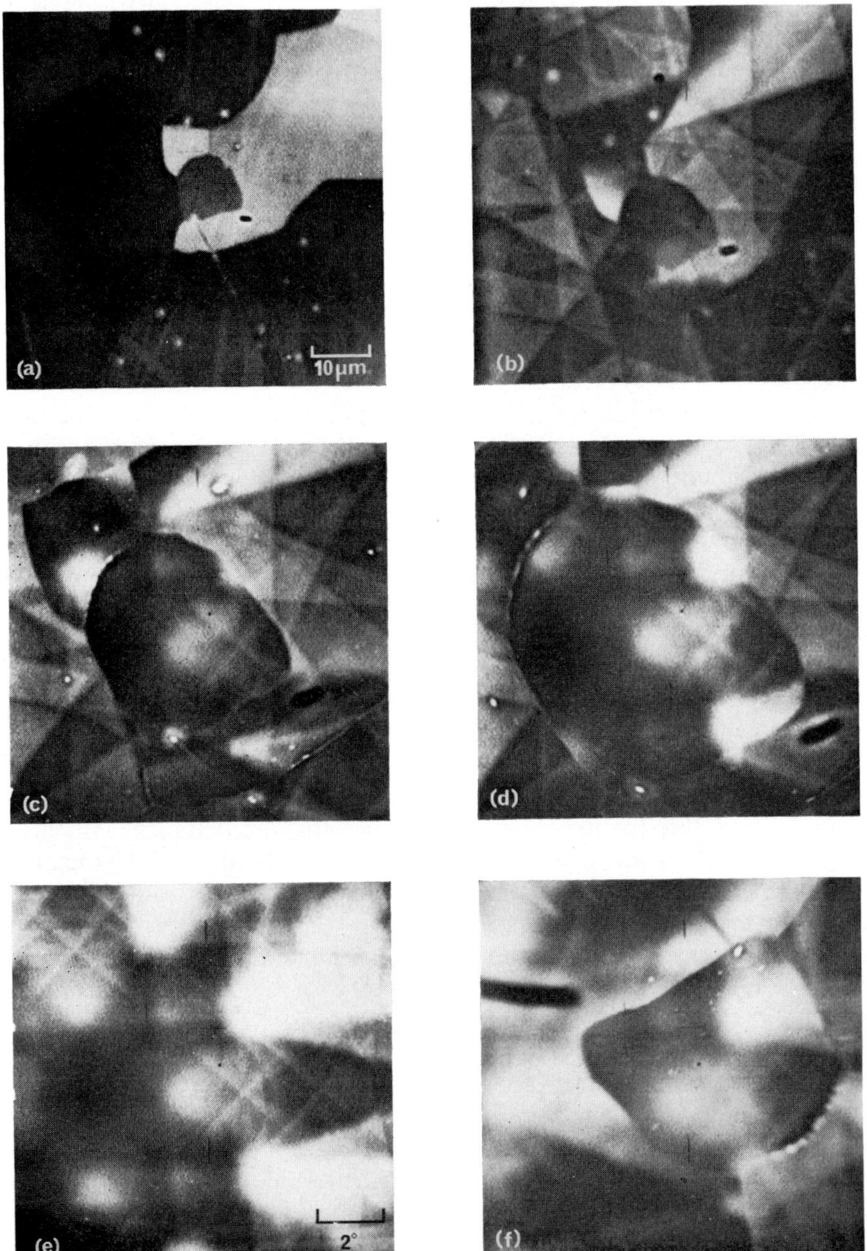

FIG. 17. Through focus series showing adjustment of microscope in SACP mode.

chosen feature as large as possible on the screen. This places the scan crossover point on the specimen surface (Figs 17(b), (c) and (d)). At the exact crossover point the channelling pattern from the selected feature will be seen (Fig. 17(e)). Further adjustment of the lens will place the crossover point the other side of the specimen and a micrograph, rotated 180° to the previous ones, will be seen (Fig. 17(f)). When the crossover point has been set with lens 3, switch back to standard micrograph mode, go to ×500 magnification and focus the image with lens 2 only. This then makes the specimen surface conjugate with both the scan point (for rocking beam operation) and the focus of lens 2 (for micrograph operation). Both the SACP and the micrograph are now optimally set up. To change from the focused micrograph to the SACP case it is only necessary to switch off the lower set of scan coils and put the magnification to ×20. The pattern will then be obtained from the object in the centre of the area observed on the micrograph.

The minimum area from which the pattern can be obtained will again be determined by the aberrations of the system. In this case, however, the limiting component will be the lens rather than the scan-coils and because the probe-forming lens is a precision component the performance will be significantly better than for the double deflection case. For a scan angle of 10° a minimum selected area of around 5–10 μm is possible. The actual value depends, as before, on both C_s and θ'. If the working distance of the lens is decreased, that is to say if the lens is strengthened, the aberrations fall and hence the minimum area is reduced. Booker and Stickler (1972) have measured the variation of performance for a Stereoscan and shown that a minimum area of around 3 μm is attainable for the shortest possible working distance. It is also possible to reduce the area by reducing θ', the scan angle, since the area is proportional to θ'^3 and even a small reduction leads to a rapid decrease in area. On instruments fitted with a "zoom" type of magnification control it is possible to set the scan angle to give the best compromise between area obtained and the ease of recognition of the pattern.

The ease and convenience of the setting up procedure for this method of selected area operation and the smallness of the area from which pattern can be obtained have combined to make the single deflection technique the most widely used of the methods available. Improvements in the design of lenses and in suitable dynamic correction circuitry make still better performance obtainable (Joy and Newbury, 1972b) with minimum areas of less than 1 μm possible. All the examples considered in this text have been obtained using this technique.

B. APPLICATION TO THE STUDY OF MICRO-CRYSTALLINE ORIENTATION

In an earlier section the way in which an ECP could be used to determine the orientation of a crystal was discussed. We will now look in more detail at the processes involved and their application to problems involving the absolute or relative orientation of micro-crystallites in a solid specimen.

The problem can be treated in two sections. There is firstly the necessity of identifying the orientation of the selected area and secondly the need to express and record this orientation in a convenient way. The most usual way of determining the orientation is by comparison with a suitable channelling map of the specimen. In the case of a cubic material this is simply a matter of generating the map of one unit triangle whereas for lattices outside of the cubic system it is necessary to form a map covering enough of the stereogram to make it possible to index any pole by operation of the symmetry elements. The SACP obtained is then matched to the map by recognizing detail. The larger the angle of scan in the pattern the easier this operation will be, as the probability of the SACP containing at least one major band will be increased. It is convenient to calibrate the record CRT in terms of degrees per centimetre for the experimental conditions used (i.e. accelerating voltage, magnification setting, lens working distance and final aperture size) by observation of an appropriate crystal. The angular width of a band on the SACP can then be measured and hence the indices of the band deduced. This will narrow the area of the map to be searched to that lying within the angular width of the SACP about the known band. The exact position of the centre of the SACP is then found by comparison of the fine detail in the pattern and map. In the case of the cubic system where the map only covers one unit triangle the SACP may be a mirror image of the detail in the map indicating that the pattern comes from the neighbouring unit triangle. For convenience in expressing orientations such a pattern would be treated as if it came from the basic unit triangle. However, when it is required to determine orientation relationships, for example in twinning processes, it is necessary to consider all the possible variants of the unit triangle from which the pattern could have come.

When the position of the centre of the SACP has been found on the map the orientation is essentially complete since the map represents a stereographic projection of the crystal and any point in the map is equivalent to a pole. The simplest way of representing the orientation of the crystal as determined by the SACP is therefore to have a portion of the stereogram for the crystal covering the area of the unit triangle map and to mark in on the stereogram the positions of the place on the map found to be the centre of the SACP. This can be done by noting the position of the point relative to the bands forming the edge of the unit triangle and the major poles in the triangle. Because it is difficult to construct a map free from distortion this procedure is best done by using the intermediate step of a carefully drawn map, as described previously, and measuring the angles between the point and several convenient bands or poles. Figure 18 gives an example of this process for an SACP taken from copper at 30 kV (Fig. 18(b)). By comparison with the appropriate map (Fig. 18(a)) the area covered by the SACP and the position of the centre of the SACP can be found as shown. The bands in the SACP were identified as (311) type bands by measurement of their width. The centre point can then be accurately placed in the drawn map (Fig. 18(c)) by noting its position relative to the edge of the bands, and the nearest poles. Finally the point can be transferred directly to the stereoprojection (Fig. 18(d)).

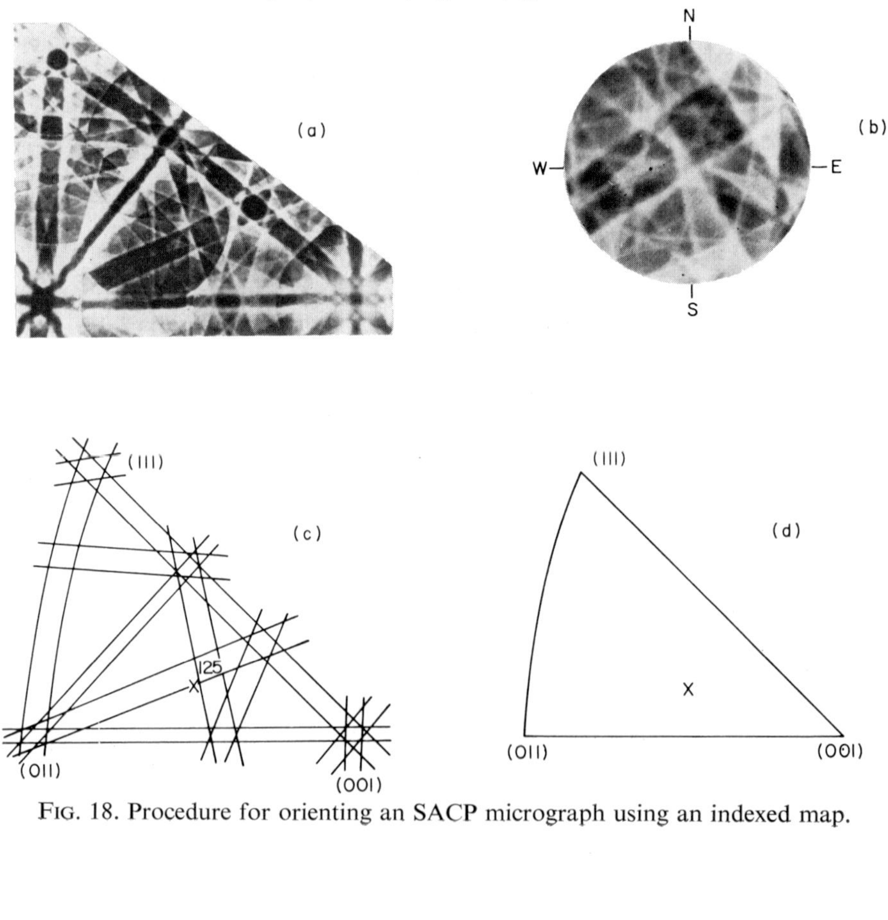

FIG. 18. Procedure for orienting an SACP micrograph using an indexed map.

The point marked in the stereogram represents the orientation of the area from which the SACP was obtained and the data could be presented in this form if required. Where the relative orientation of several points are of interest the stereographic form of presentation enables a rapid visual assessment of the situation to be made. The problem of conveniently storing the data or comparing results obtained in different experiments is however considerable with this technique because of the necessity of having to transfer points between stereographic projections. Each successive transfer will introduce some error. What is required is therefore some system for numerically representing the orientation as determined above. This can then be transformed back to a visual stereographic presentation when required but the storage and comparison of data will be much simplified.

One convenient way of doing this is to use normalized Miller indices to represent the orientation of the point. When an arbitrary section is cut through a crystal the Miller indices corresponding to the orientation of the crystal will not in general be a low value integer. For example an orientation initially determined to be near (321) might prove on more accurate analysis to be (30 19 11) or even (300 194 107). Such large numbers are unwieldy and difficult to compare with other similar numbers. This can be overcome by using normalized indices, i.e. transforming hkl to HKL where $H = h/(h^2+k^2+l^2)^{\frac{1}{2}}$ etc. The above orientations would then be (0·8018 0·5345 0·2673), (0·7793 0·4936 0·2857) and (0·7770 0·5024 0·2771), respectively. This procedure has obvious advantages. Firstly the indices represent the direction cosines of a unit vector and so relative orientations can be readily appraised. Secondly in the cubic system the cosine of the angle between poles with indices $(h_1k_1l_1)$ and $(h_2k_2l_2)$ becomes in normalized index form, $\cos\theta_{12} = H_1H_2+K_1K_2+L_1L_2$ and the angle between two poles can therefore be very rapidly computed.

If we consider the angle between the general pole HKL and (100), (010) and (001) in the cubic system we see the cosines of the three angles are H, K and L, respectively. Hence given the normalized indices of a pole its position in a stereogram can be very quickly plotted by marking in on the stereogram small circles of radius H and K about (100) and (010) respectively. Their intersection is the position of the required pole. (Note that only two circles are required since $H^2+K^2+L^2 = 1$ and hence any two given values define the third index.) Of more practical importance is the fact that this procedure can be used in reverse. The stereogram, or more conveniently the unit triangle, can be marked out with circles indicating the locus of poles with constant H, K or L. Figure 19 shows such a plot for the unit triangle stereogram and the indices of any point in the triangle read off directly. For example the point marked in Fig. 18(d) has been marked into Fig. 19 and the indices can be seen to be (0·125 0·393 0·910). The results are good to three significant figures but for the highest accuracy the indices can be renormalized finally to give more significant digits. This is particularly important when normalized indices are being used to find the angle between pairs of poles and the angles are small. This

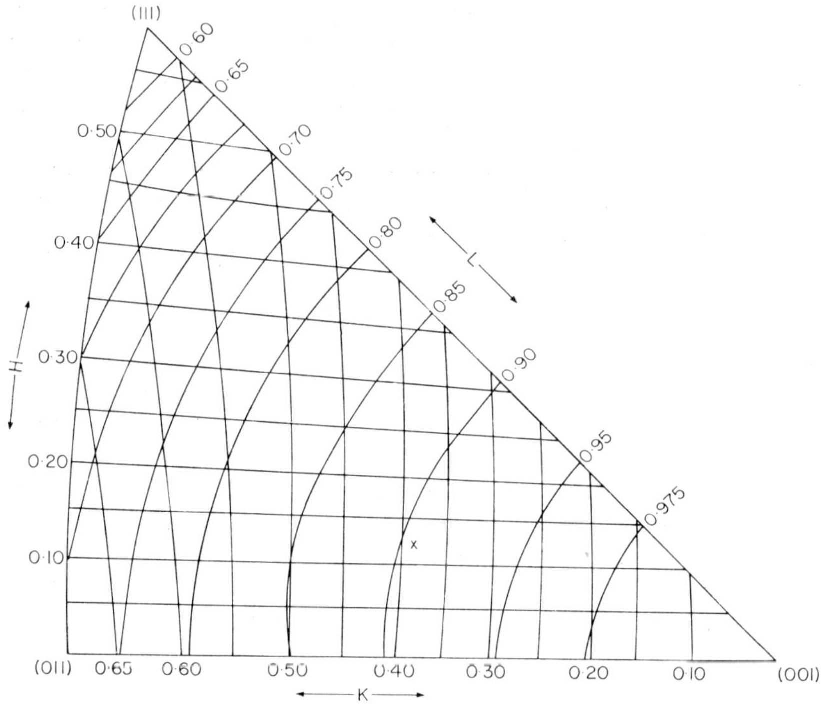

FIG. 19. Variation of normalized Miller indices in unit triangle (001) (011) (111).

technique of normalized indices offers a very handy way of storing orientation data and the numbers themselves are in a convenient form for calculation purposes. In addition the graphical method described above makes it possible to obtain quickly, or plot conversely, the normalized indices (or position) of any pole in a cubic stereogram. It is worth noting that normalized indices are also widely used as the standard way of representing Kossel orientation results and it is also conventional to use normalized Miller indices in theoretical studies of elastic properties and martensitic transformations.

There is one further way of plotting a given pole into a stereogram or conversely finding the normalized indices of a pole in a stereogram. This is due to Christian (1956) who showed that, in a cubic system stereogram of radius R, the cartesian co-ordinates of the pole HKL referred to (001) as origin are $X = RH/(1+L)$ and $Y = RK/(1+L)$. The reverse relationships are $H = 2RX/(X^2+Y^2+R^2)$, $K = 2RY/(X^2+Y^2+R^2)$, $L = (R^2-(X^2+Y^2))/(X^2+Y^2+R^2)$. If orientations are determined using the procedure described above and transferred to a stereogram of radius R then their X, Y co-ordinates can be measured and either these values or the HKL indices tabulated for further use.

Outside the cubic system the problem is more complex because of the lower symmetry involved. The X, Y notation is no longer of use but the normalized HKL indices are still valuable. A plot of HKL values for any part of a stereogram can still be constructed but it will now be necessary to draw one for each lattice parameter set.

The accuracy of the SACP technique for orientation determinations has been tested by comparison with the established Kossel technique (Joy *et al.*, 1971; Yakowitz, Chapter 13) and the accuracy for an absolute orientation shown to be better than 1° when using the techniques described above. This value is limited to some extent by the accuracy with which the specimen can be aligned normal to the optic axis. For measurements of relative orientations, for example of several grains on the same specimen, the accuracy is rather higher and ultimately limited by the divergence of the beam (about 0·15°). When a suitable map is available the SACP technique therefore provides a very rapid and accurate way of finding the orientation of small (10 μm diameters or less) areas in solid crystals. Because the SACP contrast is visible on the viewing screen it is often possible to recognize a pattern directly, without the necessity of recording it, by direct comparison with a map. Where no map is available or can be generated it may be found possible to use an accurately drawn map although the absence of grey level information makes this procedure very difficult. In such a case it may be desirable to tilt the specimen through known angles in order to bring up a pole that can be unambiguously identified and then use these angles to deduce the orientation. A computer technique has been developed (Newbury and Joy, 1971) which enables the pattern from a crystal of known lattice constant to be solved directly to yield the orientation in normalized indices HKL. This relies for its operation on the fact that a change in beam voltage will change the width of a band, or the position of a line, by an amount depending on the change in voltage and the indices (hkl) of the band. By making a choice of three bands or lines on a picture and taking two exposures at suitably separated values of accelerating voltage an accurate determination of the orientation can be made. This technique is in principle applicable to any lattice type.

The SACP orientation technique has been applied to a wide range of problems ranging from the determination of the orientation of tri-pyramid twins in silicon (Booker *et al.*, 1971) to studies of martensitic orientation relationships. One interesting application has been to the study of texture effects on polycrystalline materials. If the orientations of grains, selected at random from a polycrystal, are plotted into the unit triangle no tendency to preferential orientation is usually found.

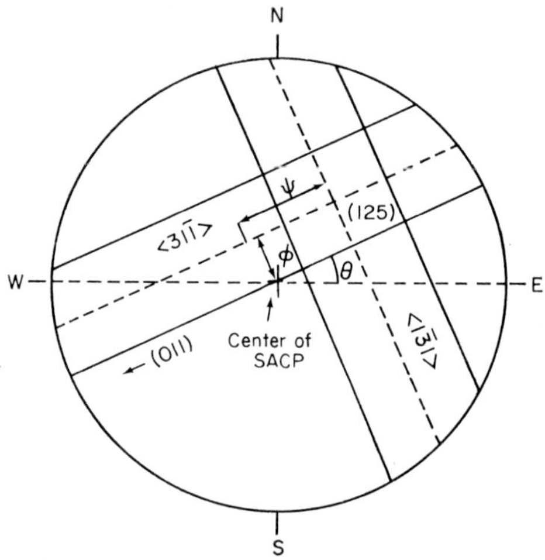

FIG. 20. Definition of angles used in orientation procedure.

If, however, one grain is taken as an origin and the orientations of all the grains touching this reference grain or lying within a specified radius of it arc plotted it is often found that these orientations cluster around each other. This result is of interest in studies of annealing and recrystallization and the SACP technique, enabling the orientation of very many grains to be obtained in a short time, is the ideal way to obtain it. In these examinations it is necessary to display the orientation of the crystal on a complete stereogram and take account of rotations in the plane of the crystal. The procedure is as follows. The lines and bands in the pattern are indexed as described above. The centre of the SACP is marked and north-south, east-west axes defined on the SACP. It is convenient to use the horizontal scan lines as the E–W axis as this ensures that all the patterns have the same initial alignment. This procedure ensures that areas of the crystal with the same orientation but with a rotation in the plane of the crystal can be distinguished. A stereographic projection is then drawn whose centre represents the centre of the SACP. The E–W, N–S axes are also marked on to the stereogram. The most prominent pole and band on the SACP are chosen as shown in Fig. 20 using the SACP indexed in Fig. 18. The angle between the chosen band and the E–W axis is θ. The band itself passes the centre at a distance ϕ and the pole is at a distance ψ from this point.

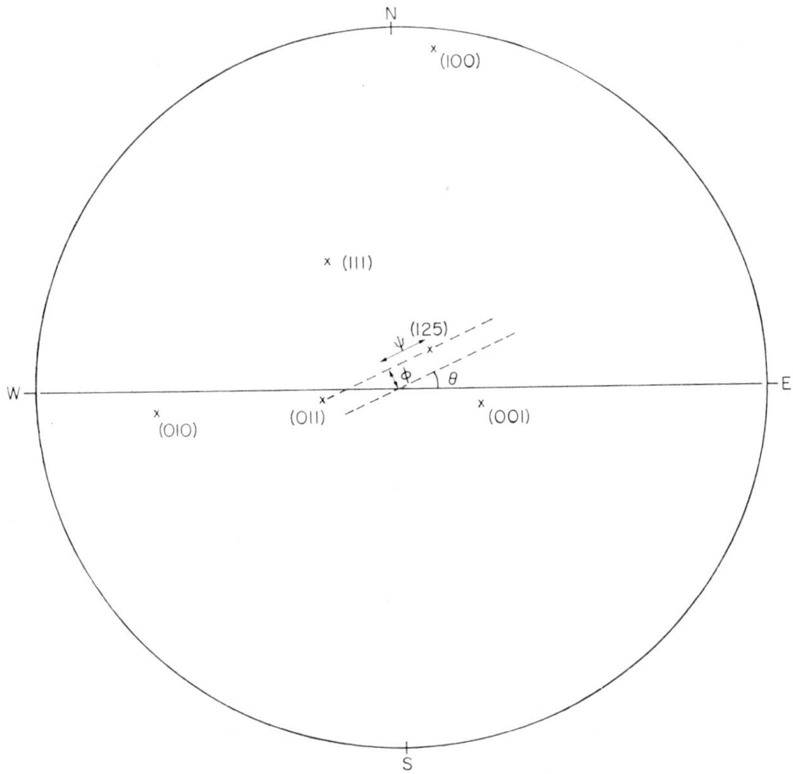

FIG. 21. Complete orientation of SACP from Fig. 18.

These three angles define the exact position of the chosen pole which can now be marked in on the stereogram as shown in Fig. 21. The major symmetry poles (001, 010, 100, etc.) can now be placed on the stereogram in their correct positions by calculating the angle between these poles and the chosen pole. A similar procedure is followed for each pattern indexed (taking care of course that each pattern recorded is identified in some way such that the correct orientation of the pattern relative to the screen axes is preserved). Because all the major symmetry poles are now marked on together with their correct placement relative to the crystal it will be possible to look for meaningful orientation relationships. A procedure of this type has been used (Ayers and Joy, 1972) to investigate the orientation relationships existing in grain boundary precipitates of Cu–Zn and Ag–Zn alloys. Similar procedures can be applied to studies of twinning axes, composition planes and martensitic transformations.

C. VARIATION OF PATTERN QUALITY WITH CRYSTALLINE PERFECTION

The applications discussed above depend on the geometry of the patterns only and give information on the lattice type, parameter and orientation. The other type of information provided by an ECP relates to the degree of perfection of the lattice.

There are two possible ways of using the SEM to study lattice imperfections. Firstly individual defects such as dislocations or stacking faults could be examined at high magnification. Contrast, in the channelling micrograph mode, will be apparent because although the beam will make a constant angle to the surface as it scans, the lattice planes will be bent in the region of the defect. The electron–optical conditions needed to see a defect in solid material in this way are very stringent. Typical requirements would be a probe size of 300 Å, a beam divergence of 10^{-3} rad and a beam current of around 10^{-9} A (Spencer *et al.*, 1972). Such conditions imply the use of a very bright electron source. However, in a thin foil the defect contrast is considerably higher and it is possible to observe contrast, in the back-scattering mode, from single defects using conventional electron sources. Results of this kind have been reported by Stern (1972) in MoS_2, and also by Woolf (1972) in similar specimens.

Examples of the type of image produced by Woolf are shown in Fig. 22. A specimen of MoS_2 of about 1000 Å in thickness was examined in the scanning transmission mode to identify the type and position of the defects present. Figure 22(a) shows dislocations imaged in this way; (b) then shows the same area observed in the back-scattering mode. Note that the same crystallographic defects can be seen. In both these micrographs the specimen has been set at the Bragg position by first obtaining a transmission selected area channelling pattern and tilting the foil to the desired orientation. In Fig. 22(c) the specimen has been tilted away from the Bragg position and the contrast has disappeared as predicted by the theory.

Secondly, large numbers of defects could be examined at low magnification. In this case normal channelling contrast will be present but it will be modified by the effect of lattice imperfections. The lattice distortions due to the presence of many defects will lower the "quality"

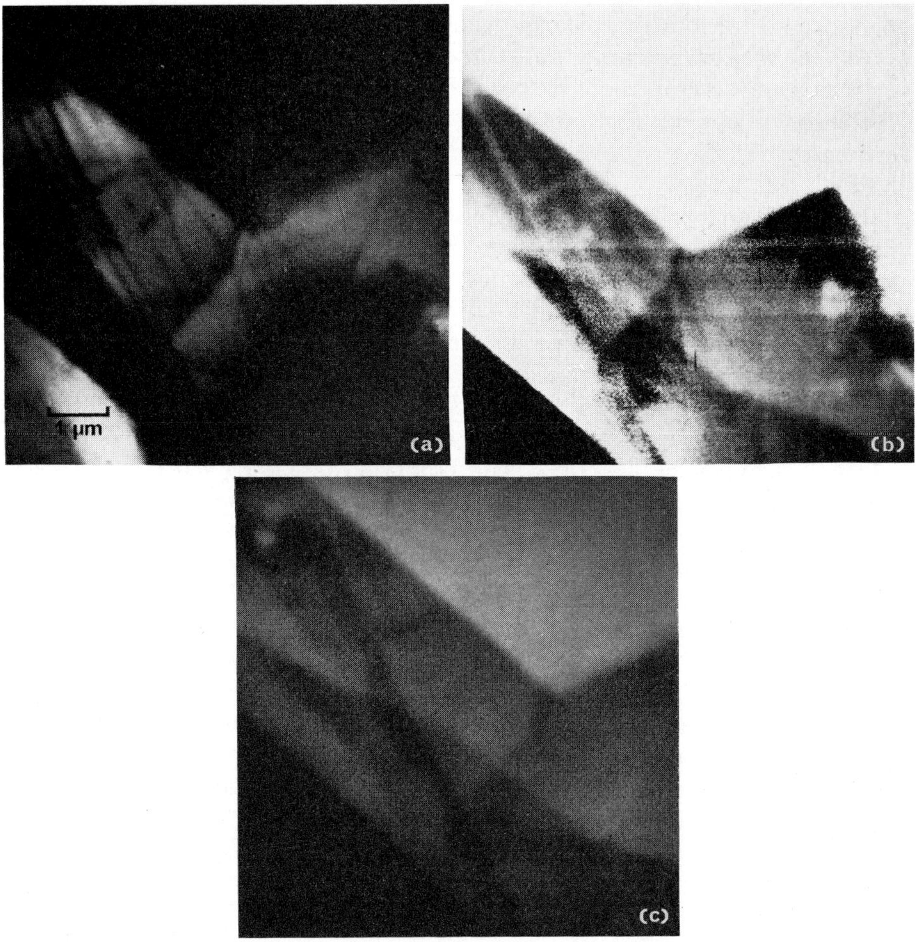

FIG. 22. Dislocation images in transmitted and back-scattered images from crystal of molybdenite. 35 kV operation. Courtesy R. J. Woolf.

of the ECP in some way. This change in quality can be assessed in several ways.

(a) *Pattern contrast.* This can be measured from the oscilloscope displaying the line trace. Care must be taken to allow for the effect of black level control and such factors as intersecting lines and contamination build up.

(b) *Line widths.* The angular widths of the bands in the pattern remains constant (at twice the appropriate Bragg angle) but the width of the lines in the pattern are found to increase. The line width is easily measured using a calibrated eye-piece to observe the SACP micrographs.

(c) *Number of high order lines*. One rapid way of assessing the pattern quality is to count the number of high order lines about a band. However, this is not a continuously variable quantity, unlike the other estimates described above, and is therefore rather limited in applicability.

It will be shown later that deviations from the expected geometrical form of an ECP gives important information about the state of the lattice, but since this effect cannot be expressed in terms of a simple parameter this does not form a convenient way of quantizing pattern quality. In general, therefore, the most useful variable with which to characterize a pattern is that of line width or contrast. It can be seen from a simple physical model that both line broadening and a fall in contrast are to be expected in a deformed lattice. A region of distortion in the lattice can be described as a combination of a local rotation of the lattice together with a change in the lattice parameter. Thus in an area where defects are present the beam is simultaneously illuminating areas of different orientation and lattice parameter. The effect is therefore to superimpose patterns of slightly different position and width. This obviously leads to a broadening of fine detail in the resultant pattern and a fall in contrast.

A considerable amount of experimental work has been carried out to find an experimental correlation between line broadening and a convenient parameter describing the state of imperfection of the crystal lattice such as percentage imposed deformation. If such a correlation can be established then it would be possible to use the SEM as a means of measuring local lattice strains by simply obtaining SACPs from area of interest and comparing the line width with a previously determined calibration curve.

Stickler *et al.* (1971) deformed samples of pure nickel, 316 stainless steel and discaloy known percentages in an Instron test machine and measured the width of the narrowest line visible in each SACP. Although there was a statistical spread in the widths measured from different patterns for each deformation step, it was found to be possible to obtain a reproducible correlation between the width of the finest line measured and the tensile elongation applied for each of the materials tested. These calibrations were then used to study a variety of phenomena and to obtain semi-quantitative values relating to the distribution of strain around an ultrasonic fatigue failure, the strain contours around a diamond pyramid indent and the deformation zone around a stress rupture in discaloy. Figure 23 shows the iso-deformation contours around this rupture plotted using the technique described above. The same workers were also able to demonstrate the effect of annealing on the quality of patterns (Stickler and Booker, 1971). This method of analysing the results is very rapid in operation since it is only required to measure the width of the finest line visible on each pattern and compare this directly with the calibration.

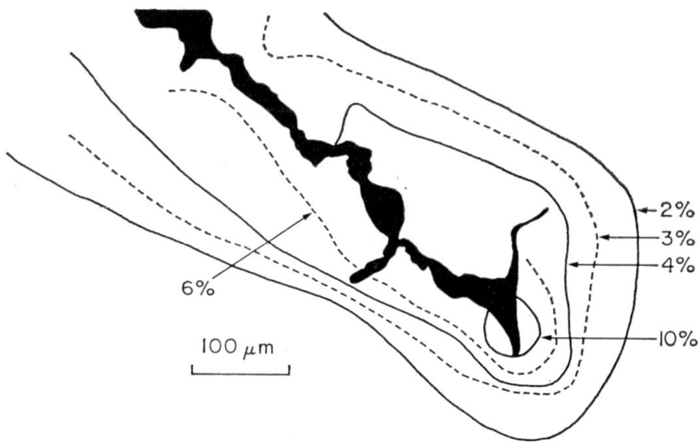

FIG. 23. Iso-deformation contours around a stress rupture in discaloy.

The difficulty with this technique is that the calibration curve must be determined for a sample of material identical with that from which it is desired to obtain data. Not only must the chemical composition be the same but the calibration samples must be annealed, prepared and deformed under conditions identical with those to be used for the test, and the electron-optical conditions used must be the same in both cases. A more serious difficulty is that the measure of percentage imposed tensile deformation is not an accurate guide to the state of the specimen, as it fails to take into account the effect of directional strains or anisotropy in the sample, or the expected effect on the pattern, since lines of different indices (*hkl*) may show different amounts of broadening as they have different natural width and are at a variety of angles to the strain imposed.

For these reasons it is rather better to characterize the specimen deformation in terms of dislocation density as this takes account of the directionality of the strain and of the anisotropy of the crystal, as well as the effect of differences in chemistry and state of annealing, etc. Stickler *et al.* (1971) measured the dislocation density of a number of test specimens using the standard TEM technique and showed that it was possible to draw an approximate universal curve of the finest detectable line width against dislocation density on which the data for all three materials examined fell. This shows clearly the advantage of this method over the percentage deformation case where each material had to have its own calibration curve.

The expected variation of line width with dislocation density has been calculated by Davidson *et al.* (1971) who found that to a reasonable approximation the pattern resolution is given by

$$\text{Pattern resolution (radians)} = 10^{-8}(\text{dislocation density (number cm}^{-2}))^{\frac{1}{2}}.$$

A comparison between the theoretical variation and the experimental data shows that the agreement is not very good except at very high dislocation densities. This discrepancy has not yet been resolved but possible explanations are that the theory does not take into account the effect of dislocations arrangements and interactions, and in practical work there is the difficulty of obtaining accurate measurements of dislocation density as, owing to the heterogeneity of deformation, this is rarely uniform. Careful experimental work is at present in progress to overcome these problems and put this technique on a sounder basis.

One other effect commonly observed on deformed materials is that the geometrical form of the ECP becomes distorted. This is especially true on materials which are deformed at a temperature above half their melting point (in absolute degrees K) (Joy and Newbury, 1971). This effect is associated with the formation in the lattice of a substructure of cells. The size of these cells depends on the exact conditions under which the deformation occurred (Garafalo, 1965) but typically they are of the order of a few microns in diameter. These cells consist of essentially perfect lattice surrounded by walls of densely packed dislocations. The nominal orientation of the subgrains is that of the grain in which they are contained but between the cells there is a slight misorientation typically of the order of 0·2° (Garafalo, 1965). When this type of structure is present the ECP will be distorted for the following reason. As the beam rocks about the surface the probe wanders around inside the selected area because of the effect of spherical aberrations in the lens (section V.A3). When the orientation inside the area is uniform this has no effect on the pattern but when an array of slightly misaligned cells is present the beam will be continuously sampling areas of different orientation and the pattern will be bent in a way depending on the intercell misorientation. Figure 24 shows two samples of deformed patterns produced in this way. The distorted pattern provides valuable evidence about the size and misorientation of the subgrain structures. If the distorted pattern is compared with an undistorted pattern of the

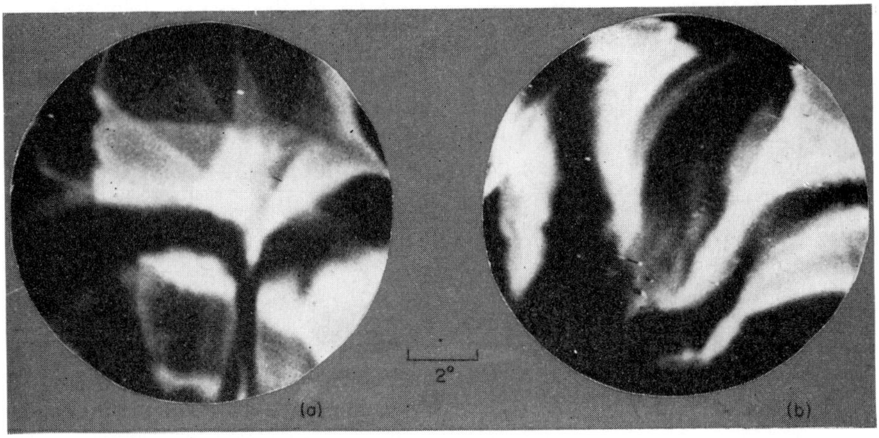

FIG. 24. Deformed channelling patterns from rolled and annealed Pb–Sn sheet.

same nominal orientation then the maximum angular offset can be measured. If the selected area is S, the average subgrain size is s and the average misorientation is Θ then the maximum offset is $\leqslant (S/s)\Theta$. If the selected area is reduced, the amount of pattern distortion is also reduced, and in the limit becomes zero when all the pattern comes from within one sub-grain. One way of achieving this is to reduce the scan angle by increasing the indicated magnification to $\times 50$ in the SACP mode. In this case the selected area will be reduced by a factor of about 8 times and consequently areas of 1 μm or so will be possible. When operating in this mode it is also possible to measure directly the offset from one sub-grain to the next since when the area selected lies over the boundary of two subgrains the pattern produced will contain a discontinuity. Therefore by measuring the area needed to produce an undistorted pattern and the offset between successive patterns formed in this way it is possible to characterize the substructure in terms of its size and misorientation. The results obtained in this way agree well with similar data obtained using the TEM but have the advantage of being far easier to produce, since no specimen thinning is needed, and many grains can be sampled, whereas in the TEM the field of view is limited.

D. USE OF CHANNELLING CONTRAST MODE

The channelling micrograph mode of operation can also be used to provide information about the nature of deformed materials. We have seen that in the case of undeformed materials each grain in a polycrystalline material exhibits a uniform level of contrast, the grey level being that found at the centre of the SACP that could be obtained from that grain. Materials which after deformation contain the non-oriented tangles of dislocations will still continue to show uniform crystallographic contrast. However, materials in which subcell formation occurs are no longer of uniform orientation and therefore the grain will appear in more complex contrast. Provided that the probe size is less than the subgrain size it is possible to reveal directly the substructure in this way as shown in Fig. 25. The exact appearance of the grain will

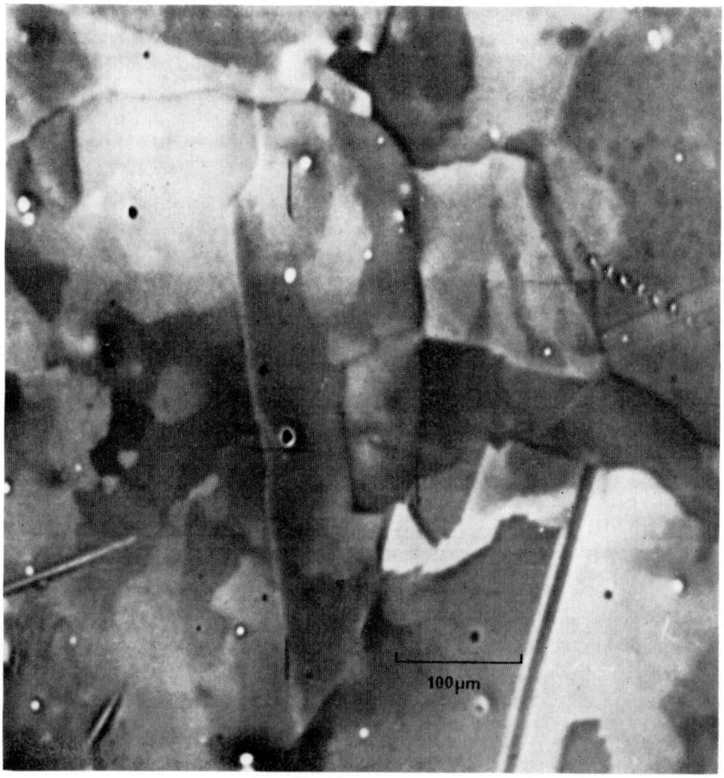

FIG. 25. Channelling micrograph of sub-grain contrast in copper deformed 10%.

depend on the diffraction conditions in operation. A slight change in the angle of incidence of the beam is sufficient to change the relative contrast of different areas of the grain. The minimum angular change necessary to cause such a change in contrast can be used as an approximate measure of the misorientation between the subgrains. One convenient way of reproducibly altering the angle of incidence of the beam to the specimen is to view the area at a medium magnification (× 100) and move the area of interest from one side of the screen to the other. The total change in incident angle is thus twice the scan angle at this magnification. If the misorientation between subgrains is entirely random the grain will have a speckled appearance, whereas if there is a progressive shift in orientation across the grain there will be bands of contrast superimposed over the speckled effect. The fact that the channelling micrograph allows such effects to be easily observed is of practical importance since to deduce the same information from TEM observations would involve measuring and plotting the relative orientation of all the subgrains in the chosen area.

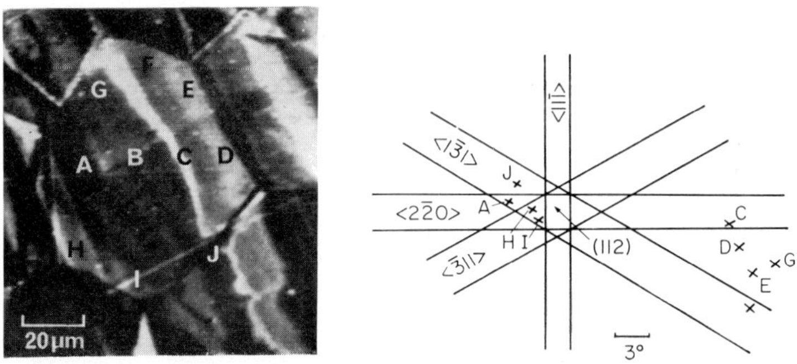

FIG. 26. Variation of orientation across a deformed grain in Fe–Si foil. Patterns were taken at the points A–J and plotted on the stereogram.

The combination of the SACP technique and the channelling micrograph mode thus enable us to make very detailed studies of deformed materials. An example combining these two techniques is shown in Fig. 26 which illustrates the contrast observed on certain grains in materials which had been rolled and annealed. The bands of contrast show that the orientation of the grain is not constant but varies from point to point. To deduce the precise way in which this occurs an SACP can be used. If the area from which the pattern comes is sufficiently small for there to be little lattice deformation across it the pattern is perfect as shown. If, however, the area is enlarged by defocusing the SACP condition (as discussed earlier) the pattern shows the magnitude of the orientation change. By taking a large number of patterns around the grain it is possible to plot out the local variation in orientation as shown on the stereogram. This shows that the contrast observed arises because the grain under observation has had its lattice effectively bent cylindrically. The annealing process has also failed to relieve the strains in this area by converting this highly strained area into smaller unstrained grains.

To summarize, the SACP and channelling micrograph modes of operation can be used to provide information about local strain distributions in solid specimens. The SACP method can be calibrated to give an indication in terms of line widths of percentage deformation or dislocation density, and the distortions and mottled contrast effects noted on SACPs and channelling micrographs can be used to give quantitative data on the substructure of the lattice.

VI. Future Trends

The most important advance in instrumentation is that enabling smaller areas to be selected. It was shown earlier that the minimum area was determined by $\frac{1}{2}C_s\theta'^3$ where θ' is the scan angle and C_s the spherical aberration coefficient of the probe lens. Since the value of θ' is determined by the smallest useful angular width of pattern acceptable, any reduction in area must be achieved by reducing C_s. Spherical aberration arises because rays making a large angle to the optic axis are deflected more strongly by the lens and hence brought to a focus on the optic axis at a point before that of the paraxial ray focus. In the case of the rocking beam technique only one ray bundle is passing through the lens at any one time and the angle which this ray makes with the optic axis is determined by the position in the scan raster. It is therefore possible in principle to correct for the effects of spherical aberrations by altering the excitation of the lens as a function of time in such a way that non-paraxial rays are less highly deflected and still come to the paraxial focus. A system for doing this has been developed by van Essen (1971). With such a system an ultimate SACP performance of better than 1 micron is possible (Joy and Newbury, 1972b) the limit being finally set by the brightness limitation of the gun and the accuracy with which the necessary correction can be set up. Improvements in brighter electron sources and more sophisticated electronic circuitry should enable areas of a few thousand ångströms to be reached.

The SACP technique is convenient to use in practice because the pattern is immediately visible on the viewing screen and the area from which the pattern has been obtained can be examined by switching back to micrograph mode. The operational convenience would be still further enhanced were it possible to view simultaneously the SACP and the area from which it was obtained. A system for doing this has been described (Joy and Booker, 1971). This relies on the fact that when the microscope is in the rocking beam mode of operation a micrograph can be obtained by slightly defocusing the lens (Fig. 16(c) and (d)). On alternate scans the lens is set to the exact SACP condition or defocused to give a micrograph, and switching circuits present the alternate images to two viewing screens with sufficient persistance to give continuous images. This arrangement makes it possible for the operator to scan a crystal to check for crystallographic features which might otherwise have been overlooked.

A third area in which advances will be made is that of pattern interpretation and orientation. The computer technique described earlier (Newbury and Joy, 1971) is in theory applicable to all crystal systems, and the practical fulfilment of this would greatly ease the problem of examining non-cubic materials.

Acknowledgements

The author is pleased to thank Dr. G. R. Booker for suggestions and helpful discussions concerning the preparation of this paper. Thanks are also due to many past and present members of the Department of Metallurgy Oxford who allowed unpublished results to be used and helped in clarifying various points, especially Drs. C. J. Humphreys and C. van Essen, Dale Newbury and Jon Spencer.

References

Ayers, J. D. and Joy, D. C. (1972). *Acta Metallurgica* **20,** 1371.

Booker, G. R. and Stickler, R. (1972). *J. Mat. Sci.* **7,** 712.

Booker, G. R., Shaw, A. M. B., Whelan, M. J. and Hirsch, P. B. (1967). *Phil. Mag.* **16,** 1185.

Booker, G. R., Joy, D. C. and Stickler, R. (1971). *EMAG*, 294.

Boyde, A. (1970). *IITRI*, **3,** 105.

Christian, J. W. (1956). *J. Inst. Metals*, **84,** 349.

Coates, D. G. (1967). *Phil. Mag.* **16,** 1179.

Davidson, S. M. (1970). *Nature, Lond.* **227,** 487.

Davidson, S. M. and Booker, G. R. (1970). *Proc. VIIth Int. Congr. Electron Microscopy, Grenoble*, **1,** 235.

Davidson, S. M., Booker, G. R. and Stickler, R. (1971). *EMAG*, 298.

Garafalo, F. (1965). "Fundamentals of creep". MacMillan, New York.

Hirsch, P. B. and Humphreys, C. J. (1970). *Proc. VIIth Int. Congr. Electron Microscopy, Grenoble*, **1,** 229.

Hirsch, P. B., Howie, A., Nicholson, R. B., Pashley, D. W. and Whelan, M. J. (1965). "Electron Microscopy of Thin Crystals." Butterworths, London.

Joy, D. C. and Booker, G. R. (1971). *J. scient. Instrum.* **4,** 837.

JOY, D. C. and NEWBURY, D. E. (1971). *IITRI*, **4,** 113.
JOY, D. C. and NEWBURY, D. E. (1972a). *IITRI*, **5,** 97.
JOY, D. C. and NEWBURY, D. E. (1972b). *J. Mat. Sci.* **7,** 714.
JOY, D. C., BOOKER, G. R., BEVIS, M. and FEARON, E. O. (1971). *IITRI*, **3,** 499.
NEWBURY, D. E. and JOY, D. C. (1971). *EMAG*, 306.
PHILLIPS, F. C. (1956). "An Introduction to Crystallography." Longmans, London.
REIMER, L., BADDE, H. G., SEIDEL, H. and BÜHRING, W. (1971). *EMAG*, 72.
RUSS, J. (1972). *Edax Editor*, **1,** 3.
SCHULSON, E. M. and VAN ESSEN, C. G. (1969). *J. scient. Instrum.* **2,** 247.
SCHULSON, E. M., VAN ESSEN, C. G. and JOY, D. C. (1969). *IITRI*, **2,** 45.
SPENCER, J. P., HUMPHREYS, C. J. and HIRSCH, P. B. (1972). *Phil. Mag.* **26,** 193.
STERN, R. (1972). *Proc. EMSA 1972* (in press).
STICKLER, R. and BOOKER, G. R. (1971). *Proc. Int. Mat. Symp., Berkeley, California*, p. 301.
STICKLER, R., HUGHES, C. W. and BOOKER, G. R. (1971). *IITRI*, **4,** 473.
THORNTON, P. R. (1968). "Scanning Electron Microscopy." Chapman and Hall, London.
VAN ESSEN, C. G. (1971). *EMAG*, 314.
VAN ESSEN, C. G. and SCHULSON, E. M. (1969). *J. Mat. Sci.* **4,** 336.
VAN ESSEN, C. G., SCHULSON, E. M. and DONAGHAY, R. H. (1970). *Nature, Lond.* **225,** 847.
VAN ESSEN, C. G., SCHULSON, E. M. and DONAGHAY, R. H. (1971). *J. Mat. Sci.* **6,** 213.
VICARIO, E. (1970). D.Phil. thesis, University of Lyon.
VICARIO, E. and PITAVAL, M. (1969). *C.r. hebd. Séanc. Acad. Sci., Paris*, **269B,** 287.
VICARIO, E., PITAVAL, M. and FONTAINE, G. (1970). *Proc. VIIth Int. Congr. Electron Microscopy, Grenoble*, **2,** 211.
WOLF, E. D. and EVERHART, T. E. (1969). *Appl. Phys. Letters*, **14,** 299.
WOLF, E. D. and HUNSPERGER, R. G. (1970). *Appl. Phys. Letters*, **16,** 526.
WOOLF, R. J. (1972). D.Phil. Thesis, University of Oxford.

ABBREVIATIONS

EMAG: Proceedings of 25th anniversary meeting of Electron Microscopy and Analysis Group, Institute of Physics (London). Held in Cambridge, July 1971.

IITRI "N": Proceedings of "Nth" annual symposium on scanning electron microscopy, Illinois Institute of Technology Research Institute, Chicago. Vol. **1** (1968), **2** (1969), **3** (1970), **4** (1971), **5** (1972).

Chapter 7

Theory of Diffraction Contrast Effects in the Scanning Electron Microscope

A. HOWIE

Cavendish Laboratory, University of Cambridge, England

Symbols

d_e = width of electron wave packet
λ = wavelength of electron
α = semi-angle of beam divergence
$\boldsymbol{\chi}$ = wave vector of incident electron
ψ = wave function describing electron
E_0 = accelerating potential
m_0 = rest mass of electron
h = Planck's constant
$\boldsymbol{\chi}'$ = wave vector of scattered electron
$\mathbf{g}, \mathbf{h}$ = reciprocal lattice vector
d = inter-planar spacing $d = 1/g$
$\mathbf{s}$ = vector of deviation from exact Bragg position
θ_B, θ_1 = Bragg angle
$\Delta\theta(= s/g)$ = angle of deviation from exact Bragg position
ξ_g = extinction distance
$b^{(j)}(\mathbf{k}^{(j)}, \mathbf{r})$ = Bloch waves, the jth one
$C_g^{(j)}$ = amplitude of g component of jth Bloch wave
$\mathbf{k}^{(j)}$ = wave vector of jth Bloch wave
$\psi^{(j)}$ = amplitude of component Bloch wave $b^{(j)}$ part of general electron wave function ψ
$q^{(j)}$ = attenuation of amplitude of jth Bloch wave
$\Lambda^{(j)}(= 1/4\pi q^{(j)})$ = penetration distance of jth Bloch wave
v = electron velocity
Δt = periodicity of thickness fringes

I_0 = intensity of transmitted electrons
I_b = intensity of back-scattered electrons
$\mathbf{b}$ = Burgers vector
$\mathbf{R}$ = displacement vector of atom from its perfect lattice position
$P^{(j)}$ = probability/unit depth of back-scattered electrons generated by the jth Bloch wave
$P(\mathbf{r})$ = periodic *fn* describing back-scattering probability in crystal
P_h = Fourier component of $P(\mathbf{r})$ corresponding to $\mathbf{h}$
P_0 = average value (zero-order Fourier component) of $P(\mathbf{r})$
V_c = volume of unit cell
e = electronic charge
m = electron mass
Z = atomic number
M_h = Debye Waller factor is $\exp(-M_h)$
σ_R = Rutherford back-scattering cross section
σ_M = Mott cross section
Λ = average attenuation distance for a plane wave
T = thickness of crystal
$\bar{I}_B$ = intensity of all back-scattered electrons from solid of thickness T, experimentally determined
$\Delta I_B = I_B(T) - \bar{I}_B(T)$
J = Incident electron current
N = Number of picture points
t_s = frame scanning time
C = contrast available
F = fraction of electrons in incident beam which enter back-scattering detector
d_p = probe diameter
B = minimum gun brightness

I. Introduction

For many applications of electron microscopy, particularly scanning electron microscopy, the wave properties of the electron are at most of marginal significance and it may be regarded as a classical particle. In actuality, however, each electron should be considered as a wave packet whose lateral extent d_e is related to the wavelength λ and semi-angle α of the optical system by the uncertainty principle.

$$d_e \gtrsim \lambda/\alpha. \tag{1}$$

Wave effects will tend to become significant when the extent of the wave packet so defined becomes comparable with or larger than the scale of the structural detail in the object studied. This will tend to be the case either when very high resolution is pursued or when rather well collimated beams are used. Historically it was in the latter kind of situation that the wave properties of the electron were first demonstrated by the observation of diffraction patterns from crystalline materials. This kind of situation also applied in the SEM experiments of Coates (1967) (see section IV below) who observed contrast patterns in a low resolution back-scattering mode. The explanation of these observations by Booker *et al.* (1967) in terms of a Bragg reflection and back-scattering mechanism already understood in conventional transmission electron microscopy (CTEM), first showed the potential importance of wave effects in the SEM. Soon afterwards the observation of lattice fringes, Fresnel fringes and diffraction contrast images of dislocations by Crewe and Wall (1970) in high resolution scanning transmission (STEM) demonstrated the significance of wave effects in the real space image.

The application of CTEM to the study of crystalline or quasicrystalline objects and their defects over the past two decades has resulted in the development of a very detailed wave theory of contrast for the interpretation of images (for reviews see Hirsch *et al.*, 1965; Heidenreich, 1964; Amelinckx *et al.*, 1970; Valdrè, 1971). A brief outline of this theory is given in section II below. The experience in the conventional instrument has amply demonstrated that the availability of such a theory in a form convenient for numerical computation is essential if quantitative data, and sometimes even if reliable qualitative results, are to be extracted from the observations. It now seems likely that a similar situation will apply in the SEM where the increased operational flexibility of the instrument allows a number of scattering and signal modes which cannot be achieved in the conventional machine. In sections III, IV and V below we examine the extent to which the existing wave theory can be applied to some of these modes of operation, pointing out the new data which may be attainable in each case and the theoretical developments which seem to be necessary if this data is to be converted into useful structural information.

II. Outline of Wave Theory for CTEM Images

A. Formation of Micrographs and Diffraction Patterns

Figure 1(a) shows schematically the ray diagram in the specimen and objective lens region of the CTEM. A parallel or nearly parallel beam emerges from the cross-over or source S and is transmitted through the specimen. The electrons in this beam can be adequately described by plane wave functions $\psi = \exp(2\pi i \boldsymbol{\chi} \,.\, \mathbf{r})$ where $\boldsymbol{\chi}$ the wave vector points in the direction of motion and, ignoring relativistic effects, has magnitude $\chi = \lambda^{-1} = (2m_0 E_0)^{\frac{1}{2}}/h$ where E_0 is the accelerating potential. In the case of crystalline specimens, strong scattered waves emerge with wave vectors $\boldsymbol{\chi}'$ given by

$$\boldsymbol{\chi}' = \boldsymbol{\chi} + \mathbf{g} + \mathbf{s} \tag{2}$$

where each such wave vector $\boldsymbol{\chi}'$ is associated with a given Bragg reflection involving a reciprocal lattice vector $\mathbf{g}$ normal to the relevant Bragg planes of spacing $d = 1/g$. The well known construction for determining the strongly excited Bragg reflections by finding the reciprocal lattice points **g** which lie close to the Ewald sphere is shown in Fig. 1(b). The small vector **s** drawn normal to **g** denotes the deviation from the exact Bragg condition $\boldsymbol{\chi}' = \boldsymbol{\chi} + \mathbf{g}$ (equivalent to $\lambda = 2d \sin \theta_B$) and corresponds to an angular deviation from the Bragg position of

$$\Delta\theta = s/g. \tag{3}$$

Since typical values of λ at electron microscope voltages are of order $d/50$ the Bragg angles are all very small and various images can easily be formed by allowing one or more of these beams to pass through the objective aperture (Fig. 1(a)). Thus a *bright field image* is formed using the zero order beam **O** alone and a *dark field image* by using a

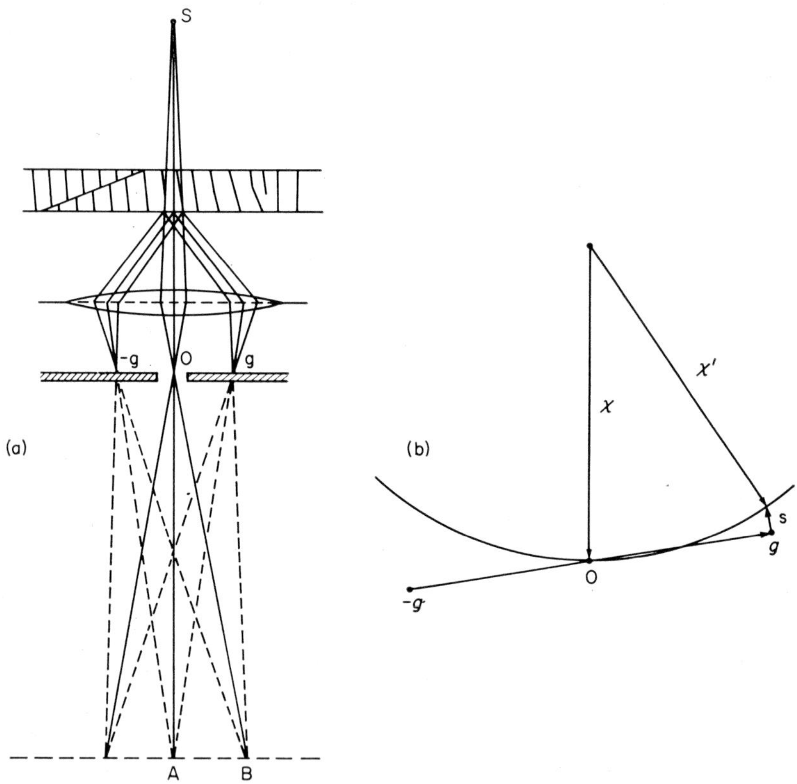

FIG. 1(a). The schematic ray diagram of the CTEM in the bright field diffraction contrast mode with the source at S. For high resolution work, a larger objective aperture admitting one or more diffracted beams **g** may be used. With the source at A and the detector at S the diagram represents a STEM system.

FIG. 1(b). Ewald sphere construction for the crystal orientation shown in 1(a). Diffraction contrast images can be obtained by selecting with the aperture the beam **O** (bright field), **g** (strong dark field) or $-\mathbf{g}$ (weak beam).

diffracted beam **g**. With a larger aperture, higher resolution can be obtained (cf. eqn (1)) and, if the zero order plus one or more diffracted beams are accepted by the aperture, their mutual interference in the image gives rise to fringes corresponding to the Bragg planes involved. In practice it is desirable to minimize the effects of the instrumental aberrations by arranging that all the waves accepted by the objective aperture make angles with the optic axis which are either as small as possible or as nearly equal as possible and this can be achieved by using tilted illumination.

By removing the objective aperture and focusing with subsequent lenses of the instrument on the plane Og rather than AB in Fig. 1(a), diffraction patterns can be obtained directly in the CTEM. *Selected area diffraction* can be obtained from a known region of the image by introducing a limiting aperture in the plane AB conjugate to the specimen. The combination of real space imaging and selected area diffraction to identify local crystal structure, orientation and diffraction conditions is of tremendous experimental value.

B. IMAGE CONTRAST EFFECTS IN PERFECT CRYSTALS

The variation of Bragg diffracted intensities with crystal thickness and orientation can be seen directly from dark field micrographs. Thus, when the orientation changes due to bending of the specimen (see Figs 2, 4 and 5 below) *bend contours* or *extinction contours* may be seen corresponding to the locus of points at the Bragg position. In a wedge crystal thickness fringes occur (Fig. 2) corresponding to oscillations of the diffracted intensity with a periodicity at the Bragg position of $\Delta t = \xi_g$ the so-called extinction distance. This phenomenon arises because of dynamical diffraction effects, i.e. multiple scattering of the electron between the states $\boldsymbol{\chi}$, $\boldsymbol{\chi}+\mathbf{g}$ and back again. The strength of this Bragg reflection process is measured by the extinction distance ξ_g. In a distance $\xi_g/2$ the diffracted amplitude builds up from zero to its maximum value of unity. The extinction distance is thus least for strong low-order Bragg reflections and typical values at 100 keV (see Hirsch *et al.*, 1965) lie in the range from 150 Å to 1000 Å. Some values at 20 keV are given in Table I. Extinction distances at other energies can readily be calculated since ξ_g is proportional to the electron velocity v. Dynamical scattering effects also determine the details of bend contours which have a width $s \sim 1/\xi_g$ corresponding, as can be seen from eqn (3) to an angular range of the crystal setting on either side of the Bragg position of $\Delta\theta_g$ given by

$$\Delta\theta_g = 1/g\xi_g. \tag{4}$$

Strong Bragg reflections thus tend to have large angular widths which may in fact in some cases be comparable with the Bragg angles themselves.

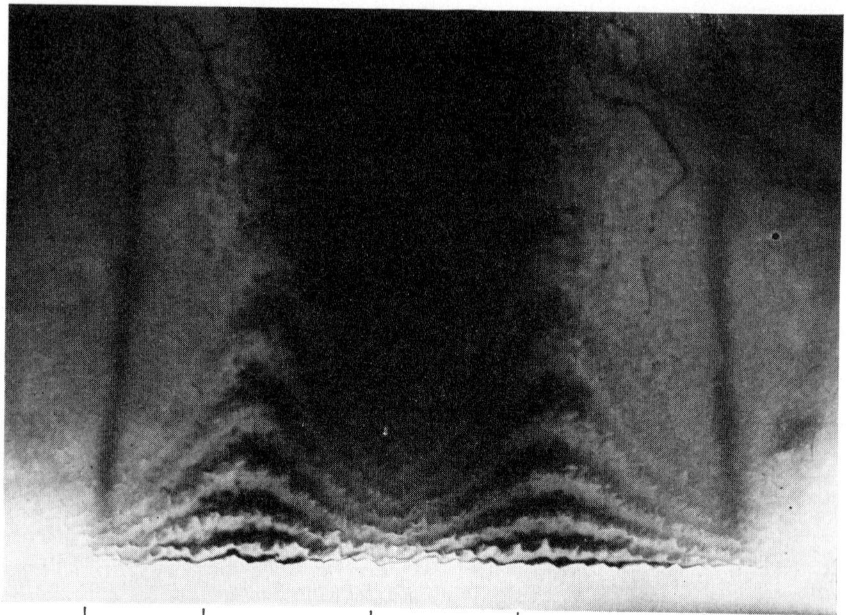

FIG. 2. Bright field image of a bent wedge crystal. The Bragg reflecting positions are indicated (cf. Fig. 4(b)). At the first order Bragg positions, $\theta = \pm\theta_1$, thickness fringes of periodicity $\Delta t = \xi_g$ occur.

TABLE I. Extinction distances ξ_g and estimated[a] Bloch wave penetration distances $\Lambda^{(1)}$ and $\Lambda^{(2)}$ for 20 keV electrons

Element	Al		Si		Cu		Au	
Reflection	111	220	220	311	111	220	111	220
ξ_g (Å)	276	525	376	668	120	206	80	145
$\Lambda^{(1)}$ (Å)	270	300	420	460	75	80	27	28
$\Lambda^{(2)}$ (Å)	860	720	1040	840	460	350	380	220

[a] These include the thermal diffuse scattering contribution (Humphreys and Hirsch, 1968; Radi, 1970) and the electronic contribution (Whelan, 1965) corresponding to scattering outside the objective aperture. The electronic contribution was reduced by a factor of 4 to take some account of screening effects and bring the data into better agreement with other calculations (see Howie and Stern, 1972).

A formal solution to the dynamical scattering problem can be obtained by constructing various combinations of plane waves linked by the Bragg reflection process which propagate self consistently in the crystal. These self-consistent solutions are known as Bloch waves and have the general form

$$b^{(j)}(\mathbf{k}^{(j)}, \mathbf{r}) = \sum_g C_g^{(j)} \exp(2\pi i(\mathbf{k}^{(j)}+\mathbf{g}) \cdot \mathbf{r}) \tag{5}$$

where the summation extends over all the Bragg reflections excited. The amplitudes $C_g^{(j)}$ and wave vectors $k^{(j)}$ for a given energy can easily be found by numerical solution of the Schrodinger equation. However the important point to note is that these Bloch waves are the fundamental solutions for the electron wave in the crystal and the general wave function ψ must be expanded in terms of them. We can thus write

$$\psi(\mathbf{r}) = \sum_j \psi^{(j)} b^{(j)}(\mathbf{k}^{(j)}, \mathbf{r}). \tag{6}$$

The amplitudes $\psi^{(j)}$ of the different Bloch waves are found by wave matching at the crystal entrance surface to the incident plane wave. In the simple case where only one Bragg reflection is excited and the crystal is at the exact Bragg position there are two Bloch waves, each of two components

$$\begin{aligned} b^{(1)} &= \frac{1}{\sqrt{2}}\{\exp(2\pi i\mathbf{k}^{(1)} \cdot \mathbf{r}) + \exp(2\pi i(\mathbf{k}^{(1)}+\mathbf{g}) \cdot \mathbf{r})\} \\ b^{(2)} &= \frac{1}{\sqrt{2}}\{\exp(2\pi i\mathbf{k}^{(2)} \cdot \mathbf{r}) - \exp(2\pi i(\mathbf{k}^{(2)}+\mathbf{g}) \cdot \mathbf{r})\}. \end{aligned} \tag{7}$$

Clearly these are nothing other than the simple standing waves familiar in multiple reflection problems. The two wave vectors have equal components normal to the Bragg planes but slightly different components $k_z^{(j)}$ in the Bragg planes. This occurs because the intensity distributions in the two waves relative to the lattice planes are rather different (see Fig. 3). $b^{(1)}$ has a large intensity in the region of low potential energy near the atomic planes and consequently has a higher kinetic energy and hence a larger wave vector $k_z^{(1)}$ than $k_z^{(2)}$. Because of these different wave vectors a beating effect with periodicity

$$\Delta t = |k_z^{(1)} - k_z^{(2)}|^{-1} = \xi_g$$

occurs as the waves travel into the crystal and this is the origin of the thickness fringes previously mentioned.

The difference in the Bloch wave intensity distributions has other important consequences, chief of which for transmission experiments is the anomalous transmission of the wave $b^{(2)}$ which avoids the region of the atom where strong large-angle elastic scattering and inelastic scattering due to excitation of atomic core electrons may occur. So far we have not considered these scattering processes specifically (since the Bragg reflection theory described above covers only low angle elastic scattering) but for purposes of transmission microscopy the main result is that electrons are scattered outside the objective aperture so that the Bloch waves are attenuated with increasing depth z in the crystal according to the equations

$$\frac{d\psi^{(j)}(z)}{dz} = -2\pi q^{(j)}\psi^{(j)}. \tag{8}$$

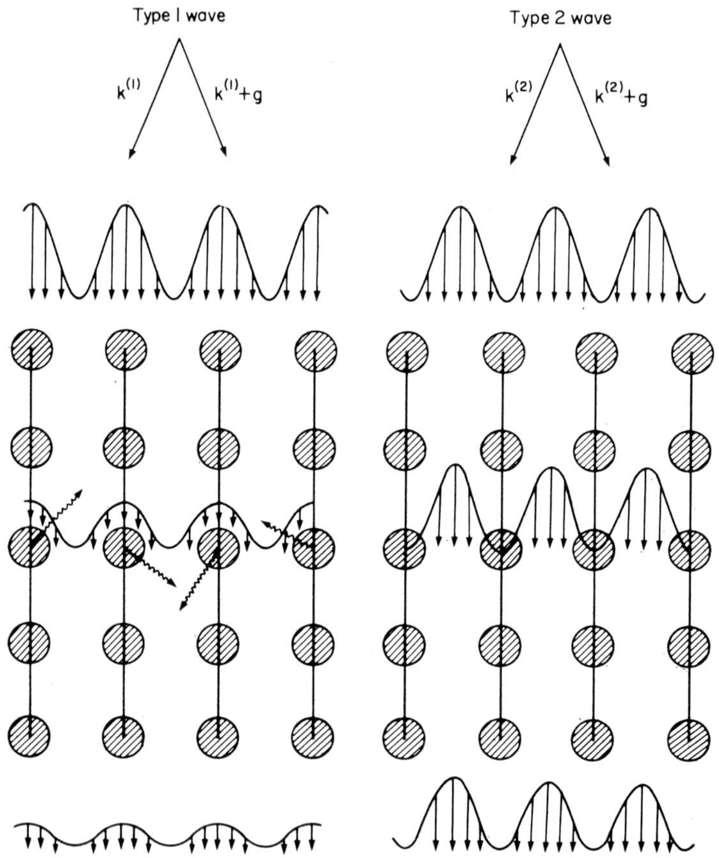

FIG. 3. Bloch wave intensities relative to the atomic planes in the two-beam case. Wave (1) is concentrated near the atoms, hence strongly scattered and poorly transmitted. Wave (2) avoids these strongly scattering regions and is well transmitted.

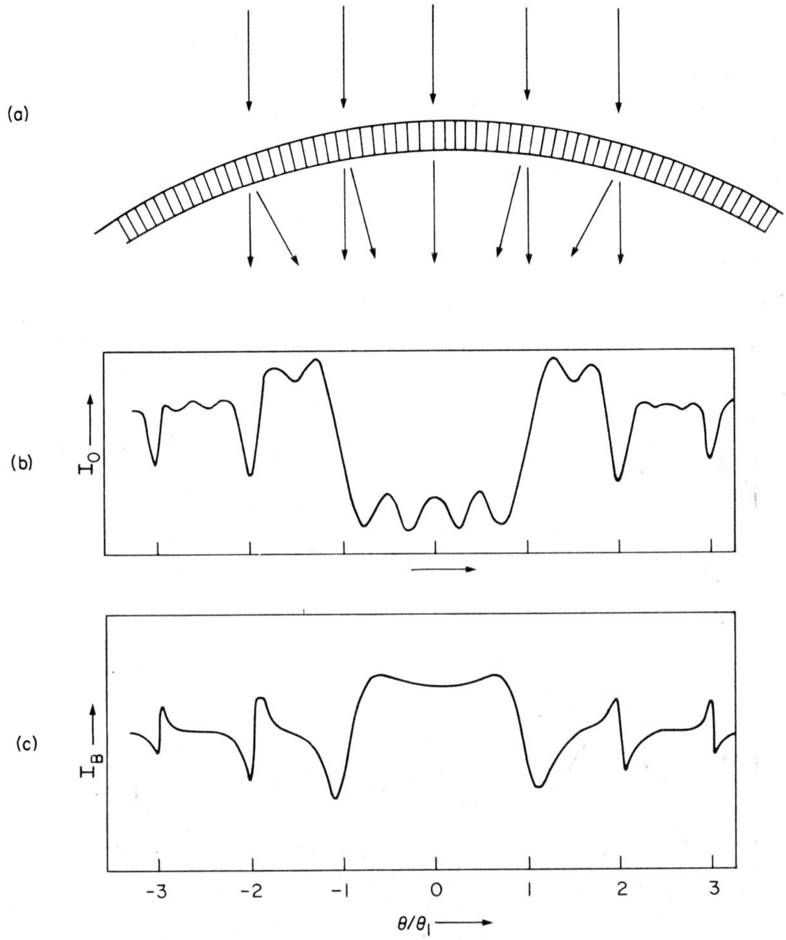

FIG. 4. For the bent crystal shown in (a) the angle of incidence θ to the crystal planes varies across the crystal giving rise to the bend contour shown in (b) where the transmitted intensity I_0 is low inside the first order Bragg positions $\theta = \pm\theta_1$ and high just outside. The corresponding back-scattered intensity I_B (or X-ray production) is shown schematically in (c).

This leads to an exponential decay $\exp(-z/\Lambda^{(j)})$ of the Bloch wave intensity $|\psi^{(j)}|^2$ where $\Lambda^{(j)} = 1/4\pi q^{(j)}$. The main contribution to $q^{(j)}$ comes from thermal diffuse scattering and recent values for various reflections have been calculated for 100 keV electrons by Humphreys and Hirsch (1968) and by Radi (1970). Since, in the Born approximation,

$q^{(j)} \propto v^{-2}$, $\Lambda^{(j)} \propto v^2$, data can readily be converted to other energies and some typical values for 20 keV electrons are given in Table I. Recently, however, it has been pointed out (Howie and Stern, 1972; Badde and Reimer, 1972) that, because of the use of Born approximation scattering amplitudes in the computation of $\Lambda^{(j)}$, the values obtained may be appreciably too small particularly in the case of high atomic number materials at lower energies ($\lesssim$ 20 keV).

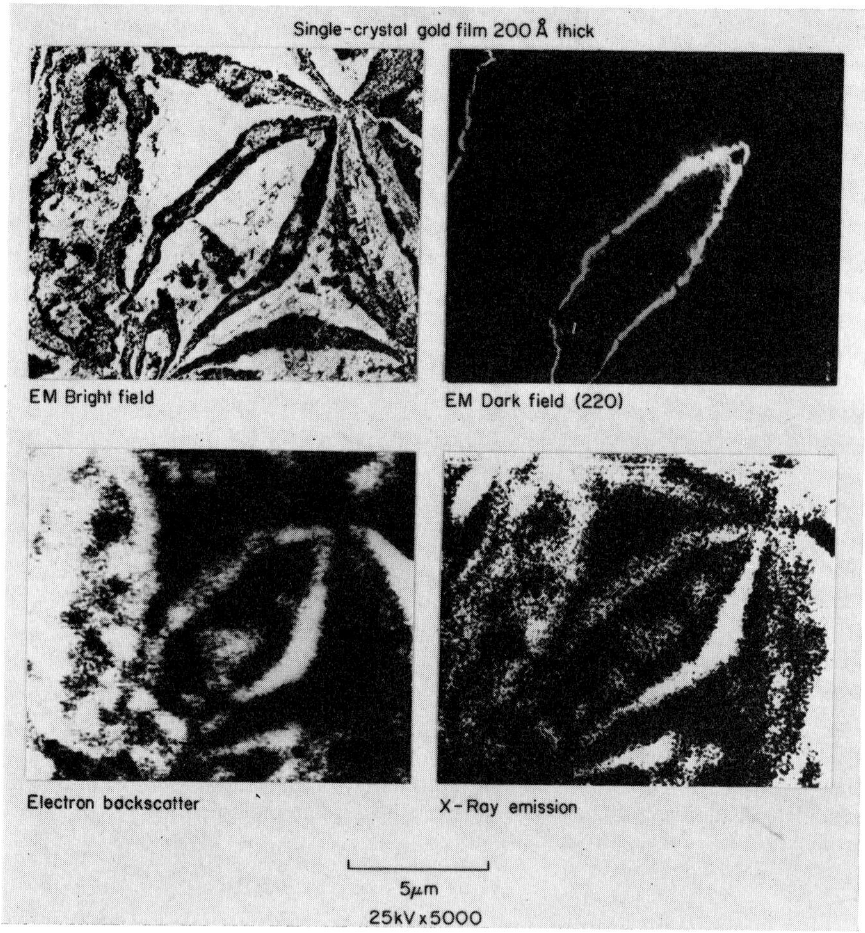

FIG. 5. Bright field, dark field, back-scattering and X-ray production intensities in the same region of a thin gold crystal showing the related dependence of all these quantities on crystal orientation. (Duncumb (1962)—courtesy *Philosophical Magazine*.)

Table I shows as expected that $\Lambda^{(2)} \gg \Lambda^{(1)}$ so that the wave which avoids the atoms is anomalously well transmitted. This wave can be strongly excited by setting the crystal just outside the Bragg position $\theta > \theta_1$ as shown in Fig. 4(a). The anomalous transmission which occurs in these specific orientations (see Fig. 4(b)) allows transmission studies to be made on comparatively thick crystals.

The concentration of the intensity distribution of the wave $b^{(1)}$ in the region of the atoms (Fig. 3) leads to anomalously high values of X-ray production and back-scattering for this wave and this is shown schematically as a function of incidence angle to the Bragg planes in Fig. 4(c). Such variations in back-scattering and characteristic X-ray production with Bragg reflection conditions were first observed experimentally by Duncumb (1962) using a specially constructed machine allowing simultaneous detection of back-scattering, X-ray emission and bright and dark field transmission (see Fig. 5). We return to these effects in section IV.

C. IMAGE CONTRAST EFFECTS IN IMPERFECT CRYSTALS

It might appear that the atomic structure of imperfect crystals and the configuration of lattice defects could be most easily found by direct resolution imaging. However, the effect of the aberrations previously noted, and also the complications which arise from the multiple scattering and can only be ignored in very thin crystals, make it extremely difficult to interpret these images since there is then no direct correspondence between the lattice fringes and the atomic planes.

Images with rather lower resolution but which are much easier to interpret can be obtained by simple bright field or dark field imaging. In the case of dislocations and other defects which tilt the lattice planes a simple estimate of the width of the images can be obtained from consideration of the local rotations of the lattice relative to the width of the Bragg peak. For electrons passing at a distance x from a screw dislocation of Burger's vector $\mathbf{b}$ the tilt $\Delta\theta$ of the lattice planes is given approximately by the expression

$$\Delta\theta = \mathbf{b}/2\pi x. \tag{9}$$

This may be equated to the misorientation $\Delta\theta$ from the Bragg position (eqn (3)) to yield an image width

$$2x = g\mathbf{b}/\pi s. \tag{10}$$

In the case of bright field or strong dark field images where $s \approx 1/\xi_g$ (eqn (4)) the image width is closely related to the extinction distance and is typically about 100 Å so that little detail can be resolved in individual dislocations although the structure of dislocation arrangements in the sample can usually be adequately observed. Much sharper images of individual dislocations can however be obtained by the weak beam method (Cockayne *et al.*, 1969) where a very weak diffracted beam with a large value of $s \gtrsim 2\times10^{-2}$ Å^{-1} is used with tilted illumination to form the image. Equation (10) then shows that images of width $\sim$15 Å can be obtained.

The effect of lattice displacements due to defects, which move an atom from its perfect crystal position $\mathbf{r}$ to a new position $\mathbf{r}+\mathbf{R}(\mathbf{r})$, can be taken into account much more quantitatively (for details see Hirsch *et al.*, 1965). It is now convenient to expand the wave function ψ in terms of Bloch waves which displace locally with the lattice

$$\psi(\mathbf{r}) = \sum_j \psi^{(j)}(z) \sum_g C_g^{(j)} \exp(2\pi i(\mathbf{k}^{(j)}+\mathbf{g}) \,.\, \mathbf{r}) \exp(-2\pi i\mathbf{g} \,.\, \mathbf{R}). \quad (11)$$

We then find that, in addition to the attenuation of the Bloch waves previously noted, the presence of the defect strain field causes transitions between the different Bloch wave states so that eqn (8) becomes

$$\frac{d\psi^{(j)}(z)}{dz} = -2\pi q^{(j)}\psi^{(j)} + 2\pi i \sum_l \psi^{(l)}(z) \exp(2\pi i(k_z^{(l)} - k_z^{(j)})z) \times \sum_g \mathbf{g} \,.\, \frac{d\mathbf{R}}{dz} C_g^l C_g^{(j)*}. \quad (12)$$

These coupled equations between the different Bloch wave amplitudes can easily be solved numerically for a given displacement $\mathbf{R}$ by starting with the values at the entrance surface $z = 0$ and integrating to $z = t$ to obtain the amplitudes of the waves leaving the crystal at each point. Clearly, however, the scattering depends on $\mathbf{g} \,.\, (d\mathbf{R}/dz)$, i.e. on the local tilt of the Bragg planes so that the simple geometrical theory of image widths described above is valid. In practice many of the properties of defect images depend in a very simple way on $\mathbf{g} \,.\, \mathbf{R}$ which may be deduced without elaborate computation. Thus the Burgers vector $\mathbf{b}$ of dislocations can be found by locating the Bragg reflections $\mathbf{g}$ for which the image vanishes corresponding to $\mathbf{g} \,.\, \mathbf{b} = 0$ and stacking fault images also disappear when Bragg planes are used which are continuous across the fault ($\mathbf{g} \,.\, \mathbf{R} = n$, an integer).

D. INELASTIC SCATTERING

The absorption model outlined above takes account of inelastic scattering of electrons outside the objective aperture. In practice, however, particularly when thicker crystals are examined, most of the electrons accepted by the aperture will also have been inelastically scattered, possibly several times, mainly by small-angle inelastic scattering processes such as plasmon excitation or single electron excitation. These processes do not involve any Bragg reflection and it can be shown (Howie, 1963; Cundy *et al.*, 1969; Humphreys and Whelan, 1969) that, at least in the case of in-focus bright or dark field images, the diffraction contrast effects are preserved unchanged. The contrast of high resolution images may, however, be affected by inelastic scattering because of chromatic aberration and also because the inelastically scattered electron image deteriorates rapidly with increasing defocus distance (Spalding, 1969; see also Metherell, 1971) as a result of the finite angle of inelastic scattering.

III. Wave Theory for STEM Images

A. The Reciprocity Theorem

Crewe and Wall (1970) demonstrated that images showing lattice fringes, dislocations and Fresnel fringes could be obtained in the STEM all bearing a striking resemblance to the images observed in the CTEM. The essential basis for this similarity may be most easily understood in terms of the reciprocity theorem of Helmholtz, as was pointed out by Cowley (1969) and by Zeitler and Thomson (1970). The theorem states that if the wave amplitude at a point $\mathbf{r}_2$ when a source is placed at $\mathbf{r}_1$ is $\psi_1(\mathbf{r}_2)$ then, for the amplitude $\psi_2(\mathbf{r}_1)$ at $\mathbf{r}_1$ when the same source is placed at $\mathbf{r}_2$, we have the result $\psi_2(\mathbf{r}_1) = \psi_1(\mathbf{r}_2)$. A proof of the theorem can be given in terms of time reversal symmetry and it holds in the presence of arbitrary fields provided that all magnetic fields are reversed in sign between the two experiments. The theorem is also valid when absorption effects occur and, even when the electrons collected have been inelastically scattered, it may be used for the wave intensities $|\psi_2(\mathbf{r}_1)|^2 = |\psi_1(\mathbf{r}_2)|^2$ provided that the energy losses involved are so small that scattering angles, focusing action of lenses etc. are not significantly changed (Pogany and Turner, 1968). An account of the application of the reciprocity principle to various problems in electron microscopy and diffraction is given by Howie (1972).

The application of the principle in the present case can be illustrated by reference to Fig. 1(a) which was previously used to describe the action of the CTEM with the electron source at S and the detector at A (a point on the photographic plate recording the image). If we now follow the theorem and place the source at A and the detector at S it can be seen immediately that we have the STEM situation used by Crewe and Wall (1970) where a relatively wide angle beam is focused to form a probe on the specimen and the detector selects the transmitted electrons emerging in a relatively small angular range. It can also be seen that the various tilted illumination methods in the CTEM mentioned above for high resolution dark field, weak beam or lattice fringe imaging are all achieved in the STEM by moving the detector to a suitable off-axis position. In general it can be said that all modes of operation in the CTEM have their counterparts in the STEM and comparison pictures have in fact been obtained in machines where both modes are possible (Shibatomi *et al.*, 1971). An example from the work of Fujimoto *et al.* (1972) is shown in Fig. 6 for the case of the Kikuchi band structure observed in transmission diffraction.

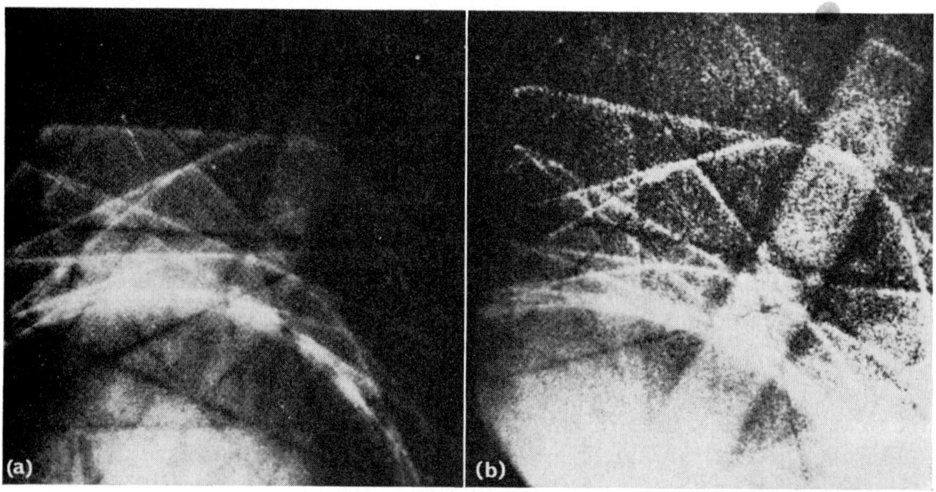

FIG. 6. Kikuchi band structure from an Si crystal observed (a) in the CTEM and (b) in the STEM. (Fujimoto *et al.* (1972)—courtesy *Zeitschrift für Naturforschung*.)

Strictly speaking, of course, the conditions required for the reciprocity theorem to apply are not exactly met in practice since in theory for instance all the apertures and accelerating electrodes should be left in place when the source and detector are interchanged. The principle clearly applies to a good approximation in the special STEM situation just described. It should be pointed out, however, that this is a rather artificial and inefficient mode of STEM operation since, as can be seen from Fig. 1(a), only a very small fraction of the electrons incident on the specimen when a large illuminating aperture is used form the signal (even less than in the CTEM). In actuality the STEM signal can be formed making use of practically all of the scattered electrons but this and the various other aspects of STEM operation which provide distinct advantages over the CTEM fall outside the scope of the reciprocity theorem whose usefulness is therefore rather strictly limited.

B. ADVANTAGES OF THE STEM OVER THE CTEM

The main advantage of the STEM over the CTEM which is relevant for discussion here is its ability to make use of electrons which have been scattered through considerable angles, with possibly significant energy loss, to form an image signal. The imaging lenses operate on the electrons before they reach the specimen. In the CTEM the chromatic and spherical aberrations of the imaging lenses, acting after the specimen, make it impossible to obtain good resolution using electrons with such a large spread in energy and angle. The reciprocity principle does not apply to inelastically scattered electrons when the energy loss is sufficient to make chromatic aberration a significant effect.

The signal from the inelastically scattered electrons can be used for various purposes. In very high resolution work such as the imaging of single atoms of heavy elements on carbon substrates, Crewe and Wall (1971) (see also chapters by Crewe in this volume) have shown that the image contrast can be significantly improved by making appropriate combinations of the small-angle inelastic scattering and the large-angle elastic scattering. The high resolution STEM may also be of advantage in weak beam work, where the low intensities and long exposure times needed cause difficulties in the CTEM. At lower resolution it may be possible to study dislocation arrangements and other defects in thicker crystals than is possible in a CTEM operating at the same voltage. The actual improvement factor which can be obtained is still in doubt and may well be a function of both the atomic number of the specimen and the voltage used. Further information on this point would not only be of practical interest but would also throw light on the basic factors which actually limit the maximum usable specimen thickness in the two instruments. At present it is difficult to isolate the effects of loss of intensity, loss of resolution due to chromatic aberration and loss of contrast due to inelastic scattering.

From the theoretical point of view the inapplicability of the reciprocity theorem to these situations is not a severe handicap. The existing CTEM theory describes the propagation of an incident plane wave through the specimen and, as pointed out by Zeitler and Thomson (1970), the case of an incident scanning probe can then be treated by Fourier transform methods. Such a method was actually used by

Howie *et al.* (1971) to compute the intensity of lattice fringes in back-scattering (see section IV below). A more significant problem which may well need attention is the kind of signal carried by the inelastically scattered electrons, especially when large energy losses or angles of scattering are involved or the image is obtained at high resolution, possibly in some out of focus condition. As discussed in section II.D, the theory developed and tested in the CTEM for the image contrast from inelastically scattered electrons is restricted to small angles of scattering and relatively small energy loss.

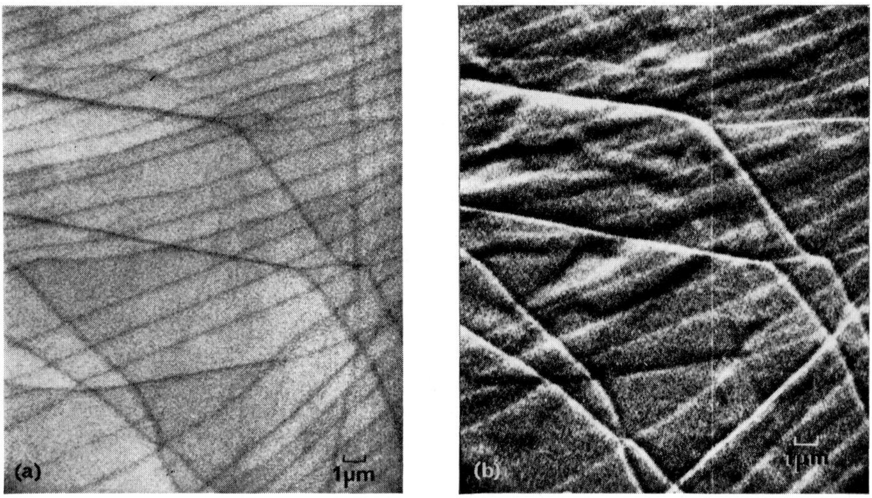

FIG. 7. Dislocations in a thin molybdenite crystal imaged in the SEM at 50 kV using (a) transmitted electrons, (b) back-scattered electrons. (Supplied by Dr. H. Hashimoto, following work of Stern *et al.*, 1972.)

Various other signals can of course be collected in the STEM to augment the information carried by the transmitted electrons. Figure 7, following work of Stern *et al.* (1972), shows dislocations in a thin molybdenite flake observed in scanning transmission (a) and back-scattered electrons (b). In principle, images from characteristic X-rays and Auger electrons may also be formed although in these cases there may be problems in obtaining an adequate count rate and it may, at least for thin specimens, be preferable to use the transmitted electron energy loss spectrum as a means of assessing local composition. In crystals the initial theoretical problem is to calculate the probability for the generation of these various signals by Bloch waves and a more detailed discussion of this is given in section IV. For thin crystals this is really the only aspect of the problem except in the case of the secondary electron signal when the difficulty of escape from even a thin specimen may need consideration.

IV. Wave Theory for the Back-scattered Electron signal

A. origin and principles of the theory

Using a low magnification scan on a single crystal, Coates (1967) observed patterns which bore a striking resemblance to the Kikuchi patterns familiar in transmission diffraction work and which moved as the crystal was rotated. The explanation of the origin of these patterns in terms of the earlier back-scattering observations of Duncumb (1962) shown in Fig. 5 was given by Booker *et al.* (1967). The back-scattered signal varies as a function of the angle of incidence which changes during the scan. These experiments opened up the possibility of practical crystallography on bulk specimens in the SEM which is discussed by Joy in this volume (Chapter 6). The role of channelling patterns in the SEM is identical to that of diffraction patterns in the CTEM. An excellent account of the early theoretical development of this whole subject, including the idea that the same contrast mechanism might be used to resolve dislocations in bulk specimens because of the local tilt of the lattice, has been given by Booker (1970).

It is interesting to point out that, quite independently, the same phenomenon had been observed and studied in low energy electron diffraction (LEED) (Laponsky and Whetton, 1959). Subsequent references to the LEED work are given by Taub *et al.* (1969) whose interpretation of the effect depends on anomalous transmission and the idea that secondaries are produced by a given Bloch wave within a distance of the surface determined by $\Lambda^{(j)}$ (see Table I). They suggest that for well transmitted Bloch waves, $\Lambda^{(2)}$ may exceed the escape distance of the secondaries so that the secondary emission is reduced. There is as yet, in fact, no very convincing independent evidence for strong anomalous transmission effects at LEED energies and this model is quite different from the one now accepted at higher energies where the primary effect is assumed to be the different probabilities $P^{(j)}$ for generation of secondaries by different Bloch waves. On this latter model, described below, an effect would still occur even if the penetration distances $\Lambda^{(j)}$ for the various Bloch waves were all equal. Various formulations of the theory differing from each other in detail and sophistication have been given by Hirsch and Humphreys (1970a,b) Vicario *et al.* (1970), Reimer *et al.* (1971), Clarke and Howie (1971a,b) and Spencer *et al.* (1972). In the last two cases the authors extended the theory to the case of imperfect crystals dealt with in IV.C below. The basis of the orientation-dependent back-scattering effect lies in the different back-scattering probabilities of the various Bloch waves mentioned previously. Since these waves decay rapidly with depth z the effect is chiefly confined to thin crystals or to the region near the entrance surface of thick crystals. We therefore first discuss the thin crystal case for both perfect and imperfect crystals in IV.B and C below. In the thick crystal case, dealt with in IV.D, separate problems occur, firstly concerning the treatment of the multiple scattering and energy loss processes familiar in this situation and secondly the question of the probe characteristics required if dislocations and other defects near the surface of bulk specimens are to be detected.

B. THEORY FOR BACK-SCATTERING FROM THIN PERFECT CRYSTALS

The probability per unit depth of generation of back-scattered primaries by a given Bloch wave described by eqn (5) is given by

$$P^{(j)} = \sum_{g,h} C_g^{(j)*} P_h C_{g-h}^{(j)} \tag{13}$$

where P_h is the Fourier component associated with the reciprocal lattice vector $\mathbf{h}$ of a periodic function $P(\mathbf{r})$ describing the back-scattering probability in the crystal. Spencer *et al.* (1972) calculated the back-scattering into the backward hemisphere using Born approximation, i.e. a Rutherford back-scattering cross section σ_R and obtained the result

$$P_h = \frac{\pi}{V_c}\left(\frac{e^2 m Z \lambda^2}{h^2}\right)^2 \exp(-M_h) \sum_j \exp(-2\pi i \mathbf{h} \cdot \mathbf{r}_j) \tag{14}$$

where the sum over j runs over the atoms at positions $\mathbf{r}_j$ in the unit cell of volume V_c and the Debye Waller factor $\exp(-M_h)$ shows the effect of lattice vibrations in smearing out $P(\mathbf{r})$ at high temperatures. The corresponding Rutherford cross section σ_R for scattering through angles $\theta \gtrsim \pi/2$ is shown in Table II together with the more accurate Mott cross section σ_M used by Reimer *et al.* (1971). The latter still varies roughly like $(Z/E_0)^2$ as suggested by eqn (14).

Using eqn (8) for the Bloch wave amplitudes and (13) for back-scattering probabilities $P^{(j)}$ we can calculate the back-scattering probability from a thin crystal of thickness t using the expression

$$I_B(t) = \int_0^t \sum_j |\psi^{(j)}(z)|^2 P^{(j)}\,dz + \int_0^t \{1 - \sum_j |\psi^{(j)}(z)|^2\} P_0\,dz \tag{15}$$

TABLE II. Estimated plane wave penetration distances Λ, Rutherford and Mott back-scattering cross sections σ_R and σ_M, X-ray production cross sections[a] σ_X and back-scattering probabilities[b] $\bar{I}_B(\Lambda) = P_0\Lambda$ at various electron energies in Al, Cu and Au

	E_0 (keV)	Λ (Å)	σ_R 10^{-20} cm^2	$\bar{I}_B(\Lambda)$ $\times 10^3$	σ_M/σ_R	σ_X/σ_R
Al	10	220	2·86	3·81	1·07	0·0016
$Z = 13$	20	420	0·74	1·88	1·00	0·0042
$\bar{I}_B(\infty)$	50	870	0·13	0·69	0·86	0·0115
$= 0{\cdot}16$	100	1700	0·04	0·40	0·70	0·0200
Cu	10	67	14·20	8·04	1·30	—
$Z = 29$	20	130	3·68	4·03	1·17	0·0190
$\bar{I}_B(\infty)$	50	300	0·65	1·64	0·99	0·0216
$= 0{\cdot}3$	100	520	0·20	0·88	0·80	0·0175
Au	10	35	105·7	21·9	1·33	—
$Z = 89$	20	50	27·4	8·12	1·56	0·0234
$\bar{I}_B(\infty)$	50	90	4·90	2·61	1·68	0·0217
$= 0{\cdot}5$	100	150	1·45	1·29	1·54	0·0170

[a] The figures refer to the K shell in Al and Cu and to the L_{III} shell in Au.

[b] The value of P_0 used corresponds to σ_R, improved values using σ_M can be obtained by multiplying by σ_M/σ_R and the corresponding X-ray production probability can be obtained from the relation $\bar{I}_X(\Lambda) = \bar{I}_B(\Lambda)\sigma_X/\sigma_R$.

in which the first term comes from Bloch waves which decay exponentially with depth into states which are assumed to behave like plane waves for back-scattering purposes and contribute the second term. This equation is limited to cases where multiple scattering and energy loss can be ignored, i.e. to crystals of thickness less than a few times Λ where Λ is the average attenuation distance for a plane wave. Values of Λ and $\bar{I}_B(\Lambda) = P_0\Lambda$ are given in Table II and show that the back-scattering in such cases is only of the order of 1%. The fractional variations with incidence angle θ due to Bloch wave effects as given by eqn (15) are very large, however (see Fig. 4(c) above). Typically an anomaly occurs over an angular region of about $\pm 1/g\xi_g$ at each Bragg position with an increase of about $\Lambda^{(1)}(P^{(1)} - P_0)$ above background for $\theta < \theta_B$ when the strongly back-scattered and strongly absorbed Bloch wave is excited and a larger dip of about $\Lambda^{(2)}(P^{(2)} - P_0)$ below background for $\theta > \theta_B$ when Bloch wave (2) is excited. These results follow from two beam theory which may not hold particularly for low-order reflections. With higher order reflections, the anomaly becomes sharper in angle, the peak and dip become more equal (because $\Lambda^{(1)}$ and $\Lambda^{(2)}$ are more similar) and the effect becomes smaller because P_g decreases. Many of the observed properties of Coates channelling patterns can be explained qualitatively by this simple model as explained by Hirsch and Humphreys (1970a,b) including the non-occurrence of lines corresponding to some Bragg reflections and the nature of the asymmetry in the lines at $\pm\theta_B$ when the Bragg planes are steeply inclined to the crystal surface. In practice, of course, most of the observations refer to thick crystals to which the extension of eqn (15) is discussed in IV.D below.

Equation (15) with constants $P^{(j)}$ derived from an appropriate function $P(\mathbf{r})$ can also be used to describe X-ray production and was in fact first introduced for this case (Hirsch *et al.*, 1962; Hall, 1966). Again a large fractional anomaly occurs but the intensity level is considerably smaller than in the back-scattering case (see Table II). The equation could probably also be used to describe the intensity of back-scattered secondaries which may be generated either by the back-scattered primaries or by the incident beam via inelastic scattering processes close to the atomic core (Hall, 1970a).

Two assumptions in eqn (15), both previously used in the anomalous absorption theory, require further investigation. It is assumed that the local back-scattering depends on the independent Bloch wave intensities rather than on the total wave intensity $|\psi(\mathbf{r})|^2$. The latter model (Hirsch *et al.*, 1962; Reimer *et al.*, 1971) leads to some additional interference terms between different Bloch waves in the expression for $I_B(t)$, and has recently been shown to give a better account of X-ray production effects in very thin crystals (Cherns *et al.*, 1973). The difference between the two models appears to be very small, however, for crystals of thickness $t \gtrsim 100$ Å. The second assumption is the plane wave behaviour of the scattered Bloch waves in producing further back-scattering. The theory of Kikuchi band formation (Hall, 1970b) shows, however, that the Bloch wave nature (i.e. strong or weak absorption and back-scattering) is to some extent preserved after diffuse scattering, particularly for the heavily back-scattered wave (1). Some account could perhaps be taken of this by using in the back-scattering theory an attenuation distance $\Lambda^{(1)}$ for this wave somewhat longer than transmission experiments suggest.

C. THEORY FOR CRYSTAL DEFECT IMAGES IN THIN CRYSTALS

The theory described in the last section can be very readily extended to the calculation of defect images observed by back-scattering from thin crystals (Clarke and Howie, 1971a,b; Spencer *et al.*, 1972). Equation (15) may still be used to obtain the back-scattered signal from each point x, y on the crystal entrance surface covered by the electron probe. The equation is integrated down a column in the z

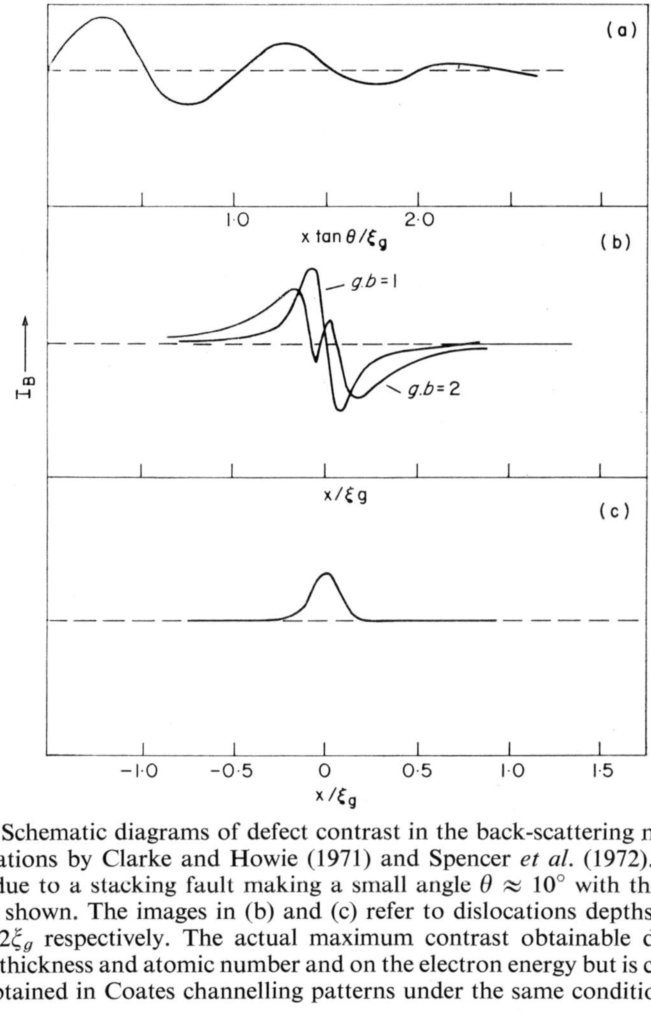

FIG. 8. Schematic diagrams of defect contrast in the back-scattering mode based on calculations by Clarke and Howie (1971) and Spencer *et al.* (1972). In (a) the contrast due to a stacking fault making a small angle $\theta \approx 10°$ with the specimen surface is shown. The images in (b) and (c) refer to dislocations depths $z \approx 0{\cdot}2\xi_g$ and $z \approx 2\xi_g$ respectively. The actual maximum contrast obtainable depends on specimen thickness and atomic number and on the electron energy but is comparable to that obtained in Coates channelling patterns under the same conditions.

direction but the Bloch wave amplitudes $\psi^{(j)}(z)$ are now obtained from the coupled eqn (12) using the known displacement field $\mathbf{R}$ of the defect. This involves using the Bloch waves defined by eqn (11) which displace locally with the lattice and, if it is assumed that the back-scattering probability $P(\mathbf{r})$ also follows the local lattice displacement, the $P^{(j)}$

quantities given by eqn (13) are the same as in the perfect crystal. The back-scattering calculations can thus be carried out by methods already familiar for the transmission case using many-beam theory if necessary. It can also be seen that the various transmission microscopy criteria for the invisibility of dislocations and stacking faults in terms of **g** . **b** and **g** . **R** will still hold for the back-scattered images.

Figure 8(a) shows that the profile of a stacking fault inclined at an angle to the crystal entrance surface is characterized by fringes running parallel to the intersection of the fault with the entrance surface. As pointed out by Spencer *et al.* (1972), the sign of **g** . **R** at the fault may be inferred from the bright or dark nature of the first fringe. Dislocations near the entrance surface display a bright-dark type of image shown in Fig. 8(b) for **g** . **b** = 1 and 2. At greater depth where only the well transmitted, weakly back-scattering Bloch wave (2) can penetrate without absorption, the presence of a dislocation can only enhance the back-scattering so that a single bright peak results (see Fig. 8(c)).

Very weak images of dislocations in the back-scattered mode were first observed in thin crystals by Clarke (1971) but considerably better images have more recently been obtained by Stern *et al.* (1972) (see also Fig. 7). So far the extent to which the observed images are in agreement with theoretical prediction has yet to be checked. Detailed studies of dislocations in thin crystals using transmission, back-scattered primaries and back-scattered secondaries will undoubtedly provide valuable information for improving the whole theory in the near future, however.

D. BACK-SCATTERING THEORY FOR PERFECT AND IMPERFECT THICK CRYSTALS

The thin crystal theory just outlined has considerable intrinsic value—firstly because it can be tested on currently available instruments and secondly because it may well be relevant for high resolution X-ray or Auger electron microanalysis work in the STEM. Its main interest, however, is as a first step to consideration of the back-scattering problem in bulk crystals where the ability to study dislocations in the back-scattering mode would be of considerable significance. Fortunately, the adoption of a semi-empirical approach allows us to avoid all the complications of the multiple scattering and energy loss problem in the depths of the crystal and extend eqn (15) to the thick crystal case fairly easily. We simply make use of the experimental data for $\bar{I}_B(T)$, the intensity of all back-scattered electrons coming from a solid of thickness T when crystalline effects are ignored. For small values of T, $\bar{I}_B = P_0 T$ but eventually $\bar{I}_B$ flattens off to a constant value characteristic of the bulk which is rather insensitive to the incident electron energy but increases with Z. Some values of $\bar{I}_B(\infty)$ for the case of normal incidence are given in Table II. $\bar{I}_B$ also varies slowly with the incident angle θ and in cases of oblique incidence the appropriate value of $\bar{I}_B$ should be used in the argument which follows.

We divide the sample into the thin entrance region of thickness t and the remaining underlying regions. To a good approximation, all of the back-scattering from the underlying regions penetrates the thin entrance layer so that the total back-scattering $\bar{I}_B$ in the non-crystalline case consists of a part $P_0 t$ from the upper layer and a part $\bar{I}_B - P_0 t$ from the underlying regions generated by an intensity $1 - P_0 t$ transmitted from the thin upper region. In the case where crystalline effects occur, the back-scattering from this upper region is $I_B(t)$ (eqn (15)) and an intensity $1 - I_B(t)$ is transmitted so that the back-scattering from the underlying regions is $(\bar{I}_B - P_0 t)\,(1 - I_B(t))/(1 - P_0 t)$. Adding this to $I_B(t)$ from the upper region and neglecting second order terms, we obtain for the back-scattering in the thick crystal case the expression

$$I_B(T) = \bar{I}_B(T) + (1 - \bar{I}_B(T)) \sum_j \int_0^T |\psi^{(j)}(z)|^2 (P^{(j)} - P_0)\,\mathrm{d}z. \tag{16}$$

This expression can be used over the whole range of thickness, incident angle and energy. For thin crystals at near normal incidence it reduces, apart from a negligible second order term, to eqn (15). For very thick crystals the upper limit in the integral may be taken as infinity if required since all the Bloch wave intensities decay exponentially with z in characteristic distances $\Lambda^{(j)}$. It can be seen that the signal to background ratio for both channelling patterns and defect images contains the factor $(1 - \bar{I}_B(T))/\bar{I}_B(T)$ and hence, as an alternative to using $\bar{I}_B$, one can discuss the contrast of defect images or lattice planes relative to that of channelling patterns since both decrease in the same way with increasing crystal thickness (Howie *et al.*, 1971; Clarke and Howie, 1971a,b).

A more detailed theory of multiple scattering has been given by Hirsch and Humphreys (1970a,b) and by Spencer *et al.* (1972) in an effort to avoid the use of empirical data. So far, however, their theory ignores the effects of energy loss and in the end they have to use data for the electron range which then yields an equation equivalent to (16). The multiple scattering does, however, lead to a further contrast mechanism not included in eqn (16) since, as a result of medium-angle scattering ($\theta < \pi/2$) of the Bloch waves in the entrance regions of the crystal, the subsequent back-scattering in deeper layers will be altered. This effect, considered semi-empirically by Reimer *et al.* (1971) and in more detail by Booker *et al.* (1973), appears to lead to an increase of the signal $\Delta I_B = I_B(T) - \bar{I}_B(T)$ by a factor of about 2·5 in bulk crystals. Equation (16) could probably be revised to include this.

Taking account of these effects we can use the data of Table II to obtain some rough working estimates of the available contrast $C \approx 2{\cdot}5\, \bar{I}_B(\Lambda)(1 - \bar{I}_B(\infty))/\bar{I}_B(\infty)$ in bulk crystals. Typical values are only a few per cent so that a rather high incident probe current J must be used if the image is to be detected against the statistical noise with a reasonable number of picture points N and frame scanning time t_s. Using the values quoted by Booker (1970) we find

$$J = 10^{-17}\, N/C^2 F t_s,\ \mathrm{A} \tag{17}$$

where F is the fraction of electrons in the incident probe actually entering the back-scattering detector and might be in the order of 0·1. With $N = 10^6$ points per frame and $t_s = 10$ sec (both fairly ambitious figures) and $C = 5\%$ a probe current of 4×10^{-9} A would be required, for instance. Reference to Fig. 8 shows that if the full image contrast is to be obtained this current must be delivered into a probe of diameter d_p and illumination semi-angle α specified by

$$d_p \lesssim \xi_g/5 \tag{18}$$

$$\alpha \lesssim 1/g\xi_g. \tag{19}$$

As discussed by Booker (1970), Clarke and Howie (1971b) and Spencer *et al.* (1972), these conditions imply a certain minimum gun brightness B in view of the relation

$$B = 0{\cdot}4\, J/d_p^2\alpha^2. \tag{20}$$

Typical values for 20 keV operation might be $d_p = 100$ Å, $\alpha = 5 \times 10^{-3}$ rad leading to a value of $B = 5 \times 10^7$ A cm^{-2} sr^{-1} necessitating the use of a field emission gun. These rather stringent requirements could of course be somewhat relaxed if longer scanning times or fewer picture points were used. The use of energy selective detectors (Wolf and Everhart, 1969) might enable the contrast C to be raised by a factor of 5 with a corresponding decrease in the factor F in eqn (17). B could then be reduced by a factor of 5.

If larger values of α are used than those specified by eqn (17) the contrast C would be reduced by a factor of roughly $1/\alpha g\xi_g$ since only the angular component of rotation out of the Bragg planes is relevant. Equations (17) and (20) then show that the increased current available is just that required to detect this reduced contrast level. A similar argument applies, at least for the dislocation images of the type shown in Fig. 8(c), if the spot size is increased beyond the value specified in eqn (18) since once again the reduction in the contrast by a factor of about $\xi_g/5d_p$ is compensated by the increased current available. This latter course would lead to undesirable loss of resolution, however, and images of the type shown in Fig. 8(b) might become undetectable. The reduction in contrast following such increases of α or of d_p might also reduce the visibility of diffraction contrast images relative to the topographical contrast.

In using the same value of d_p in eqns (18) and (20) we have assumed that aberration effects can be ignored. As indicated by Clarke and Howie (1971b), however, spherical aberration and also the diffraction limit (eqn (1)) may become significant at energies E_0 lower than, say, 10 keV. The use of higher voltages gives an increase in brightness proportional to v^2 and also allows dislocations to be detected at greater depths below the surface. However, as pointed by Spencer *et al.* (1972), the decrease in contrast at higher voltages means that the current required (eqn (17)) increases very considerably.

The theory developed here can also be applied to the X-ray production problem in thick crystals. In this case, however, Spencer (1972) has shown that the situation is somewhat more complicated since the X-ray production in the deeper parts of the crystal can be influenced by the changes in the electron back-scattering in the surface layers. Bragg reflection anomalies of about 1% are predicted but these may actually reverse in sign relative to the thin crystal effect because of the interaction with back-scattering.

V. Glancing Angle Scanning Microscopy

The theory of the preceding section and the experimental information now available indicate that in the back-scattering mode at near-normal incidence structural information can be obtained from the region within a few hundred Å of the surface of bulk crystals. For many problems of surface physics and chemistry, however, it could be desirable to obtain the information from a region much closer to the surface and the use of more oblique incidence conditions is a possible way to achieve this (Wells, 1971). Analogous techniques for surface studies were employed many years ago in the conventional electron microscope operating in a reflection mode (Halliday, 1965) but proved to be of limited value mainly because of the restricted field of view in focus at any one time. This difficulty could in principle be overcome in the SEM by the use of dynamic focusing during the scan.

From the diffraction point of view the obvious step is to make use of Bragg reflection from planes parallel or nearly parallel to the crystal surface (see Fig. 9). Because of refraction effects, Bragg reflection occurs at a grazing angle θ_i less than θ_B (the Bragg angle), and the lowest order Bragg reflections are frequently inaccessible. With reflections like 333 or 511, however, and accelerating potentials $E_0 \approx 10$ keV, incident angles of $\theta_i \approx 5\times10^{-2}$ rad can be used. The angular range of Bragg reflection is still correctly given by eqn (4) and for those reflections $\Delta\theta_B \approx 10^{-3}$ rad. Typical experimental values of the Bragg reflected intensity are 5×10^{-2} (Colella and Menadue, 1972).

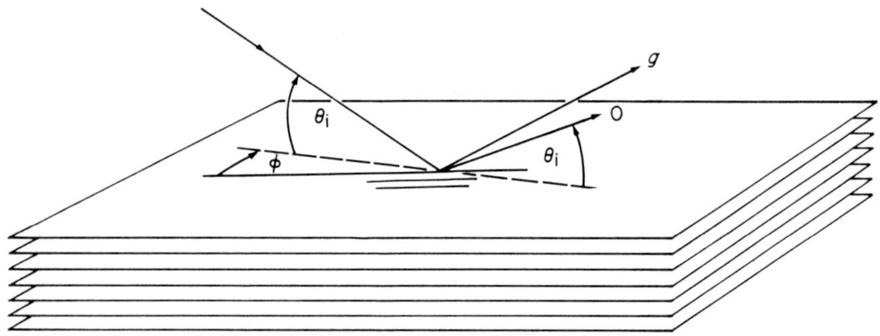

FIG. 9. Schematic ray diagram for the glancing angle mode. The incidence and emergence angles θ_i are exaggerated for clarity and are the same for all the beams. Diffraction from planes normal to the surface gives rise to a Bragg reflected beam **g** whose intensity varies with the azimuthal angle ϕ. Bright field or dark field scanning images could be formed using the **O** and **g** beams.

In the glancing angle Bragg condition the penetration of the elastically scattered waves is usually not more than 3 or 4 atomic layers below the surface. Even without the scanning facility, therefore, the use of electron beams of energy from a few keV upwards at oblique or even grazing angles of incidence is of value in surface analysis either by X-ray studies (Sewell *et al.*, 1969) or by the method of Auger spectroscopy. In principle it should be possible to exploit these non-scanning techniques further since, if the incidence angle θ_i is kept constant and the aximuthal angle ϕ varied, Bragg reflections from planes cutting the crystal surface at steep angles can also be excited (see Fig. 9). The consequent formation of Bloch wave intensity distributions such as those already discussed for transmission diffraction (Fig. 3) could then lead to anomalies in X-ray production and Auger emission as a function of ϕ which might enable the crystallographic position of various atomic species to be located. Some interesting variations in Auger emission which may have been due to this effect have indeed been observed in preliminary experiments carried out by J. M. Wilson and R. M. Stern (personal communication).

Diffraction studies of surfaces, whether by LEED or by the use of faster electrons at glancing angles, mainly provide information about the periodicities in the surface and provide very little easily interpretable information about surface imperfections or irregularities for which a real space picture is essential. A. J. Craven (private communication) has examined the possibility of scanning at glancing angles with a 10 keV beam. Using a field emission gun system, a 200 Å spot can be formed at a working distance of ~ 8 cm with a beam convergence less than the Bragg peak width $\Delta\theta_B$. As a result of the glancing angle used the resolution would thus be 200 Å in one direction but only 4000 Å in the other direction. Scanning over an area of 10 μm $\times$ 100 μm would not change the incidence angle by more than $\Delta\theta_B$ but for a larger area of scan a double deflection system might be necessary to keep θ_i constant. A real space image with all the bright and dark field facilities and related diffraction pattern would be extremely valuable even with this rather poor resolution. The diffraction theory presented in earlier sections can be modified for the glancing angle situation in the perfect crystal case (Colella, 1972) but the case of imperfect crystals has yet to be examined. The difficulties arising from the glancing angle of incidence could be overcome if the electron optical problem of forming a small probe with low energy electrons could be solved. The problem of developing an adequate theory to interpret scanning data from imperfect crystals in the LEED range however would clearly be a more formidable challenge than any of the other developments of wave theory recounted here.

References

AMELINCKX, S., GEVERS, R., REMAUT, G. and VAN LANDUYT, J. (eds) (1970). "Modern Diffraction and Imaging Techniques in Material Science." North-Holland, Amsterdam.

BADDE, H. G. and REIMER, L. (1972). *Proc. EMCON Vth European EM Congress*, 440–441, Institute of Physics, London.

BOOKER, G. R. (1970). *In* "Modern Diffraction and Imaging Techniques in Material Science." (S. Amelinckx, R. Gevers, G. Remaut and J. Van Landuyt, eds), pp. 553–653. North-Holland, Amsterdam.

BOOKER, G. R., SHAW, A. M. B., WHELAN, M. J. and HIRSCH, P. B. (1967). *Phil. Mag.* **16,** 1185–1191.

BOOKER, G. R., JOY, D. C., SPENCER, J. P. and HUMPHREYS, C. J. (1973). *Proc. VIth Scanning Microsc. Symp.* (*Chicago*), pp. 251–258.

CHERNS, D., HOWIE, A. and JACOBS, M. H. (1973). *Z. Naturf.* **28a,** 565–571.

CLARKE, D. R. (1971). *Phil. Mag.* **24,** 973–979.

CLARKE, D. R. and HOWIE, A. (1971a). *Proc. EMAG EM Conf.* 302–305, Institute of Physics, London.

CLARKE, D. R. and HOWIE, A. (1971b). *Phil. Mag.* **24,** 959–971.

COATES, D. G. (1967). *Phil. Mag.* **16,** 1179–1184.

COCKAYNE, D. J. H., RAY, I. L. F. and WHELAN, M. J. (1969). *Phil. Mag.* **20,** 1265–1270.

COLELLA, R. (1972). *Acta crystallogr.* **A28,** 11–15.

COLELLA, R. and MENADUE, J. F. (1972). *Acta crystallogr.* **A28,** 16–21.

COWLEY, J. M. (1969). *Appl. Phys. Letters* **15,** 58–59.

CREWE, A. V. and WALL, J. (1970). *Optik*, **30,** 461–474.

CREWE, A. V. and WALL, J. (1971). *Proc. EMSA*, 24–25.

CUNDY, S. L., HOWIE, A. and VALDRÈ, U. (1969). *Phil. Mag.* **20,** 147–163.

DUNCUMB, P. (1962). *Phil. Mag.* **7,** 2101–2105.

FUJIMOTO, F., KOMAKI, K., TAKAGI, S. and KOIKE, H. (1972). *Z. Naturf.* **27a,** 441–444.

HALL, C. R. (1966). *Proc. R. Soc.* **A295,** 140–163.

HALL, C. R. (1970a). *Phil. Mag.* **21,** 1075–1080.

HALL, C. R. (1970b). *Phil. Mag.* **22,** 63–72.

HALLIDAY, J. S. (1965). *In* "Techniques for Electron Microscopy" (D. Kay, ed.), pp. 525–545. Blackwells, Oxford.

HEIDENREICH, R. D. (1964). "Fundamentals of Transmission Electron Microscopy." Interscience, New York.

HIRSCH, P. B., HOWIE, A. and WHELAN, M. J. (1962). *Phil. Mag.* **7,** 2095–2100.

HIRSCH, P. B., HOWIE, A., NICHOLSON, R. B., PASHLEY, D. W. and WHELAN, M. J. (1965). "Electron Microscopy of Thin Crystals". Butterworths, London.

HIRSCH, P. B. and HUMPHREYS, C. J. (1970a). *Proc. IIIrd Scanning Mic. Symp., Chicago*, 449–455.

HIRSCH, P. B. and HUMPHREYS, C. J. (1970b). *Proc. VIIth Int. E. M. Congress* **1,** 229–230.

HOWIE, A. (1963). *Proc. R. Soc.* A **271,** 268–287.

HOWIE, A. (1972). *Proc. Vth European E.M. Congress*, 408–413, Institute of Physics, London.

HOWIE, A., LINNINGTON, P. F. and TOMLINSON, P. N. (1971). *Phil. Mag.* **23,** 1559–1565.

HOWIE, A. and STERN, R. M. (1972). *Z. Naturf.* **27a,** 382–389.

HUMPHREYS, C. J. and HIRSCH, P. B. (1968). *Phil. Mag.* **18,** 115–122.

HUMPHREYS, C. J. and WHELAN, M. J. (1969). *Phil. Mag.* **20,** 165–172.

LAPONSKY, A. B. and WHETTON, N. R. (1959). *Phys. Rev. Letters* **3,** 510–512.

METHERELL, A. J. F. (1971). *In* "Advances in Optical and Electron Microscopy" (R. Barer and V. E. Cosslett, eds). Vol. 4, pp. 263–360.

POGANY, A. P. and TURNER, P. S. (1968). *Acta crystallogr.* A **24,** 103–109.

RADI, G. (1970). *Acta crystallogr.* A **26,** 41–56.

REIMER, L., BADDE, H. G., SEIDEL, H. and BÜHRING, W. (1971). *Z. angew. Phys.* **31,** 145–151.

SEWELL, P. B., MITCHELL, D. F. and COHEN, M. (1969). *In* "Developments in Applied Spectroscopy" **7A** (Groves and Perkus, eds), pp. 61–79. Plenum, New York.

SHIBATOMI, K., YAMANOTO, T. and KOIKE, H. (1971). *Proc. EMSA*, 26–27.

SPALDING, D. R. (1969). Ph.D. thesis, University of Cambridge.

SPENCER, J. P. (1972). *Proc. Vth European E.M. Congr.* 478–479, Institute of Physics, London.

SPENCER, J. P., HUMPHREYS, C, J. and HIRSCH, P. B. (1972). *Phil. Mag.* **26,** 193–213.

STERN, R. M., ICHINOKAWA, T., TAKASHIMA, S., HASHIMOTO, H. and KIMOTO, S. (1972). *Phil. Mag.*, 1495–1499.

TAUB, H., STERN, R. M. and DVORYANKIN, V. F. (1969). *Phys. Stat. sol.* **33,** 573–577.

VALDRÈ, U. (ed.) (1971). "Electron Microscopy in Material Science." Academic Press, New York and London.

VICARIO, E., PITAVAL, M. and FONTAINE, G. (1970). *Proc. VIIth Int. E.M. Congr.*, **2,** 211–212.

WELLS, O. (1971). *Appl. Phys. Letters* **19,** 232–237.

WHELAN, M. J. (1965). *J. appl. Phys.* **36,** 2099–2110.

WOLF, E. D. and EVERHART, T. E. (1969). *Appl. Phys. Letters* **14,** 299–300.

ZEITLER, E. and THOMSON, M. G. R. (1970). *Optik*, **31,** 258–280; 359–366.

Chapter 8

Quantitative Conductive Mode Scanning Electron Microscopy

D. B. HOLT

Metallurgy Department, Imperial College of Science and Technology, London, England

Symbols

α = linear absorption coefficient
A = atomic weight in grams
A^1 = constant in equation (50)
B^1 = constant in equation (50)
β = factor appearing in the p–n junction characteristic and depending upon the type of recombination occurring
b = ratio of the electron mobility to the hole mobility
C^1 = constant in equation (47)
C = the velocity of light
D = minority carrier diffusion coefficient
d = radius of the beam energy dissipation volume
Δ = the ratio of the number of secondary emission electrons to the number of incident beam electrons
δ = the total fraction of the incident electrons that are lost by emission ($= \eta + \Delta$)
ε = dielectric constant
E = electric field
E_0 = the energy of an incident beam electron ($= qV_b$)
E_c = the diffusion field ($= kT/qL_p$)
E_{eff} = the effective energy of the incident electrons
e_i = the average energy required to generate a hole-electron pair (the "ionization energy")
η_{cc} = charge collection efficiency of a barrier, the fraction of generated pairs that are separated ($= I_m$ actual$/I_{\max}$)
η = the ratio of primary emitted (back-scattered) electrons to incident electrons

η_q = "electron voltaic quantum efficiency" $\left(= \frac{qV_b}{e_i}\left(1-\frac{V_0}{V_b}\right)\right)$

$F(X_0)$ = function appearing in the theory of the bulk e.v.e. and given by equation (59) or (60)

G = generation factor, the number of hole-electron pairs produced per incident electron ($= E_{\text{eff}}/e_i$)

g' = total generation rate ($= GI_b/q$)

g = gain in the barrier e.v.e. ($= I/I_b$)

$G(X_0)$ = function appearing in the theory of the bulk e.v.e. and given by equation (64)

$G_\delta(X_0)$ = function appearing in the theory of the bulk e.v.e. and given by Figure 41(b)

I = current delivered to the external (amplifier) load in the barrier e.v.e.

I_b = the current in the incident electron beam

I_e = total current of electrons emitted from the specimen

I_s = the specimen current i.e. the "absorbed electron current" ($I_b = I_s + I_e$)

I_j = the forward current through a p–n junction

I_0 = reverse saturation current of a p–n junction

$I_{\max}$ = the maximum possible value of the barrier e.v.e. current ($= GI_b$)

Im actual = the highest value of the barrier e.v.e. current obtained on scanning across the barrier ($= \eta_{cc}GI_b$)

I_{sc} = diode short circuit current

Ig = the current generated in the barrier by electron bombardment ($= I_{sc}$ when the beam is incident on the junction i.e. $x = 0$)

J = current density

k = Boltzmann's constant

$\overline{K}$ = ratio of the average energy of back-scattered electrons to the beam energy

L = minority carrier diffusion length

Lp = diffusion length of holes in n-type material

Le = effective diffusion length determining the volume in which the conductivity is increased by electron bombardment ($\simeq L_p/5$ for n-type Ge according to Munakata (1967a)

Ld = downstream diffusion length (defined for equation (81))

Lu = upstream diffusion length (defined for equation (81))

l = length of the specimen between contacts

Mo = rest mass of an electron

μ_p = mobility of holes

n = minority carrier density

n_i = intrinsic carrier density

N = number of hole-electron pairs collected per incident electron ($= \eta_{cc}G$)

N_A = acceptor atom concentration

Ni = impurity concentration in atoms cm^{-3}

P_p = photomultiplier response current due to cathodoluminescence generated in p-type GaP

Pn = photomultiplier current due to n-type GaP cathodoluminescence

P_{in} = beam input power density in μW cm^{-2}

$p(\xi)\,d\xi$ = the energy lost by the electron beam in penetrating from a depth ξ to $\xi + d\xi$

Δp = number of excess carriers generated in n-type material

po = equilibrium hole concentration in the absence of the beam

q = the charge on an electron

R = electron range (depth of penetration)

r = external load resistance
R_j = junction resistance
R_{sh} = junction shunt resistance
R_s = junction series resistance
R_0 = resistance of the rod or filament from one contact to the other of constant cross section and just including the region of bombardment enhanced conductivity (figures 38 and 43)
ρ = resistivity
s = surface recombination velocity
S = cross section of the specimen
S_0 = cross section of the rod containing the region of bombardment enhanced conductivity (Fig 38 and 43)
$\Delta\sigma$ = small change of conductivity due to carriers generated by electron bombardment
T = absolute temperature
τ = minority carrier lifetime
τ_p = hole lifetime
V = diode voltage
V_b = electron beam accelerating voltage
Vc = chemical contribution to the bulk e.v.e. voltage
V_d = Dember contribution to the bulk e.v.e. voltage
V_D = Diffusion voltage of a p–n junction
V_g = guard ring voltage
V_0 = "dead voltage" obtained from empirical curves of η_q versus V_b
V_{oc} = diode open circuit voltage (barrier e.v.e.)
V_r = the reverse bias voltage applied to a diode
V_s = voltage applied to the specimen
W = depletion region width
Z = atomic number of the material
Z_d = depth of the centre of the energy dissipation sphere (Fig. 16)

I. Introduction

A. TYPES OF CONDUCTIVE MODE SIGNAL

In the conductive mode of scanning electron microscopy the signals that are used are the currents or voltages that are induced in specimens by the electron beam bombardment.

Three physically distinguishable types of signal are employed and as a confusing variety of terms are used in the literature, some time will be spent in defining the three distinct types of signal. These are illustrated in Fig. 1. The specimen current in Thornton's (1968) terminology, is the leakage current from the beam to earth via the specimen. It is referred to by some workers as the "absorbed electron" current. This term is descriptive of the fact that by conservation of mass or charge we must have:

$$I_b = I_s + I_e \tag{1}$$

where I_b is the beam current, I_s is the specimen current and I_e is the total emissive current consisting of primary or back-scattered electrons, secondary electrons and intermediate-energy tertiary electrons including the Auger electrons. We will consider the significance of this relation for the specimen current type of conductive mode micrograph below.

The second term that is used in conductive mode SEM micrography is that of charge collection signals. These currents flow round circuits connecting two contacts to the specimen. The use of the term "charge collection current" as the general description for all the phenomena giving rise to signals in circuits like those illustrated in Figs 1(b) and (c)

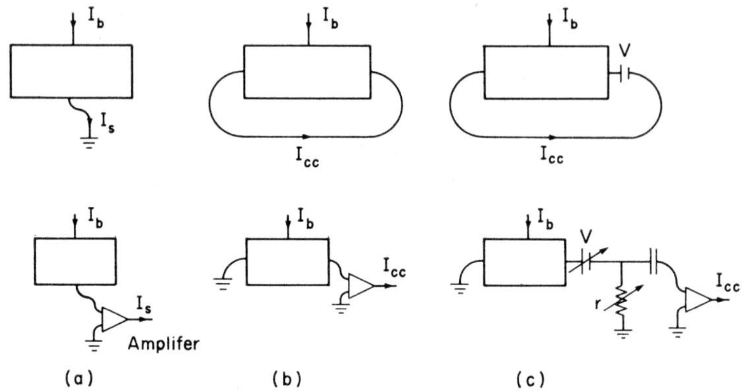

FIG. 1. The three types of conductive mode signal. In each case the upper diagram is schematic, and indicates only the essential physical nature of the signal, while the lower diagram indicates the type of circuit that is used to extract the signal. The amplified signal may then be displayed as a video or Y-modulation micrograph or recorded as a line scan trace. (a) The leakage current from the beam, I_b, to earth via the specimen is known as the specimen current, I_s (b) and (c): When a closed circuit is made between contacts to two separate points on the specimen surface, charge collection currents, I_{cc}, may be obtained. (b) In the absence of any external voltage source, charge collection currents will flow only if the specimen exhibits electron voltaic effects (barrier or bulk electron voltaic effects) to generate electromotive forces in the specimen and drive currents round the external circuit. (c) If external biasing is applied, charge collection current variations (contrast) may be obtained, even in the absence of electron voltaic effects, due to β-conductivity. (The term "electron beam induced conductivity", EBIC or BIC, has been indiscriminately applied to both barrier electron voltaic effect signals and β conductivity and could equally well logically apply to specimen currents. Its use, therefore, cannot be recommended).

follows the practice adopted in relation to semiconductor radiation counters (Dearnaley and Northrop, 1966) such as the Li-drifted Ge and Si detectors that are used in energy dispersive X-ray spectrometer attachments to scanning electron microscopes (see Chapter 12). The charge collection terminology is again that previously adopted by Thornton (1968). The phrase is descriptive of the fact that the currents arise when the negatively charged current carriers (electrons) move toward and collect near one contact while the positively charged carriers (holes) collect near the other contact.

B. THE INFLUENCE OF THE AMPLIFIER CIRCUIT

In the other modes of operation of the SEM the importance of the characteristics of the detector is obvious. The analogous factors in the case of the conductive mode are not so clearly recognized, however.

The observed signal depends not only on the physical effect induced in the specimen by electron bombardment but also on the electrical parameters of the circuit that is used to detect the signal. For example we shall see in the case of the barrier electron voltaic effect that, depending on the resistance of the amplifier circuit, the signal may appear as voltage with no current, the open circuit voltage; as a current with no voltage, the short circuit current; or as both a current and a voltage. The "detector" circuit for the conductive mode of scanning electron microscopy in general consist of the specimen, one or two contacts and the amplifier. It may include additional biasing batteries, resistors, isolating capacitors, etc., for example as indicated in Fig. 1(c). Moreover, the specimen in general will not be a simple ohmic resistance. It may have an appreciable capacitance as well, depending on its dimensions and dielectric constant. The contacts may not be ohmic. The time constant of the whole circuit is often not negligible. This means that the amplifier cannot follow rapid changes in the signal and because of this appropriate scan rates may have to be adopted to avoid signal distortion.

Care must therefore be taken to establish the equivalent circuit characteristics of the specimen with its contacts and of the detector circuitry. Only then will it be possible to calculate back from the observed signal (contrast) to obtain the physical properties of the specimen. Thus intelligent attention to the detector characteristics is as much a necessary precondition for quantitative interpretation of the conductive mode of scanning electron microscopy as it is for any other mode.

C. THE TYPES OF MATERIALS TO WHICH THE CONDUCTIVE MODE CAN USEFULLY BE APPLIED

The conductive mode of operation of the scanning electron microscope has particularly attracted people interested in semiconducting materials and devices because in this field the electrical properties which are directly observed are the properties that are of practical importance. A number of forms of contrast occur in semiconductors which yield valuable information related to the functioning of solid state electronic devices and these effects have tended to monopolize attention in this field.

Moreover, the quantitative work with which we are concerned here has naturally been concerned only with the most developed and best understood semiconducting materials, that is with silicon, germanium, GaAs and GaP.

It is obvious that the largest charge collection effects are likely to occur in those materials in which the densities of charge carriers introduced by electron bombardment are greater than or equal to the densities present in equilibrium in the absence of the electron beam. The mechanism by which the incident beam generates charge carriers in semiconductors is indicated in Fig. 2. Electron–electron collisions "ionize" the atoms of the crystal by raising electrons from a filled state in the valence band to an empty state in the conduction band. This leaves a hole in the valence band.

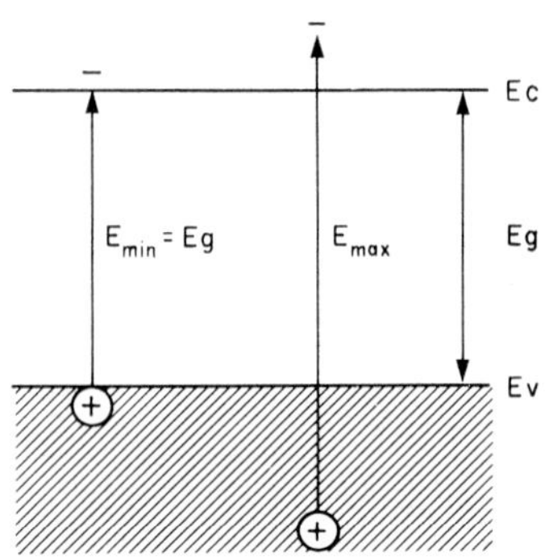

FIG. 2. Generation of hole-electron pairs by electron bombardment of semiconductors or insulators involves raising electrons by collision from a state in the filled valence band below E_v, to an empty state in the conduction band above the minimum energy E_c. This may be done with the absorption of any amount of energy from $E_{\min}$ which is equal to the forbidden energy gap $E_g = E_c - E_v$ up to some maximum value. The average energy required for hole-electron pair generation is denoted by e_i (see Table I).

The order of magnitude of the density of charge carriers introduced into a semiconductor by electron bombardment can be calculated as follows. The increase in the number of each kind of charge carrier per unit volume due to the bombardment, Δn, will be the number generated per second times the carrier lifetime, τ, divided by the volume in which the carriers occur, V, i.e. $\Delta n = GI_b\tau/qV$ where I_b is the beam current and q is the charge on an electron so that I_b/q is the number of electrons incident per second, G is the generation factor, i.e. the number of hole-electron pairs generated per incident electron which will be discussed in section III.A below. V can be calculated as follows. The volume in which the beam energy is dissipated in silicon is approximately a sphere tangent to the surface (Bresse, 1972). The radius of this sphere is half the depth of penetration, R, of electrons in silicon. In addition the minority carriers will spread out into a further volume determined by the carrier diffusion length L. Thus the volume containing the carriers generated has a radius $R/2+L$. For a 10 kV beam, $R = 1{\cdot}5\ \mu m$ in silicon (Bresse, 1972). Let $L = 1\ \mu m$ and τ be 2×10^{-3} sec. G for these conditions is $2{\cdot}8\times10^{3}$ hole-electron pairs per incident electron. Let $I_b = 10^{-10}$ A. Then Δn will be $1{\cdot}5\times10^{21}$ hole-electron pairs cm^{-3}. This is an upper limiting value and will be less for specimens with shorter minority carrier lifetimes τ: A similar calculation by Wittry and Kyser (1964) for GaAs with a lifetime of $\tau = 10^{-9}$ sec and $L = 3\ \mu m$ gave, for a beam of 30 kV and 1 μA, Δn, $= 3\times10^{16}\ cm^{-3}$. Lifetimes in GaAs can be as long as 10^{-3} sec and correspondingly higher hole-electron pair densities will result. These electron bombardment generated carrier densities can thus be equal to or greater than typical equilibrium majority carrier densities in semiconductors which range from say $10^{14}\ cm^{-3}$ for light doping to 10^{18}–$10^{19}\ cm^{-3}$ for degenerate doping. They are small, however, compared to the carrier densities in metals which are one or more electrons per atom, i.e. 10^{22}–$10^{23}\ cm^{-3}$. Moreover, while electron bombardment will raise valence electrons to higher states in a metal, it cannot produce holes as the outermost band is not filled in metals. Thus carriers of opposite sign cannot be generated and charge collection effects cannot occur in metals. In addition, local electrical effects will be screened by the free electron gas in metals. Thus the conductive mode of the SEM cannot be expected to produce so much useful information when applied to metals as it can when applied to semiconductors.

The term semiconductor is not used in a narrowly defined, exclusive sense in this context, however. Hole-electron pair generation and charge collection effects can occur in virtually any non-metallic solid whether relatively conducting or insulating and whether crystalline or glassy in structure. Thus, while up to the present time work in this field has been concentrated on a few well developed semiconductors, the conductive mode of the SEM is capable of much wider application.

D. CONTRAST MECHANISMS IN THE CONDUCTIVE MODE

The specimen current arises from the electrons absorbed in the specimen and contrast arises from the variations in the factors determining the number of electrons emitted, i.e. the electron emission coefficients.

As indicated in Fig. 1, a distinction must be made between charge collection currents occurring in the presence and in the absence of an external biasing voltage. When an external bias is applied, this provides the necessary e.m.f. to drive the current through the external detection circuit. Contrast can then arise through variations in the conductivity increment induced in the specimen by the electron bombardment. We will follow Munakata (1968a,b, 1972) and refer to this as β-conductivity. β in this context has the same meaning as in "β-rays", i.e. it refers to the electron irradiation that produces the conductivity increment.

The effects of electron bombardment on the electrical properties of semiconducting or insulating materials can be employed in electronic devices, especially opto-electronic devices such as TV camera tubes (vidicons etc.) and image storage devices. Research workers in this field tend to use the acronyms EBIC (electron beam induced conductivity), BIC (beam induced conductivity) or EBC (electron bombardment conductivity) indiscriminately for *all* the electrical effects of electron bombardment. (For a review of this work see Kazan and Knoll, 1968.) For this reason their use in the SEM field would tend to obscure the important distinctions between the different contrast mechanisms occurring in conductive mode microscopy. So, although EBIC would appear logically to refer to "β-conductivity", its use cannot be recommended.

In the absence of an external voltage, charge collection currents will only flow if an internal e.m.f. is produced in the specimen by electron bombardment. Such internal potential differences are produced by "electron voltaic effects". There are two distinct electron voltaic effects (e.v.e.s). These are the barrier e.v.e. and the bulk e.v.e. In both cases a built-in electric field in the interior of the specimen acts to separate the holes from the electrons generated by the incident beam. This results in a difference in the electrochemical potential (i.e. the free energy or Fermi level) of electrons at the two contacts. This energy difference provides the e.m.f. to drive the signal through the detecting amplifier. Barrier e.v.e.s arise from the built-in electric fields at electrical barriers like *p–n* junctions and Schottky barrier contacts. Bulk e.v.e.s arise from the built in fields due to non-uniform impurity distributions in the "bulk" of the semiconducting material. The two forms of contrast thus provide information on the position and properties of barriers and of impurity striations and similar inhomogeneities in the bulk.

II. Specimen Current Micrographs

Equation (1) shows that the specimen current is complementary to the total emissive current since the beam current is constant. Therefore, in cases in which the specimen current signal and the emissive mode signal are *proportional to* I_s *and* I_c *respectively*, contrast in the two modes is simply complementary. In general the specimen current signal is indeed simply equal to the specimen current times the amplifier gain. The variations in the specimen current signal therefore are equal in magnitude and opposite in sign to the variations in the emissive current. The number of back-scattered electrons varies with the atomic number of the material so "atomic number" compositional contrast occurs in I_e. The number of secondary electrons varies with the angle between the incident beam direction and the normal to the specimen surface so topographical contrast also is contained in I_e. In the emissive mode these forms of contrast can be observed separately by using detectors which respond either to primary or secondary emitted electrons. In specimen current micrographs both forms of contrast are observed. Moreover, beam spreading in the interior of the specimen makes resolution in the specimen current micrographs worse than in the secondary electron micrographs. The advantages of great depth of focus for studying surface topography and of the ready availability of atomic contrast in specimen current microscopy nevertheless make this technique useful for comparison with X-ray micrographs in electron probe microanalyser studies of complex alloy structures (Goldstein, 1969).

Often the emissive mode signals are *not* proportional to the emitted current, however. This is because a variety of forms of "collection contrast" occur in secondary electron micrographs. The video voltage displayed to form such micrographs is the result of amplifying the current arising from the fraction of the emitted electrons that produce a response in the detector. It is the variations in the fraction of the emitted electrons that are collected by the detector that are responsible for collection contrast. Examples of collection contrast include voltage contrast and magnetic contrast. Thus when collection contrast is present, the contrast in the emissive mode micrographs (especially the secondary electron micrographs) and the specimen current micrographs will not be complementary, despite eqn (1), because the ratio of secondary electron signal to emitted current varies from point to point.

Collection contrast is important only in semiconductor and magnetic materials. In electrically inert materials the two types of micrograph, specimen current and emissive mode, are complementary and the specimen current micrograph contains no new information. Under these circumstances, therefore, the secondary electron micrograph is superior because the resolution in the conductive mode micrographs is lower due to beam spreading in the solid.

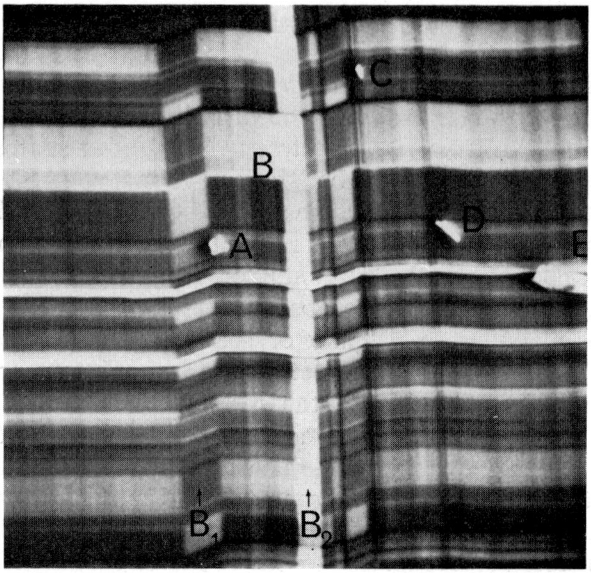

(a)

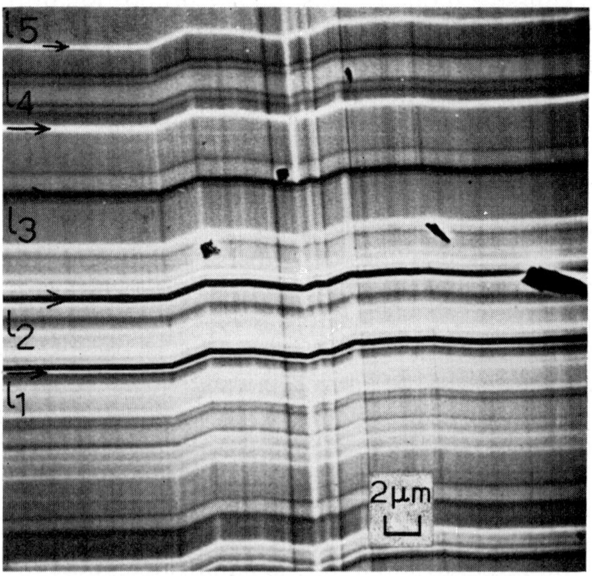

(b)

FIG. 3(a) and (b). [See caption opposite.]

A specimen of a semiconducting material in which the contrast in the specimen current and the emissive mode secondary electron micrographs was not complementary is shown in Fig. 3. Atomic number contrast is not encountered in relatively pure and homogeneous semiconducting materials. The non-complementarity of Figs 3(a) and 3(b) is therefore probably due to the occurrence of voltage contrast.

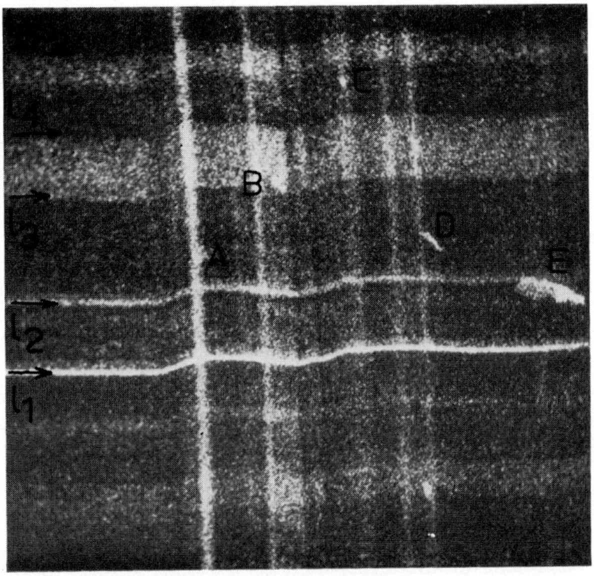

(c)

FIG. 3. Striations in a platelet of ZnS grown from the vapour phase (a) Secondary electron emissive mode micrograph (b) specimen current micrograph (c) cathodoluminescence micrograph. The contrast of many features in (a) and (b) is complementary, black in one micrograph corresponding to white on the other. Examples are the linear markings lettered l_1 to l_5. Contrast in the polytype bands B_1 and B_2 however is not complementary in (a) and (b). Contrast in band B_1 alternates from dark to light from top to bottom in (a) but is uniform in (b). Band B_2 is much whiter than adjacent bands in (a), but it has the same signal strength as the adjacent bands in (b). Dust particles marked A to E are of assistance in identifying corresponding points in the micrographs. (After Holt and Culpan, 1970.)

By adding $I_s = I_b - I_e$ to the secondary electron signal, the contrast due to variations in I_e could be cancelled in secondary electron micrographs, leaving only the collection contrast.

III. CHARGE COLLECTION MICROGRAPHS

We will first deal with the electron voltaic effects which are the only source of signals in the absence of external bias voltages (Fig. 1(b)), and then treat the β-conductive effects which can be obtained with an applied bias (Fig. 1(c)).

A. THE BARRIER ELECTRON VOLTAIC EFFECT

The barrier electron voltaic effect is the generation of voltages by electron bombardment of electrical barriers in semiconducting or insulating materials. The term "barrier electron voltaic effect" emphasizes the close analogy with the better known barrier photovoltaic effect employed in solar cells. The phrase barrier electron voltaic effect was introduced by Ehrenberg *et al.* (1951) and Rappaport (1954) who first observed the phenomenon. The effect has been studied in both

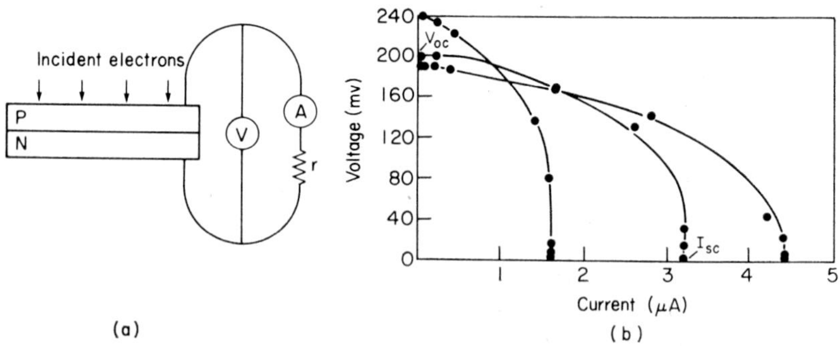

FIG. 4. The barrier electron voltaic effect in a semiconductor *p–n* junction diode. Electron bombardment in the experimental situation shown in (a) generates a potential difference between the contacts to the diode which will cause an external current to flow. (b) The diode current–voltage characteristics of a number of Si *p–n* junctions (after Rappaport, 1954). As the external load resistance r is reduced to zero, the current rises to the maximum value I_{sc}, the "short circuit current". For very large load resistances the voltage rises toward the maximum possible value V_{oc}, the "open circuit voltage".

p–n junctions and Schottky barriers. The essential experimental fact about the barrier e.v.e. is shown in Fig. 4(b). The effect can thus be made manifest as a maximum current, the short-circuit current I_{sc}, but no voltage, if the detector circuit has a negligible resistance. The effect will be detected as a maximum voltage, the open-circuit voltage V_{oc}, but no current, when the detector circuit input resistance is effectively infinite. For intermediate detector circuit resistances intermediate values of current and voltage will be observed. For this reason it is undesirable to refer to the signal produced as a voltage, a current or a beam-induced conductivity. It is best to employ the full phrase barrier electron voltaic effect signal or contrast.

The ratio of the current I delivered to the amplifier, to the incident electron beam current I_b which is called the gain g, generally increases with the reverse bias voltage applied to the diode. The form of this increase is shown in Fig. 5(b) for the case of a gold-silicon Schottky barrier (rectifying contact). The gain also depends upon the depth of the barrier below the entry surface for this plan-view geometry, in which the bombarding electrons are incident in a direction normal to the plane of the barrier, as shown in Fig. 5(c).

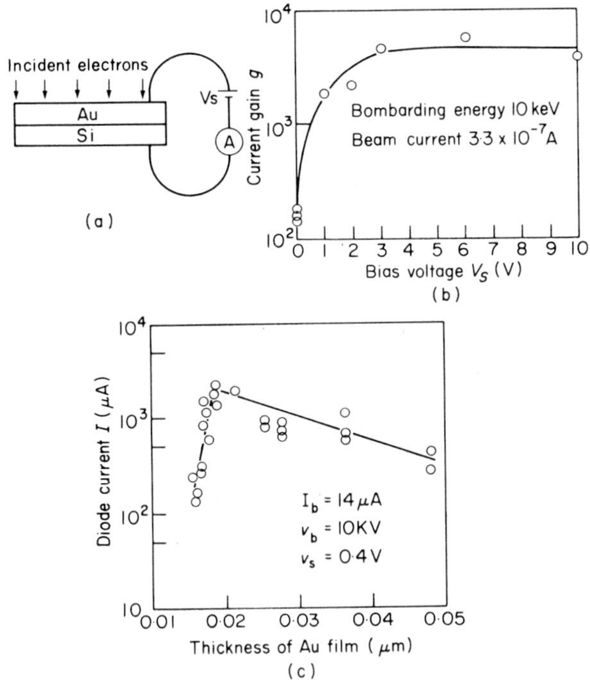

FIG. 5. The barrier electron voltaic effect in a rectifying Au–Si contact. (a) The experimental arrangement. (b) The variation of current gain with reverse bias voltage. (c) The dependence of diode current on the thickness of the gold layer. (After Miyaji and Miyazaki, 1963.)

In this section the essential physics of the barrier electron voltaic effect will next be outlined before proceeding to the analytical treatments in subsequent sections. The basic physical mechanism underlying the barrier electron voltaic effect is the separation, i.e. the "collection" by the electric field of the barrier of the pairs of oppositely charged electrons and holes generated by the electron beam bombardment. In more detail this occurs as follows.

TABLE I. Experimental Values for the Average Energy Required to Generate a Hole-Electron Pair

Material	e_i (eV)	Reference
Si	3·6	Fabri *et al.* (1963)
	3·6	Baldinger *et al.* (1964)
	4·2[a]	Rappaport *et al.* (1956)
Ge	2·84	McKenzie and Bromley (1959)
	4·4[a]	Rappaport *et al.* (1956)
GaAs	4·6[a]	Wittry and Kyser (1965)
	6·8[a]	Kalibjian and Mayeda (1971)
GaP	7·8	Goldstein (1965)
InSb	0·42	Ivakhno and Nasledov (1965)
CdS	7·3	Van Heerden (1957)
	7·8[a]	Munakata (1966a)
	9·3[a]	Lappe (1959)
CdTe	5·0	Vavilov *et al.* (1964)
PbS	2·0	Smith and Dutton (1958)

[a] Indicates a value obtained in experiments involving electron bombardment. As noticed by Lappe (1959) these values tend to be larger than those obtained in α particle or photon bombardment studies. According to Klein (quoted by Jones *et al.*, 1973) $e_i = E_g + (9/5)E_g + r(h\omega_p)$ where E_g the energy gap is the minimum potential energy required to create a hole-electron pair, the second term represents residual kinetic energy and the third term represents energy dissipated as a number r of phonons of energy $h\omega_p$ where h is Planck's constant and ω_p is the angular frequency of the phonons.

The electrons of the incident beam expend some of their energy in ionizing the atoms in the solid. In terms of the band theory of semi-conducting or insulating crystals this means that electrons are raised from the filled valence band to the empty conduction band as shown in Fig. 2. The number of hole-electron pairs produced per electron in the beam is called the generation factor G which may be written

$$G = \frac{E_{\text{eff}}}{e_i} \tag{2}$$

where E_{eff}, the effective energy of the incident electrons is less than E_0, the beam energy, because some of the energy of the beam is emitted and back-scattered with the emissive current I_e, and e_i is the average formation energy of hole-electron pairs. e_i is greater than E_g, the forbidden energy gap of the material, because some pairs are created with energy separations greater than E_g and because some of the hole-electron pairs are lost by recombination before they can be detected. e_i is generally several times greater than E_g. Experimental values of e_i for a number of materials are listed in Table I.

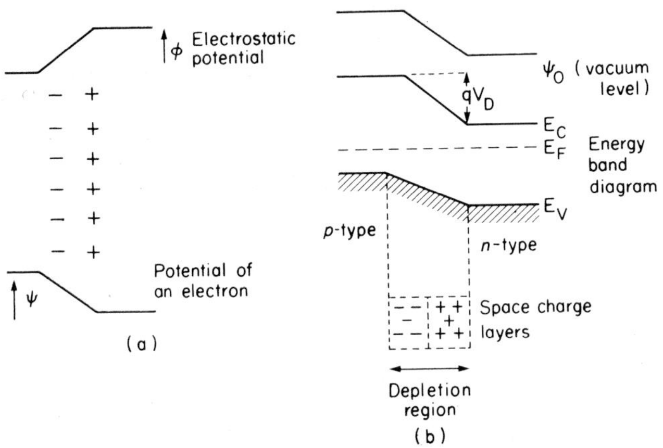

FIG. 6. Electrical barriers are esssentially dipole layers. (a) Dipole layer showing the variation of electrostatic potential ϕ and of the energy of negatively charged electrons ψ. (b) *p–n* junction showing the energy band diagram above and the positive and negative space charge regions making up the depletion region, below. The space charge regions constitute the dipole layer responsible for the "diffusion voltage" V_D and the consequent "built-in-field" of the junction. (c) Schottky barrier at a metal-semiconductor contact. The dipole layer in this case consists of a surface charge on the metal (shown as negative, this being the usual case in practice) and a space charge region extending some distance into the semiconductor. The depletion region thus lies entirely in the semiconductor.

All electrical barriers are basically dipole layers as is indicated in Fig. 6. In the well known case of the *p–n* junction, for example, the requirement that the Fermi level E_F must be constant across the junction, taken together with the fact that E_F is near E_c in the *n*-type material but near E_v in the *p*-type material, means that the bands must be bent as shown in Fig. 6(b). Physically this is brought about by the occurrence in the space charge region of positive and negative space charge layers on either side of the junction plane. The space charge region is "depleted" of free charge carriers because the Fermi level there is far from both band edges. The space charge dipole layer produces the change in potential, qV_D, where V_D is the "diffusion voltage". The change in potential is manifested in the energy band bending and the depletion region constitutes the high resistivity barrier. Schottky barriers at metal-semiconductor contacts are similar as shown in Fig. 6(c).

The barrier electron voltaic effect arises through the action of the built-in field, in separating holes and electrons which are generated in or near the depletion region, as shown in Fig. 7. (On energy band diagrams electrons tend to fall and positively charged holes tend to rise since by convention the electrostatic potential of electrons is plotted increasing upwards. This is the opposite of the potential as defined in terms of a unit positive charge in elementary electrostatics.) This results in the flow of a current generated in and near the junction of magnitude I_g. The separated holes and electrons cannot simply be eliminated by current flow back through the barrier, as this contains the non-conducting depletion region and the potential step opposes this current flow. The separated hole-electron pairs, however, produce a non-equilibrium potential difference V across the barrier and this, unlike the equilibrium potential difference V_D, can lead to current flow in an external circuit. This can perhaps best be understood in terms of the Fermi level which is the locus of the free energy of the electrons. As a result of the electron bombardment the Fermi energies in the p- and the n-type material cease to be equal as indicated in Fig. 7(a) and external work can be done. The energy for this is of course derived from the kinetic energy of the bombarding electrons. The two extreme cases are shown in Fig. 7(b) and (c). If the diode is short circuited no voltage can exist between the p- and n-regions so $V = 0$ and the maximum possible external current flows. This is the short circuit current, I_{sc}. At the opposite extreme, if the diode is open circuited no current can flow so $I = 0$ and the maximum possible voltage V_{oc}, is developed between the p- and n-regions. In this case the flow of electron bombardment generated "minority carrier" electrons from the p- to the n-side of the junction is balanced by the drift current of the numerous free electrons ("majority carriers") from the n-type material over the potential barrier at the junction. This potential barrier is reduced from the equilibrium value qV_D by the amount qV_{oc} where V_{oc} is the open circuit voltage. Thus this "forward" drift current has the value corresponding to the application of an external bias of V_{oc} in the forward direction. Similarly, the current of bombardment generated minority carrier holes from the n-type side is just balanced by the forward drift current of majority carrier holes from the p-side as shown in Fig. 7(c).

The large gains that occur in the barrier electron voltaic effect (Fig.

5(b)) are easily explained in terms of eqn (2). Beam accelerating potentials are tens of thousands of volts and thus E_{eff} is of the order of magnitude of 10^4 eV whereas e_i is a few times Eg, that is e_i values are a few electron volts, and the number of electron hole pairs generated per incident electron is of the order of magnitude 10^3–10^4. This provides

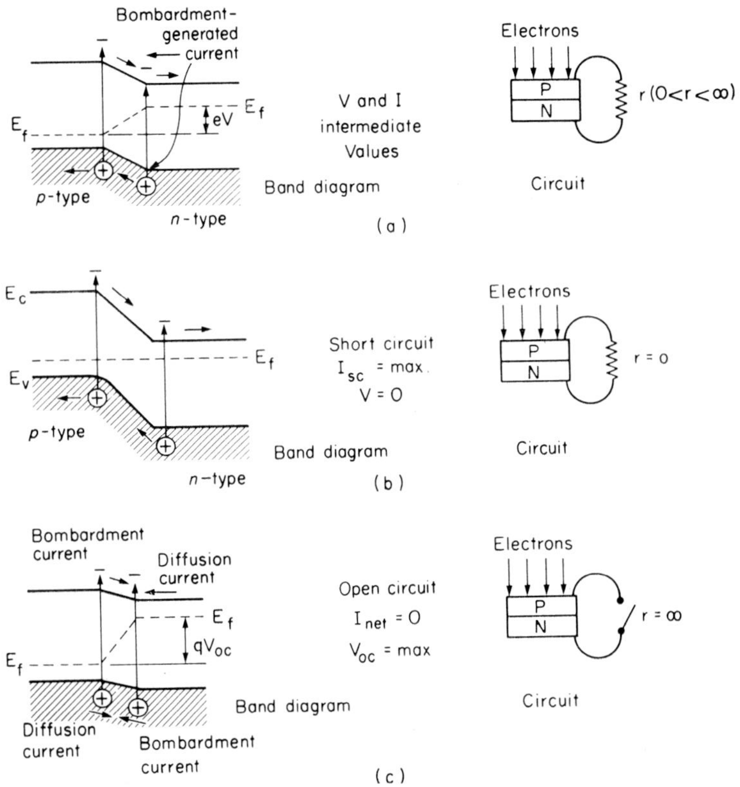

FIG. 7. The barrier electron voltaic effect at a *p–n* junction. Electron bombardment generates hole-electron pairs which are separated by the built-in electric field due to the dipole layer at the junction. (a) For intermediate values of the detector circuit resistance r, values of current and voltage are obtained that are intermediate between the open circuit and short circuit values. (b) With the diode short circuited, the separated hole-electron pairs recombine as a result of current flow round the external circuit and no externally detectable voltage is produced ($V = 0$) but the current is a maximum ($I = I_{sc}$). (c) With the diode open circuited no net external current can flow ($I = 0$). The electron and hole currents induced by the electron bombardments are equal and opposite to the electron and hole currents due to the diffusion of majority charge carriers over the potential step at the junction. This potential step is reduced by the open-circuit voltage, V_{oc}, which is developed across the junction.

the upper limit to the current gain for if all electron-hole pairs are separated by the barrier field, i.e. if the collection efficiency of the barrier is 100%, then the gain is given by

$$g_{max} = \frac{I_{max}}{I_b} = \frac{GI_b}{I_b} = G \tag{3}$$

where g_{max} is the maximum possible value of the gain and I_{max} is the maximum possible value of the charge collection current, i.e. the current that results when all the hole-electron pairs generated, GI_b, are separated by the built-in field. The charge collection efficiency, η_{cc} of the barrier is given by

$$\eta_{cc} = \frac{I}{I_{max}} = \frac{I}{GI_b} \tag{4}$$

where I is the actual current supplied by the specimen. The charge collection efficiency is thus the fraction of the generated carriers that is collected. Figures 5(b) and (c) show that the charge collection efficiency of a barrier can be increased by reverse biasing (b) and can be changed by altering the barrier geometry (c). The effect of reverse biasing arises through increasing the width of the depletion region, since the carriers generated within this region are all collected by the field. The effect of the barrier position arises through varying the position of the depletion region in relation to the volume within which the carriers are generated so that a greater or lesser fraction of this volume and hence of the carriers generated occurs where they will be efficiently collected.

The fact that the current gain in the barrier electron voltaic effect can be so large means that this type of SEM signal is large and readily observed. The large gains that are possible aroused interest in possible device applications of the effect among workers in solid state electronics. Consequently the barrier electron voltaic effect was theoretically analysed in considerable detail and the theories were experimentally checked in the main developed semiconducting materials, Si, Ge and GaAs. In barrier electron voltaic cells which are also known as beta voltaic batteries, beta particles from suitable radioactive isotopes are used to bombard the junction to produce electrical power (Rappaport, 1954; Pfann and van Roosbroeck, 1954; Rappaport *et al.*, 1956; Kalibjian and Mayeda, 1971). A little work has also been done on other possible device applications including switches (Brown, 1962), fixed pattern generators for use in television (Miyazaki *et al.*, 1966) and photoemissive devices (Abraham *et al.*, 1966).

Having introduced the basic facts and ideas regarding the barrier electron voltaic effect above, we give in the next section an algebraic analysis based on the work originally done on betavoltaic batteries.

(1) *The Current–Voltage Characteristic of the Barrier Electron Voltaic Effect*

The situation in an electron-bombarded *p–n* junction with the geometry shown in Fig. 4(a) is represented electrically by the equivalent circuit of Fig. 8(a). Because the current–voltage characteristics of *p–n* junctions are not linear the junction resistance R_j is not constant. However, it is assumed that $R_{sh} \gg R_j$ where R_{sh}, the specimen shunt resistance, is due to surface and junction leakage, and $R_j \gg R_s$ where R_s is the series resistance of the *n*- and *p*-type material. Only the simplified equivalent circuit of Fig. 8(b) then needs to be considered. The equations to this equivalent circuit are (Pfann and van Roosbroeck, 1954):

$$I_g = I_j + I \tag{5}$$

$$V = Ir \tag{6}$$

and

$$I_j = I_0[\exp(qV/\beta kT) - 1]. \tag{7}$$

Equation (5) expresses Kirchoff's Law and (6) expresses Ohm's Law. Equation (7) is the standard expression for the forward current in a rectifying *p–n* junction, where I_0 is the reverse saturation current of the diode, V is the forward voltage applied, k is Boltzmann's constant, T is the temperature in degrees absolute and β is a constant that is equal to or greater than one. Experimentally, values from $\beta = 1$ through $\beta = 2$ and higher are found. $\beta = 1$ is the value predicted by the simplest theory of *p–n* junction rectification in which it is assumed that only drift and diffusion currents flow. $\beta = 2$ is predicted by the theory including linear recombination in the depletion layer (Sah *et al.*, 1957) and non-linear recombination can be invoked to explain the occurrence of a wide range of higher values of β. In fact (Kleinman, 1956): "It can be shown that if the recombination follows some power v of the injected density . . . the forward characteristic of a simple *p–n* junction is of the form $\exp(qV/2(v+1)kT)$", i.e. $\beta = 2(v+1)$. A cynical summary is that large values of β indicate bad junctions in undeveloped materials.

The current I_j arises because, as shown in Fig. 7(a), the separation of the electron bombardment generated hole-electron pairs by the junction field, which constitutes the current I_g, alters the electrostatic potentials on either side in the sense corresponding to lowering the potential step, that is, the current I_g produces forward biasing. The current given by eqn (7) is then basically due to the increased diffusion of the majority carriers from either side over the energy barrier which has been lowered by an amount qV.

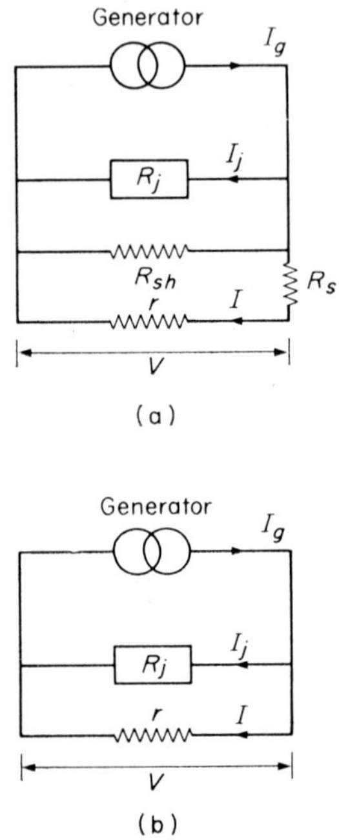

FIG. 8. Equivalent circuits for an electron bombarded *p–n* junction and detector. The constant current generator represents the generation of the current I_g due to hole-electron pairs separated by the junction. R_j is the effective impedance of the junction through which a forward current I_j flows in the opposite direction to the electron beam generated current I_g. R_s is the series resistance due to the resistance of the *p* and *n* type material, and of the contacts. R_{sh} is the parallel shunt resistance due to surface and junction leakage. r is the (detector cricuit) load resistance and I is the externally observable diode current, flowing when a voltage V is observed. (a) Full equivalent circuit. (b) Simplified circuit corresponding to the case that $R_s \ll R_j$ and $R_{sh} \gg R_j$. This is the assumption made in deriving the theory. (Pfann and Van Roosbroeck, 1954; Rappaport *et al.*, 1956.)

Combining eqns (5) and (7) we have

$$I = I_g - I_o[\exp(qV/\beta kT) - 1]. \tag{8}$$

This is the current–voltage (I–V) relation for the *p–n* junction electron voltaic effect and its form is that shown in Fig. 4(b). The diode short circuit current is the value of I for $V = 0$ (Fig. 7(b)) and substituting into (8) shows that

$$I_{sc} = Ig. \tag{9}$$

At the other extreme the maximum voltage, the open circuit value V_{oc}, corresponds to $I = 0$. Substituting into eqn (8) leads to

$$V_{oc} = \frac{\beta kT}{q} \ln \{(I_{sc}/I_o)+1\}. \tag{10}$$

Three quantities are readily obtained from the experimental I–V relation which is measured by varying r from 0 to ∞ with constant I_b and V_b. These are I_{sc}, V_{oc} and I_0. I_0 is determined from the slope of the I–V curve at $V = 0$ since from eqn (8):

$$\left(\frac{\partial I}{\partial V}\right)_{v=0} = -\frac{q}{\beta kT} I_0$$

i.e. the diode saturation reverse current

$$I_0 = -\frac{\beta kT}{q}\left(\frac{\partial I}{\partial V}\right)_{v=0} \tag{11}$$

and β can be obtained from the slope of the I–V curve plotted on semilog paper. Thus the barrier electron voltaic effect in plan view, that is with the geometry of Fig. 3(a), can be used to determine the nature of p–n junctions.

The treatment given above is based on the simplification of the equivalent circuit from that of Fig. 8(a) to that of Fig. 8(b). Therefore care is required in the application of the theory. R_j is defined to be

$$R_j = \frac{\partial V}{\partial I}. \tag{12}$$

Combining this with eqn (7) gives

$$R_j = \frac{\beta kT}{qI_0} \exp(-qV/\beta kT). \tag{13}$$

This has its maximum and minimum values under short circuit and open circuit conditions (see Fig. 3(b)). On short circuiting, $V = 0$ and

$$R_{j_{max}} = \left(\frac{\partial V}{\partial I}\right)_{v=0} = \frac{\beta kT}{qI_0}. \tag{14}$$

For open circuit conditions $I = 0$ and $V = V_{oc}$

$$R_{j_{min}} = \frac{\beta kT}{qI_0} \exp(-qV_{oc}/\beta kT)$$

and substituting for V_{oc} from eqn (10):

$$R_{j_{min}} = \frac{\beta kT}{qI_0}\left(\frac{I_{sc}-1}{I_0}\right). \tag{15}$$

Using the experimentally determined values of I_{sc} and I_0 the values of $R_{j_{max}}$ and $R_{j_{min}}$ can be calculated and used to check the validity of the simplification of the equivalent circuit from that of Fig. 8(a) to that of Fig. 8(b) for any particular specimen. This is not usually necessary for *p–n* junctions in a well developed material like Si but in newer and less well behaved materials it is important.

If the short circuit current I_{sc} is recorded this will give a measure of the collection efficiency of the barrier via eqn (4) if G is known. This can generally be increased by reverse biasing (Fig. 4(b)). The way in which this occurs will be dealt with in the next section. However, in general the collection efficiency will be less than 100%. Therefore the experimental values of the current gain (eqn (3)) even when maximized by reverse biasing will give values of G which are smaller than the theoretical value. That is, the experimental values lead to values of e_i on applying eqn (2), which are larger than the values of Table I. This is the form in which the comparison of theory and experiment has been expressed in SEM work, as we shall see in the next section.

The theory given above applies to steady state (d.c.) conditions. When the signal, I or V, varies account must be taken of the capacitance of the junction, of the bulk material in the specimen, and of the contacts, all of which contribute to the time constant of the detection (amplifier) circuit. The junction capacitance is voltage dependent and so will vary with the applied bias. The minority carrier lifetimes also contribute to limit the rise-times of the observed signal (Mann *et al.*, 1961; Miyazaki *et al.*, 1966; Abraham *et al.*, 1966). It is necessary to test for such effects by varying the line scan speed and determining whether the form of the signal variation changes. A speed must be selected that is below the limit at which these effects become apparent so that the shape of the Y modulation or pen-chart trace is independent of the scan speed.

Essentially the same treatment as that given above for *p–n* junctions applies also to Schottky barriers except that in eqn (7) which now represents the ideal Schottky barrier characteristic $\beta = 1$. To get this ideal behaviour it is necessary to employ a guard ring electrode surrounding the circular Schottky contact.

(2) *Experimental Determinations of Generation Factors and Charge Collection Efficiencies*

The simplest possible assumption that can be made is that the maximum value of the diode current is given by

$$I_{max} = \frac{I_b E_o}{e_i}. \tag{16}$$

This has been used in non-scanning work by Miyazaki *et al.* (1966) and Abraham *et al.* (1966). The assumption here is that the whole of the energy of the incident beam I_bE_0, is dissipated in generating hole-electron pairs at a cost of e_i per pair. This approximation ignores the energy loss involved in the emission current which was taken into account in eqns (2) and (3) by employing an E_{eff}, that was less than E_0, the energy in electron volts of the incident electrons of the beam.

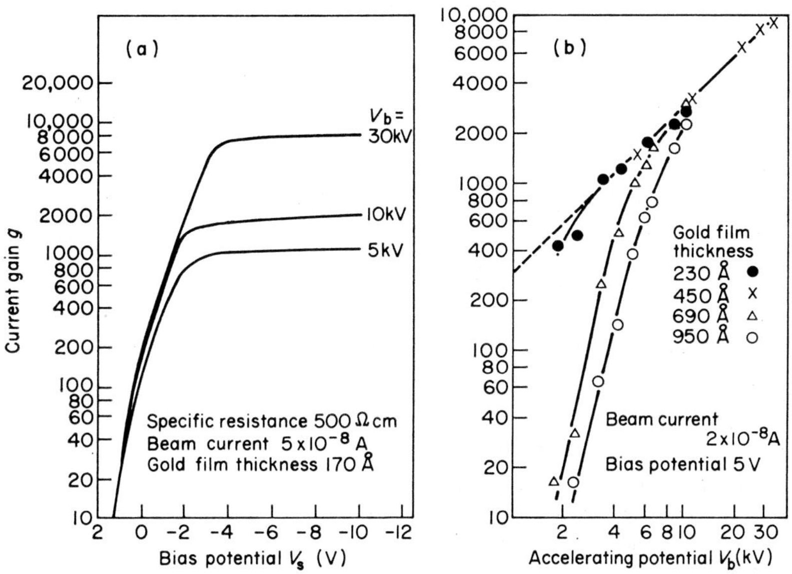

FIG. 9. Barrier electron voltaic effect in Au–Si Schottky diodes (after Miyazaki *et al.*, 1966). (a) Current gain as a function of the diode reverse bias and the beam accelerating voltage V_b. (b) Current gain as a function of beam voltage and thickness of the gold film. The broken line represents the simplest theoretical relation

$$G = qV_b/e_i = E_0/3.5.$$

Equation (16) also assumes that all hole electron pairs are separated (100% collection efficiency), i.e. that the only losses are those due to the theoretically calculable minimum rate of hole-electron pair recombination intrinsic to the material of the diode. This will not lead to errors so long as the material is free of recombination centres and the whole of the volume in which the beam energy is dissipated and the hole-electron pairs are generated, occurs in the depletion region of the barrier. Miyazaki *et al.* applied a reverse bias to widen the depletion region and saturate the gain as shown in Fig. 9(a) in work on Au–Si Schottky diodes. Then on the above assumption the gain for the maximum diode current is, combining eqns (3) and (16);

$$g_{\max} = G = \frac{E_0}{e_i} = \frac{qV_b}{e_i} \tag{17}$$

where V_b is the accelerating voltage of the beam. The agreement between this oversimplified expression (E_0 not E_{eff} being used) and experiment was nevertheless good at large values of V_b as Fig. 9(b) shows. The reason that the experimental values fall below the theoretical line for low values of V_b is that a portion of the beam energy is dissipated in the initial gold layer (Fig. 5(a)) and does not produce hole-electron pairs in the depletion region. This fraction of the beam energy falls both relatively and absolutely as V_b increases. The energy loss in the metal film is also greater, the greater the thickness of the metal, as Fig. 9(b) illustrates.

A method for correcting for the loss of beam energy in the emissive current was introduced by Czaja (1966) and used in a modified form also by Bresse (1972). Czaja, following Sternglass (1954) writes

$$\delta = (I_b - I_s)/I_b = \eta + \Delta \tag{18}$$

where η is the fraction of incident electrons lost by back-scattering, i.e. with energies greater than 50 eV and Δ is the fraction of electrons lost by secondary emission, i.e. with energies less than 50 eV and δ is the total fraction of electrons emitted. Czaja argued that if all the incident electrons that did not penetrate into the specimen were elastically reflected, i.e. with the full beam energy, E_0, it would be correct to rewrite eqn (3) as

$$I_{\max} = I_g = GI_s\,. \tag{19}$$

Because Czaja worked with *p–n* junction diodes in which the space charge region thickness was much less than the diameter of the energy depletion volume the charge collection efficiency was substantially less than 100%. We therefore rewrite eqn (19) as

$$I_{\max} = NI_s \tag{20}$$

where N is the number of hole-electron pairs separated per electron in the beam. In view of eqn (20) Czaja plotted his diode current as a function of I_s. Czaja found on measuring the diode current for S_i *p–n* junctions as a function of I_s that a simple relation like eqn (20) was only obtained if a reverse bias was applied, as Fig. 10 shows. He therefore measured the diode current under reverse bias, not short circuit, conditions. The measured values of the diode current, $I_{\max}$, plotted as a function of I_s shown in Fig. 11 then gave experimental values for N (eqn (20)).

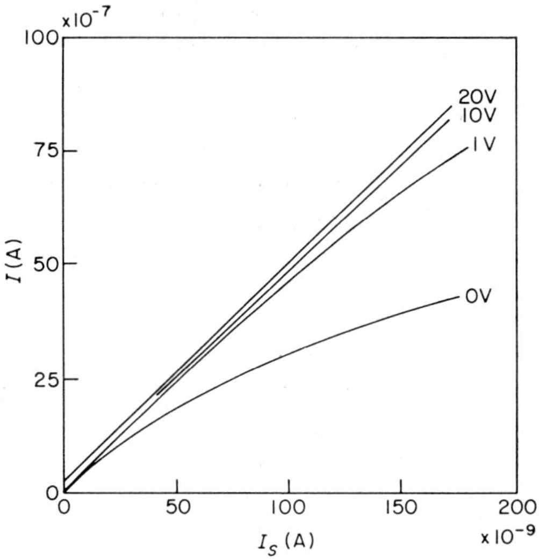

FIG. 10. The variation of diode current I as a function of the specimen current I_s for various reverse bias voltages in a Si p–n junction. The non-linearity of the relation between I and I_s at zero bias (short circuit conditions) would lead to varying values of the generation factor or of the charge collection efficiency. (After Czaja, 1966.)

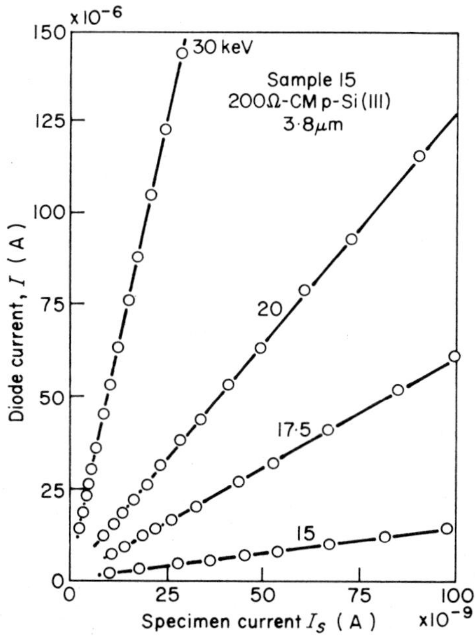

FIG. 11. Diode current as a function of specimen current at different beam energies E_0. The specimen was a diffused Si p–n junction with the junction 3.8 μm below the beam entry surface. (After Czaja, 1966.)

The values so obtained required correction because in fact the incident electrons which give rise to secondary emitted electrons and consequently do not contribute to the sample current still lose almost all their energy within the sample by producing hole-electron pairs. Moreover the incident electrons which are back-scattered also contribute to the generation factor G since they are back-scattered at energies less than the incident value E_0. Therefore Czaja corrected the measured value to obtain

$$N_{corr} = N - N_{b.s.} - N_{s.e.} \tag{21}$$

where $N_{b.s.}$ and $N_{s.e.}$ are the numbers of pairs generated by electrons which also give rise to back-scattered and to secondary electrons respectively. Following Sternglass (1954) Czaja wrote the average energy of the back-scattered electrons as

$$E_{b.s.} = \overline{K}E_0 \tag{22}$$

where $\overline{K} = 0{\cdot}45 + 2\times10^{-3}\times Z$. (This was only obtained for $0{\cdot}2$ keV $\leqq E_0 \leqslant 32$ keV by Sternglass but Czaja assumed it to apply up to 40 keV.) From eqns (18) and (22) we can write

$$N_{b.s.} = \eta[(1-\overline{K})E_0/e_i^*] \tag{23}$$

where Czaja assumed e_i^*, the effective formation energy for pairs at the average energy of back-scattered electrons ($\simeq E_0/2$), could be written as

$$e_i^*(E_0) = \frac{E_0/2}{N(E_0)-N(E_0/2)}. \tag{24}$$

Czaja also wrote

$$N_{s.e.} = G\Delta. \tag{25}$$

Combining eqns (25), (23) and (21) gives

$$N_{corr} = N(1-\Delta) - \eta[(1-\overline{K})E_0/e_i^*]. \tag{26}$$

Czaja's experimental determination of $\delta = \eta + \Delta$ together with Sternglass's value of η as shown in Fig. 12 made it possible to obtain N_{corr}. Czaja then compared this experimental value with a theoretical value of N, which is the product of G, the generation factor, and η_{cc}, the charge collection efficiency calculated as follows.

The geometry of Czaja's diodes was as shown in Fig. 13. The number of carriers collected is the sum of three contributions

$$N = N_1 + N_2 + N_3$$

where

$$N_1 = \frac{1}{e_i}\int_0^{x_1} p(\xi)e^{(\xi - X_1)/L_1}\,d\xi \quad (n\text{-type}) \tag{27}$$

$$N_2 = \frac{1}{e_i}\int_{x_1}^{x_2} p(\xi)\,d\xi \quad \text{(space charge region)} \tag{28}$$

$$N_3 = \frac{1}{e_i}\int_{x_2}^{N} p(\xi)e^{(x_2 - \xi)/L_2}\,d\xi \quad (p\text{-type}) \tag{29}$$

where R is the maximum depth of pair-production, i.e. the electron range perpendicular to the surface, and is a function of E_0, and $p(\xi)\,d\xi$ is the energy lost by the electron beam in penetrating from a depth ξ to $\xi + d\xi$ perpendicular to the surface. The total beam energy is given by

$$E_0 = \int_0^{R(E_0)} p(\xi)\,d\xi. \tag{30}$$

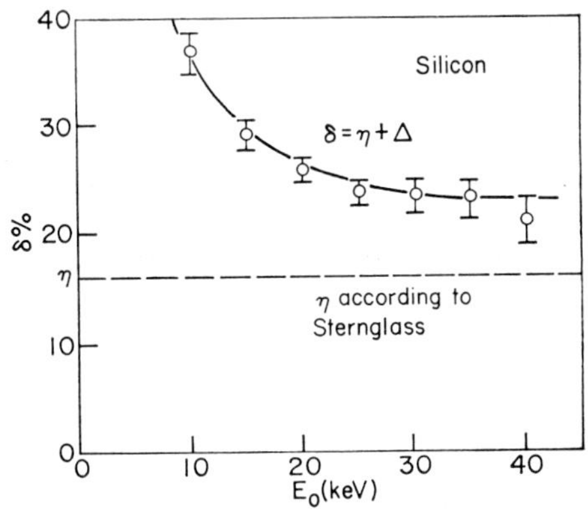

FIG. 12. The experimental values of the total emission coefficient δ for Si. (After Czaja, 1966.)

The reason for the exponential variation with L_1 and L_2 in the n- and p-type regions is that those carriers generated at a distance from the depletion region of $\xi - X_1$ or $X_2 - \xi$ must first diffuse to the depletion region before the junction field can accelerate them across the barriers, i.e. "collect" them. As indicated in Fig. 14, it is only the minority carriers in both types of material that will be collected. The exponentials in eqns (27) and (29) arise from the solution of the diffusion equation under steady state conditions as will be shown in section III.A(3) below, dealing with edge-on scanning observations on the barrier electron voltaic effect.

Czaja (1966) evaluated eqns (27)–(29) using two different exponential functions for $p(\xi)$ based on the work of Makhov (1960a,b and c). The agreement between Czaja's N_{corr} values from eqn (26) and his theoretical N values from eqns (27)–(29) in the case of a diode in which diffusion induced slip (this will be described below) had not occurred is shown in Fig. (15). To carry out the calculation, Czaja assumed $e_i = 3{\cdot}60$ eV per hole-electron pair which is the accepted value (Fabri *et al.*, 1963;

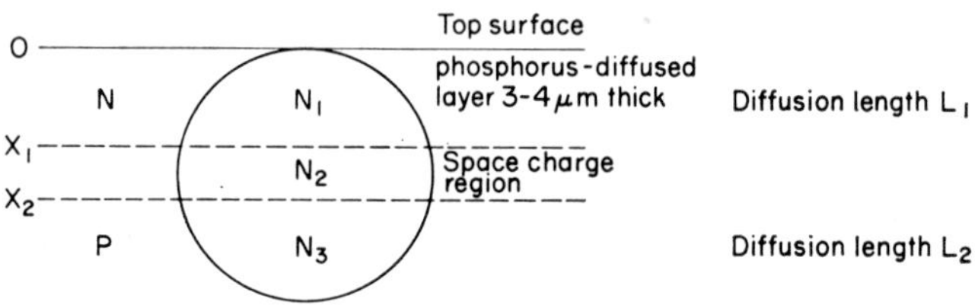

FIG. 13. The three zones of charge collection for the p–n junctions in Czaja's (1966) work. N_1 carriers were collected from pairs generated in the n-type material of diffusion length L_1, N_2 from those generated in the depletion region and N_3 from those generated in the p-type material of diffusion length L_2.

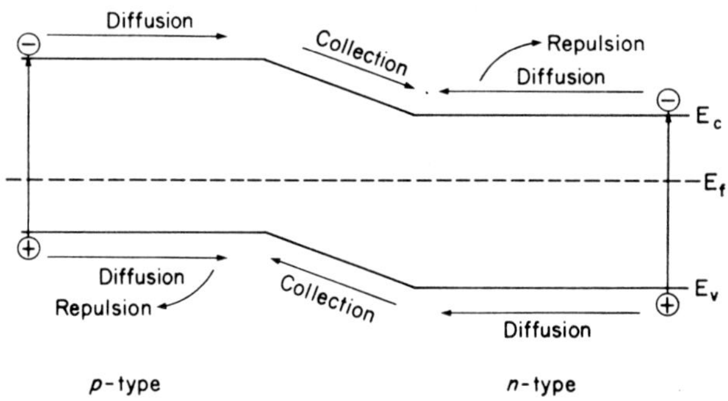

FIG. 14. When hole-electron pairs are generated at a distance from the p–n junction, the carriers must first diffuse to the junction before they can be "collected" i.e. accelerated across the barrier and so separated from the other member of the pair, by the built-in field. From both sides it is only the minority carriers, that is the holes from the n-type and electrons from the p-type material, which will be collected.

Baldinger *et al.*, 1964), and treated the minority carrier diffusion lengths L as adjustable parameters to obtain the best agreement between the theoretical and experimental values of N. In this way values for L were obtained. (Actually Czaja found that he had to assume, as others working with similar diodes previously had had to do, that L_1 had two different values, one for an initial surface layer of depth ξ_0 and a second one for depths $> \xi_0$.) The use of these results for the study of contrast due to diffusion induced slip will be discussed in the next section.

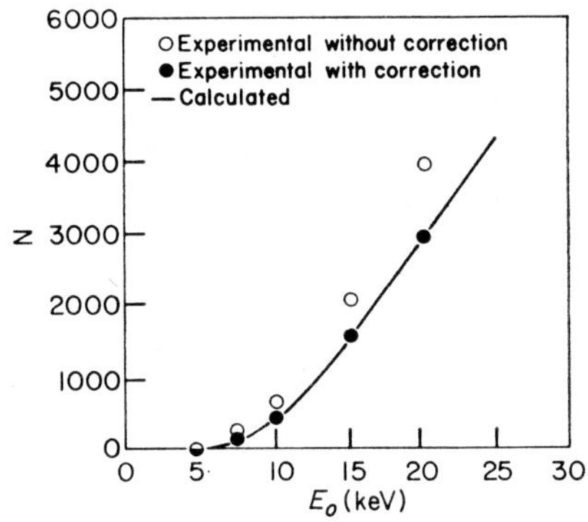

FIG. 15. Charge collected per electron in the specimen current versus the beam energy for a Si *p–n* junction. (After Czaja, 1966.)

Bresse (1972) used an approach similar to that of Czaja to study Schottky diodes of Al on Si. Bresse based his analysis on the use of the penetration theory of Kanaya and Okayama (1970) together with the Archard (1961) diffusion model. The creation zone or energy dissipation volume in the silicon under beam bombardment is then spherical and is centred on the diffusion depth ξ_d and the maximum depth of hole-electron pair creation is R which is the diameter of the sphere. Bresse (1972) following Kanaya writes

$$\rho R = 2{\cdot}76\times 10^{-11} A Z^{-8/9} E_0^{-5/3} \left(\frac{1+E_0/2M_0C^2}{1+E_0/M_0C^2}\right)^{5/3} \tag{31}$$

and

$$\xi_d/R = 1/(1+\gamma) \tag{32}$$

where ρ is the density of the material in g cm^{-3}, A is the atomic weight in g, Z is the atomic number, $\gamma = 0{\cdot}187Z^{2/3}$, M_0 is the rest mass of an

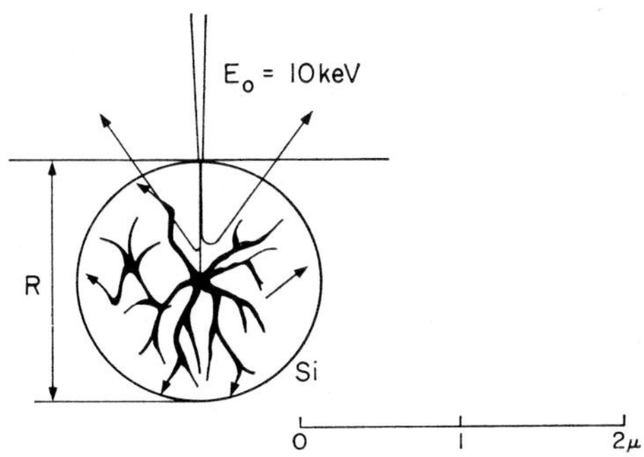

FIG. 16. The form of the energy-dissipation volume in Si on the Kanya and Okayama (1970) theory. (After Bresse, 1972.)

electron and C is the velocity of light. In this model the ratio in eqn (32) is independent of E_0. The following values are obtained for $E_0 = 10$ keV for Si: $R = 1{\cdot}5$ μm and $\xi_d/R = 0{\cdot}48$. The form of the energy dissipation volume in Si is therefore that of a sphere of radius R which is approximately touching the free surface ($\xi_d/R \simeq 0{\cdot}5$) for all beam potentials in the range 10–100 keV, on this model, as shown in Fig. 16.

Bresse (1972) chose to calculate the generation factor G (the number of hole-electron pairs created by one incident electron of energy E_0) in terms of the energy of the beam of current I_b (not of I_s as did Czaja (1966)) and so had a simpler expression after correcting for the back-scattered electrons only:

$$G_{\text{theoretical}} = \frac{E_0}{e_i}(1-\bar{K}\eta) = \frac{E_{\text{eff}}}{e_i}\,. \tag{33}$$

Bresse was able to test the relation of eqn (31) between R and E_0 by reverse biasing his Schottky diodes which were of the form shown in Fig. 17. The width of the depletion region W_m varies as

$$W_m = 4{\cdot}2\times 10^7 (V_r/N_i)^{\frac{1}{2}} \tag{34}$$

where V_r is the reverse bias voltage and N_i is the impurity concentration in the Si in atoms cm^{-3}. When the diode current reached its maximum value as a function of V_r, the corresponding value of W_m from eqn (34) gave a measurement for R but the accuracy of the measurements was not high.

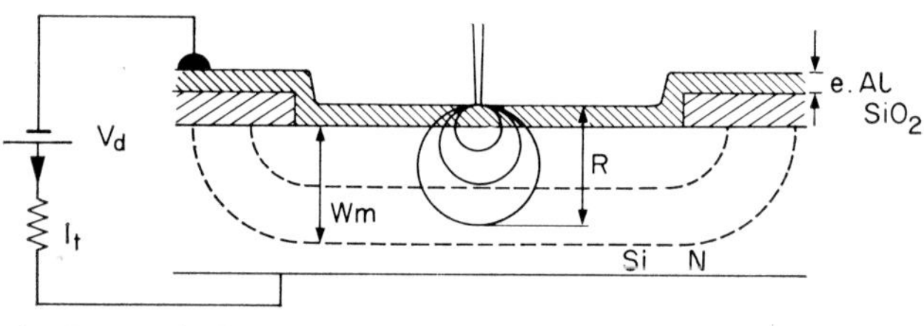

FIG. 17. The form of Si Schottky diodes used in Bresse's determinations of the generation factor. (After Bresse, 1972.)

The measured current I is made up of two components

$$I = I_g + I_r \tag{35}$$

where I_r is the reverse current of the diode.

For good Shottky diodes (and rectifying p–n junctions) I_r is small and virtually constant. The beam energy dissipated in penetrating the Al contact will not contribute to the diode current since the space charge region in the Al is very small ($\simeq$ 50 Å) so Bresse used contacts of thickness very small compared to R ($+ \simeq 0{\cdot}1\ \mu$m whereas $R \geqslant 1{\cdot}5\ \mu$m) in order to reduce this collection loss.

For complete collection of the pairs the maximum value for the experimentally measured generation factor is given by

$$G_m = \frac{I_{\max}}{I_b}. \tag{36}$$

The proportionality of the generation factor to the beam energy was confirmed as shown in Fig. 18. The agreement was within 5% for $E_0 = 15$ keV. The experimental value was always the lower of the two because the carriers created in the aluminium contacts were not collected. For each beam energy the experimental value G_m was substituted into the theoretical expression (33) to obtain a value for e_i. Bresse's results were nearer the accepted value (Baldinger *et al.*, 1964) of 3·60 eV than those of Czaja (1966) which are also listed in Table II.

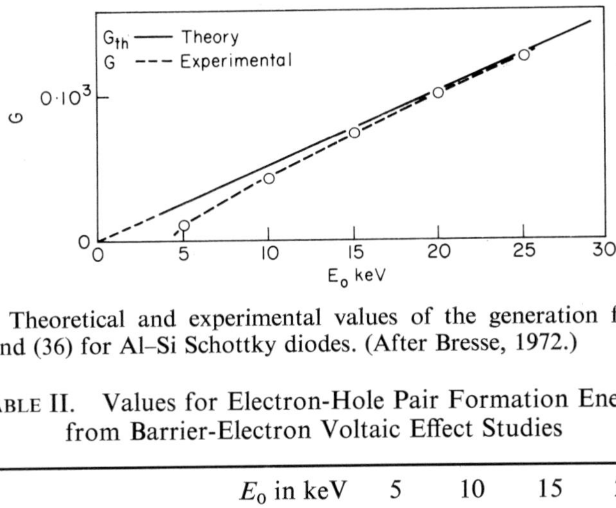

FIG. 18. Theoretical and experimental values of the generation factor (from eqns (33) and (36) for Al–Si Schottky diodes. (After Bresse, 1972.)

TABLE II. Values for Electron-Hole Pair Formation Energy from Barrier-Electron Voltaic Effect Studies

	E_0 in keV	5	10	15	20	25
Bresse (Si Schottky Diodes)	e_i in eV	8·9	5·5	3·74	3·66	3·63
Czaja (Si *p–n* junctions)	e_i in eV	29·8	9·5	4·22	2·96	

Kalibjian and Mayeda (1971) used a different, device-oriented approach to the problem of calculating the generation factor and the barrier charge collection efficiency. Following the treatment of the barrier photovoltaic effect given by Johnson (1957) they write

$$V_{oc} = \frac{\beta kT}{q} \ln \left\{ \frac{1+(\Delta p/po)_N}{1+(\Delta p/po)_P} \right\} \tag{37}$$

where $\Delta p/po$ is the ratio of excess hole concentration due to the generation of carrier pairs by the beam to the equilibrium hole concentration in the absence of the beam. The subscripts N and P denote the side of the junction being considered.

For moderate and low input power densities this reduces to

$$V_{oc} = \frac{\beta kT}{q} \ln \left\{ \left(\frac{\Delta p}{po} \right)_N + 1 \right\} \tag{38}$$

since po in the p-type material is large and $(\Delta p/po)_P \ll 1$.

This they rewrite as

$$V_{oc} = \frac{\beta kT}{q} \ln \{ 6{\cdot}25 \times 10^{12} n_q \alpha \tau_p p_{in} / V_b \eta_i^2 + 1 \} \tag{39}$$

where p_{in} is the input power density in $\mu W\,cm^{-2}$, η_q is the fractional electron voltaic quantum efficiency, α is the electron absorption coefficient in cm^{-1} (assuming that electrons are absorbed so that the beam intensity decreases with depth of penetration Z as $e^{-\alpha Z}$, where α is a linear absorption coefficient for electrons), τ_p is the hole lifetime in seconds, and n_i is the intrinsic carrier concentration in carriers cm^{-3}.

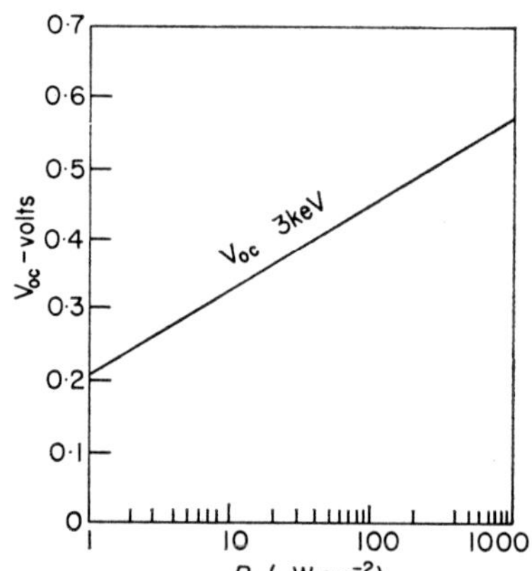

FIG. 19. Open circuit voltage as a function of input power for the electron voltaic effect in GaAs p–n junctions. (After Kalibjian and Mayeda, 1971.)

The quantum efficiency is defined to be

$$\eta_q = \frac{qV_B}{e_i}\left(1 - \frac{V_0}{V_b}\right) \tag{40}$$

where V_0 is an extrapolated dead voltage from the η_q versus V_b curve. Thus η_q is an empirically corrected generation factor G (see eqn (2)). Hence in the notation of Kalibjian and Mayeda

$$I_{sc} - I_b = \eta_q I_b\,. \tag{41}$$

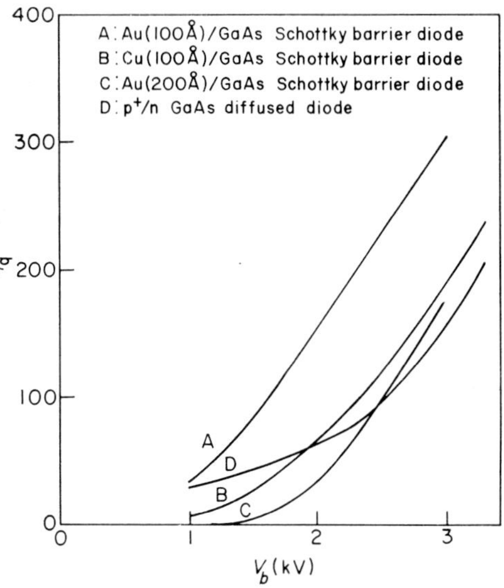

FIG. 20. Variation of the "quantum efficiency" i.e. the empirically corrected generation factor of eqn (40) with beam voltage for GaAs diodes. (After Kalibjian and Mayeda, 1971.)

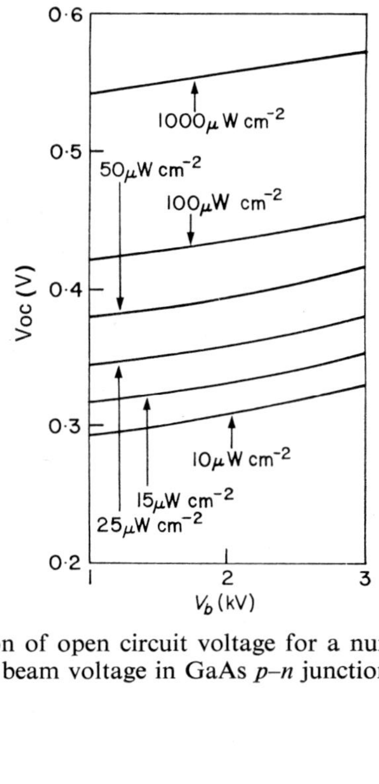

FIG. 21. The variation of open circuit voltage for a number of values of input power as a function of beam voltage in GaAs p–n junctions. (After Kalibjian and Mayeda, 1971.)

Kalibjian and Mayeda made measurements on both p–n junctions and Schottky diodes of GaAs. They found that as predicted by eqn (39), the open circuit voltage V_{oc} was proportional to the log of the input power p_{in}, as shown in Fig. 19. The variation of the quantum efficiency (generation factor) η_q with beam voltage was found to be as shown in Fig. 20, hence the necessity to correct for V_0. The variation of V_{oc} with V_b is shown in Fig. 21 for diffused p–n junctions and in Fig. 22 for Schottky barrier diodes with Cu and Au contacts.

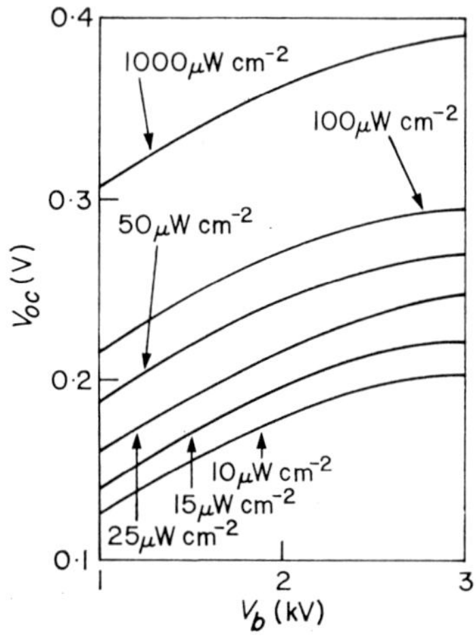

FIG. 22. The variation of open circuit voltage with beam voltage at several input power levels for Cu–GaAs Schottky barrier diodes. (After Kalibjian and Mayeda, 1971.)

The assumption that the injected hole density in the p-type material is negligible in comparison with the equilibrium hole density that was necessary to simplify eqn (37) to eqn (38) is not generally valid in SEM work. The values of V_b used in the SEM are an order of magnitude higher than those of Kalibjian and Mayeda and the current densities in the finely focused beam are higher also. Moreover, the empirical correction for a dead voltage V_0 of eqn (40) and Fig. 20 has not been found necessary in SEM work. This may be because these corrections are small relative to the much larger beam potentials used in the SEM and the accuracy of the work done hitherto has not warranted such detailed corrections. It may also be due to the fact that GaAs was used in these measurements. The dead voltage presumably corresponds to the value required to give values of penetration R greater than the depth of the dead layer (Wittry and Kyser, 1966; see also chapter 10) due to surface states. The surface state density, etc. and hence the depth of the dead layer, will depend on the material and on the surface treatment, hence the variations in V_0 of Fig. 20.

The efficiency of collection of the hole-electron pairs, that is, the fraction of the carriers created that are collected depends on the nature of the barrier. In particular it depends on the width of the depletion region as can be seen from the case treated by Czaja (eqns (27) to (29) and their dependence upon the width $X_2 - X_1$). The charge collection efficiency will also depend in principle on the field and diffusion lengths in the depletion region.

In the simplest treatments the efficiency is made to depend only on the width of the depletion region by assuming a collection efficiency of 100% within this region. This was implicit in eqn (28), for example. By reverse biasing, the width of the depletion region can be increased. This is the mechanism of the increase in the gain with reverse bias voltage up to a saturation value corresponding to the theoretical limit of $g_{max} = G$ and $\eta_{cc} = 1$, shown in Fig. 5(b). By scanning across the depletion region it is possible to measure the width by observing the distance over which the gain has its constant, maximum value, as will be discussed in section III.A(4). This is, of course, only possible if the width of the depletion region is greater than the probe size, i.e. greater than the diameter of the energy dissipation volume which determines the resolution of the SEM in the conductive mode. The efficiency of collection, however, in principle also depends on the electric field outside the depletion region. Czaja (1966) assumed in eqns (27) and (29) that minority carriers that were generated outside the depletion region would only reach the depletion region by random diffusion with the characteristic diffusion lengths. However, if there is a long range electric field outside the depletion region, the minority carriers will be accelerated toward the junction. That is, in addition to diffusion there will be a "drift" contribution to the numbers of minority carriers that are collected.

The reason for the presence of these fields is a general one and this is also important in connection with the bulk electron voltaic effect to be discussed in section III.B. In extrinsic semiconductors the position of the Fermi level in the forbidden energy gap depends on the doping. In *p*-type material E_f is near E_v (see, e.g. Fig. 6(b)) and in *n*-type material E_f is near E_c in equilibrium. In fact the position of E_f varies relative to the band edges whenever the type or concentration of the impurity atoms varies. The Fermi energy (free energy of electrons) must be everywhere the same in equilibrium, however. This requires that the bands bend, i.e. that the electrostatic potential varies. This is a more general statement of the same argument that showed that band bending and internal electrical fields must occur at *p–n* junctions and Schottky barriers.

For example an exponential doping profile gives rise to a field which is constant throughout the region, whereas a Gaussian doping profile gives a field which increases linearly as the junction is approached. Hoff and Everhart (1970) have calculated the collection efficiency of Gaussian *p–n* junctions for a number of sets of boundary conditions.

Gaussian concentration profiles are the type produced by depositing one of the dopants on the surface of the semiconducting material at a lower temperature and diffusing it into the material in a higher temperature anneal. The consequence is that there is a long-range linearly varying field outside the depletion region of the form illustrated in Fig. 23 and the occurrence of an additional drift contribution increases the charge collection efficiency of the junction above the value given by a calculation like that of eqns (27)–(29).

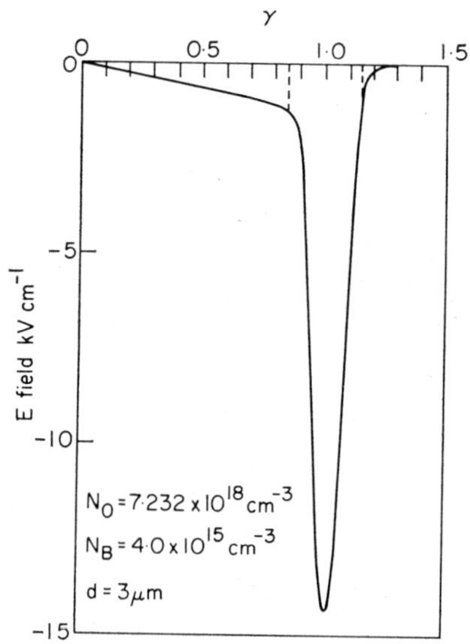

FIG. 23. The thermal equilibrium electric field versus distance from the free surface normalized to the depth of the junction for a Gaussian-diffused *p–n* junction. (After Hoff and Everhart, 1970.)

No experimental tests of the expressions obtained by Hoff and Everhart (1970) have been published, but in his Ph.D. thesis, P. Hoff did make some measurements that were in reasonable agreement with the theory (T. E. Everhart, personal communication).

This work does indicate the way in which barrier electron voltaic effect observations can in principle be analysed to yield information on the electrical form of the barrier under examination.

(3) *Studies of the Barrier Electron Voltaic Effect using Scanning in Plan View*

In one of the earliest of all conductive mode observations it was found that when the barrier electron voltaic effect signal was displayed to form an SEM micrograph of diffused *p–n* junctions in Si in which diffusion-induced slip had occurred, the contrast took the form of grids of dark lines crossing the field of view in crystallographic directions (Lander *et al.*, 1963), as shown in Fig. 24. Czaja and Patel (1965) had put forward the hypothesis that this contrast effect was due to the dislocations introduced by the slip that resulted from the diffusing in of impurities with marked atomic size mismatches with the Si. These dislocations, it was suggested, acted as the sites of an increased electron-hole recombination rate. This enhanced recombination in the slipped region reduced the number of minority carriers collected from the diffused *n*-type region of Fig. 13, and due to the reduction of N_1 of eqn (27) due to an effective reduction in L_1, the diode current I was reduced and so the screen brightness was reduced when the electron beam scanned across a dislocation line.

FIG. 24. Plan view scanning electron micrograph using the barrier electron voltaic effect reverse-biased current as video signal. This Si *p–n* junction had pronounced diffusion-induced slip. (After Czaja, 1966.)

Evidence to support the view that the dark lines seen in barrier electron voltaic effect plan view micrographs were due to dislocations was provided in two ways. Firstly, Czaja and Wheatley (1964) showed that the patterns of lines in micrographs like Fig. 24 and the patterns of lines observed in the same areas when a reliable dislocation etch had been employed were the same. This type of correlation was repeated in a number of cases also by Czaja and Patel (1965). Further support for this hypothesis was produced by indenting the top, *n*-type layer of previously perfect Si *p–n* junctions giving featureless barrier electron voltaic effect micrographs. When these specimens had been annealed, a process which is known to result in a pattern of dislocations around the indentation, the barrier e.v.e. micrographs were observed to show a pattern of dark loops originating at the indentation. These were of the form that dislocation loops would be expected to take. Etching confirmed that pitting occurred at the ends of the loops, that is at the points where the dislocations reached the surface of the specimen.

Care must be taken in interpreting dark lines on plan view barrier e.v.e. micrographs. Neve and Thornton (1966) found that microcracks could also give rise to dark lines. These were, however, generally narrower and more uniform in contrast along their length than those due to dislocations. In this work, too, etching was used to identify the defects observed in the SEM micrographs.

In 1966 Czaja carried out the analysis recounted in the previous section in which curve fitting involving eqns (27)–(29) was used to obtain values for the minority carrier diffusion lengths L_1 and L_2 for perfect Si diffused junctions, that is junctions which did not exhibit diffusion induced slip effects in barrier e.v.e. contrast. Repeating this for the reduced barrier e.v.e. signal at the dark lines of a micrograph like that of Fig. 24 gave a value for the diffusion length due to enhanced recombination at dislocations. From this in turn he calculated a lifetime and a capture radius for recombination at the dislocations. We will not give this analysis in detail because (1) it was based on relatively few results of limited accuracy, and (2) it involved the use of an early theory of the electrical effects of dislocations in semiconductors and (3) the effective assumption that each dark line in micrographs like Fig. 24 is a dislocation. These features are the lines of intersection of slipped planes and the *p–n* junction, i.e. they each contain large numbers of dislocations which the conductive mode of the SEM cannot resolve. However, Czaja's (1966) analysis led to a value for the recombination cross section for dislocations in *n*-types silicon that was not unreasonable and lent further support to the hypothesis that the observed contrast was due to enhanced recombination due to dislocations.

No further work on dislocation contrast in barrier e.v.e. micrographs has been published although this problem has been the subject of additional research for some years (G. R. Booker, personal communication).

An important practical problem is that while defects can readily be observed by many methods and their presence can be correlated with device failure, it has not been possible to establish one-to-one correspondences between structural defects and changes in electrical behaviour. This is because dislocations and stacking faults do not always have harmful effects. It has been suggested, therefore, that the electrical effects are not due to the structural imperfections themselves but to chemical impurity precipitates which sometimes, but not always, occur at the defects. An important step forward in the study of this type of problem was taken by Ravi *et al.* (1973) in the case of stacking faults produced during the surface oxidation of silicon.

They examined 2-ohm cm *n*-type epitaxial layers of Si 20 μm thick which had been oxidized in steam at 1100°C and then boron diffused to produce diodes $0{\cdot}1 \times 0{\cdot}1$ cm with a junction depth of 4 μm. The diffusion temperature was 1250°C with an integrated boron concentration of 3×10^{16} cm^{-2}.

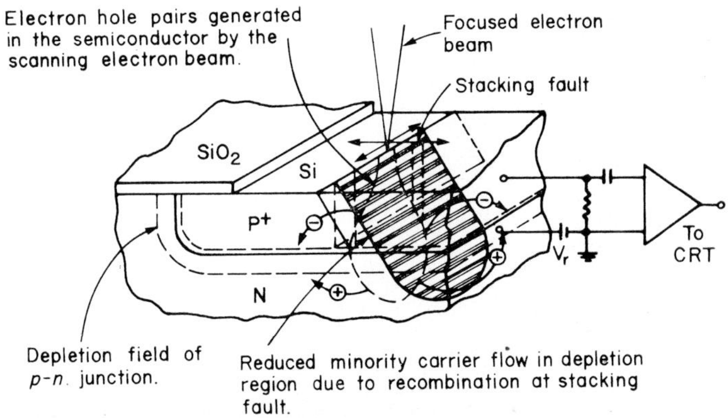

FIG. 25. Schematic diagram showing a stacking fault produced by surface oxidation and running down to intersect a *p–n* junction parallel to the surface. The detection circuit shown on the right makes it possible to bias the junction (by V_r) without the d.c. diode current reaching the amplifier (due to the capacitor). The "capacitatively coupled" signal, however, is detected. (After Ravi *et al.*, 1973.)

Ravi *et al.* (1973) used transmission electron microscopy (TEM) to determine the size and nature of the stacking faults and whether they did or did not have associated precipitates. An SEM was used to scan the *p–n* junctions in plan view and observe the defect contrast in barrier e.v.e. micrographs. The current–voltage characteristics of the junctions were measured and related to the numbers and types of stacking faults in each diode.

The geometry of the faults which ran into the interior of the devices from the surface on {111} planes to intersect the *p–n* junctions is shown in Fig. 25. The detection circuit allowed the diodes to be biased while being examined in the SEM using the barrier e.v.e. Recombination at the stacking faults was presumed to be the reason for the contrast observed in unbiased specimens at stacking fault sites. The dark contrast in the top micrograph in Fig. 26 is due to enhanced recombination, the subsequent bright region is due to the capacitor in the detection circuit (Fig. 25). The dark dots at the top and bottom edges of the stacking fault resulted from the presence of second phase particles precipitated on the partial dislocations bounding the fault and directly observed in the TEM.

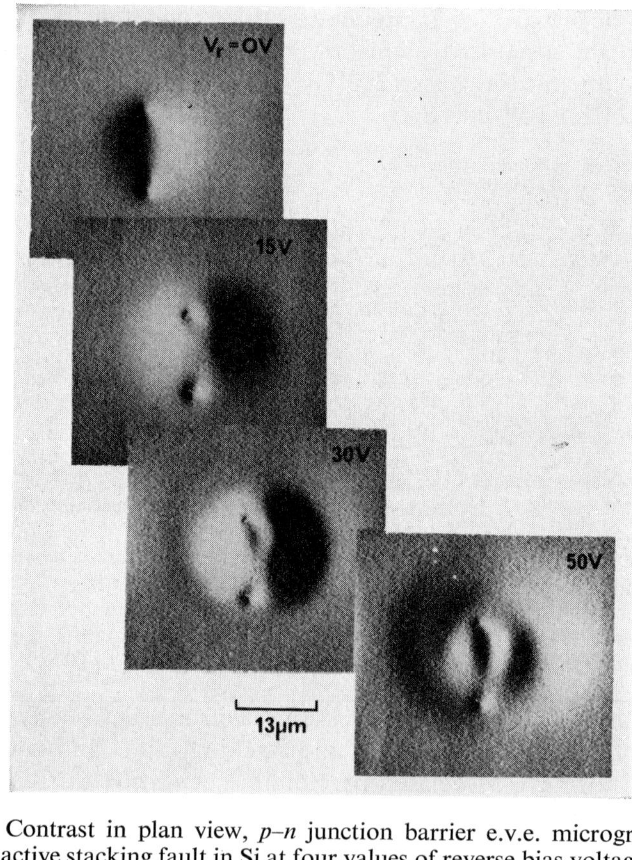

FIG. 26. Contrast in plan view, *p–n* junction barrier e.v.e. micrographs of an electrically active stacking fault in Si at four values of reverse bias voltage. The scan direction was from left to right. (After Ravi *et al.*, 1973.)

It was found that the stacking faults could be divided into electrically active and electrically inactive on the basis of their barrier e.v.e. contrast with a reverse bias voltage applied to the junction. The electrically active faults were those which for a reverse bias applied above some threshold voltage gave rise to an *increase* in the barrier e.v.e. current greater than 10% of its average background value. This arbitrary detection limit was set to give a signal gain greater than the noise level and so define the threshold (reverse) voltage. The reversal of the contrast on crossing the threshold can be seen at a high magnification in Fig. 26. The micrograph with a bias voltage $V_r = 0$ shows recombination contrast (dark initially in each scan from the left) whereas the micrograph with $V_r = 15$ V (above threshold) shows current enhancement (bright initially, i.e. on the left). This contrast reversal in a circular zone surrounding the fault is characteristic of the current enhancement effect. With further increased bias annular contrast zones appeared, as can be seen in the lower micrographs, for which Ravi *et al.* offered no explanation.

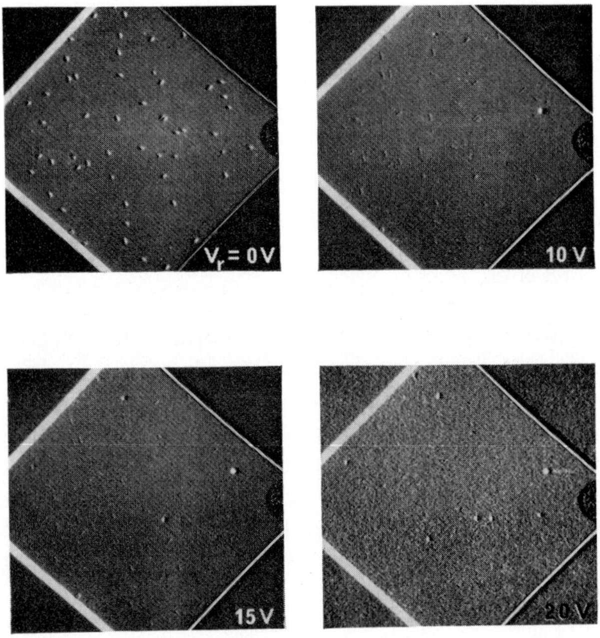

FIG. 27. Barrier e.v.e. micrographs of a Si diode at four values of the reverse bias voltage. (After Ravi *et al.*, 1973.)

Using a low magnification whole diodes could be observed as in Fig. 27 and the numbers of electrically active faults counted and their thresholds determined. For example, the micrograph in Fig. 27 with $V_r = 0$ shows that a total of fifty-five faults were present. The $V_r = 10$ V micrograph shows that there was one active fault with a threshold below this bias voltage. The 15 V micrograph shows two more faults active, and the $V_r = 20$ V micrograph shows five more faults to be active. Thus there were a total of eight electrically active faults with thresholds between about four and 20 V in this diode.

It was found that the lowest reverse bias threshold voltage for any stacking fault in a diode was inversely proportional to the log of the reverse bias current divided by the reverse bias voltage obtained for low voltages from the current voltage characteristic as shown in Fig. 28.

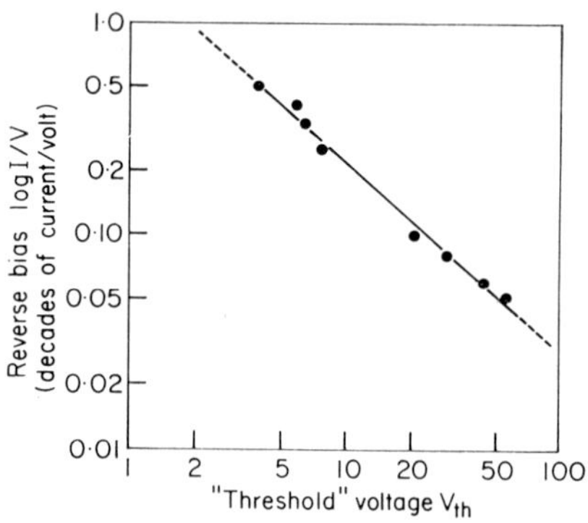

FIG. 28. Reverse biased diode characteristics in decades of current per volt versus the minimum threshold voltage for electrically active stacking faults in the diode. (After Ravi *et al.*, 1973.)

Ravi *et al.* found that the threshold voltage for electrically active stacking faults was dependent on the length of the fault as shown in Fig. 29. The upper limit of these measured values of threshold (70 V) is about 70% of the nominal avalanche (breakdown) voltage for these diodes. Extrapolation indicates that electrically inactive faults are those with lengths greater than about 30 μm which have thresholds above the diode breakdown value. The lower limiting value (about 4 V) corresponds to the detection limit set by the noise level in the detection and display system.

TEM studies showed that the stacking faults were "decorated" with precipitates, especially at the bounding partial dislocations. The stacking faults as indicated in Fig. 25 were roughly semicircular areas so that their measured "lengths" (parallel to the surface) were twice their depths measured in the fault plane. Thus the relation between the threshold voltage and length (Fig. 29) is also a relation between the threshold voltage and the depth of the stacking fault relative to the constant 4 μm depth of the *p–n* junctions in the Si diodes studied by Ravi *et al.* (1973). These authors concluded that the smaller faults were more electrically active, i.e. they introduced greater excess reverse currents into the diode. They presented models to account for the electrical effects of the faults based on the elastic strain fields of the stacking faults and the distortions produced in the junctions by the presence of the faults. These models are speculative and fall outside the scope of this review, however.

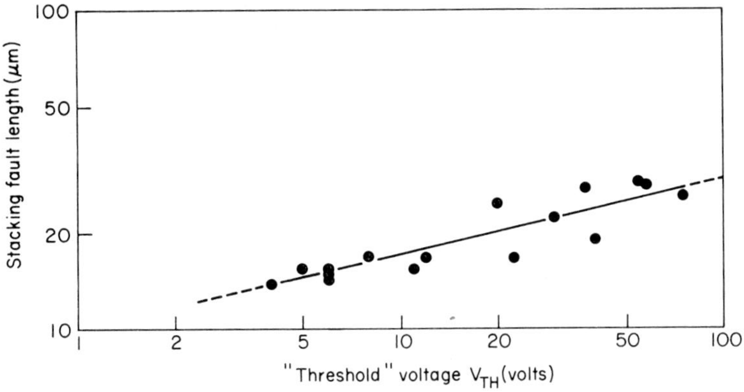

FIG. 29. Threshold voltage of electrically active stacking faults versus fault length. (After Ravi *et al.*, 1973.)

(4) *Studies of the Barrier Electron Voltaic Effect using Scanning normal to the Barrier*

When the incident electron beam is scanned along a line normal to the plane of the *p–n* junction over a surface normal to the junction plane as shown in Fig. 30, the problem is to calculate the variation of the diode current with the distance X from the beam to the junction. This was first done by Kyser and Wittry (1964) and Wittry and Kyser (1965) as follows.

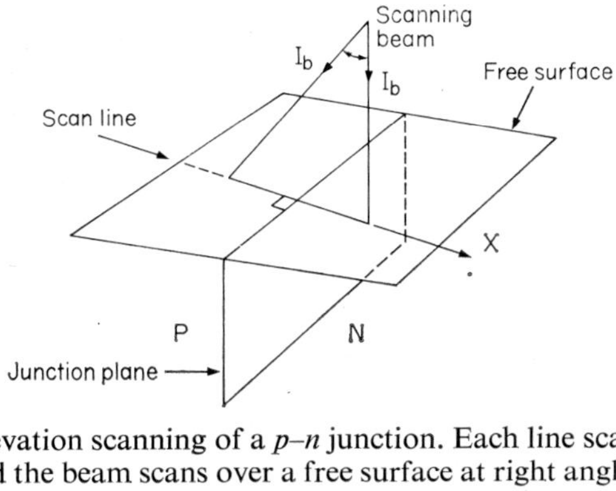

FIG. 30. Side elevation scanning of a *p–n* junction. Each line scan is made normal to the junction and the beam scans over a free surface at right angles to the junction.

Assuming that there is no electric field in the material except at the junction the minority charge carrier density n obeys the "diffusion" or "continuity" equation

$$D\nabla^2 n + GI_b - n/\tau = 0 \tag{42}$$

where D is the minority carrier diffusion coefficient, and τ is the minority carrier lifetime. Since the generation rate GI_b is zero except in the relatively small energy dissipation volume under the incident beam, throughout which it can be treated as approximately constant, for present purposes this second term may be omitted. Rearranging and remembering that

$$L^2 = D\tau \tag{43}$$

where L is the minority carrier diffusion length, eqn (42) may be rewritten as

$$\nabla^2 n - n/L^2 = 0. \tag{44}$$

Assuming that in effect the electron bombardment introduces a point source of carriers the situation has spherical symmetry and the appropriate solution of the equation is $n = \alpha e^{-r/L}/r$ where α is a constant of dimensions cm^{-2}. The number of minority carriers that reach the junction and are therefore accelerated across it per second (see Fig. 14), i.e. the number that are "collected" is given by the value of $-D\,\partial n/\partial X$ at the junction at $X = 0$, integrated over the plane of the junction. The result is that the charge collection current is given by

$$I(X) = 2\pi\alpha D e^{-X/L}. \tag{45}$$

By calculating the current through a hemisphere of radius d in the limit as d decreased, Wittry and Kyser (1964) showed that $2\pi\alpha D$ is the strength of the source, GI_b, the number of minority carriers generated per unit time. Hence

$$I = GI_b e^{-X/L} = I_{max} e^{-X/L}. \tag{46}$$

For a source of *finite* size, assuming that the generation rate is constant, G_0, inside a spherical energy dissipation volume and zero everywhere else, Kyser and Wittry (1964) obtained the solution

$$n = \frac{\alpha e^{-r/L} + \beta e^{+r/L}}{r} + \tau G_0.$$

By setting the flux of excess carriers out of a sphere of radius r_1 equal to the number of carriers produced in this volume per second minus the number that recombine per second, it can be shown that $\alpha = -\beta$.

Hence the solution can be rewritten as

$$n = \frac{C^1 \sinh(r/L)}{r} + \tau G_0 \tag{47}$$

where C^1 is a constant. By differentiating, the value of r for which there is a maximum carrier concentration is found to be $r = 0$. At $r = 0$ the excess carrier concentration is given by $\tau G_0 + C^1/L$.

Higuchi and Tamura (1965), basing themselves on the analogy with the long established method of measurement in which a light spot is scanned across the *p–n* junction, in effect used eqn (46) to measure the minority carrier diffusion lengths in Si. They did this by measuring the distance from the junction to a point where the barrier electron voltaic effect current fell to $1/e$ of the value at the junction. They found that with this technique it was possible to measure very short diffusion lengths and to observe any variations in the diffusion length over large areas of the diode.

This method for determining L cannot be recommended, however. The radius of the energy dissipation volume, i.e. the effective radius of the probe, depends on the beam voltage. The range of values of X on either side of the junction for which the probe will overlap the depletion region will therefore increase as V_b increases. The distance from the junction to the point at which the barrier e.v.e. current falls to $1/e$ of its peak value thus will also be voltage dependent and not in general equal to L. The slope of the linear portion of a plot of the logarithm of the barrier e.v.e. current against X should be used to determine L.

Munakata (1966a) observed the barrier electron voltaic effect current on scanning across a grain boundary in CdS. He plotted his measurements on semilog paper and it is clear that his results had the exponential form of eqn (46), as can be seen in Fig. 31. Extrapolating the semilog plot of the diode current values on either side of a junction up to the intersection yields a value for $I_{\max} = GI_b$ (eqn 46)). The actual maximum value of the current is less than this due to the fact that the charge collection efficiency is less than 100%. Thus (Fig. 32)

$$I_{m\ \text{actual}} = \eta_{cc} G I_b$$

and

$$\frac{I_{m\ \text{actual}}}{I_{\max}} = \eta_{cc}. \tag{48}$$

Moreover

$$\frac{I_{\max}}{I_b} = G = \frac{E_{\text{eff}}}{e_i}. \tag{49}$$

Munakata (1966a) used eqn (49) to estimate e_i for CdS to be 7·8 eV, which was in reasonable agreement with the other published values of 7·3 eV (van Heerden, 1957) and 9·3 eV (Lappe, 1959).

The actual form of the diode current as a function of X for a GaAs laser diode is shown in Fig. 33. Wittry and Kyser (1965) suggested that

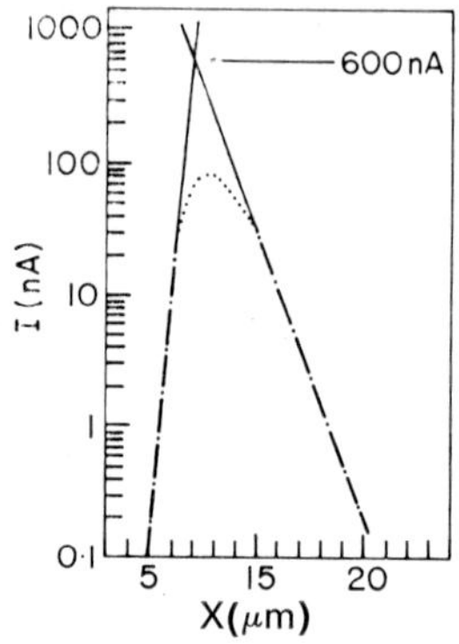

FIG. 31. Barrier electron voltaic effect short circuit current versus distance scanned across a grain boundary in CdS. (After Munakata, 1966a.)

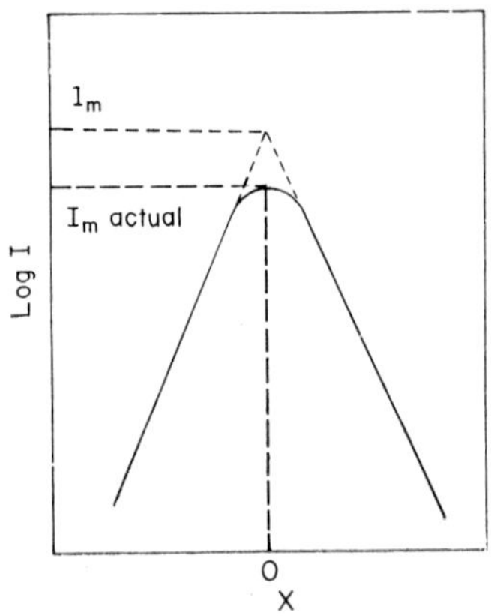

FIG. 32. Barrier electron voltaic effect current versus distance scanned across a p–n junction showing the extrapolated and actual values of the current produced with the beam incident on the junction.

this curve on the n-type side could best be represented by an equation of the form

$$I = A^1 \exp(-X/L_1) + B^1 \exp(-X/L_2) \tag{50}$$

where A^1 and B^1 are constants and the diffusion length L_2 is approximately an order of magnitude greater than L_1. They found similar values of L_1 at electron beam voltages of both 15 and 30 kV and assumed L_1 to be the minority carrier diffusion length. They suggested that the second term was due to infrared radiation reaching the junction which would make $L_2 = \alpha^{-1}$ where α is the optical absorption coefficient. However, their measured value for L_2 of 30 μm which would make $\alpha = 300\ \text{cm}^{-1}$ did not agree with the value for the absorption coefficient for the cathodoluminescent spectral peak which is about $1500\ \text{cm}^{-1}$.

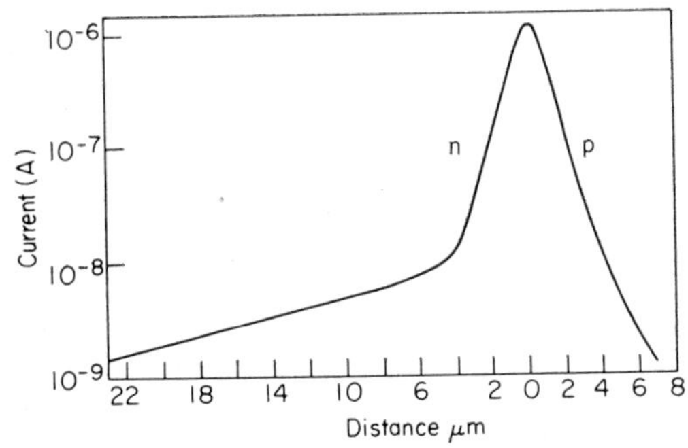

FIG. 33. Barrier electron voltaic effect current versus distance from the p–n junction on scanning across a GaAs laser diode. (After Holt and Chase, 1973.)

Micrographs can be produced by using the diode current for large values of X as video signal (Holt and Chase, 1968) and it was found that the contrast in charge collection micrographs of this type and in the luminescent mode micrographs of the same specimens was identical (Holt and Chase, 1973). This lent support to Wittry and Kyser's (1965) hypothesis as it indicated that both types of signal arise from the same fundamental process, which at large distances from the junction is the emission of light. This is observed directly in the cathodoluminescent micrograph and indirectly as a self-detected current at the p–n junction in the charge collection micrograph, i.e. some photons generated by the beam reach the junction where they are absorbed to recreate hole-electron pairs which are separated by the junction field. Holt and Chase found that signals could be detected for large values of X (0·5 mm and more) showing that L_2 is large and α therefore small. The reason for this is that the initial cathodoluminescence spectrum is rapidly modified, for increasing values of X, by strong absorption of the light of high photon energies. The spectral peak of the radiation transmitted as far as the junction is therefore of a lower energy and the value of L_2 obtained from the $I(X)$ curve relates to this modified peak. This is dealt with quantitatively in Chapter 10. It was found that the values of α to be expected on this basis for different values of X were in good agreement with the experimental values of $1/L_2$. In order to be detected at the p–n junction, a considerable proportion of the incident light must be absorbed within a minority carrier diffusion length on either side of the junction plane. The absorption coefficient data (Chapter 10) shows that despite being little absorbed in the n-type material, the light reaching the junction will still be strongly absorbed in the Zn doped p-type material. Moreover, the absorption coefficients in GaAs can increase by up to ten times with compensation (Lucovsky *et al.*, 1965) as occurs on the p-side of the junction of Fig. 33. For radiation emitted in the p-type material, self absorption would be very high,

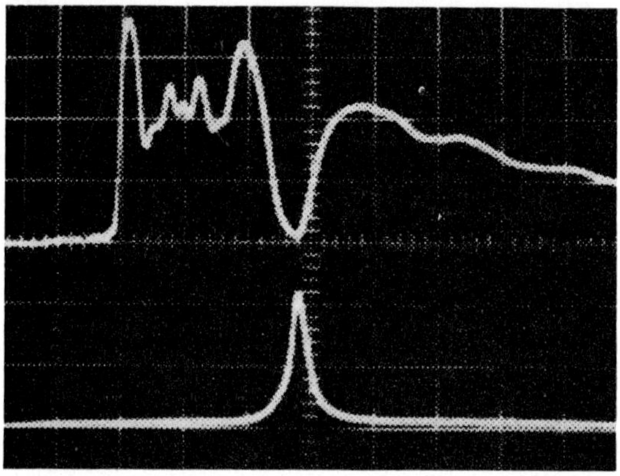

FIG. 34. Line scans of the infrared cathodoluminescence signal (upper trace) and of the barrier e.v.e. short circuit current (lower trace) for a GaAs laser diode. (After Wittry and Kyser, 1965.)

little would reach the junction and little of that would be absorbed at the junction. It was found experimentally that no long-range component corresponding to L_2 occurred on the p-side. Thus the signal on the p-side varied in accordance with eqn (46) while that on the n-side varied as eqn (50) but with L_2 itself increasing with increasing values of X. Values for the minority carrier diffusion lengths and the currents A^1, B^1, etc., that are obtained in this way are reasonable and in agreement with the results of other methods (Wittry and Kyser, 1965; Holt and Chase, 1973). The effect of self-absorption on the effective value of α is discussed in Chapter 10.

Charge collection and cathodoluminescence are competing processes in GaAs. Charge collection at the p–n junction involves separating hole-electron pairs. The separated holes and electrons cannot then recombine radiatively to give out an infrared photon. The inverse relation of the two types of signal was demonstrated by Wittry and Kyser by comparing line scan records of the two signals detected simultaneously as shown in Fig. 34. The charge collection process which is competitive with the cathodoluminescence emission process is of course that in which minority carriers diffuse to the junction to be collected. That is, it is the process represented by the first term of eqn (50). The second term representing the contribution to the charge collection current due to self-detection of the cathodoluminescence is of course directly proportional to the cathodoluminescence. Therefore, the barrier e.v.e. current micrographs at large values of X, when the second term predominates exhibit the same contrast as do the cathodoluminescence

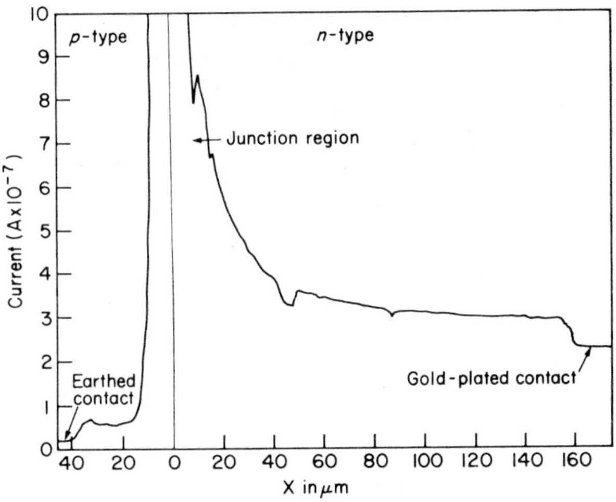

FIG. 35. Barrier electron voltaic effect current recorded at a higher amplifier gain than in Fig. 28, versus distance from the *p–n* junction of a GaAs laser diode showing dot-and-halo contrast (see Chapter 10). (After Holt and Chase, 1968.)

micrographs. These contrast effects arise from non-uniform distributions of radiative (or non-radiative) recombination centres. Such distributions cause variations in the number of photons produced by the recombination of the constant number of hole-electron pairs produced by the electron beam, GI_b, with consequent diminution in the number of photons reaching the *p–n* junction. These effects are responsible for the local variations in the barrier e.v.e. current shown in Fig. 35 recorded at a higher amplifier gain setting than Fig. 33 and for a specimen exhibiting Te segregation contrast as shown in Fig. 36. Te atoms have a 12% larger radius than the As atoms they replace and consequently have a strong tendency to segregate to dislocation lines as a Cottrell atmosphere (Shaw and Thornton, 1968; Holt and Chase, 1973). This "dot-and-halo" form of contrast is discussed in greater detail in Chapter 10.

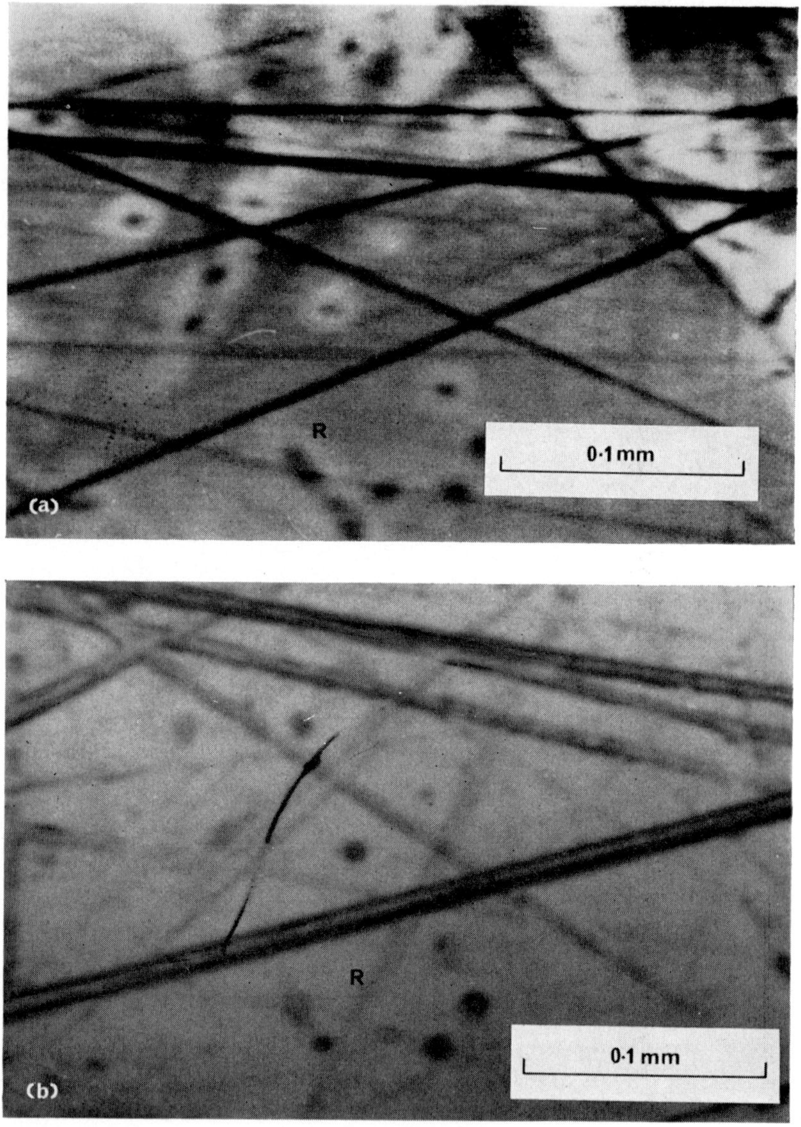

FIG. 36. (a) Barrier electron voltaic effect micrograph of the mirror face of a Te-doped GaAs laser diode. The *p–n* junction is at the top of the picture. (b) Optical micrograph of the same area after etching. (After Holt and Chase, 1968.)

Studies of a similar type have been carried out on GaP electroluminescent diodes (Calverley and Wight, 1970; Lidgard, 1972; Hackett *et al.*, 1972; Holt *et al.*, 1973). The form of the barrier e.v.e. current for a line scan normal to the *p–n* junction is shown in Fig. 37. The rather special conditions which have to be satisfied for a contribution to the barrier electron voltaic effect current to arise from the self-detection of cathodoluminescence as discussed above and in Chapter 10

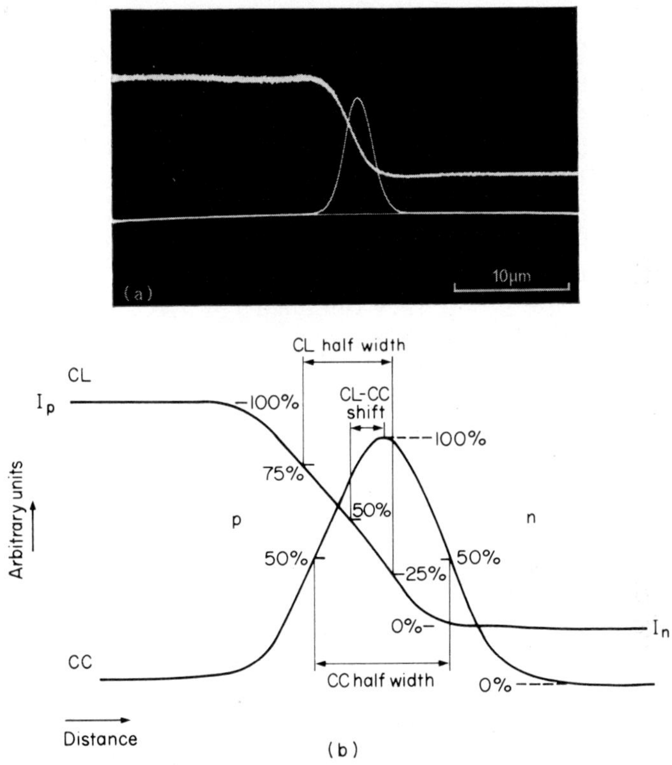

FIG. 37. (a) Charge collection (barrier electron voltaic effect) and cathodoluminescence signals for a line scan across a GaP *p–n* junction. This is a photograph of the experimental Y-modulation signal traces. (b) Diagram of (a) giving the definition of the mid-point of the cathodoluminescence curve and of the CL–CC shift. (After Holt *et al.*, 1973.)

are not satisfied in the case of GaP. The form of the barrier e.v.e. current as a function of X is that of eqn (46) and not of eqn (50) in the present case, therefore, on both the *n*- and the *p*-sides of the junction. The cathodoluminescence signal recorded in Fig. 37 is simply the photomultiplier response to the total emission spectrum. It falls from a high value for the strongly red-emitting p-type material to a low value for the weakly green-emitting n-type material. The 50% cathodoluminescent signal value can be defined as shown in Fig. 33(b). This point is displaced from the maximum position on the charge collection trace as shown in Fig. 37(b) by a distance known as the CL–CC shift. This parameter has been found to correlate with the electroluminescent efficiency of the GaP (Calverley and Wight, 1970; Holt *et al.*, 1973). The values of L and of e_i that are derived from the analysis of barrier e.v.e. current line scan traces across GaP p–n junctions are found to vary with the beam voltage V_b in some cases (Lidgard, 1972; Holt *et al.*, 1973) but not in others (Hackett *et al.*, 1972). This behaviour is not currently understood.

Bresse and Lafeuille (1971) carried out a more complex calculation than that of Kyser and Wittry (1964), of the diode current as a function of the beam–junction distance. The model of Kanaya and Okamoto (1970) was used and two cases considered: that for negligible surface recombination (surface recombination velocity $s = 0$) and that for $s = \infty$. For the case $s = 0$, they obtained $I = I_b G \exp(-X/L)$ which is the same as eqn (46). For the case $s = \infty$, the result was obtained that the effect of surface recombination is that on a semilog plot instead of I *vs* X being a straight line of constant slope, I *vs* X is a curve that is concave upward, as shown in Fig. 39. (The same conclusion was derived previously by van Roosebroek (1955) for the closely analogous barrier photovoltaic effect for p–n junctions.)

Bresse (1972) carried out an experimental test of this result using a Schottky diode with a guard ring, as shown in Fig. 38. By altering the guard ring voltage V_g the surface recombination velocity s can be changed from $s = 0$ to $s = \infty$. Figure 39 plots the results that he obtained for $E_0 = 15$ keV, $R = 2{\cdot}6\ \mu$m, $V = 0$ V and $V_g = +5$ V which makes $s = 0$, and for the same conditions except $V_g = -5$ V which makes $s = \infty$. The general trend of the results is in agreement with the form of the predicted curves. According to Bresse the difference for the case $s = \infty$ between the theoretical and the experimental values, could be explained by the shape of the junction which was not a plane perpendicular to the free surface but as shown in Fig. 38.

Hackett (1972) developed a method to avoid the effect of surface recombination in line scan measurements of the minority carrier diffusion length. He angle-lapped the specimen as shown in Fig. 40 so that the surface scanned was at an angle $0 \ll 90°$ to the p–n junction. The effect of scanning along X is then to alter the height of the point of impact of the beam on the surface above the junction. In effect the scan becomes one normal to the plane of the junction. Hackett made use of a result of Wittry and Kyser that for a uniform large-area electron-beam excitation of minority carriers near the surface $z = 0$ of a semi-infinite semiconductor, the steady-state one-dimensional carrier distribution $\Delta p(z)$ approaches the asymptotic form

$$\Delta p(z) = Ce^{-z/L} \tag{51}$$

several diffusion lengths away from the surface. This result is independent of the surface-recombination velocity and the excitation

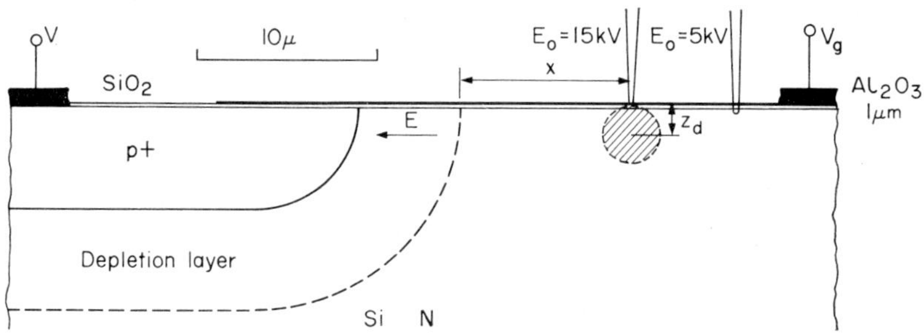

FIG. 38. The form of silicon guard ring device used by Bresse (1972) to measure the barrier electron voltaic effect current as a function of the distance X for surface recombination velocities $s = 0$ and $s = \infty$. (After Bresse, 1972.)

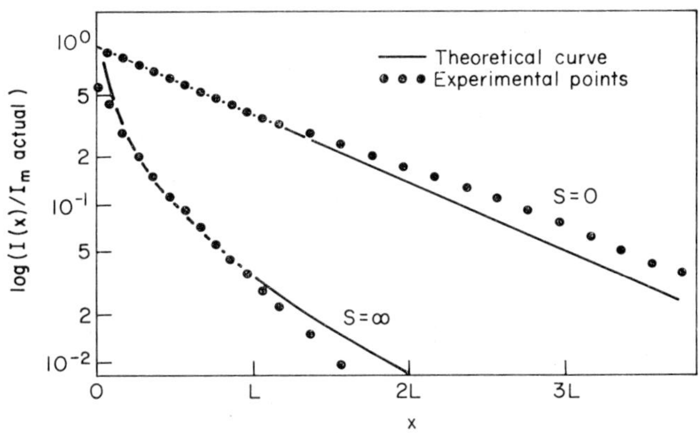

FIG. 39. Comparison of the theoretical and experimental results for the variation of the barrier electron voltaic effect current in the device shown in Fig. 38. (After Bresse, 1972.)

distribution, provided that the depth of the excitation is comparable to the minority-carrier diffusion length L. Hence under the circumstances of Fig. 40

$$I(X) = I_0 \exp(-X \sin\theta/L). \tag{52}$$

Hackett used this technique to measure L in an efficient GaP red-emitting electroluminescent junction.

This method is more time consuming than the normal line scan method of Fig. 30 because of the necessity for angle-lapping the specimen and it destroys the device. However, in cases in which surface recombination is suspected to be playing a role, it may be adopted as a check on the reliability of the simpler method.

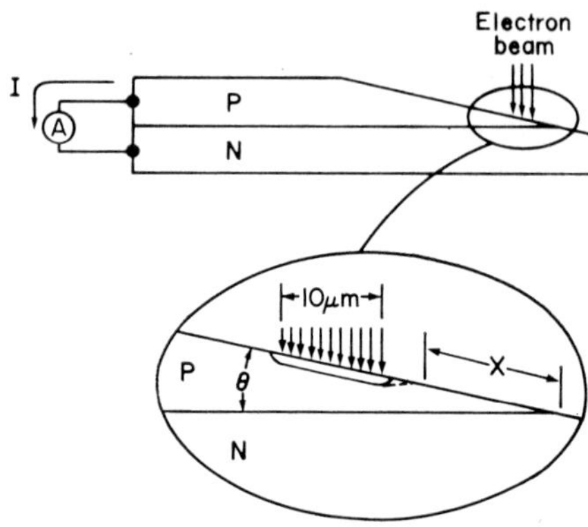

FIG. 40. Angle-lapped specimen configuration for determining the minority carrier diffusion length (in the *p*-type material). A large diameter electron beam is used as shown in the enlarged lower diagram, and scanned over the surface to vary *X*. (After Hackett, 1971.)

A different criticism of the method of Fig. 30 was raised by Zimmermann (1972). He stated that when the beam is incident near the junction (i.e. for $X <$ a few times L) the efficient collection of the minority carriers by the junction disturbs the carrier diffusion and the simple exponential relation (eqn 46) will not be observed. This effect he considers more important than surface recombination for small X. If the material on one side (or both!) of the junction is too narrow, there may be no range of values of X in which the exponential variation of $I(X)$ occurs and L cannot then be measured by this method.

Zimmermann pointed out that if the beam position is kept constant then when the beam is switched off the barrier e.v.e. current will decay as

$$I(t) = I_0 \exp(-t/\tau) \tag{53}$$

where I_0 is the initial steady state value of $I(X)$, t is the time from switch-off and τ is the carrier lifetime, and this relation holds at rather smaller values of X than required for the validity of eqn (46). Zimmermann showed that lifetime measurements could be made in silicon high-power diodes and thyristors with a spatial resolution of about three carrier diffusion lengths, using a pulsed 40 keV beam of 10 μm diameter. Lifetimes larger than 0·3 μsec could be measured. It was found that in one diode the lifetime varied by a factor of four over a distance of a few millimetres.

Hunter *et al.* (1973) used both the Zimmermann pulsed beam method and the normal line scan method to measure L and τ independently on a number of silicon device structures.

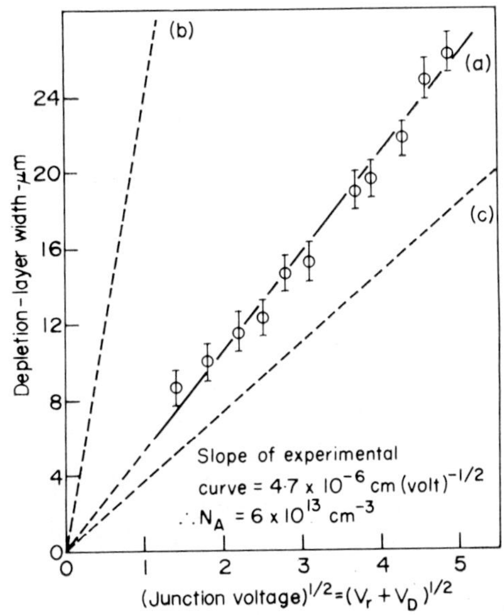

FIG. 41. Measured values of the depletion region width as a function of the reverse bias voltage V_r for a Si *p–n* junction. (After MacDonald and Everhart, 1965.)

To summarize, the barrier electron voltaic effect line scan trace method in many cases provides a convenient method for determining minority diffusion lengths over a wide range of values. Hwang *et al.* (1971) using GaP specimens in which the *p–n* junctions met the surface at an angle were able to measure diffusion lengths as short as 0·13 μm with an accuracy of 15%. In addition, by observing line scans at a number of positions along the junction, it is possible to observe whether the diffusion lengths (and the junction charge collection efficiencies from $I_{m\ \text{actual}} = \eta_{cc} G I_b$) are everywhere the same. The normal line scan method can be combined with the pulsed beam method of Zimmermann (1972) to obtain the carrier lifetime as well.

The efficiency of collection of the hole-electron pairs, that is the fraction of the carriers created that are collected, depends on the nature of the barrier. If the width of the depletion region is greater than or

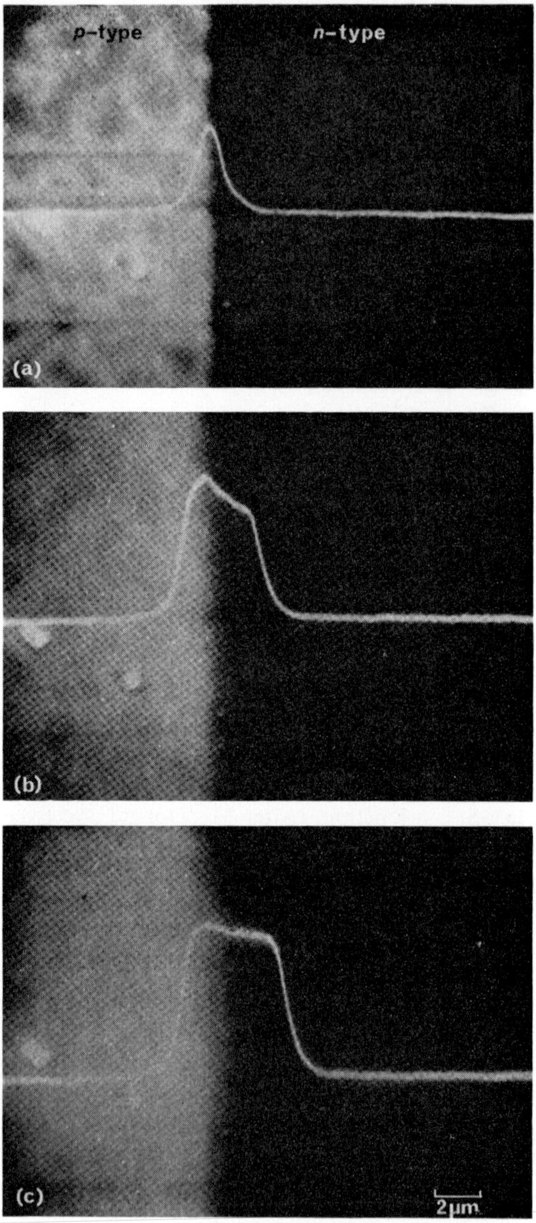

FIG. 42. Secondary electron emission micrographs showing voltage contrast at a *p–n* junction in $ZnSe_yTe_{1-y}$. On each micrograph is superimposed the barrier e.v.e. current line scan trace. (a) Zero bias. (b) 10·5 V reverse bias. (c) 22·5 V reverse bias. (After Aven *et al.*, 1972.)

equal to the radius of the energy dissipation volume, a constant maximum value of the collected diode current $I_{m \text{ actual}}$ will be recorded while the beam scans across the depletion region. The width of the depletion region varies with the applied voltage. Reverse biasing increases the diffusion voltage across the *p–n* junction (Fig. 6(b)) and this leads to an increase in the width of the depletion region as

$$W = \left[\frac{2\varepsilon}{qN_A}(V_r + V_D)\right]^{\frac{1}{2}} \tag{54}$$

where W is the depletion region width, ε is the dielectric constant of the material, N_A is the acceptor atom concentrator on the *p*-side, which is assumed to be of much higher resistivity than the *n*-type side of the junction, and V_r is the reverse bias voltage. MacDonald and Everhart (1965) used this scanning method to obtain a direct measurement of W. Their measurements of W in a *p–n* junction of the n^+–p type to which eqn (54) applies are shown in Fig. 41. The experimental results gave the acceptor density as 6×10^{13} cm^{-3} with a sensitivity that is indicated by the dashed slopes corresponding to $N_A = 10^{13}$ and 10^{14} cm^{-3}.

A related observation was made on *p–n* junctions in $ZnSe_yTe_{1-y}$ by Aven *et al.* (1973), as shown in Fig. 42. This is a set of secondary electron emission micrographs showing voltage contrast at a *p–n* junction under three different biases. Superimposed on each is a barrier e.v.e. current line scan trace. It can be seen how under increasing reverse bias voltages the depletion region spreads into the *n*-type material giving a widening plateau in the barrier e.v.e. current trace as in the work of MacDonald and Everhart. The observations of Fig. 42 were used to estimate the zero bias width of the depletion region in order to discuss non-radiative recombination in this region under forward bias conditions. This constituted a part of a major analysis of the cathodoluminescence from $ZnSe_yTe_{1-y}$ diodes in relation to the electroluminescent efficiency of these devices by Aven *et al.* This work is discussed in Chapter 10.

B. THE BULK ELECTRON VOLTAIC EFFECT

The bulk electron voltaic effect is the generation of a voltage across a semiconducting specimen by electron bombardment of a region of non-uniform resistivity. The physical origin of this effect can be illustrated by means of Fig. 43 which is derived from a diagram given by Tauc (1962), for the bulk photovoltaic effect. The following account of the physical origin of the bulk electron voltaic effect is also closely based on Tauc's treatment. Localized resistivity variations in semiconductors arise through variations in the local densities of the donor and acceptor impurities. Consequently in such regions the position of the Fermi level alters relative to the valence and conduction bands. Since in equilibrium the Fermi level, which is the electrochemical potential, i.e. the free energy of electrons, must be constant the bands must bend as shown in Fig. 43(c), so that localized electric fields exist. When localized electron bombardment occurs in such a region, a bulk electron voltaic effect voltage is generated by two mechanisms so that

$$V = V_c + V_d \tag{55}$$

where V_c is the "chemical" contribution and V_d is the Dember voltage contribution. The chemical contribution arises as follows. The hole-electron pairs generated between points b and c as shown in Fig. 43(a)

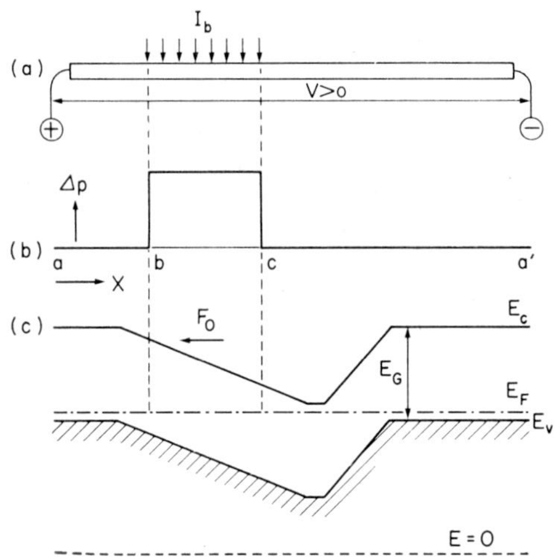

FIG. 43. The physical situation involved in the bulk electron voltaic effect. (a) The irradiated sample. (b) The distribution of the additional charge carrier density Δp generated by electron bombardment. (c) The energy levels of the semiconducting material in the absence of the beam.

and (b) are separated by the field F_0. In the open circuit condition electrical charges build up which alter the field F_0 until the current of "collected" (separated) pairs ceases. The voltage V_c so generated is directly measurable. In the case shown in Fig. 43 the end a acquires a positive charge and the end a' acquires a negative charge. The other component of the bulk electron voltaic effect e.m.f. arises at the boundaries of the electron beam. There is a concentration gradient of injected carriers dn/dx at these boundaries and both holes and electrons will therefore diffuse away from the beam. The mobilities of the holes and electrons in general will not be equal, however, and so the charge carriers of one sign will diffuse further into the material away from the beam. This does not occur symmetrically at the two sides of the beam in the presence of non-uniform doping. This asymmetric separation of charge carriers of opposite sign sets up the Dember voltage.

The theoretical treatment of the bulk photovoltaic effect given by Tauc (1962) was modified by Munakata (1966b, 1967a, 1967b) to give an analysis of the bulk electron voltaic effect. Munakata (1966b) neglected the chemical voltage on the grounds that the field is usually not so high as to alter the diffusive flow of the carriers. He therefore (Munakata, 1967b) used Tauc's expression for the Dember voltage in the case of weak irradiation (i.e. the case that the number of carriers generated per unit volume $\Delta n \ll n_0$) to write for an n-type semiconductor

$$\Delta V = \frac{kT}{q}\frac{2}{b+1}\Delta\sigma\frac{\mathrm{d}\rho}{\mathrm{d}X}\Delta X \tag{56}$$

where k is Boltzmann's constant, T is the absolute temperature, b is the ratio of the electron mobility to the hole mobility, $\Delta\sigma$ is the small change in the conductivity caused by the charge carriers created by the electron beam, ρ is the resistivity and X is the distance along the specimen. $\Delta\sigma$ is assumed to be restricted to a region ΔX (the energy dissipation volume plus the volume into which the generated carriers diffuse). $\Delta\sigma$ and ΔX are assumed to be constant for all positions of the beam along the specimen length. Integration of ΔV then gives $\rho(\Delta V)$ defined by

$$\rho(\Delta V) = \int \Delta V\,\mathrm{d}X \propto \Delta\sigma\,\Delta X \int \frac{\mathrm{d}\rho}{\mathrm{d}X}\,dX \propto \rho \tag{57}$$

which is proportional to the resistivity. In germanium, with which Munakata worked, ΔX was determined by the diffusion length for minority carriers. This diffusion length therefore governs the errors and the resolving power of the method.

To obtain an analytical solution for $\rho(\Delta V)$ by integrating the expression in eqn (56) as indicated above, it is necessary to introduce explicit expressions for ΔX, $\Delta\sigma$ and $\mathrm{d}\rho/\mathrm{d}X$. Munakata (1967a) sets $\Delta X = 2L_e$ where L_e is an effective diffusion length giving the volume in which excess hole-electron pairs occur around the point of impact of the beam, as shown in Fig. 44. According to Munakata, $L_e \simeq L_p/5$ where L_p is the diffusion length for holes in n-type material. If the diameter of the electron beam is greater than $2L_e$ the equations are modified in a minor way as discussed below.

To obtain values for $\Delta\sigma$, it is first necessary to obtain values for the densities of injected (excess) carriers in all parts of the specimen as shown in Fig. 44, i.e. to the left of the beam irradiated volume of length ΔX, in this volume and to the right of it. These expressions for Δp, the number of excess holes (and the equal number of excess electrons)

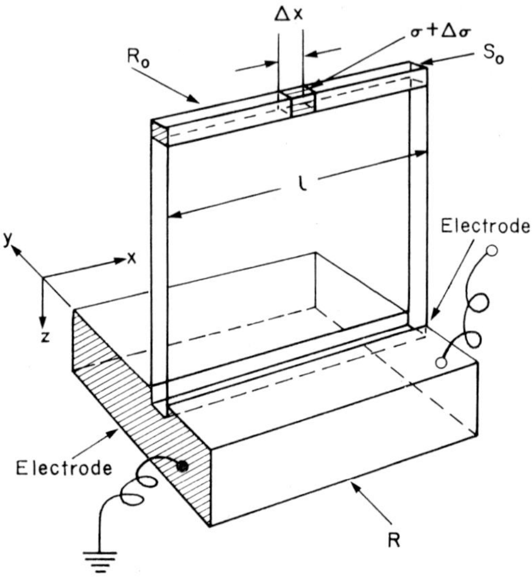

FIG. 44. The effective volume of the probing electron beam in Munakata's theory of the bulk electron voltaic effect. (After Munakata, 1968b.) The same arrangement of the specimen was used in Munakata's (1968b) treatment of β-conductivity.

are obtained as solutions of the equation for diffusive flow of the carriers (eqn (42)) with appropriate boundary conditions. The result is that

$$\Delta\sigma\,\Delta X = \Delta\sigma 2L_e = \frac{q}{kT}\frac{b+1}{2}GI_b\frac{L_p^2}{S}[F(X_0)] \tag{58}$$

where S is the cross section of the specimen and

$$F(X_0) = 1-\frac{\sinh((l-X_0)/L_e)+\sinh(X_0/L_e)}{\sinh(l/L_e)} \tag{59}$$

where l is the length of the specimen between ohmic contacts, and X_0 is the distance of the electron beam from the first contact at $X = 0$. When the radius of the energy dissipation volume d is comparable with L_e, the effective diameter of the volume containing injected excess carriers should be $2(L_e+d)$ and $F(X_0)$ is given by

$$F(X_0) = 1-\frac{\sinh(d/L_e)}{(d/L_e)}\times\frac{\sinh((l-X_0)/L_e)+\sinh(X_0/L_e)}{\sinh(l/L_e)}. \tag{60}$$

However, the effect of d is not serious when d/L_e is less than 1, especially when l/L_e is large.

When l/L_e is large $F(X_0)$ is equal to 1 except at the contacts where it falls steeply to zero (Munakata, 1966b, 1967a), as shown in Fig. 45.

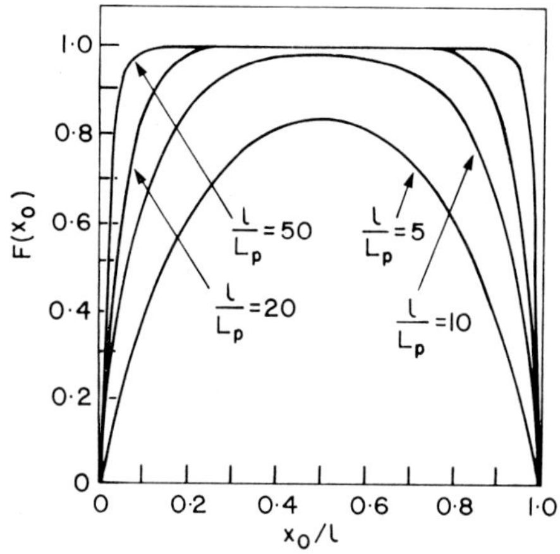

FIG. 45. The form of the function $F(X_0)$ of eqn (59). When the specimen length is $\gg L_p$ the influence of the contacts in reducing the number of injected carriers Δp, the altered conductivity $\Delta\sigma$ and hence the induced voltage decrement ΔV is confined to a narrow end effect. For relatively short specimens, however, the effect occurs throughout. (After Munakata, 1966b.)

Substituting eqn (58) into (56) gives

$$\Delta V = \left(\frac{\mathrm{d}\rho}{\mathrm{d}X}\right) 2GI_b \frac{L_p^2}{S}[F(X_0)]. \tag{61}$$

If $\mathrm{d}\rho/\mathrm{d}X$ is constant and given by

$$\frac{\mathrm{d}\rho}{\mathrm{d}X} = \frac{\rho_2 - \rho_1}{l}. \tag{62}$$

Then substituting into eqn (61) and integrating from $X = 0$ to X_0 gives

$$\rho(\Delta V) = (\rho_2 - \rho_1) 2GI_b \frac{L_p^2}{S}[G(X_0)] + C \tag{63}$$

where

$$G(X_0) = \left(\frac{X_0}{l}\right) + \left(\frac{L_e}{l}\right) \frac{\cosh((l - X_0)/L_e) - \cosh(X_0/L_e) - \cosh(l/L_e) + 1}{\sinh(l/L_e)} \tag{64}$$

and C is a constant of integration taken to be zero in Munakata's (1967a) discussion. The functions $\rho(X)$ and $G(X_0)$ are shown in Fig. 46.

To treat the resolving power of this technique Munakata considered the case, shown in Fig. 47, of a resistivity variation given by

$$\rho = \rho_1 + \Delta\rho\left[1 - \cos\left\{2\pi\left(\frac{X - l/2}{\lambda}\right)\right\}\right] \tag{65}$$

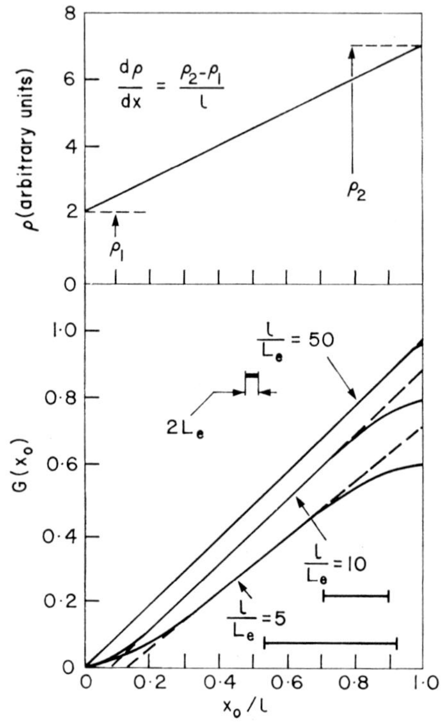

FIG. 46. (a) Linear resistivity variation and (b) integrated values of induced voltages. For small values of L_e the form of the voltage due to the bulk electron voltaic effect exactly reproduces the original resistivity variation. (After Munakata, 1967b.)

where $\Delta\rho$ is the amplitude of the resistivity peaks separated by a distance λ. Substituting into (61) and integrating, Munakata (1966) obtained

$$\rho(\Delta V) = \Delta\rho 2GI_b \frac{L_p^2}{S} \frac{1}{1+(2\pi(L_e/\lambda))^2}[G_\delta(X_0)] + C_\delta \tag{66}$$

where $G_\delta(X_0)$ is a complicated function given by Munakata (1966b, 1967a), and shown graphically in Fig. 47(b) and C_δ is a constant of integration. In the case of a regular periodic variation in the resistivity, the bulk electron voltaic effect voltage reproduces the form of the variation except near the electrodes (Munakata, 1967b). In the case of a step in the resistivity (a linear rise from ρ_1 to ρ_2 over a short distance) the bulk e.v.e. voltage consists of a peak (Munakata, 1966b). This is, of course, the case of a barrier and the result is essentially that of the barrier electron voltaic effect. Munakata's treatment shows that when obtaining the minority carrier diffusion length from a plot of the log of the barrier e.v.e. voltage care must be taken for if the value of L_p is $>l/10$ the result will be erroneous because of the end effect, that is because of the drop in $F(X_0)$ shown in Fig. 45, near the contacts.

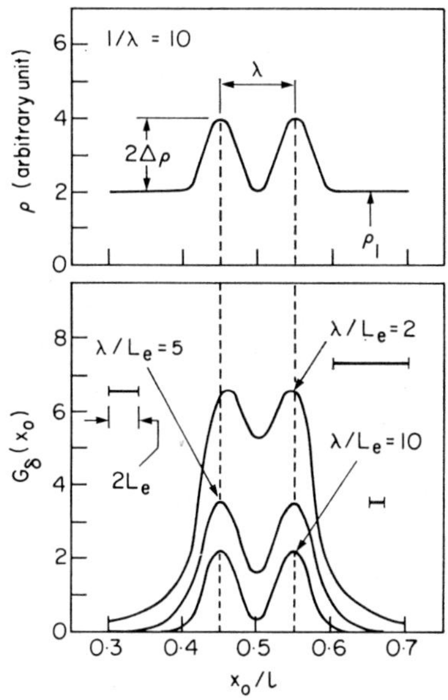

FIG. 47. Case of a specimen containing two nearby resistivity peaks leading to a treatment of resolving power (a) resistivity and (b) functional form of the voltage due to the bulk electron voltaic effect. When $\lambda/L_e > 10$ the voltage has the same form as the resistivity. (After Munakata, 1967b.)

Munakata (1967b) used an electronic integrating circuit to obtain $\rho(\Delta V)$ from the incoming signal, ΔV, and displayed this as a Y-modulation trace on a synchronously scanned C.R.O. By calculating the variation of resistivity $\rho(\Delta V)$ corresponding to a particular change of signal ΔV the relative scale of the output can be fairly readily calibrated. However, to obtain an absolute value for the resistivity, the constant of integration C_δ has to be determined by an independent measurement. Munakata showed that the values for ρ *vs* X obtained in this way were in good agreement with the results obtained by conventional mechanical single-probe measurements.

Equation (61) shows that the sensitivity of the method, i.e. the magnitude of ΔV, the signal, depends strongly on the minority carrier diffusion length L_p. The sensitivity of the method is therefore good in well developed materials such as Ge in which L_p may be large. In such cases the effective diffusion length L_e, which determines the spatial resolution of the method, was found to be only a few tenths of L_p, so that good resolutions were also obtained. In materials such as GaAs and CdS, however, the small values of L_p result in voltage changes ΔV which are too small to allow this method to be used. To overcome this limitation Munakata developed a second technique which employs an effect that he referred to as β-conductivity and which is usable for crystals having short minority carrier diffusion lengths.

C. β-CONDUCTIVITY

β-Conductivity is observed when the bombarded specimen has an external d.c. biasing voltage applied as in Fig. 1(c). The electron beam creates free charge carriers and this increases the conductance of the specimen in the bombarded region. The voltage across the specimen will therefore decrease if the bias is supplied by a constant current source, or alternatively an increased current will flow if a constant voltage source is employed. The voltage decrease or the current increase is then used as the β-conductivity signal.

(1) *Munakata's Constant Current Theory*

Munakata (1968a, 1968b) treated the case in which a constant current source is used and the voltage is detected. The β-conductivity $\Delta\sigma$ is constant but its effect is position dependent because the voltage decrement is dependent on the relative magnitudes of the electron beam induced conductivity and the original conductivity. The contribution of the β-conductivity proper to the voltage decrement has to be separated from the contributions due to the bulk electron voltaic effect and that due to the specimen current. Munakata (1968a, 1968b) showed the β-conductive signal to be proportional to the square of the local electric field in the material and thus to the square of the local resistivity if the current density is constant.

Munakata (1968b) again treated the specimen as consisting of a filament or rod of resistance R_0, which contains the energy dissipation volume of length ΔX and cross section S_0, and the remainder of the specimen of resistance R, as shown in Fig. 44. The equivalent circuit of the specimen and measurement circuit for this work is as shown in Fig. 48. Munakata's (1968b) analysis of this arrangement is as follows:

Let σ and E be the conductivity and electric field in the rod. The current density J will be given by

$$J = \sigma E = \frac{V_s}{R_0} \frac{l}{S_0} \tag{67}$$

where V_s is the applied voltage. The effect of the electron beam is to change the conductivity in the energy dissipation volume by $\Delta\sigma$ and the field locally by δE. Then

$$J + \Delta J = (\sigma + \Delta\sigma)(E - \delta E). \tag{68}$$

Since the bar is connected to a constant current source $\Delta J = 0$ and therefore

$$\Delta\sigma E \simeq \delta E \sigma. \tag{69}$$

The decrease in the voltage between the ends of the bar is therefore given by:

$$\delta V_0 = \int_0^l \delta E \, \mathrm{d}X \simeq E \frac{\Delta\sigma}{\sigma} \Delta X = \frac{l}{\sigma^2} \frac{V_s}{R_0} \frac{1}{S_0} \Delta\sigma \, \Delta X. \tag{70}$$

The actual observable decrease in the specimen voltage V_s must be calculated using the equivalent circuit of Fig. 48. The voltage decrease δV detected by the resistor r is then found to be

$$\delta V = \frac{rR_t}{r + R_t} \frac{\rho^2}{R_0^2} \frac{\Delta\sigma \, \Delta X}{S_0} V_s = \frac{rR_t}{r + R_t} E^2 \frac{S_0}{V_s} \Delta\sigma \, \Delta X \tag{71}$$

where ρ is the resistivity and R_t the total specimen resistance, i.e.

$$R_t = \frac{R_0 R}{(R_0 + R)}.$$

Equation (71) shows that the β-conductive signal is proportional to E^2 or ρ^2 as stated above. If R_0, S_0, $\Delta\sigma$ and ΔX are constant as the beam is scanned along the X direction $(\delta V)^{\frac{1}{2}}$ gives the resistivity.

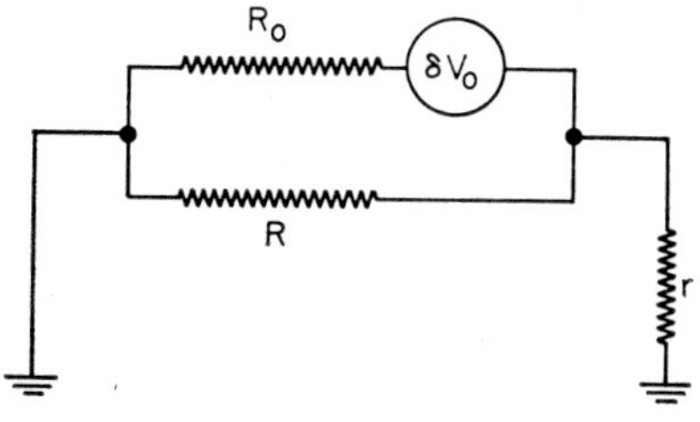

FIG. 48. The equivalent circuit of the arrangement shown in Fig. 44. $R_t = R_0 R/(R_0 + R)$. (After Munakata, 1968b.)

When the resistivity varies along the Y or Z directions as well as the X direction, R_0 and hence J are not constant along the X axis (the line scanned). In this case the β-conductive voltage signal does not give the resistivity distribution but the field distribution along the beam path.

As in the bulk electron voltaic effect case, the resolution is limited by the size of the incident beam (d) and the diffusion length for minority carriers. The formula for $\Delta\sigma\Delta X$ in eqn (71) is that obtained previously (eqn 58). Thus the end effect due to the presence of ohmic contacts, contained in the function $F(X_0)$ shown in Fig. 45 occurs in the β-conductive signal too.

Using a more general approach Munakata (1968a) obtained for the total change in the terminal voltage of the small bar alone (i.e. neglecting the effect of the remainder of the specimen in parallel with the irradiated bar):

$$\Sigma V_0 = I_s R_1 + \Delta\sigma\,\Delta X \rho^2 J + \Delta\sigma\,\Delta X \frac{kT}{q}\frac{2}{b+l}\frac{\mathrm{d}\rho}{\mathrm{d}X} \tag{72}$$

where R_1 is the resistance to earth from the point of impact of the beam, i.e. R_1 is a fraction of R_0 the resistance of the whole bar. The first term on the right hand side of eqn (72) is the ohmic drop due to the specimen current flowing to earth through part of the bar and the contact at one end. The second term is the β-conductive signal δV calculated above (eqns (71) and (58)). The third term is the bulk e.v.e. voltage.

The actual change of the specimen voltage ΣV in view of the equivalent circuit of Fig. 48 is given by

$$\Sigma V = \frac{r}{r+R_t}\frac{R_t}{R_0}(\Sigma V_0). \tag{73}$$

Of the three terms contributing to the observable change in the terminal voltage across the specimen only the β-conductive voltage signal δV can be externally controlled by adjusting the biasing current density J. The method is of special interest as noted above for those materials in which the bulk e.v.e. is small owing to small minority carrier diffusion lengths. The fact that the β-conductive signal can be set to a convenient value, within limits, by adjusting the external current source is a practical advantage of this technique. The fact that the β-conductive signal is proportional to the square of the resistivity whereas the bulk e.v.e. is proportional to the gradient of the resistivity simplifies the analysis of the signal in the present case.

Munakata (1968b) used a "rooter" to produce the signal for display in the SEM. That is, he employed an amplifier which for an input δV gives an output which is proportional to $(\delta V)^{\frac{1}{2}}$ and thus to the resistivity.

He demonstrated that the method could be used successfully for both Ge and CdS, that is for materials with both large and small minority carrier diffusion lengths.

Munakata (1972) showed that the spatial resolution of the β-conductive technique for determining local variations in the electrical field could be increased by chopping the incident electron beam at frequencies in the kHz range. This effect occurs since the resolution with a continuous beam depends on L_e which depends on L_p. The effective lifetime for minority carriers generated by the pulsed beam, however, decreases as the frequency of the pulses increases. The experimental observations on Ge supported this prediction.

(2) *Gopinath's Constant Voltage Theory*

Recently Gopinath (1970), and Gopinath and de Monts de Savasse (1971) treated the situation in which the voltage applied is kept constant and the current change due to electron bombardment is used as β-conductive signal. The treatment gave expressions for the spatial variation, from anode to cathode, of the signal as a function of the bias voltage. Both high and low bias field cases were treated whereas Munakata's treatment considered only the small bias field case. In Gopinath's work both linear and annular specimen geometries were considered. We will summarize first the linear theory.

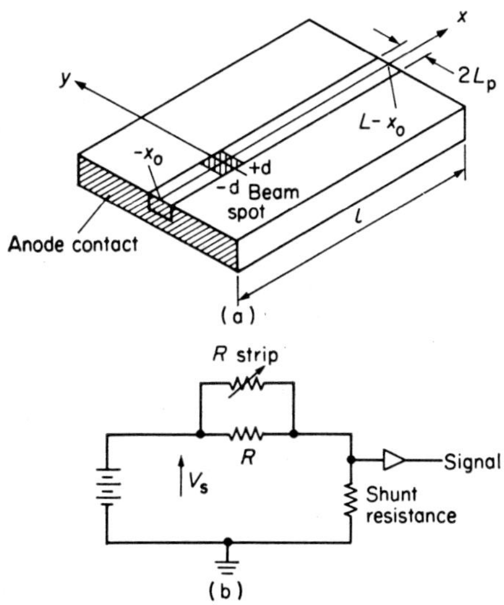

FIG. 49. (a) Geometry of the linear devices with the irradiated strip with origin at the centre of the beam spot. (b) The equivalent circuit of the device. The shunt resistance is much smaller than R. (After Gopinath, 1970.)

In the absence of the beam Ohm's Law states for the scanned rod of Fig. 49(a):

$$R_0 I = V_s \tag{74}$$

where V_s is the applied specimen voltage. Beam bombardment alters this to

$$(R_0 - \Delta r)(I + \Delta I) = V_s + \Delta V.$$

But since a constant voltage source is used $\Delta V = 0$ and therefore

$$\Delta r I \simeq \Delta I R_0 \tag{75}$$

(i.e. assuming $\Delta r \Delta I$ to be negligible). Therefore combining eqns (74) and (75)

$$\Delta I \simeq \frac{\Delta r(X_0)}{R_0^2} V_s. \tag{76}$$

For an n-type specimen the effect of the bombardment is to generate Δp excess minority carriers and produce a change in conductivity

$$\Delta\sigma(X) = q\Delta p(X)(1+b)\mu_p \tag{77}$$

where μ_p is the hole mobility. Now

$$R_0 = \int_{-X_0}^{L-X_0} \frac{\mathrm{d}X}{\sigma S_0}$$

and the effect of the beam is to reduce this resistance of the scanned rod to

$$R_0 - \Delta r(X_0) = \int_{-X_0}^{L-X_0} \frac{\mathrm{d}X}{(\sigma + \Delta\sigma) S_0}. \tag{78}$$

Substituting (78) and (77) into (76) gives

$$\Delta I = \left\{\frac{V_s q(1+b)\mu_p}{R_0^2 S_0}\right\} \int_{-X_0}^{L-X_0} \frac{\Delta p(X)}{\sigma^2}\,\mathrm{d}X. \tag{79}$$

To obtain $\Delta p(X)$ for insertion into (79) Gopinath solved the continuity equation assuming that there is no space charge in the specimen and that the minority carrier distribution is uniform across the rod, i.e. no variation with y:

$$\frac{\mathrm{d}^2\Delta p}{\mathrm{d}X^2} - \frac{(E/E_c)}{L_p}\frac{\mathrm{d}\Delta p}{\mathrm{d}X} - \frac{\Delta p}{L_p^2} = \frac{g'\tau_p}{L_p^2} \tag{80}$$

where $E_c = (kT/qL_p)$ is the diffusion field. The boundary conditions applied were

(i) $\Delta p = 0$ at $X = -X_0$ and at $L - X_0$, i.e. ohmic contacts,
(ii) Δp is continuous at $\pm d$, the spot boundaries (Fig. 44(a)), and
(iii) $\mathrm{d}\Delta p/\mathrm{d}X$ is continuous at $\pm d$.

The solution of (80) has the form:

$$\begin{aligned} \Delta p &= A_1 \exp\left(\frac{X}{L_u}\right) + B_1 \exp\left(\frac{-X}{L_d}\right) & -X_0 \leqslant X \leqslant -d \\ &= A_2 \exp\left(\frac{X}{L_u}\right) + B_2 \exp\left(\frac{-X}{L_d}\right) + \frac{g\tau_p L_d L_u}{L_p^2} & -d \leqslant X \leqslant d \quad (81) \\ &= A_3 \exp\left(\frac{X}{L_u}\right) + B_3 \exp\left(\frac{-X}{L_d}\right) & d \leqslant X \leqslant L_0 - X_0 \end{aligned}$$

where

$$L_d = \frac{2L_p}{\{(E/E_c)^2 + 4\}^{\frac{1}{2}} - E/E_c}$$

and

$$L_u = \frac{2L_p}{\{(E/E_c)^2 + 4\}^{\frac{1}{2}} + E/E_c}$$

are referred to as the "downstream diffusion length" and the "upstream diffusion length" respectively.

For the low-field case, $L_d \simeq L_u \simeq L_p$, the results of Gopinath's analysis are essentially similar to Munakata's treatment.

In the high field case, i.e. for $E/E_c \gg 1$, and assuming that $L_p \ll L$ so that end effects are not dominant, the integral in equation (79) reduces to

$$\int_{-X_0}^{L-X_0} \frac{\Delta p}{\sigma^2}\,\mathrm{d}X \simeq \text{constant}\,(L - X_0 + 3D). \quad (82)$$

This is only valid if σ is constant.

Physically this means that for high applied fields (E/E_c large) the minority carriers (holes—since we are considering n-type material) drift only towards the cathode, diffusion towards the anode being negligible. In this case the signal (current) will vary linearly with distance from the cathode in accordance with (82) and reach its peak value near the anode as shown in Fig. 50(b) and (c) for $E/E_c = 10$.

For intermediate values of the applied field Gopinath obtained the calculated variations of current (signal) with X shown in Fig. 50 for various assumed values of E/E_c by substituting the values of Δp from eqn (81) into eqn (79). By comparing the shape of the β-conductive current line scan trace across a device of the form shown in Fig. 49(a) for different bias levels with the calculated curves of Fig. 50, an estimate of the value of L_p can be obtained. By examining line scans at various positions y across the specimen any non-uniformities in the field can be located. By decreasing the bias the bulk electron voltaic effect signal can be obtained giving the variations in the resistivity (which is assumed constant in the calculations of Fig. 50).

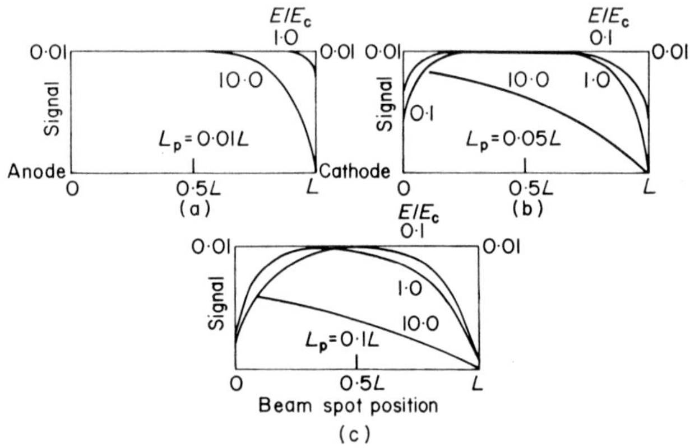

FIG. 50. β-conductivity current signal as a function of the position of the spot along a linear device as shown in Fig. 49(a). The results are normalized by setting $g'\tau_p = 1.0$, spot width $2d = 0.01$ and the ordinate is multiplied by

$$\frac{\sigma^2 R_0^2 S_0}{V_s q(1+b)\mu_p} \qquad \text{(eqn 79)}$$

σ is assumed constant with X. (After Gopinath, 1970.)

Gopinath (1970) reported observations on a GaAs Gunn oscillator of the linear geometry analysed above and found the results in agreement with the theory. He was able to obtain a consistent set of estimates for E_c, L_p and τ_p from the form of the line traces.

Gopinath and de Monts de Savasse (1971) extended the theory to devices with annular geometry. This required a simple two-dimensional treatment to predict the spatial distribution of the signal in annular devices. The theory predicts a geometry-dependent contrast at small applied bias when diffusion predominates, and a field dependent contrast at larger bias when drift dominates. Again SEM observations on GaAs transferred electron devices with annular geometry were reported and these supported the theory. They pointed out that additional contrast will appear due to any local defects that affect the generation and recombination rate. The generation rate $g'(= GI_b)$ in the continuity eqn (42) and (80) appears as a multiplication factor in

Δp. The local recombination rate in the energy dissipation volume determines the number of carriers that diffuse or drift away from the spot and therefore the size of the region of enhanced "β-conductivity". If a localized region has an infinite recombination rate no carriers will drift outward and therefore the region will appear dark on the micrograph regardless of the magnitude of the biasing field. This is analogous to the effect of dislocations on the barrier electron voltaic effect studied by Czaja (1966) as outlined previously.

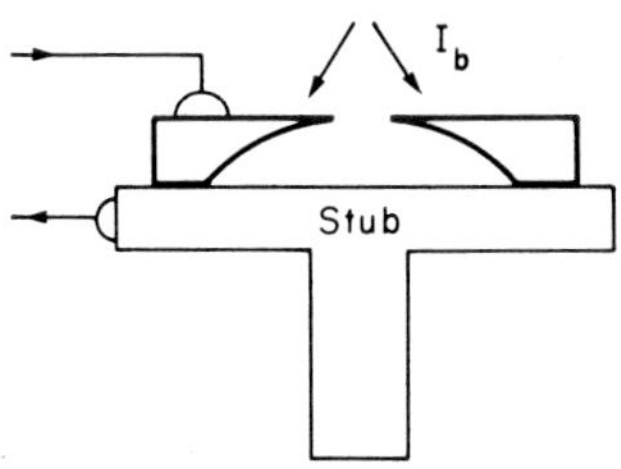

FIG. 51. A standard transmission electron microscope specimen mounted for examination by β-conductivity observations in the SEM. (After Shaw and Booker, 1969.)

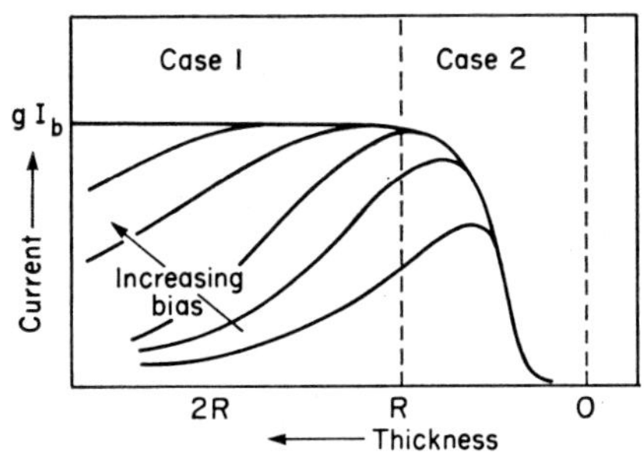

FIG. 52. β-conductive current as a function of the distance of the beam from the centre of the hole in the specimen of Fig. 51, and as a function of the biasing field. (After Shaw and Booker, 1969.)

For defect analysis studies, Shaw and Booker (1969) considered the β-conductivity effect in specimens of the standard thinned disc form used in transmission electron microscopy as shown in Fig. 51. They reported in graphical form as shown in Fig. 52 the results of an analysis (which was not, however, given in full) of the variation of the current signal with R the distance from the centre of the specimen. The bias field falls with increase in R because a constant voltage is applied between the top and bottom of the specimen but the thickness increases with R. Preliminary observations were reported that were in reasonable agreement with expectations. This method which makes possible a direct comparison with the transmission electron microscope results should prove useful in identifying defects with observed conductivity contrast properties.

REFERENCES

ABRAHAM, J. M., WOLFGANG, L. G. and INSKEEP, C. N. (1966). *In* "Advances in Electronics and Electron Physics" (L. Marton, ed.), Vol. 22**B**, pp. 671–682. Academic Press, London and New York.

ARCHARD, G. D. (1961), *J. appl. Phys.* **32,** 1505–1509.

AVEN, M., DEVINE, J. Z., BOLON, R. B. and LUDWIG, G. W. (1972). *J. appl. Phys.* **43,** 4136–4142.

BALDINGER, E., GUTMANN, J. and MATILE, G. (1964). *Z. angew. Math. Phys.* **15,** 90–93.

BRESSE, J. F. (1972), SEM 1972. *Proc. Vth Ann. SEM Symposium IITRI, Chicago*, 105–112.

BRESSE, J. F. and LAFEUILLE, D. (1971). *In* "Electron Microscopy and Analysis." Conf. Series No. 10, 1971, pp. 220–223. Institute of Physics, London.

BROWN, A. V. (1962). *I.E.E.E. Trans. El Devices* **ED–10** (1962), 8–12.

CALVERLEY, A. and WIGHT, D. R. (1970). *Solid St. Electronics.* **13,** 382–385.

CASEY, H. C. and KAISER, R. H. (1967). *J. electrochem. Soc.* **114,** 149–153.

CHASE, B. D. and HOLT, D. B. (1973). *Phys. Stat. sol.* **a19,** 467–478.

CZAJA, W. (1966). *J. appl. Phys.* **37,** 4236–4248.

CZAJA, W. and WHEATLEY, G. H. (1964). *J. appl. Phys.* **35,** 2782.

CZAJA, W. and PATEL, J. R. (1965). *J. appl. Phys.* **36,** 1476–1482.

DEARNALEY, G. and NORTHROP, D. C. (1966). "Semiconductor Counters for Nuclear Radiations." Spon, London.

EHRENBERG, W., LANG, C. S. and WEST, R. (1951). *Proc. phys. Soc.* **64A,** 424.

FABRI, G., GATTI, E. and VELTO, V. S. (1963). *Phys. Rev.* **131,** 134–136.

GOLDSTEIN, B. (1965). *J. appl. Phys.* **36,** 3853–3856.

GOLDSTEIN, J. I. (1969). In "Electron Probe Microanalysis" (Advances in Electronics and Electron Physics Suppl. 6, A. J. Tousimis and L. Marton, eds), pp. 245–290. Academic Press, London and New York.

GOPINATH, A. (1970). *J. Phys. D Appl. Phys.* **3,** 467–472.

GOPINATH, A. and DE MONTS DE SAVASSE, T. (1971). *J. Phys. D Appl. Phys.* **4,** 2031–2038.

HACKETT, W. H. (1972). *J. appl. Phys.* **43,** 1649–1654.

HACKETT, W. H., SAUL, R. H., DIXON, R. W. and KAMMLOTT, G. W. (1972). *J. appl. Phys.* **43,** 2857–2868.

HIGUCHI, H. and TAMURA, H. (1965). *Jap. J. appl. Phys.* **4,** 316–317.

HILL, D. E. (1964). *Phys. Rev.* **133,** A 866–872.

HOFF, P. and EVERHART, T. E. (1970). *IEEE Trans. El. Devices* **ED–17,** 458–465.

HOLT, D. B. and CHASE, B. D. (1968). *J. Mater. Sci.* **3,** 178–182.
HOLT, D. B. and CHASE, B. D. (1973). *Phys. Stat. sol.* **a20,** 135–144.
HOLT, D. B., CHASE, B. D. and CENSLIVE, M. (1973). *Phys. Stat. sol.* **a 20,** 459–467.
HOLT, D. B. and CULPAN, M. (1970). *J. Mater. Sci.* **5,** 546–556.
HUNTER, D. R., PAXMAN, D. H., BURGESS, M. R. and BOOKER, G. R. (1973). *SEM 73 Proc. Conf. Newcastle*, pp. 208–213. Institute of Physics, London.
HWANG, C. J., HASZKO, S. E. and BERGH, A. A. (1971). *J. appl. Phys.* **42,** 5117–5119.
IVAKHNO, V. N. and NASLEDOV, D. N. (1965). *Sov. Phys. solid St.* **6,** 1651–1655.
JONES, G. A. C., NAG, B. R. and GOPINATH, A. (1973). *Proc. VIth Ann. SEM Symposium, IITRI, Chicago*, 309–316.
JOHNSON, E. O. (1957), R.C.A. *Review*, **18,** 556–577.
KALIBJIAN, R. and MAYEDA, K. (1971). *Solid St. Electronics* **14,** 529–540.
KANAYA, K. and OKAYAMA, S. (1970). *In Microscopie Electronique* 1970, *Grenoble*, 159–160. Soc. Franc. Micros. Electron, Paris.
KAZAN, B. and KNOLL, M. (1968). "Electronic Image Storage", pp. 33–51, and pp. 114, 147–154, 237, 360–368. Academic Press, New York and London.
KLEINMAN, D. A. (1956). *Bell System. Tech. J.* **35,** 685–706.
KYSER, D. F. and WITTRY, D. B. (1964). *In* "The Electron Microprobe" (T. D. McKinley, K. F. J. Heinrich and D. B. Wittry, eds), pp. 691–714. Wiley, New York.
LANDER, J. J., SCHREIBER, H., BUCK, T. M. and MATHEWS, J. R. (1963). *Appl. Phys. Letters* **3,** 206–207.
LAPPE, F. (1959), *Z. Phys.* **154,** 267–285.
LIDGARD, G. (1972), *Solid St. Electronics* **15,** 159–164.
LUCOVSKY, G., VARGA, A. J. and SCHWARZ, R. F. (1965). *Solid St. Commun.* **3,** 9–13.
MACDONALD, N. C. and EVERHART, T. E. (1965). *Appl. Phys. Letters*, **7,** 267–269.
MANN, H. M., HASLETT, J. W. and LIETZ, G. P. (1961). *I.E.E. Trans. Nucl. Sci.* **NS 8,** 151–156.
MCKENZIE, J. M. and BROMLEY, D. A. (1959). *Phys. Rev. Letters* **2,** 303–305.
MAKHOV, A. F. (1961a), *Sov. Phys. Solid St.* **2,** 1934–1941.
MAKHOV, A. F. (1961b), *Sov. Phys. Solid St.* **2,** 1942–1944.
MAKHOV, A. F. (1961c), *Sov. Phys. Solid St.* **2,** 1945–1951.
MIYAJI, K. and MIYAZAKI, E. (1963). *Jap. J. appl. Phys.* **2,** 129–130.
MIYAZAKI, E., MAEDA, H. and MIYAJI, K. (1966). *In* "Advances in Electronics and Electron Physics" (L. Marton, ed.), Vol. 22A, pp. 331–339. Academic Press, London and New York.
MUNAKATA, C. (1965). *Jap. J. appl. Phys.* **4,** 697.
MUNAKATA, C. (1966a). *Jap. J. appl. Phys.* **5,** 1251.
MUNAKATA, C. (1966b). *Jap. J. appl. Phys.* **5,** 756–763.
MUNAKATA, C. (1967a). *Microelectronics and Reliability*, **6,** 27–33.

MUNAKATA, C. (1967b). *Jap. J. appl. Phys.* **6,** 963–971.
MUNAKATA, C. (1968a). *Jap. J. appl. Phys.* **7,** 1051–1055.
MUNAKATA, C. (1968b). *J. Phys. E Sci. Instr.* **1,** 639–642.
MUNAKATA, C. (1972). *J. Phys. D: Appl. Phys.* **5,** 1000–1004.
NEVE, N. F. B. and THORNTON, P. R. (1966). *Solid St. Electronics* **9,** 900–901.
PFANN, W. G. and VAN ROOSBROECK, W. (1954). *J. appl. Phys.* **25,** 1422–1434.
RAPPAPORT, P. (1954). *Phys. Rev.* **93,** 246–247.
RAPPAPORT, P., LOFERSKI, J. J. and LINDER, E. G. (1956). *R.C.A. Rev.* **17,** 100–128.
RAVI, K. V., VARKER, C. J. and VOLK, C. E. (1973). *J. electrochem. Soc.* **120,** 533–541.
SAH, C. T., NOYCE, N. and SHOCKLEY, W. (1957). *Proc. I.R.E.* **45,** 1228–1243.
SHAW, A. M. B. and BOOKER, G. R. (1969). *In* "Scanning Electron Microscopy 1969". *Proc. IInd Ann. SEM Symposium, IITRI, Chicago*, pp. 459–463.
SHAW, D. A. and THORNTON, P. R. (1968). *J. Mater. Sci.* **3,** 507–518.
SMITH, A. and DUTTON, D. (1958). *J. Opt. Soc. Am.* **48,** 1007–1009.
STERNGLASS, E. J. (1954). *Phys. Rev.* **95,** 345–358.
TAUC, J. (1962). "Photo and Thermoelectric Effects in Semiconductors", pp. 98–112. Pergamon, Oxford.
THORNTON, P. R. (1968). "Scanning Electron Microscopy." Chapman and Hall, London.
VAN HEERDEN, P. J. (1957). *Phys. Rev.* **106,** 468–473.
VAN ROOSBROECK, W. (1955). *J. appl. Phys.* **26,** 380–391.
VAVILOV, V. S., NOLLE, E. L., EGOROV, V. D. and VINTOVKIN, S. I. (1964). *Sov. Phys. solid St.* **6,** 1099–1104.
WITTRY, D. B. and KYSER, D. F. (1964). *J. appl. Phys.* **35,** 2439–2442.
WITTRY, D. B. and KYSER, D. F. (1965). *J. appl. Phys.* **36,** 1387–1389.
WITTRY, D. B. and KYSER, D. F. (1966). *J. phys. Soc. Japan* **21** (*suppl.*), 312–316.
ZIMMERMANN, W. (1972). *Phys. St. sol.* **a12,** 671–678.

Chapter 9

Cathodoluminescence

M. D. MUIR and P. R. GRANT

Department of Geology, Royal School of Mines, Imperial College, London, England

List of Symbols and Abbreviations

implied = *not* defined in the text

CRT = Cathode ray tube
SEM = Scanning electron microscope
e.h.t. = Extra high tension (implied)
$R(\lambda)$ = Count rate read for wavelength
λ = Wavelength
$K(\lambda)$ = Total response factor
k = Partial response factor
$Q(\lambda)$ = Photomultiplier quarter efficiency
$F(\lambda)$ = Collected fraction of radiation reaching monochromator
$A(\lambda)$ = Self absorption coefficient rate
$\phi(\lambda)$ = Number of photons of a given wavelength generated per second
ϕ = Quantum yield
η = mirror efficiency
τ_L = Transmission coefficient of perspex light pipe
τ_F = Transmission coefficient of fibre light pipe
°K = Degrees Kelvin
BB = Blue band
RB = Red band

S = Light sum
J = Peak height (subscript indicating the peak)
CL = Cathodoluminescence/cathodoexcitation
UV = Ultraviolet/photoexcitation
$^5D_3$ = Third split state of fifth d orbital (implied)
T = Temperature
p = Density of hole-electron pairs
g = Generation rate of free electrons by electron beam
σ = cross-section governing capture of a free hole by a free electron
N = Concentration of excitonic molecules
n = Concentration of free excitons
α = Capture coefficient from free excitons.
ϕ° = Natural quantum yield
ϕ_i = Quantum yield (subscript indicating relation process leading to radiative transition)
IC = Internal conversion
IS = Intersystem crossing
$T_1 \leftarrow S_0$ = Absorption
$T_1 \rightarrow S_0$ = Emission
K_{IC}^T = Rate constant for internal conversion in a triplet state
K_{IC}^S = Rate constant for internal conversion in a singlet state
ϕ_F(sol) = Fluorescence quantum efficiency in solution
ϕ_F(cryst) = Fluorescence quantum efficiency in solid
e^- = electron
e^+ = hole
e^0 = Exciton
e_r^+ = Relaxed hole
h = Planck's constant
r = frequency
τ = Decay time at time t
τ_0 = Total decay time
I = Luminescence intensity
I_0 = Initial luminescence intensity
t = time
τ = Mean decay time
τ^0 = Natural decay time
τ_i (where $i = F$ for fluorescence, P for phosphorescence) = decay of component indicated by $_i$
E_n = Energy of line emitted when recombination is via a donor and acceptor separated by distance r_n.
E_G = Gap energy
E_D = Binding energy of the electron to the donor
E_A = Binding energy of the hole to the acceptor
ε = Low frequency dielectric constant
r_n = Distance between donor and acceptor
$P(r)\,dr$ = Probability of finding a donor acceptor pair with a separation between r and r and $dr(r_n)$
α = Coulomb interaction
q_1 = Charge on donor
q_2 = Charge on acceptor
F_n = Radiative transition probability
Rd = Half the Bohr radius of the shallowest centre
F_e = Fraction of centres excited
M = Constant
P = Constant
a_0 = Threshold of impurity conductive upon excitation
LO = Longitudinal optical
E_a^i = Activation energy
A = Constant
k = Boltzmann's constant
X, Y = coordinate axes
p.m. = photomultiplier
sr = steradian
L_0, L_2 = emissions from radiative transitions
E = energy
δE = small increment of E
ν = frequency

K_F = rate constant for fluorescence decay
$\parallel$ = parallel to
$\perp$ = perpendicular to
P = degree of polarization
ERL = electron recombination luminescence
HRL = hole recombination luminescence

I. Introduction

The production of X-ray and electron emission from targets bombarded by an electron beam is a well-known phenomenon. The phenomenon of light emission from such targets is less well understood. Nonetheless, everyone is familiar with oscilloscopes and television, and the production of light from the screens of these scanning beam devices. It was, and in fact, still is, the oscilloscope as an experimental instrument that has yielded much of the empirical information about phosphors, which are light emitting materials. With such a system, it is convenient to manipulate an electron beam, and analyse the light spectrum produced when the beam strikes the phosphor. By the use of suitable pulsing devices, the rise and decay times of phosphors can be determined.

The beam of a scanning electron microscope (SEM) can be readily manipulated with respect not only to its position, but also with respect to its voltage, current, and in some cases pulse frequency. An obvious drawback in most SEMs is the inaccessibility of the specimen once vacuum operating conditions are reached. These problems will be discussed along with the subsequent analysis and detection of the light component of the total radiation emitted. Regardless of the type of instrument used—whether SEM or electron microprobe—it is necessary for better appreciation of the phenomenon of cathodoluminescence to have some understanding of the physics of electron-stimulated light; this will be discussed in the latter part of this chapter.

II. Instrumentation used in Cathodoluminescence Studies

The simplest system in which cathodoluminescence may be examined is a cathode ray tube (CRT), and it was, in fact, the detection of luminescence excited by electrons that led Thomson to his discovery of "cathode rays" or electrons. In a CRT, the electron gun is situated in a vacuum tube and the electron beam (which may be modulated in any of a number of ways) impinges upon a phosphor covering a glass or other transparent screen. The luminescence produced is viewed after it has passed through the glass screen and its characteristics (intensity, wavelength, decay times, etc.) measured.

Recent work on cathodoluminescence has been carried out in electron microscopes, as well as in the conventional CRTs. The electron probe family of instruments, the electron probe microanalyser, and the scanning electron microscope are particularly suitable for cathodoluminescence work. The so-called electron luminescence microscope (Le Poole *et al.*, 1968) was specifically designed for the study of cathodoluminescence properties of minerals. Recently, electron gun attachments have been manufactured for light optical microscopes (Sippel, 1968). This has initiated a whole new field of optical microscopy. Effectively the optical cathodoluminescence microscope is a CRT with the glass side of the screen forming the specimen stage of an ordinary optical microscope (Fig. 1; after Sippel, 1968). It is not capable of any improvement in resolution over the optical microscope and is, therefore, rather limited in application, although it has already made some valuable contributions in the field of mineralogy.

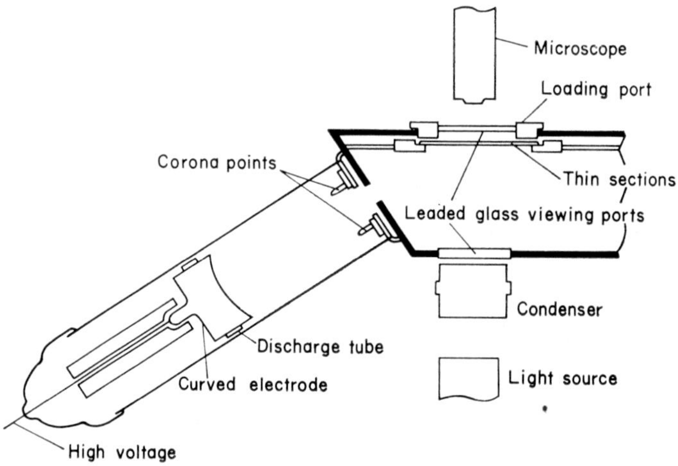

FIG. 1. Schematic diagram of Sippel's optical/cathodoluminescence microscope. (From Sippel, 1968.)

The scanning beam instruments are obviously powerful tools in the study of cathodoluminescence itself, and in the investigation of the cathodoluminescent properties of materials. The instruments are analogous to CRTs, with an electron source, an evacuated column, and a specimen surface which acts as a phosphor. All reported work in the literature thus far carried out using such instrumentation has involved reflected cathodoluminescence only, but with the advent of scanning transmission instruments, or even with the use of a transmission stage in one of the present generation of instruments, the detection of transmitted cathodoluminescence becomes feasible. In Sippel's (1968) optical cathodoluminescence microscope only transmitted cathodoluminescence is utilized.

In the scanning beam instruments, the accelerating potential of the electron gun (employing a tungsten filament as a source in most instruments in use at present) may be varied from 1 kV to 30 kV in the scanning electron microscopes, and from 1 kV to about 50 kV in the electron probes. In some instruments, the accelerating potential is continuously variable, while in others only a fixed number of accelerating potentials is available, e.g. 3 kV, 15 kV, and 25 kV. High accelerating potentials are not particularly desirable for cathodoluminescence work because they decrease the resolution of the cathodoluminescence micrograph, and will increase saturation effects. On insulating specimens, high beam voltages will also increase specimen charge and, therefore, decrease the number of primary electrons reaching the specimen surface, with a consequent loss of cathodoluminescence efficiency.

All the scanning beam instruments use one or more electromagnetic lenses to focus the beam and it is possible to calculate the spot size of the beam on the specimen surface (Smith, 1972; see also Chapter 1 of this volume). By calculating the excitation volume, it is possible to determine the resolution of the cathodoluminescence image. Obviously we have here a conflict between obtaining optimum resolution and maximum intensity. In order to obtain the best resolution, it is necessary to use the finest spot size available for any instrument, but in order to achieve the greatest intensity of output, it is necessary to use a beam current of as high a value as possible so long as this is less than the saturation or damage level for the material under investigation. A high beam current is usually obtained by reducing the lens currents, and hence increasing the spot size. The two conditions are manifestly incompatible and a compromise has to be sought, which usually results in a sacrifice of resolving power. This is because cathodoluminescence intensity is generally very low, and in order to minimize detection problems conditions must be allowed to favour the production of the most intense luminescence output. Usually, therefore, in our own instrument (Cambridge Stereoscan Mark IIA), the accelerating potential for cathodoluminescence work is set at 20 kV or less: the first condenser lens is set at about 0·3–0·4 A: the second condenser lens is switched off: and the final lens ammeter usually reads about 0·65 A. The beam current meter reading will be of the order of 175 μA. (Although this is not to be considered as the beam current at the specimen, it is the beam current at the electron gun.) In order to measure the beam current at the specimen, it is necessary to undertake a series of measurements with a Faraday cage on the specimen stage. The conditions thus obtained are, therefore, similar to those used in electron probe microanalysis and the specimen current is usually of the same order, 1×10^{-7} A.

The electron luminescence microscope of Le Poole *et al.* (1968) has an electron gun whose accelerating potential can be varied up to 30 kV, and a single electromagnetic lens for focusing. It also has stigmators, and X and Y deflectors to provide easy beam tilt. Clearly this instrument is not designed to give optimum resolution and the reason for this is that the cathodoluminescence is detected optically using a polarizing optical microscope. Unlike Sippel's system, however, it can detect reflected cathodoluminescence. The spot size of the electron beam is quoted as 30 μm or larger, and the maximum magnification obtainable with their optical microscope is $\times 300$. It is, therefore, more strictly comparable with Sippel's instrument than any of the scanning beam instruments and we will not consider it further here.

III. Detection of Cathodoluminescence

Obviously, if the equipment being used has an optical column, the luminescence can be viewed directly (Sippel, 1968; Le Poole *et al.*, 1968). It is undeniable that there is a considerable psychological advantage in the investigator being able to see the distribution of colour in his specimens for himself, and to be able to correlate them directly with familiar optical images of the same specimen. Most electron probe microanalysers are equipped with an optical microscope which is fixed to a port in the specimen chamber, and a great deal of work has been published using this technique (Long and Agrell, 1965; Smith and Stenstrom, 1965; Davey, 1966; Goni and Remond, 1969; Remond *et al.*, 1970).

Unfortunately, most commercially available scanning electron microscopes do not have attached optical microscopes and it is usually a geometric impossibility to fit an optical tube to any part of the specimen chamber and be able to view the specimen. This in itself is not necessarily a drawback since, using an optical detection system, resolution and magnification are limited to their usual light optical values.

McMullan (1952) suggested that it should be possible to obtain scanning cathodoluminescence images and McMullan and Smith (Smith, 1956) succeeded in obtaining such images using a photomultiplier tube in a scanning electron microscope. The type of system used was described by Davey (1966) who has applied it to an electron probe microanalyser. The light emitted from the specimen surface was collected by a reflecting objective of the type used in ultraviolet microscopy. The specimen was slightly tilted with respect to the beam so that the light had to pass through a minimum distance in the specimen. (This latter fact is important because cathodoluminescence is produced at the surface or *just below it* and unless the specimen is transparent or almost transparent, light produced below the surface will be lost to the collector.) The light collected was then passed through a fused silica window, and subsequently could be passed through absorption filters. In Davey's suggested system, the light could then be reflected by a mirror into an optical column, or into a photomultiplier tube operated in a photon counting mode. The output from the photomultiplier tube could be used to produce a scanning image in the same way as that produced by a secondary electron detection system, and cathodoluminescence images can be photographed in a similar way. This, of course, produces a micrograph illustrating the distribution only of the total cathodoluminescence intensity in black and white. By interposing filters before the mirror, Davey was able to record separate micrographs of narrow bands of the total emission, and to compare the spatial distribution of the bands. In particular, he investigated the distribution of the orange emission of a Mn activated ZnS phosphor with respect to the distribution of the blue emission.

Davey used an ingenious method for visually displaying decay times of his material. He applied pulses to the electron beam which deflected it from the condenser aperture and prevented it from reaching the specimen. The photomultiplier output pulses were used to modulate the beam of an oscilloscope synchronized to the electron beam pulse generator and a visual estimate of the decay time could then be made.

The technique of observing luminescence in the SEM was described by Thornton (1968), and the first experimental results were published by Williams and Yoffe in 1968, in a study of stacking faults in ZnSe single crystals. In their first paper, these authors used a perspex light pipe to collect the luminescence from the specimen and pass it to the photocathode of a photomultiplier tube, the output from which was used to form an image of the total luminescence output of the specimen. Later, Williams and Yoffe (1969a) used a quartz light pipe rather than a perspex one, and noted that "care was taken to avoid energetic electrons striking the light pipe and producing a confusing background emission". This latter point is important, because many substances in the SEM specimen chamber are potentially luminescent. Lenses and light pipes made of perspex give off bright green luminescence, and even although they are not in the line of the incident beam, primary electrons are scattered from every surface in the specimen chamber, and may cause lenses and light pipes to luminesce. Furthermore, the specimen chamber must be kept clean, because Bond *et al.* (1974) have shown that many commonly used vacuum oils are luminescent, and furthermore, many dust particles are brightly luminescent.

Most of the work in published SEM cathodoluminescence literature has been carried out using the type of light collection system proposed by Williams and Yoffe, with minor modifications. Some systems rely upon a lens (of quartz or perspex) placed in front of the photomultiplier tube window, in an effort to collect the maximum amount of light.

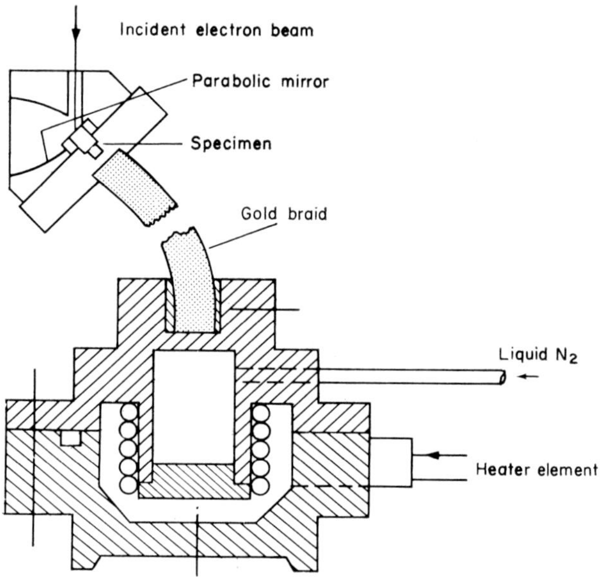

FIG. 2. Diagram of the hot and cold stage with a parabolic mirror for the focusing of cathodoluminescence. (As used by Jones *et al.*, 1973.)

Others have no lens intervening between the specimen and photomultiplier tube photocathode, and it is difficult in practice to see any difference between micrographs obtained with or without a collecting lens. However, if a lens or light guide is to be used it should have transmission characteristics which will encompass the full range of luminescence wavelengths to which the specimen will give rise. These are not necessarily easy to anticipate and for this reason, more than for any other, it is probably better not to use any light collection system. Remond *et al.* (1972) commented that because quartz is luminescent under electron beam excitation, for the light collection system they use an undoped plastic which has been shown in practice to be non-luminescent.

Several authors have proposed the use of a parabolic mirror behind the specimen to facilitate collection. The specimen is placed at the focus of the mirror, and in theory, a maximum amount of light is reflected towards the collector. Muir *et al.* (1971) described the use of a parabolic mirror for cathodoluminescence spectroscopy, and variants on this have been used by Jones *et al.* (1973) (see Fig. 2) who used a parabolic mirror to investigate cathodoluminescence of semiconducting devices; the mirror was mounted on a probe cooled to liquid nitrogen temperatures. Bond *et al.* (1974) have also used a parabolic mirror, and found that it improved the detection of cathodoluminescence by a factor of twenty.

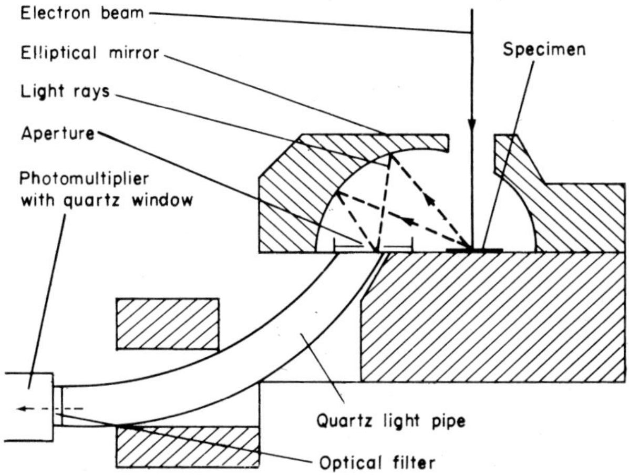

FIG. 3. Elliptical mirror configuration of cathodoluminescence detection. (From Hörl and Mügschl, 1972.)

An ingenious improvement on the parabolic mirror was employed by Hörl and Mügschl (1972a,b). They enclosed their specimens in an elliptical mirror (see Fig. 3) with the specimen at one focus. A small hole in the mirror allowed the beam to pass through and a light pipe was placed at the other focus to lead the light to a photomultiplier tube, and very efficient collection can be obtained. A similar system has been used by Carlsson, L. and van Essen, C. G. (personal communication), and with the use of fibre optics spectroscopy as well as image formation could be carried out.

Remond *et al.* (1972) also stressed one of the major pitfalls in the use of a photomultiplier tube to detect cathodoluminescence. This is that the arrival of any energy packet over a certain threshold value at the photocathode may stimulate electron emission which would cascade along the tube and be recorded as if it were a light photon. In the specimen chamber of a scanning electron microscope, as we have already established (Chapter 1), a proportion of the primary electrons is elastically reflected from the surface of the specimen, and may strike any part of the specimen chamber—including the photomultiplier tube window. The photomultiplier tube is unable to distinguish between these high energy primaries and the lower energy photons of cathodoluminescence, and the primaries are, therefore, an ineradicable source of noise in the cathodoluminescence signal. X-ray photons, too, may produce photomultiplier noise.

There is a further problem in that if the intensity of the cathodoluminescence output is low, as it very frequently is, in order to produce a cathodoluminescence image on the SEM viewing screen, it is necessary to utilize the very high gain capability of the photomultiplier tube. This clearly amplifies both the photomultiplier signal and the noise, and the majority of the signal may be that produced by the arrival of primary electrons at the photocathode. Thus, it is clearly easy to produce a false cathodoluminescence image which is actually a primary electron image. This can be confirmed by switching the electron detection system to show only primaries, and comparing this image with the putative cathodoluminescence image. If they appear to be identical, then the "cathodoluminescence" image is probably not being formed by light photons, but by high energy primaries. The images may actually be identical, but it is difficult to be sure of this without determining the wavelength of the light emitted. Secondary electrons and X-ray photons are unlikely to make any noticeable contribution to the production of a false image, but they will contribute to the noise.

Naturally, if the cathodoluminescence image shows some features which do not correspond with the morphological features of the primary electron image, these can be interpreted as being due to light emission (Muir, 1970). If in addition filters are inserted in the path of the light, in front of the photomultiplier tube window, and if these give different distribution images for different wavelengths, again the signal can be interpreted as being cathodoluminescence from the specimen.

Despite this, one of the major difficulties in the study of cathodoluminescence in the SEM, where there is no optical viewing system, is being sure that the signal produced is genuine, and not produced by electrons. The only way in which this can be unequivocally demonstrated is by the use of some device which removes the light from the specimen chamber and enables it to be seen and subjected to wavelength analysis.

Finally, it is obviously important to select the photomultiplier tube used for cathodoluminescence work so that the wavelength response of the photocathode is suitable. For general purposes, the spectral response of the tube should cover as wide a wavelength range as possible. The S20 type tubes (produced by most manufacturers) give a good detection efficiency over most wavelengths. For specialized work, or work in the infrared, other tubes have to be used.

IV. Cathodoluminescence Spectroscopy

Davey's system (1966) used in the electron probe microanalyser can be adapted for use in the scanning electron microscope, and indeed Williams (1968) and Williams and Yoffe (1969a,b) did this. An instrument capable of wavelength selection was designed and used as early as 1960 (Davoine *et al.*, 1960), but despite this early interest, comparatively little published work is available on cathodoluminescence spectroscopy carried out in the SEM. It is useful to consider two forms of collection of cathodoluminescence signal: non-dispersive, when the total signal is collected, usually for image formation, and intensity measurements; and dispersive, where the signal is split into its constituent lines, and spectra, or rise and decay times are recorded.

In order to carry out such dispersive measurements, a transparent window has to be inserted in the wall of the specimen chamber. Not all commercial instruments are suitable for this type of adaptation, but in most models, access ports are incorporated in the specimen chamber for the attachment of accessories such as X-ray spectrometers, and these can usually be fitted with a window. The choice of material for the window will naturally depend upon the wavelengths which are to be analysed. The window should be able to transmit all wavelengths without strong absorption bands at any part of the spectrum and in practice fused silica, "Spectrosil", and other such materials have satisfactory transmission characteristics. In any event, transmission characteristics of the window used should be measured using a standard lamp, and calculated among the errors of the detection system utilized. Most window materials give satisfactory transmission of ultraviolet and visible light, but are opaque (or have strong absorption edges) in the infrared. Very little work has yet been done on infrared emission, but it appears to be a field with considerable potential value, particularly in semi-conductor research, and some of the materials which transmit in the infrared possess sufficient mechanical strength to withstand the specimen chamber vacuum. Undoubtedly, this is an area where considerable research effort will be expended in the future.

Plastic and quartz fibre optics have recently been used with great success in transmitting the light from the specimen to the entrance slit of the monochromator (Carlsson, L. and van Essen, C. G., personal communication; Steyn and Holt, 1973). This prevents light loss due to misalignment of the mirror, although there is a considerable light loss during transmission along the fibre bundle, and a rather severe light loss at any joins which have to be made in the fibre bundle. However, this appears experimentally to be much less than that caused by misalignment.

The monochromator should be placed as close as possible to the transparent window, and ideally, should be bolted on to the specimen chamber. In many commercial instruments, however, this is a physical impossibility, because the plinth upon which the SEM electron optical column is supported is frequently remarkably large in size. If the monochromator has to be at some distance from the specimen, then the light path from the window to the entrance slit should be totally enclosed so as to exclude all ambient light. The choice of monochromator for this work is not critical provided that three criteria are satisfied, namely:

(a) The wavelength resolution of the monochromator should be as good as possible, and probably a discrimination of 10 Å or less should be available.

(b) The monochromator should be able to analyse all wavelengths that can be transmitted by the window.

(c) The monochromator should be able to record a spectrum in a relatively short time, say 2–3 min, so that if continuous exposure to the electron beam causes alteration to the specimen, with subsequent alteration of cathodoluminescence output, then at least not too much of this will have taken place during the recording of the spectrum, and the beginning and end of the spectrum will be more or less comparable.

The next important part of the analytical system is the photomultiplier tube at the output from the monochromator. Davey (1966) suggested that this photomultiplier tube should be operated in the photon counting mode, since this is particularly suitable for low light intensities. Muir *et al.* (1971) described the use of such a system and compared it with the conventional d.c. amplification applied to photomultiplier tubes in spectroscopy. They found that d.c. amplification provided a suitable signal, with reasonable signal/noise ratios at high light intensities, but that operation in the photon counting mode was essential for the normal low light intensities encountered in cathodoluminescence work. Briefly, the difference in operation between the two systems is this: using d.c. amplification, the photomultiplier tube is operated at a high gain using a high voltage supply; instabilities are easily caused by small

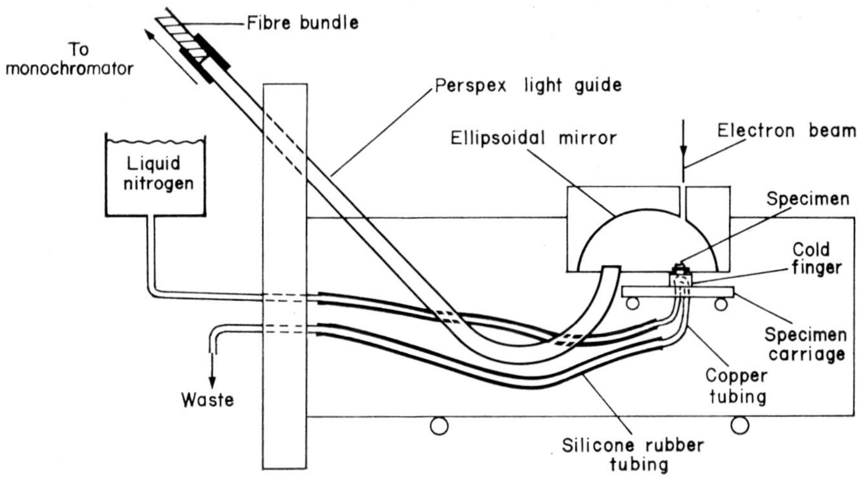

FIG. 4. The combined cold stage and elliptical mirror for the collection of cathodoluminescence. (As described by Steyn and Holt, 1973.)

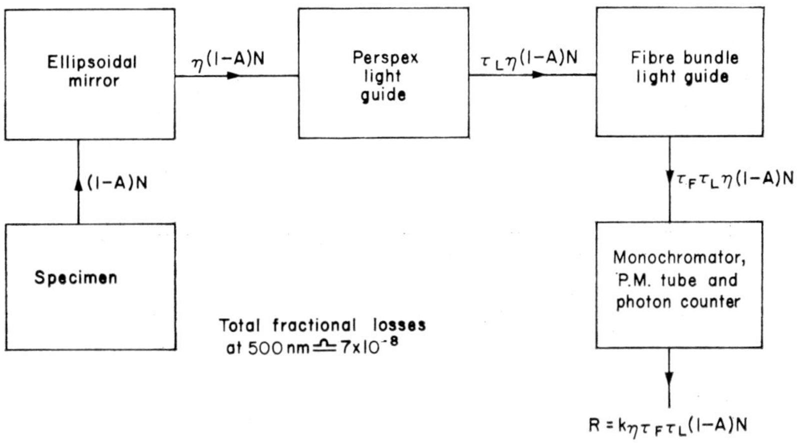

FIG. 5. Block diagram indicating the stages and losses between stages in the collection and analysis of cathodoluminescence. (From Steyn and Holt, 1973.)

supply voltage or temperature changes; in the photon counting mode, the p.m. tube is run at low gain, but is coupled closely with a fast high-gain amplifier, making the p.m. tube signal much more stable, and less affected by external factors. The signal/noise ratio is greatly improved, and a discriminator can be used to cut out noise due to photomultiplier tube events which are not generated by incident photons.

So far quantitative results from cathodoluminescence spectroscopy in the SEM are few, but Steyn and Holt (1973) measured the fraction of

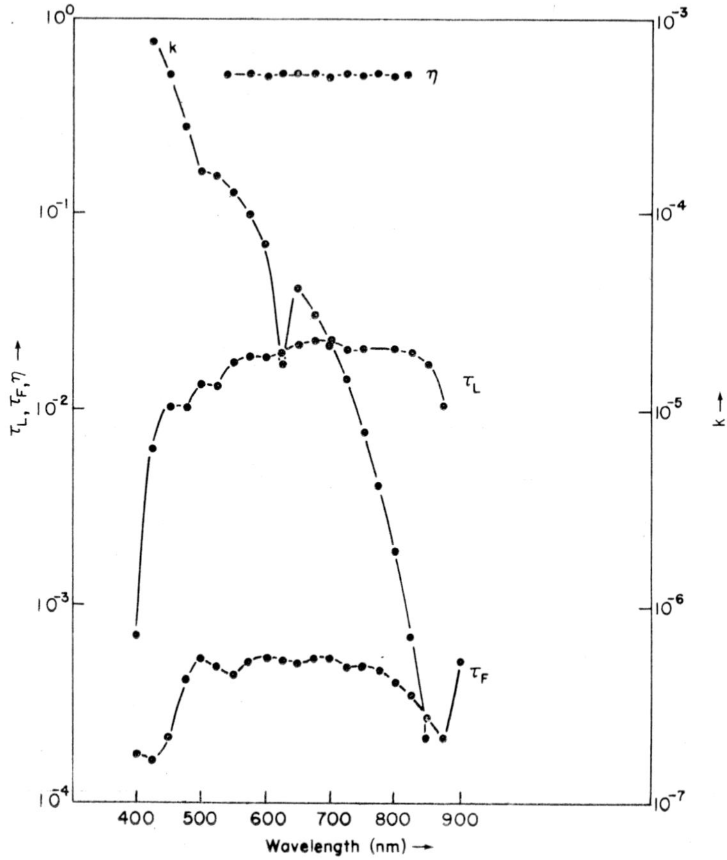

FIG. 6. A graphical representation of the losses involved in the collection and analysis of cathodoluminescence spectra. τ_L = transmission coefficient for the perspex light guide; τ_F = transmission coefficient for the fibre light guide; η = mirror efficiency; K = combined efficiency of the monochromator, photomultiplier and photon-counter. (From Steyn and Holt, 1973.)

the light collected and transmitted at each step in a system consisting of an ellipsoidal mirror and light guide (Fig. 4), and monochromator, photomultiplier tube, and photon counter (Fig. 5). The individual loss factors for each stage in the system of Fig. 5 were measured as a function of wavelength as shown in Fig. 6. The overall instrumental response factor K resulting from all these losses in series is shown for a monochromator slit width of 1·8 mm in Fig. 7. To calculate these losses, Steyn and Holt used the relationship:

$$R(\lambda) = (Q\lambda)E(\lambda)F(\lambda)[1-A(\lambda)]\phi(\lambda) \quad (1)$$

where ϕ = the quantum yield (= the number of photons of a given wavelength λ generated per second) = N in Fig. 5.

R = the count rate read on the photon counter

Q = the photomultiplier quantum efficiency

E = the transmissive efficiency of the monochromator

F = the fraction of quanta that escape from the specimen that are collected, and then delivered into the entrance slit of the monochromator

A = the total loss correction due to self-absorption etc. in the material

and

$1-A$ = the fraction of quanta generated in the specimen that escape.

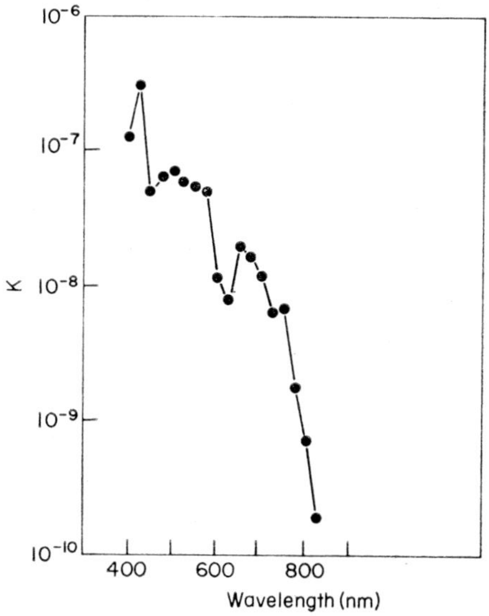

FIG. 7. The total instrument response factor (K) with a monochromator slit width of 1·8 mm; in this case the entrance slit and the exit slit are coupled and equal. (From Steyn and Holt, 1973.)

The total instrument response factor $K(\lambda)$ is the product of Q, E, and F. F can be determined as the product of the mirror efficiency η, and the transmission coefficient of the perspex light pipe, and that of the fibre optics τ_L and τ_F respectively.

The output from the photomultiplier tube can be treated in a number of ways. For instance, the movement of the grating in the monochromator can be stopped at a single band, and this signal can be fed back into the amplification system of the SEM to provide a display of monochromatic cathodoluminescence. Williams and Yoffe (1969b) illustrated such micrographs in their work on ZnSe platelets. They photographed the same area using the signal at 440 nm, and then the signal at 448 nm. The areal distribution of the luminescence at these two different wavelengths is quite distinct, but because the wavelengths are so close to one another, it would have been extremely difficult to carry out such work using a filter system. Yoffe *et al.* (1973), using the same system, gave further results from other materials.

If the output from the photomultiplier tube is amplified and passed through a ratemeter, then comparative intensities at different wavelengths can be measured and recorded using either a pen chart recorder or a multichannel analyser with the spectrum recorded on a CRT. It is convenient to have some kind of tape facility for storage and processing of the data.

V. Practical Problems Encountered

Many practical problems bedevil the study of cathodoluminescence. There are several reasons for this, but the most important factor is that the detection of cathodoluminescence requires the isolation of a comparatively small number of low energy photons from a large number of high energy primary electrons. The ratio of signal to noise is therefore low and may be made even lower by the noise contribution of secondary electrons and X-ray photons, and introduced instrumental noise. Detection equipment chosen should, therefore, attempt to increase rather than further decrease the signal to noise ratio. Most cathodoluminescence detection systems use d.c. amplification on the photomultiplier tube output, which, because of the low signal strength, increases the noise produced by the photomultiplier tube itself. Thus, the intensity modulation of the cathodoluminescence signal displayed on the SEM CRT screen contains a high proportion of noise from the primary beam and the photomultiplier tube. This problem can be alleviated by operating the photomultiplier tube in the photon counting mode, which eliminates much of the tube's contribution to the noise in the signal, and by using the Hörl and Mügschl elliptical mirror which cuts down on primary beam noise.

In cathodoluminescence work it would be possible, using commercially available signal processors, to subtract the primary electron signal (collected by an electron detector) from the signal collected by the cathodoluminescence detector, which consists of a mixture of primary electron and cathodoluminescence signals. The complete grey level range of the SEM would therefore be available for displaying the cathodoluminescence signal, and the primary electron contribution virtually eliminated.

Statistical noise in cathodoluminescence images can only be reduced by taking micrographs using very long exposure times, which will integrate the noise. This is not, however, suitable for all materials, because in some cases, the cathodoluminescence output deteriorates if the beam remains on a single spot for any length of time. This drawback is frequently encountered in organic materials (Pease and Hayes, 1966); and so with materials such as this, some statistical noise may have to be tolerated in the subsequent illustration. Thus the major factor in detecting and illustrating cathodoluminescence is to improve the signal to noise ratio, while the greatest problem in recording cathodoluminescence spectra is low emission intensity from most specimens.

Obviously, to obtain the maximum amount of information from any low intensity signal, the collection of that signal must be as efficient as possible. Cathodoluminescence collection systems have been described previously in this paper, and will not be further referred to here. First of all, it is essential that the specimen chamber and all the equipment within it should be scrupulously clean, and free from dust and vacuum oil. Secondly, since it is impossible to prevent back-scattered primaries from striking any or all of the materials inside the specimen chamber, all these materials should have their cathodoluminescence spectra recorded (if they have any). These emission spectra can then be used in correcting the emission spectrum of the specimen. Thirdly, and vitally, all ambient light should be rigorously excluded from the system. Bond *et al.* (1974) have discovered that light emitted from the tungsten filament of their SEM can make a significant contribution to the luminescence detected.

One of the most effective ways of increasing luminescence efficiency is by lowering the temperature. Williams (1968) and Williams and Yoffe (1969a,b) chilled their specimens by mounting them on a small movable liquid nitrogen-cooled stub. Hörl (1972b) illustrated the cathodoluminescence emission from solid nitrogen at a temperature of 4·8°K, using a liquid helium-cooled stage. Shay *et al.* (1970) carried out a series of experiments on the cathodoluminescence efficiency of $ZnSiP_2$ at low temperatures (not in an SEM). They noted that above 50°K, the luminescence efficiency begins to drop off very rapidly, and decay times become very short. Below 50°K, the luminescence efficiency increases as the temperature falls, and the output at 10°K is several orders of magnitude greater than the output at 100°K. The 100°K output is considerably greater than that at room temperature. This will be discussed further in the next section of this chapter.

A further possible way of increasing luminescence efficiency is by increasing the brightness of the electron source—in other words, by producing more electrons at the specimen surface without increasing the accelerating potential of the beam. This latter is undesirable, because it tends to produce specimen charging, and saturation effects in the specimen. A description of the various types of possible electron source is given in Chapter 1 of this volume, and the factors important for cathodoluminescence work are summarized in Table I of this present chapter. Although the field emission source in theory gives greater brightness for any corresponding accelerating potential, it has been shown (Chapter 1 of this volume) that at high beam currents, and large spot sizes, the lanthanum hexaboride gun gives the maximum power. However, so far, no cathodoluminescence work has been carried out using these alternative electron sources.

TABLE I. Comparison of the Tungsten Hairpin Filament, Lanthanum Hexaboride Rod, and Field Emission Source Types of Electron Gun, Showing Brightness at Different Accelerating Potentials and Lifetimes

Type of electron source	Brightness $A\ cm^{-2}\ sr^{-1}$	Accelerating potential	Lifetime (hours)
Tungsten hairpin[a]	10^4–10^5	20 kV	30
LaB_6[a]	10^6–10^7	20 kV	250[c]
F.E.S. tungsten tip[b]	10^7	5 kV	1000[c]

[a] Ahmed (1972). [b] Veneklasen and Siegel (1972).
[c] Depending largely on vacuum conditions at the gun.

VI. Cathodoluminescence Mechanisms

Although there is a large body of literature referring to luminescence in general, there are very few publications referring specifically to cathodoluminescence. One of the objects of this paper was to try to find out if any general conclusions could be drawn. However, although there are many mechanisms proposed in specific cases, none of these appears to be of general validity.

VII. Comparison of Cathodoluminescence with Photoluminescence

There is no basic difference between the effects caused by an electron beam or an ultraviolet source as an excitation agent for stimulating light emission (cathodoexcitation and photoexcitation respectively;

producing cathodoluminescence and photoluminescence respectively). The emission spectra produced are qualitatively similar. Dittmann *et al.* (1968) show the resemblance between ultraviolet (Fig. 8, top), and electron beam (Fig. 8, bottom) emission spectra of a ZnS–Cu activated phosphor, at room temperature and at liquid nitrogen temperature.

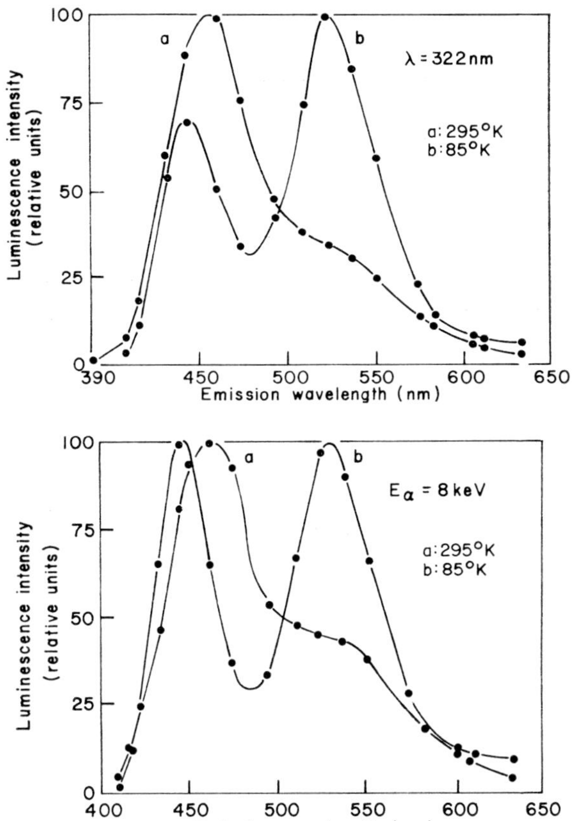

FIG. 8. (After Dittmann *et. al.*, 1968)

Cathodoluminescence, like photoluminescence may be considered, as the luminescence resulting from the relaxation of two excited states. In studies of the luminescence of organic materials a distinction is made between fluorescence and phosphorescence, although these terms are obsolete in the field of the luminescence of inorganic materials.

Radiative depopulation of an excited singlet state gives rise to the fluorescence component, whereas the phosphorescence component is due to radiative depopulation of the triplet state (see Fig. 9, modified after Becker, 1969). Because the energy between the first triplet state and the ground state is invariably less than that between the first singlet state and the ground state, emission from the former is almost always at longer wavelengths than from the latter. The phosphorescent decay is inherently slower than the fluorescent, and whether or not a phosphor is fluorescent or phosphorescent is controlled by whichever relaxation component is dominant.

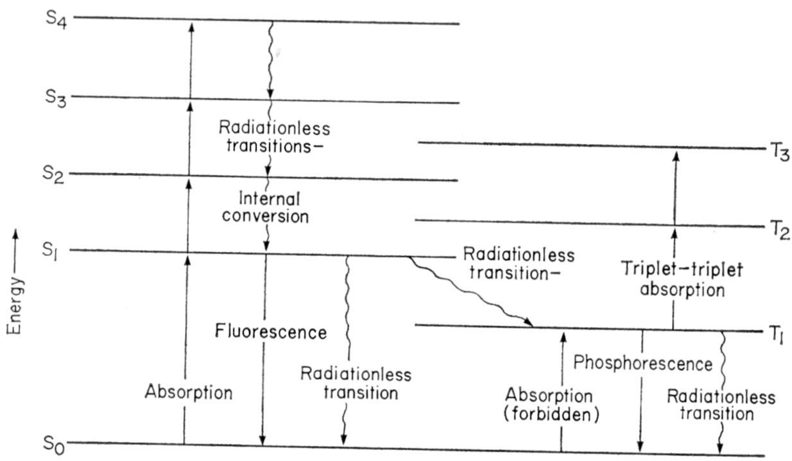

FIG. 9. (After Becker, 1969.)

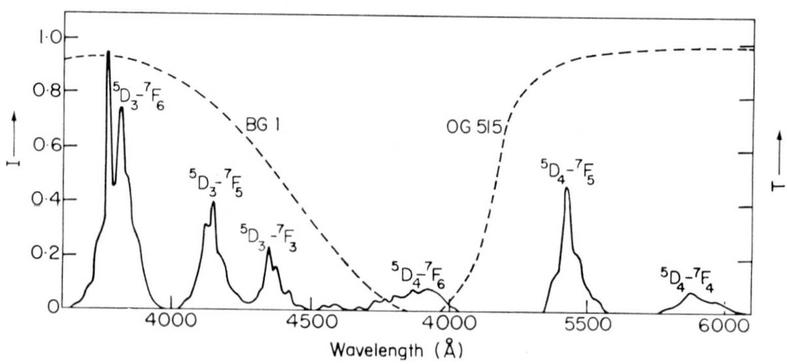

FIG. 10. (After Fijnaut, 1971.)

Cathodoluminescence spectra can therefore be regarded in a similar way to photoluminescence spectra, and the same forms of interpretation can be applied. For example, a cathodoluminescence emission spectrum of $Sr_3(PO_4)_2$–Tb activated, has been annotated using conventional photoluminescence notation (see Fig. 10, after Fijnaut, 1971).

Apart from the whole spectrum, the most important characteristics of cathodoluminescence are the rise and decay times for individual component lines of the spectrum. Pastrnak and Karel (1968) examined the excitation of manganese activated A1N using an ultraviolet and an electron beam source. They found that the shape of the spectra were

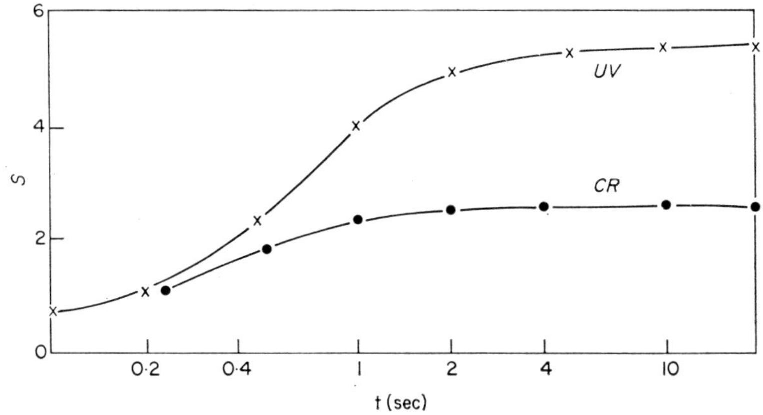

FIG. 11. Curve indicating the rise time and total luminous output for a phosphor with continuous cathodo-, and photo-excitation (*CR* and *UV*, respectively) from time $t = 0$; S = light sum. (From Pastrnak and Karel, 1968.)

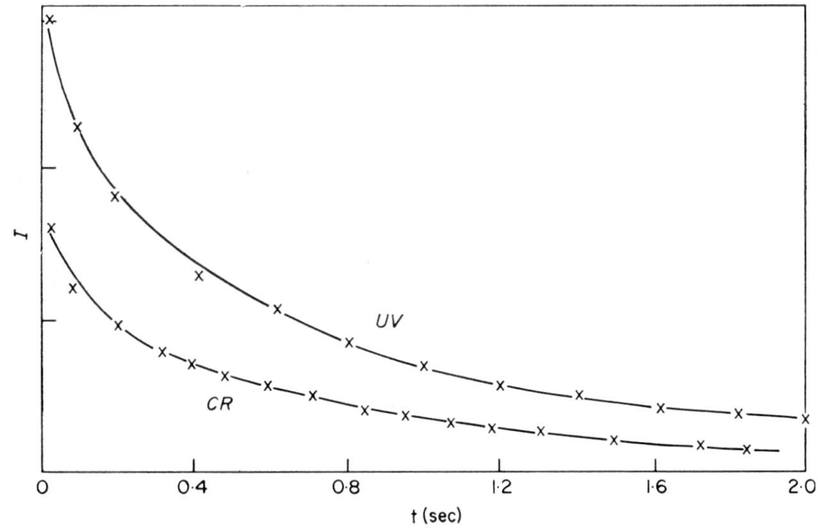

FIG. 12. Decay curves—a comparison of the relaxation times subsequent to the cessation of excitation by an electron beam (*CR*), and ultraviolet light (*UV*). (From Pastrnak and Karel, 1968.)

identical with a blue band (BB) produced by the unactivated phosphor, as well as a red band (RB) from the manganese activated phosphor. Investigation of the decay processes revealed that irrespective of whether photoexcitation or cathodoexcitation were used, the phosphor has rise time characteristics (Fig. 11), and fast decay components and slow decay components (Fig. 12) which are very similar for the two excitation sources. The crosses on the curves of Fig. 12 represent the blue band and, the solid lines the red band.

With weak cathodo- and photoexcitation, the light sums (S) for the blue band and for the red band retain the same ratios. With higher excitation intensity, the blue band becomes dominant. Figure 11 shows the parallelism of the curves for photo- and cathodoluminescence once a steady state has been achieved. If we compare the ratios of the heights (J) of the different bands [$(J_{RB}/J_{BB})_{CL}$ for cathodoexcitation, and $(J_{RB}/J_{BB})_{UV}$ for photoexcitation], we find that

$$\left(\frac{J_{RB}}{J_{BB}}\right)_{CL} = 14{\cdot}5 \tag{2}$$

and

$$\left(\frac{J_{RB}}{J_{BB}}\right)_{UV} = 13{\cdot}5. \tag{3}$$

The ratio of the light sums is quoted by Pastrnak and Karel as $S_{UV}/S_{CL} = 2{\cdot}3$ for both the blue and the red band. Thus the emission under cathodoexcitation is less than that under photoexcitation for similar energy input in the system AlN activated by Mn. However, these relations only hold for low intensity excitation.

It has been shown by Tolstoy and Abramov (1968) that at higher intensities the cathodoluminescence increases less rapidly than linearly with increasing excitation energy. The authors describe this phenomenon as non-linear quenching. Their diagram summarizing the effect is shown in Fig. 13. Unless care is taken to see that the excitation intensity is more or less comparable in the case of cathodoexcitation and photoexcitation, i.e. equal in power per unit area, the usually higher excitation intensities in the former will give relatively low quantum yields.

Specifically in the case of cathodoluminescence, though there is, no doubt, an optimum value for the incident beam density with respect to the emission intensity, Fijnaut (1971) showed that for the $^5D_3$ emission from $Sr_3(PO_4)_3$–Tb activated phosphor, the decay time was independent of the incident beam current energy and density, over the range of 2·2–8·0 kV, and 10^{-11}–10^{-8} A cm^{-2} respectively. It is worth noting, however, that most of Fijnaut's work was carried out using the

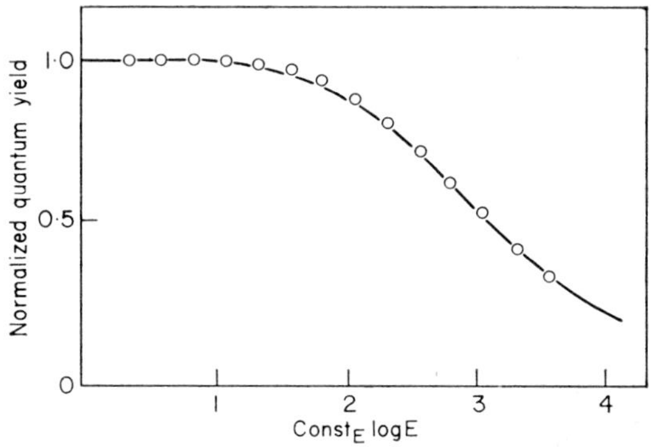

FIG. 13. A graph demonstrating the effect of excess intensity of excitation (E) on the luminescence yield. (After Tolstoy and Abramov, 1968.)

lower beam current (i.e. 10^{-11}) so that saturation of the excited energy levels of the Tb^{3+} ions was not achieved.

Bril *et al.* (1968), working on the same phosphor, noted a delay in emission following pulsed cathodoexcitation. This suggested to Fijnaut (1971) the existence of a "cascade" process in the initial stages of

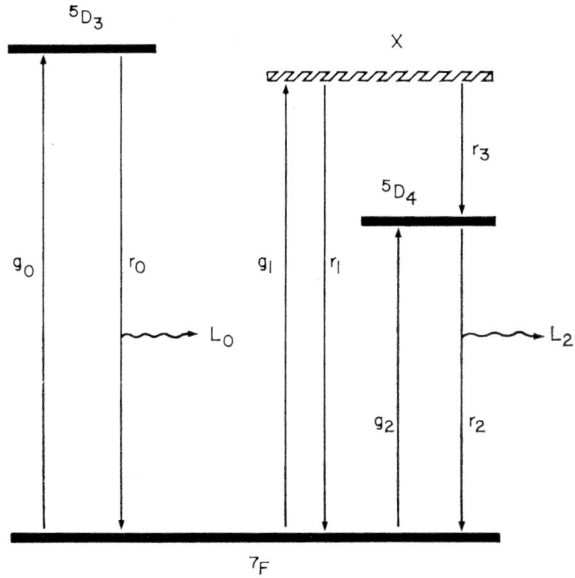

FIG. 14. Energy level diagram illustrating the cascade effect as described by Fijnaut (1971). Here an excited state (X) loses energy non-radiatively to excite a lower energy state which may relax radiatively. The specimen in this case was a $Sr_3(PO_4)_2$–Tb phosphor; g_i, r_i, where $i = 0, 1, 2$, are excitations and transitions, respectively. (After Fijnaut, 1971.)

excitation. By this, we understand him to suggest that de-excitation resulting from the cathodoexcitation may in some cases create "secondary" luminescence—excited by secondary electrons, X-ray photons, ultraviolet and visible photons, infrared radiation etc., and that any or all of these processes may contribute to the final emission. Thus using electron beam excitation, all the other excitation and radiative de-excitation processes could potentially be observed. Figure 14 illustrates this point in a simple three state situation where X is an unknown level, L_0 and L_2 are radiative emissions. In a limited number of cases, none of the other excitation processes is sufficiently energetic to stimulate luminescence emission. Terol and Aguino (1968) describe some rare earth activated phosphors which emit luminescence under cathodo-excitation but not photoexcitation.

VIII. Luminescence in Organic and Inorganic Systems

In inorganic materials, luminescence is often produced by radiative electronic transitions in the vicinity of some form of luminescence centre. These centres may represent impurity ions within the lattice, or they may be defects in the crystal lattice. Luminescence produced by impurities is described as extrinsic, and that produced by lattice relaxation is described as intrinsic. The term "luminescence centre" describes a wide variety of mechanisms, and although it is useful as a blanket term, particularly where the exact mechanism is unknown, it is of little value in modern interpretations of luminescence phenomena.

In organic systems, luminescence is produced largely as a result of electronic transitions between molecular orbitals.

A. Organic Systems

On excitation to any state above the first excited singlet, relaxation may occur back to the lowest excited singlet state. From this point, *fluorescence* emission to the ground state or non-radiative transitions may occur. A transition to the triplet state may take place and, if the triplet state is occupied, be followed either by non-radiative de-excitation to ground state or by the emission of *phosphorescence* (where the spin of the orbital states from which the emission originates differs from the de-excited state), or by both processes (see Fig. 9, after Becker, 1969).

There are four measurable characteristics associated with molecular emission. The energy, E, of the photons of any wavelength of the fluorescence spectrum is given by the relation:

$$E = h\nu \tag{4}$$

where h = Planck's constant, ν = the frequency of the emitted radiation.

The energy of the excited state with respect to the ground state is determined by the vibronic (*vibr*ational electr*onic*) band, or orbital, with the same energy as the emission peak. This represents the transition from the zero vibrational level of the excited state to the zero vibrational level of the ground state. In any other case, vibrational energy would have to be considered as a part of the total emitted (see Fig. 15, after Becker, 1969). The second characteristic of excitation, its decay time,

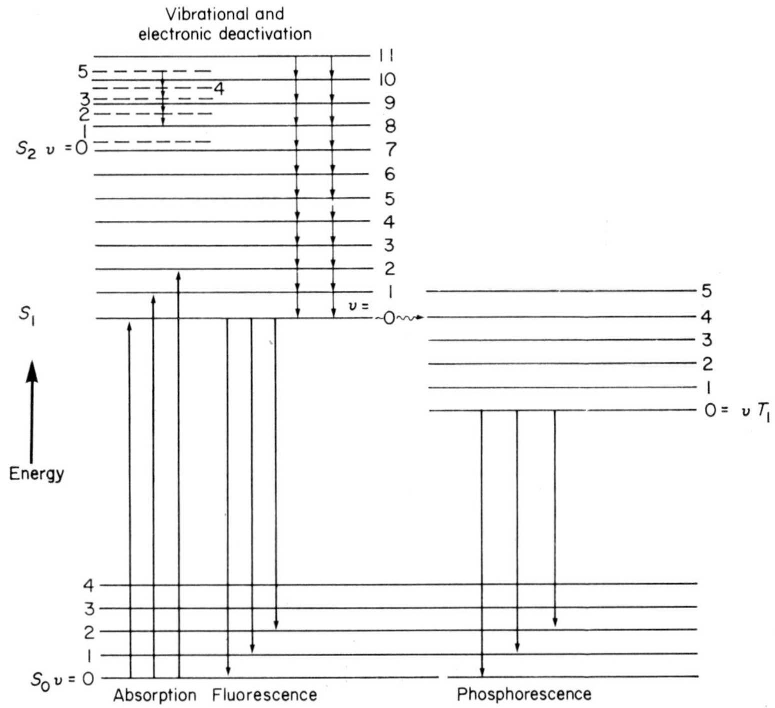

FIG. 15. Schematic energy level diagram showing electronic states (S_i = singlet, T_i = triplet) and vibrational levels (v). (From Becker, 1969.)

is short and difficult to measure. Luminescence intensity decays on the removal of the excitation source, exponentially with time:

$$I = I_0 e^{-t/\tau} \tag{5}$$

where I_0 = initial intensity at an arbitrary time zero, I = intensity at a later time t, τ = the mean decay time of the emission process. This will be the natural decay time τ^0 only in the absence of quenching processes when $t = \tau$, then

$$I = I_0 \cdot \frac{1}{e} \cdot \tag{6}$$

Now, the rate constant K, can be related to the decay time τ, for fluorescence as follows:

$$K_F = \frac{1}{\tau_F}. \tag{7}$$

Also, there is a relationship between τ_F, i.e. the observed decay time, and the natural decay time τ_F^0:

$$\tau_F = \tau_F^0 \phi_F^0 \tag{8}$$

again in this case for fluorescence. ϕ_0^F is the natural quantum yield. The overall decay observed has a time τ which is the sum of the decay times for all the relaxation processes involved. Becker (1969) treats authoritatively with these aspects of the underlying physico-chemical principles.

The quantum yield of the emission spectra is difficult to measure but can be defined as:

$$\phi_F = \frac{\text{number of fluorescence quanta emitted}}{\text{number of quanta absorbed to a singlet excited state}} \tag{9}$$

$$\phi_P = \frac{\text{number of phosphorescence quanta emitted}}{\text{number of quanta absorbed to a triplet excited state}} \tag{10}$$

because the direct absorption $T_1 \leftarrow S_0$ is forbidden, excitation to the triplet state occurs via a singlet state and a subsequent non-radiative transition. When the total band gap energy is that of the photons emitted, quantum yields from these states may be termed natural (or in Becker's terminology "intrinsic"—but we shall not use this term here, to avoid confusion with the semi-conductor usuage).

Where no photochemical reaction takes place:

$$\Sigma_i \phi_i^0 = \phi_F^0 + \phi_P^0 + \phi_{IC}^0 = 1 \tag{11}$$

where ϕ_{IC}^0 = the quantum yield for non-radiative transitions (often equivalent to internal conversions, and hence the subscript IC) between electronic states. ϕ_{IC}^0 may be small, in which case:

$$\phi_F^0 + \phi_E^0 = 1. \tag{12}$$

Fluorescence and phosphorescence are thus complementary processes.

ϕ_F^0 is related to the rate constants as follows:

$$\phi_F^0 = \frac{K_F}{K_F + K_{IC}^S + K_{IS}} = K_F \tau_F \tag{13}$$

where K_{IC}^S = the rate constant for internal conversion, and hence non-radiative relaxation for the first excited singlet. K_{IS} = the rate constant for intersystem crossing, i.e. between singlet and triplet states.

For phosphorescence:

$$K_{IS} \gg K_F + K_{IC}^S$$

and therefore:

$$\phi_P^0 = \frac{K_P}{K_P + K_{IC}^T} = K_P \tau_P^0 \quad (14)$$

where K_{IC}^T = the rate constant for internal conversion from the lowest triplet state. If K_{IS} is not considerably greater than $K_F + K_{IC}^S$, then

$$\phi_P^0 = \frac{K_{IS}}{K_F + K_{IC}^S + K_{IS}} \frac{K_P}{K_P + K_{IC}^T}. \quad (15)$$

The quantum yield for intersystem crossing is:

$$\phi_{IS}^0 = \frac{K_{IS}}{K_{IS} + K_F + K_{IC}^S}. \quad (16)$$

Polarization is a further measurable emission characteristic. Using photoexcitation with oriented molecules, such as occur in crystals, particular transitions may become more probable when the incident radiation is polarized in a particular direction.

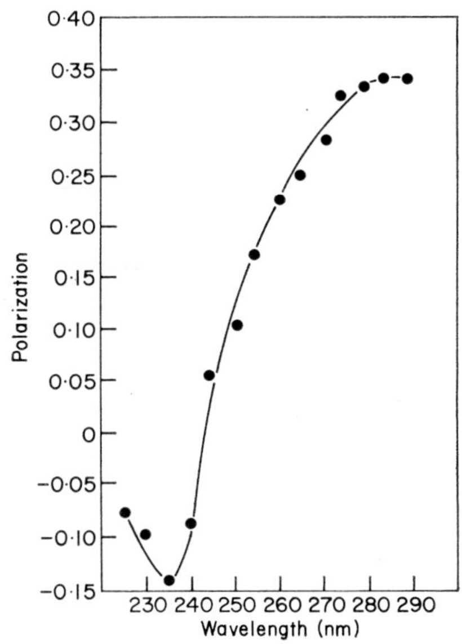

FIG. 16. (After Weber, 1960)

The direction of the polarization of the emission with respect to a crystal of known orientation can be used to characterize absorption bands when observed as a function of the wavelength of excitation. Figure 16 (after Weber, 1960) shows how the variation of the excitation wavelength affects the polarization of the emission from phenol in propylene glycol in the solid state at $-70°C$. For some hydrocarbons, phosphorescence is polarized perpendicular to the plane of the rings (see Becker, 1969, for further discussion of particular systems).

In general, the degree of polarization can be defined as follows:

$$P = \frac{I_{\parallel} - I_{\perp}}{I_{\parallel} + I_{\perp}} = \frac{(3\cos^2\alpha) - 1}{(\cos^2\alpha) + 3} \tag{17}$$

where $I_{\parallel}$ and $I_{\perp}$ are the intensities of the observed parallel and perpendicular components respectively, and α is the angle between the emission transition dipole and the particular absorption band. α is commonly 90° or 0°, so P has a maximum value of $+\frac{1}{2}$, and a minimum value of $-\frac{1}{3}$.

All four of the above emission characteristics, energy, decay time, quantum efficiency, and polarization, are observable also in inorganic systems, although they do not generally depend on transitions between molecular orbitals, and the mechanisms involved depend on the nature of the host material, e.g. whether it is an ionic crystal or a covalent complex. However, organic substances can, in certain cases, behave as ionic solids. Albrecht *et al.* (1968) reported on the luminescence of *N*,*N*,*N*′,*N*′-tetramethyl paraphenylenediamene (TMPD) in 3-methylpentane at 77°K. At this temperature, the solution has been chilled to a rigid amorphous solid. The TMPD exhibits biphotonic ionization (i.e. two photons are produced for every incident photon) in the near ultraviolet, and this luminescence is accompanied by photocurrents. A similar phenomenon was described by Geacintov *et al.* (1968) who distinguished between the quantum efficiency of fluorescence in solutions (ϕ_F(sol) and ϕ_F(cryst)). ϕ_F(sol) is generally independent of excitation wavelength, and for some aromatic substances, ϕ_F(cryst) does not vary much with the excitation wavelength (they give as an example anthracene), but is more likely to be affected by the condition of the crystal surface. Tetracene, however, does show that ϕ_F(cryst) depends on the excitation wavelength (Geacintov *et al.*, 1968). The ϕ_F(cryst) for tetracene begins to decrease as the excitation energy reaches about 3 eV, and higher voltages, and this event coincides with the appearance of a bulk generated photocurrent. This current is due to bulk ionization, i.e. creation of free electrons and holes, and above 3 eV takes over from the production of vibrationally excited neutral singlet states. Ionization competes with the radiative exciton formation process and the fluorescence efficiency decreases. In solution, none of the above processes takes place, and hence Geacintov *et al.* (1968) conclude that crystalline, not molecular processes are involved here.

B. INORGANIC PROCESSES—LUMINESCENCE MECHANISMS

A variety of mechanisms has been proposed for emission processes in a large number of inorganic phosphors. They include hole-electron pair processes, exciton processes, donor-acceptor pair mechanisms etc., all of which will be discussed below.

1. *Hole–electron Pair Processes*

Excitation by high energy particles causes a transfer of electrons from the valence band to the conduction band with a consequent formation of hole–electron pairs. If the electrons and holes have an energy of two to three times that of the interband gap, the electrons and holes may, in their turn, create hole–electron pairs. This non-equilibrium energy distribution of charge carriers undergoes thermalization in 10^{-11} sec. The number of charge carriers in an energy interval (E) to $(E+\mathrm{d}E)$ depends on the (energy) level density in the band, and in semi-conductors, this increases according to $\sqrt{E}$ for a simple parabolic band model. Thermal losses account for approximately 50 % of the energy of the hole–electron pairs, and typically only 25–30 % of the excitation energy is recovered as cathodoluminescence emission (Levshin, 1968). Excitation energy close to the energy of interband transitions will cut down the thermal losses, but will promote exciton formation with further loss of luminescence efficiency.

After this initial hole-electron pair formation process, a quasi-equilibrium distribution of charge carriers is established. The majority of hole–electron pairs fill traps which are produced by activator ions or other types of lattice defect. Recombination of electrons and holes takes place by band-to-band transitions, as well as at luminescence and quenching (non-radiative recombination) centres. At low temperatures, the shallowest traps are most rapidly filled, and in the case of certain ZnS phosphors, they form a kind of continuum at the bottom of the conduction band, thus effectively narrowing the forbidden energy gap. If the temperature is allowed to rise (from 17°K to 77°K in the example cited by Levshin, 1968), electrons and holes migrate from the shallow to deeper levels. If the phosphor in such a case has several activators at different levels with respect to the valence band, then different emission spectra can be observed at different temperatures.

Lushchik (1970) in a survey of luminescence processes in alkali halides distinguishes between electron recombination luminescence (ERL) where electrons recombine with ionized luminescence centres, and hole recombination luminescence (HRL) where holes recombine with atomic activator centres. ERL is accompanied by the production of conductivity, and the electrical currents can be controlled by electric fields. HRL is not accompanied by current production, because in the alkali halides, holes are less mobile than electrons, and the emission from HRL is much more akin to that of the intrinsic emission of the host lattice than to that of ERL emission.

Tunnel recombination plays an important role in HRL of alkali halides. The final stage of HRL is tunnel trapping of an electron e^-, from an uncharged atomic centre A^0, by an approaching hole, e^+, resulting in a positively charged atomic centre A^+, and a photon $h\nu$, where h is Planck's constant, and ν is the frequency of the photon. Because of the low mobility of holes in alkali halides, and because atomic centres have large effective radii, the probability of tunnel trapping of an electron by a hole is greater than the probability of an electron and hole recombining to excite the activator. The recombination probability for tunnel luminescence is independent of temperature (Lushchik, 1970).

2. *Exciton Processes*

Excitons are defined as mobile uncharged quanta of crystal excitation. They may be bound or paired with more or less closely spaced particles (Levshin, 1968). If an electron is given the symbol e^-, and a hole, e^+, then an exciton can be represented as e^0. For the latter three quanta, Lushchik (1970) distinguishes between their excited or relaxed states; e.g. an excited hole may be represented by e^+, and a relaxed hole by e_r^+. The properties of excited and relaxed holes are different, and similarly, the energy of excitons in an excited state differs from that of the relaxed state (e_r^0).

At low temperatures, relaxed holes are immobile, leading to a sharp decrease in the efficiency of recombination luminescence as the temperature is reduced. Unrelaxed holes can move to luminescence centres even at low temperatures (by tunnel trapping). Relaxed (or bound) exciton emission also drops off at low temperatures, but unrelaxed (or free) excitons can excite luminescence at low temperatures.

The binding energy of bound excitons is great (e.g. 23·0 meV in $ZnSiP_2$ from Raman scattering measurements; see Shay *et al.*, 1970), and in consequence, in systems where there is a large amount of bound exciton formation, there is a decrease in luminescence efficiency. Free excitons produce distinctive luminescence at low temperatures, and two may combine to form loosely aggregated molecules, which may also give rise, at very low temperatures, to characteristic luminescence.

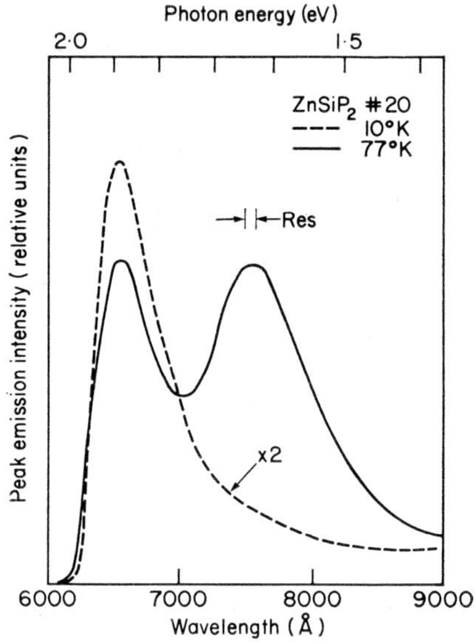

FIG. 17. Cathodoluminescence spectra of $ZnSiP_2$ at different temperatures. The electron beam current was the same for both measurements; neither spectrum was corrected for Sl response of the photomultiplier. (From Shay *et al.*, 1970.)

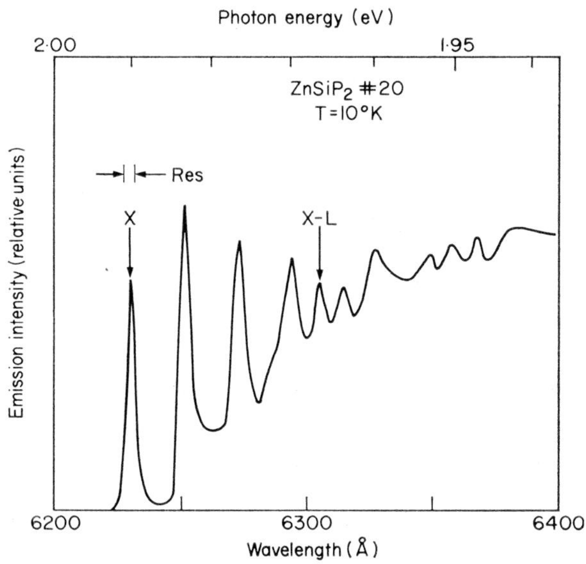

FIG. 18. Recombination peaks for excitons and phonons in $ZnSiP_2$. (From Shay *et al.*, 1970.)

Excitonic processes are important under conditions of cathodo-excitation. For example, Shay *et al.* (1970) have studied the luminescence from a $ZnSiP_2$ crystal at 77°K. This has a spectrum with two peaks, at 660 nm and 760 nm (see Fig. 17). A series of sharp lines occurs on the short wavelength side of the 660 nm peak (Fig. 18).

The two most likely origins for the sharp line spectrum are:

(a) radiative recombination of electrons on donors with holes on acceptors (donor–acceptor pair recombination); or

(b) radiative recombination of bound excitons with many phonon (non-radiative energy losses) replicas.

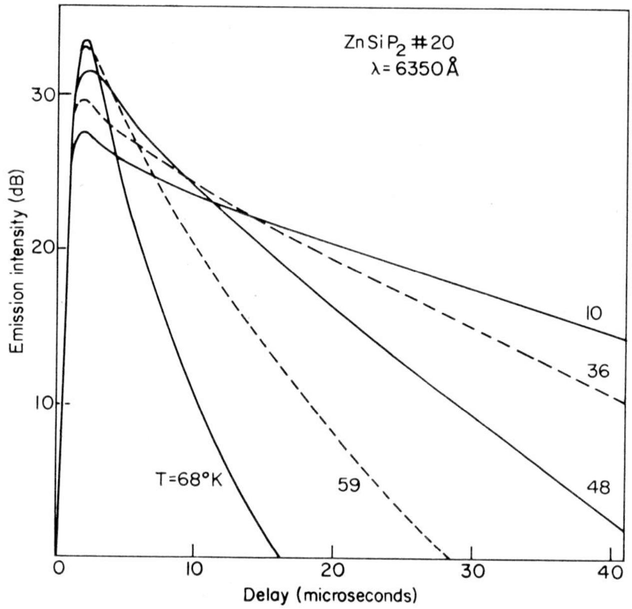

FIG. 19. (After Shay *et al.*, 1970.)

According to Shay *et al.* (1970), the two types of spectra can be distinguished by examining the time-resolved spectra of the sample. In the case of $ZnSiP_2$, the decay time for all lines is identical, and the time dependence of the decay is exponential. The decay time of donor–acceptor pair recombination lines and bands is variable over the spectrum (see Thomas *et al.*, 1965). The decay of the entire spectrum with a single exponential constant is consistent with recombination of a bound exciton with phonon replicas.

The decay of the 660 nm peak becomes significantly more rapid above 25°K (Fig. 19). For a temperature range of 25–50°K, the radiative efficiency is more or less constant. Therefore, although with rising temperature the decay time becomes shorter, the luminescence becomes more intense. Cuthbert and Thomas (1967) suggested that a similar phenomenon observed in GaP is due to thermal population of an excited state of the bound exciton having a shorter lifetime in this temperature range. Above 50°K, the radiative efficiency begins to decrease and decay becomes faster in $ZnSiP_2$. The luminescence is thermally quenched and experimental results fit the curve given by Shay *et al.* (1970) from the relationship

$$\tau = \tau_0/[1+A \exp(E_a/kT)] \quad (18)$$

where τ = the decay time at a time t

τ_0 = the total decay time

E_a = the activation energy

A = a constant, k is Boltzmann's constant, and T is the temperature in °K.

The regularly spaced sharp lines in the cathodoluminescence spectrum of $ZnSiP_2$ are attributed to radiative recombinations of a bound exciton with the simultaneous emission of 0, 1, 2, . . . phonons of energy of about 6·4 meV. This energy is too little for lattice phonons in $ZnSiP_2$ as determined by Raman scattering spectra, and is therefore probably a local mode associated with an impurity in the lattice. If this is so, then additional lines representing replication by $\pm n$ lattice phonons should be emitted simultaneously with the lines mentioned above due to the radiative recombination of a bound exciton. Such lattice phonon replicated lines do occur between the regularly spaced bound exciton replica lines. Shay *et al.* (1970) were able to identify the no-phonon line. The separation between this line and the other lattice phonon replicated lines corresponds to an energy of 23·0 meV, or in other words, the binding energy for excitons in this system as determined by Raman spectroscopy (the lines referred to above can be seen in Fig. 18).

Another recent example of the application of cathodoexcitation to the study of excitons is given in Radford *et al.* (1972). Using ultraviolet stimulation, they found that the blue emission of CdS was dominated by two bound exciton lines, and the only indication of the presence of free excitons in the spectrum was the zero (or no) phonon free exciton line which was present as a shoulder on the bound exciton emission. Using cathodoexcitation at 100 kV, Radford *et al.* (1972) found that two lines dominate the emission spectrum. These are 39 meV apart, which is approximately equal to the energy of a longitudinal optical (LO) phonon in CdS. The lines are separated by multiples of the LO phonon energy from a line at 485·65 nm given by excitons of the series (Γ_9–Γ_7) with a quantum number of $n = 1$. The two principal lines in the spectrum represent the first and second phonon replicas of the Γ_9–Γ_7 zero phonon exciton line. These are free exciton lines. Some smaller peaks in the spectrum represent bound exciton lines. Excitons of kinetic energy E are distributed in the exciton band according to a Maxwell–Boltzmann distribution, which is related to an effective temperature T. Thus the exciton effective temperature should be different from that of the lattice. The one phonon emission line was compared with theoretical values for emission at various effective temperatures. The best fit is at 42°K or 62°K. The two phonon line gives the best fit for 44°K. The surface temperature of the specimen was 13 ± 2°K,

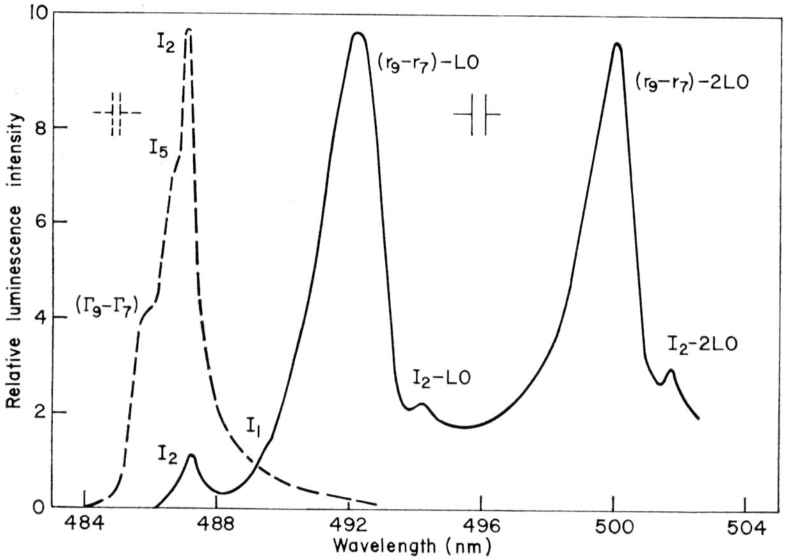

FIG. 20. Photoexitation (dotted) compared with cathodoexitation (solid). (From Radford *et al.*, 1972.)

measured with a thermocouple, although the specimen was mounted on a liquid helium cold finger. Thus, Radford *et al.* (1972) were able to determine that the effective temperature for free excitons in CdS is 45°K, which is higher than the measured lattice temperature. Bound excitons have effective temperatures that are the same as the lattice. In photoluminescence, all observed excitons are bound excitons, with an effective temperature the same as that of the host lattice. In cathodoluminescence, free excitons are formed, and Radford *et al.* consider that holes and electrons produced by cathodoexcitation have a high kinetic energy and are very abundant. The high effective temperatures indicate that relaxation of electrons and holes to lattice temperatures is very slow in cathodoluminescence. Figure 20 compares the photo- and cathodo-excited spectra.

Cuthbert (1970), using hyperpure silicon, described how excitonic molecules may be formed by the binding together of two free excitons into a free complex with small binding energy. In 1966, Haynes described a model for excitonic molecules in pure silicon at low temperatures where radiative emission is produced by recombination of an electron and hole within the excitonic molecule. The remaining electron and hole are ejected into the conduction and valence bands. Cuthbert modified the above model, allowing for non-radiative recombinations of excitonic models.

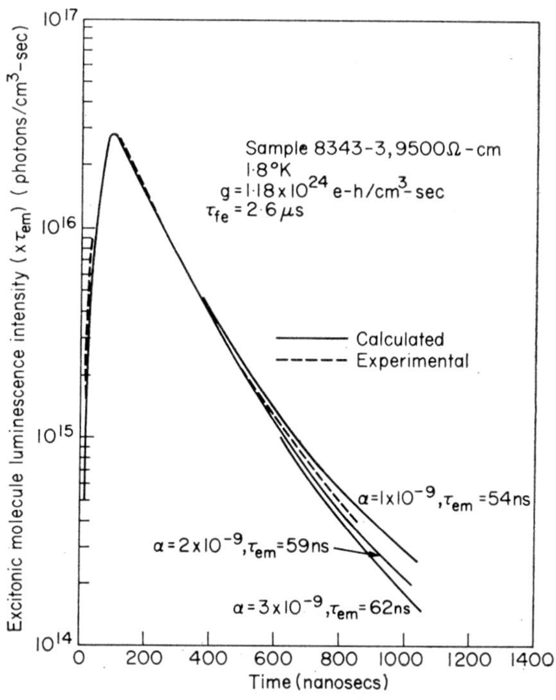

FIG. 21. (After Cuthbert, 1970.)

Experimentally, Cuthbert used 300 kV electrons, and calculated that they produced 10^{17} hole-electron pairs cm^{-3} per 86 nsec pulse at 1·8°K in intrinsic silicon. At low temperatures, the intrinsic luminescence of Si is composed of a superposition of free exciton and excitonic molecule bands. The use of time resolved spectroscopy permits a distinction being drawn between the decay times of free excitons and excitonic molecules (Figs 21 and 22). In decay curves for excitonic molecules, there is a post-excitation peak due to a build-up of the excitonic molecule density (Fig. 21) by the combining of pairs of free excitons to form excitonic molecules. If it is assumed that the time constant for the formation of free excitons from hole-electron pairs is small, then, in effect, excitons can be regarded as being directly produced by the electron beam. According to Cuthbert's results, this assumption is valid for Si at liquid helium temperatures.

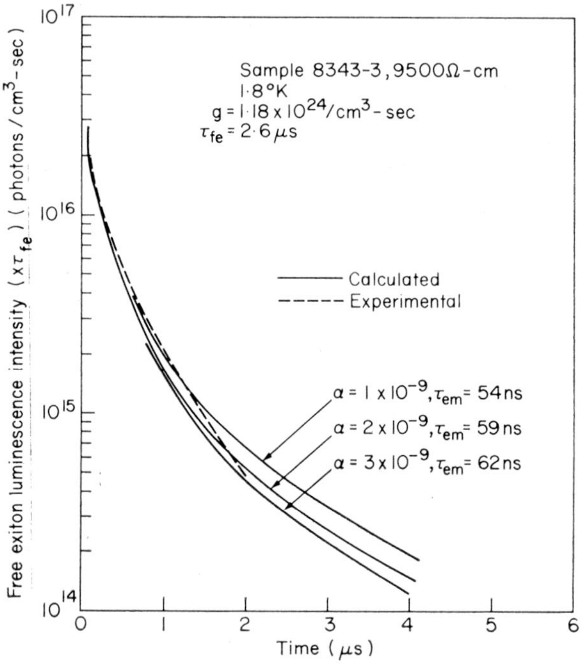

FIG. 22. Free exciton luminescence in hyperpure Si. (After Cuthbert, 1970.)

The equation governing the density, p, of hole-electron pairs during the exciting pulse is:

$$\frac{\mathrm{d}p}{\mathrm{d}t} = g-[\sigma+\nu]p^2 \tag{19}$$

where g = the generation rate of free electrons by the beam, finite for $0 \leqslant t \leqslant 86$ nsec

σ = cross-section governing the capture of a free hole by a free electron

ν = the frequency of the emitted radiation.

Since $p = 0$ at $t = 0$, then the equation can be solved for p:

$$p = (g/[\sigma+\nu])^{\frac{1}{2}} \tanh (g[\sigma+\nu])^{\frac{1}{2}}t. \tag{20}$$

After the generation of hole-electron pairs ceases, the $1/e$ decay time of the hole-electron pair density is:

$$\tau_{1/e} = (e-1)/(g[\sigma+v])^{\frac{1}{2}}. \tag{21}$$

Therefore, substituting Cuthbert's experimental values in the case of Si: $\tau_{1/e}$ = about 2 nsec. There would be no observable post excitation peak for free excitons, since Cuthbert's equipment was capable of measuring only time intervals greater than 4 nsec (compare Figs 21 and 22). If the concentration of excitonic molecules $= N$, and the concentration of free excitons $= n$, and the capture coefficient of excitonic molecules from free excitons $= \alpha$, and the time of the pulse $= t$, then:

$$\frac{dn}{dt} = g - \alpha n^2 - \frac{n}{\tau_{fe}} + \frac{N}{\tau_{em}} \tag{22}$$

τ_{fe} = the lifetime for free excitons due to radiative and non-radiative decay combined,

τ_{em} = the lifetime for excitonic molecules due to radiative and non-radiative decay combined.

It is assumed that σ for direct creation of excitonic molecules is small. Hence

$$\frac{dN}{dt} = \tfrac{1}{2}\alpha n^2 - \frac{N}{\tau_{em}}. \tag{23}$$

Each transition results in the liberation of a dissociated hole-electron pair. It is assumed that these particles regroup into excitons so that one free exciton becomes available for each excitonic molecule that decays. The radiative decay time of an exciton bound into an excitonic molecule should not be very different from that of a free exciton because of the weak bonding between the constituent excitons. However, in practice this is not true. In Cuthbert's experiments the excitonic molecules decayed about four times as fast as the free excitons. The rapid decay time can only be accounted for by Auger, non-radiative recombination processes, and these are analogous to Auger recombinations of bound excitons decaying at neutral donors in silicon. Henry and Nassau (1970) described the lifetimes of bound excitons in CdS, where the edge emission of CdS at liquid helium temperature consists of donor-acceptor pair recombinations in the green, and decay of excitons bound to neutral donors and acceptors in the blue. They managed to measure the lifetimes of these lines and found that they approximated very closely to the radiative lifetime for the I_2 line (an exciton bound to a neutral donor), and concluded that because of the close agreement between the values, the I_2 line was represented in their spectrum. Because its lifetime agreed so well with the calculated value for the expected lifetime, they concluded that the non-radiative Auger effect is negligible for the I_2 line.

Donor–acceptor Pair Mechanisms

Donor–acceptor pairs which may lie at any distance from each other are associated luminescent centres in semi-conducting phosphors (Williams, 1968). Shionoya (1968) gives an example of pair emission using time-resolved spectroscopy in an investigation of the luminescence of ZnS-type crystals. He finds that the peak shifts if the power per unit area is decreased (Fig. 23). He explained this by a donor–acceptor pair model:

$$E_n = E_G-(E_D+E_A)+e^2/\varepsilon r_n \tag{24}$$

where E_n = the energy of the line emitted when recombination takes place via a donor and an acceptor with separation r_n

$e^2/\varepsilon r_n$ = the Coulomb interaction of the pair

e = the electron charge

ε = the low frequency dielectric constant

E_G = the gap energy

E_D and E_A are the binding energies of the electron to the donor and of the hole to the acceptor respectively.

Therefore, the larger r_n is, the smaller becomes E_n. We should see a shift to a lower emission photon energy for the following reason: the transition probability for this type of emission also becomes smaller with

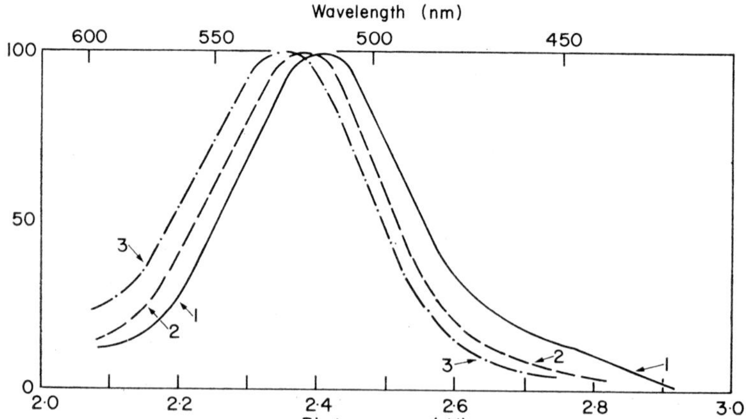

FIG. 23. (After Shionoya, 1968.)

increasing pair distance. At high excitation energies, the pairs with large r_n will be saturated since the transition probability for widely spaced pairs is small whereas the pairs with small r_n will not be saturated (because the transition probability is higher). Therefore, as the excitation energy is increased, the large r_n donor–acceptor pairs remain saturated, but more of the smaller r_n donor–acceptor pairs are excited and decay radiatively. This leads to a relative increase in the intensity of the lines corresponding to smaller values of r_n, i.e. higher energies.

The donor–acceptor pair mechanism has been dealt with more exhaustively by Hagston (1971), in a study of the theory of the energy shift of donor–acceptor pair emission with varying excitation energy. He develops a theory to take account of flash (or pulsed) excitation and continuous excitation. He suggests that the probability of finding a donor–acceptor pair with a separation distance between r and $r + \mathrm{d}r(r_n)$ is:

$$P(r)\,\mathrm{d}r \propto 4\pi N r^2\,\mathrm{d}r \exp[\alpha/r] \exp[-\tfrac{4}{3}\pi N r^3] \tag{25}$$

where

$$\alpha = q_1 q_2/\varepsilon kT \tag{26}$$

ε = the dielectric constant of the medium
N = the total concentration of the donors
q_1 = the charge on donors, q_2 = the charge on acceptors } $q_1 = q_2 = e$ (see eqn (24))
k = Boltzmann's constant
T = absolute temperature.

Such a pair emits light of the energy E_n, in accordance with eqn (24), and the pair possesses a radiative transition probability F_n where:

$$F_n \propto r_n^{2m} \exp[-r_n/R_d] \tag{27}$$

where R_d = half the Bohr radius of the shallowest centre, which Hagston assumes to be the donor
m = a number close to zero.

It is necessary to find an expression for the fraction of centres (f_e) in the excited state initially, separated by a distance r_n. This can be determined by the rate equation:

$$\frac{\mathrm{d}f_e}{\mathrm{d}t} = (1 - f_e) g \sigma(r_n) - f_e F_n \tag{28}$$

where g = the excitation intensity
$\sigma(r_n)$ = the capture cross section for pairs separated by distance r_n

In flash (or pulsed) excitation, excitation processes are complete before recombination processes begin to act, $f_e F_n$ is absent from eqn (28), and applying the boundary condition that $f_e = 0$ when $t < 0$, then because the intensity (g) has to be replaced by $g\delta(t)$, a delta function = one at $t = 0$, the equation becomes:

$$f_e = 1 - \exp\left[-g\sigma(r_n)\right]. \tag{29}$$

In continuous excitation, the left-hand side of eqn (28) is zero and:

$$f_e = [1 + F_n/g\sigma(r_n)]^{-1}. \tag{30}$$

As g tends to zero:

$$f_e \rightarrow g\sigma(r_n)/F_n. \tag{31}$$

From eqn (24), we can derive the distribution of the emission intensity immediately after the flash (or pulse):

$$\mathrm{d}E \propto \mathrm{d}r_n/r_n^2 \tag{32}$$

and if we gather together the results from eqns (25), (27), (29), and (32), it is possible to show that the emission intensity per unit energy at energy E, $I(E)$ is given by:

$$I(E) \propto r_n^4 \exp\left[\alpha/r_n\right] r^{2m} \exp\left[-r_n/R_d\right] \exp\left[-\tfrac{4}{3}\pi N r_n^3\right]\{1 - \exp\left[-g\sigma(r_n)\right]\}. \tag{33}$$

The emission peak energy is given by maximizing $I(E)$, and leads to the balance equation

$$-\frac{1}{R_d} - \frac{\alpha}{r_n^2} - 4\pi N r_n^2 + \frac{p\beta + 4 + 2m}{r_n} = 0 \tag{34}$$

where

$$\beta = \frac{g\sigma(r_n)\exp\left[-g\sigma(r_n)\right]}{1 - \exp\left[-g\sigma(r_n)\right]}$$

and equals 0 when $g\sigma(r_n) = \infty$ and equals 1 when $g\sigma(r_n) = 0$.

If the balance eqn (34) is satisfied under a given set of conditions, then altering these conditions such that an effectively positive contribution is added to the left hand side, then the emission peak will move to a new position corresponding to a larger r_n. If the alteration in conditions causes a negative contribution to the left-hand side of eqn (34), then r_n will become smaller.

Equations (32) to (34) apply to flash (or pulsed) excitation. In the case of continuous excitation, the only difference lies in the contribution of $(1/f_e)\partial f_e/\partial r_n$ to eqn (34). As $g \rightarrow \infty$, i.e. at the limit of high excitation, $f_e = 1$ for all r_n in both cases, and hence $(1/f_e)\partial f_e/\partial r_n$ is zero.

At this limit, both modes of excitation give an emission which peaks at the same energy. Assuming no preferential pairing, and making $m = 0$, the position of the peak corresponds to:

$$r_n = 4R_d. \tag{35}$$

This has been established experimentally for GaP (Thomas *et al.*, 1965). High excitation energy in this case can be considered as that energy required to saturate all centres. In the case of weak excitation energy, i.e. where $g \to 0$, the contribution of $(1/f_e)\partial f_e/\partial r_n$ is $(+p/r)$ for flash excitation, whereas for continuous excitation, the contribution is:

$$\frac{1}{\sigma}\frac{\partial \sigma}{\partial r_n} - \frac{1}{F_n}\frac{\partial F}{\partial r_n} = \frac{p}{r_n} - \frac{2m}{r_n} + \frac{1}{R_d}. \tag{36}$$

Since m is a number close to zero, in continuous excitation, an additional positive term has been added to the balance eqn (34) compared with the case for flash excitation. Hagston (1971) has interpreted this to mean that in the limiting case of weak intensity, the emission peak following flash excitation occurs at a higher energy (smaller r_n) than it would in the case of continuous excitation. If $\alpha = m = 0$, then:

$$r = (4+p-\delta)R_d \tag{37}$$

where $\delta = 4\pi N r^3$. This is for the case of flash excitation. For continuous excitation:

$$r = (1+\tfrac{1}{4}p)^{\frac{1}{3}}(N)^{-\frac{1}{3}}. \tag{38}$$

p is a number experimentally found to lie between 2 and 3 for GaP. For emission following flash excitation, the variation of the emission peak between the limiting cases of excitation ($g \to 0$, and $g \to \infty$) involves only a very narrow range of r_n. In the case of GaP, Thomas *et al.* (1965) found that a reasonable value for $p-\delta$ is 2, and hence r_n varies between $4R_d$ and $6R_d$. In the case of continuous excitation, however, it is clear that the larger the value of N, the smaller will be r_n. Because as N increases, a point is eventually reached where impurity banding becomes significant, if we take this point as an upper limit for N, then, following Hagston, who uses a combination of theory and experiment, it is possible to derive the equation

$$N^{\frac{1}{3}}a_0 = 0{\cdot}2 \tag{39}$$

a_0 is the value of a material-dependent factor for the onset of impurity conduction. Combining eqns (38) and (39), it is possible to show that

$$r_n \gtrsim 2^{\frac{2}{3}}5R_d. \tag{40}$$

The minimum value for r_n is approximately $8R_d$, and this is in agreement for GaP (Thomas *et al.* (1965) require a value of $10R_d$), and for CdS (Colbow (1966) requires a value of $8R_d$).

Therefore in the case of flash excitation, the variation in r_n with g is much less than in the case of continuous excitation, and Hagston (1971) uses this to explain the observed variations in peak shift in donor–acceptor pair mechanisms for luminescence.

IX. Configurational Co-ordinate Models

Butler (1968) classified the processes of activation in a series of phosphors whose lattices are oxygen dominated, such as, for example, barium trisilicate, or calcium fluophosphate. He found that activators could be divided into five classes:

1. *Primary activators:* where luminescence arises from transitions which are allowed in the free ion. Examples are: Ga^+, In^+, Tl^+, Ge^{2+}, Sn^{2+}, Pb^{2+}, As^{3+}, Sb^{3+}, Sn^{3+}, B^{3+}, Cu^+, Ag^+.
2. *Secondary activators:* where transitions are forbidden in the free ion. Examples are Mn^{2+}, Fe^{3+} (Fe^{2+} is a poison because the close grouping of energy levels allows ready radiationless energy loss, i.e. quenching).
3. *Lanthanide activators:* where excitation is an allowed transition but emission a forbidden transition in the free ion. Examples are Eu^{3+}, Tr^{3+}, Gd^{3+}.
4. *Covalent complex activators:* in these the ligand fields are very strong and the free ion energy cannot be correlated directly with the excitation and emission bands. An example is Mn^{4+}.
5. *Double activation:* two activators are required for this process and the mechanisms are not fully understood. An example is calcium fluophosphate activated jointly by antimony and manganese.

Butler (1968) has proposed a general hypothesis for activation in this group of compounds (the so-called oxyphosphors). He suggests that the various components of both the excitation and emission bands can be derived from discrete energy levels of the activator ion which is substitutional. Using a configurational co-ordinate model, the potential energy of the activator can be plotted as a function of the distance between the activator ion and the ions which surround it in the crystal lattice. The radius of ions in a crystal lattice is not precisely measurable, but varies from one such ion to another, depending upon the ion's exact site in the lattice. In Fig. 24a, the circles enclosing the ions enclose 90 % of the electron cloud. In Fig. 24b, the circles enclose 95 % of the electron cloud. In the second diagram, there is considerable

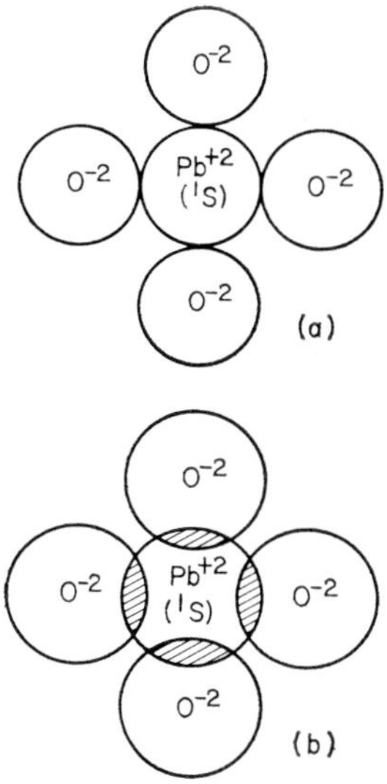

FIG. 24. Simplified electron density distribution contours: (a) 90% contour for Pb^{2+} in four fold co-ordination with oxygen, (b): as (a), but contouring at 95%, showing considerable overlap of the "clouds". (From Butler, 1968.)

overlap with the secondary ligand field. If the Pb ion were to be excited, then the volume and distribution of the overlap would be different yet again.

The configurational co-ordinate model shown in Fig. 25 shows potential energy curves for Pb^{2+}. Curve A represents the ^1S state of s^2 configuration, B the ^3P state of sp configuration without inclusion of

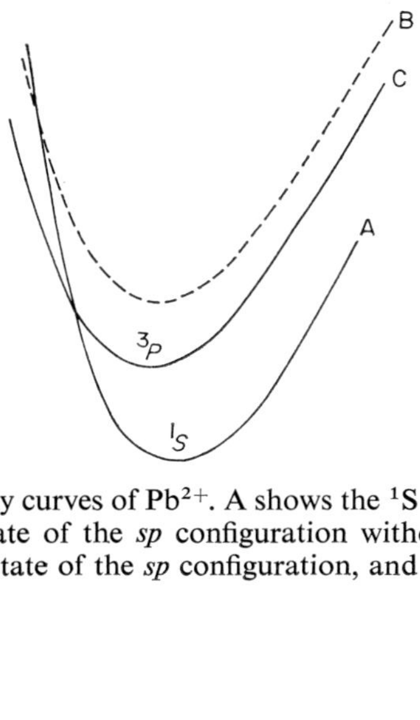

FIG. 25. Potential energy curves of Pb^{2+}. A shows the ^{1}S state of the s^2 configuration; B shows the ^{3}P state of the sp configuration without inclusion of overlap energy; C shows the ^{3}P state of the sp configuration, and includes overlap energy. (After Butler, 1968.)

overlap, and C, the ^3P state including the overlap. When the difference in overlap is included, as in curve C, the effect of the surrounding ions has been to lower the energy of the excited state to less (and in some cases much less—Butler quoted 50% less) than the normal unexcited state. The combination of the lowering of the crystal field and the free ion value determines the energy gap between the normal and excited state and the amount of lowering undergone by the crystal field will be a function of the properties both of the activator ion and of the ligand field in which it is situated. For examples of the action of the different types of activator ion, see Butler (1968).

Similar mechanisms were proposed for ZnS type II–VI compounds in a lengthy paper by Birman (1968). He had been concerned about the lack of reliability and predictability of luminescence theory and had developed the Ligand–Field-Band theory that he used to predict and interpret certain luminescence emissions from ZnS with different activators. He was able to offer both theoretical and practical proof of his predictions.

Engleman and Barnett (1970) and Barnett and Engleman (1970) have further developed the quantitative theory of luminescent centres in a configurational diagram model, and they have applied their theoretical results to a study of Cr^{3+} in oxides. It is important to note here that their theoretical relationships can be applied as readily to triplet–singlet transitions in organic molecules. For further information, the reader is referred to these two articles.

X. Quenching

In organic systems, this may occur as a result of several processes. In the case of metallo-organic complexes containing paramagnetic ions, the high spin momentum caused by such ions creates a strong non-uniform magnetic field of a strength comparable with that created by a centre with a high atomic number; spin-orbit coupling can thus be expected in such complexes and this increases the probability of singlet–triplet transitions so that it dominates the $S_1 \rightarrow S_0$ probability and fluorescence is quenched. Phosphorescence may be emitted and would be the sole emission under such circumstances.

Such complexes are common and include several important naturally occurring substances such as the porphyrins and chlorophylls. Fluorescence results on such complexes show clearly that a sharp distinction can be drawn between non-transition metal ion and transition metal ion derivatives. None of the latter has exhibited fluorescence irrespective of the state of oxidation of the metal. Furthermore, a general decrease in fluorescence intensity can be related to increasing atomic number of the metal, although the relationship is not simple, and exceptions occur which appear to depend on the ionic radius of the metal which may cause geometric distortions.

Where essentially complete spin-orbit coupling exists in organometallic complexes, even expected phosphorescence may not occur because other factors become dominant. For further details of quenching in fluorescence and phosphorescence in organometallic complexes, see Becker (1969).

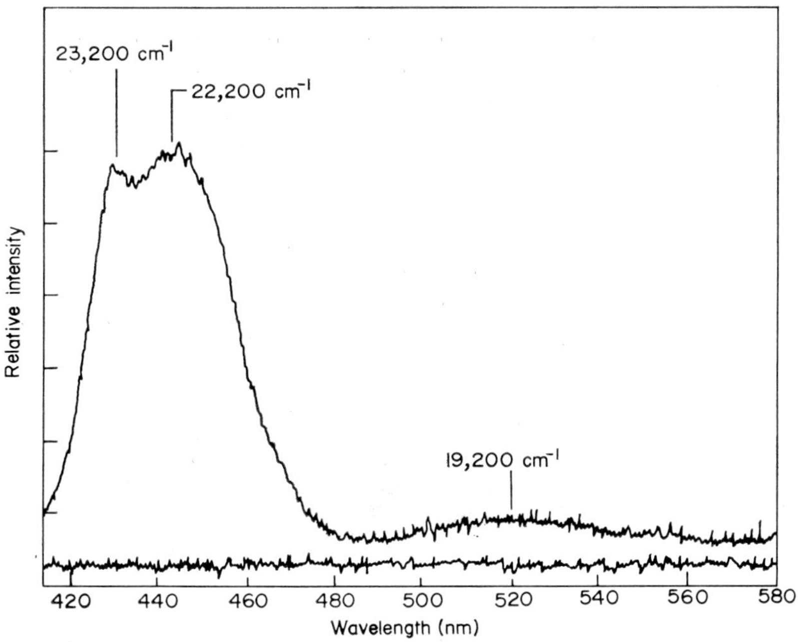

FIG. 26. The fluorescence spectrum of $LiAl_5O_8$–Fe^{3+} activated. Stoichiometrically the $LiAl_5O_8$ is in the inverse spinel form, with the Fe^{3+} in the octahedral sites. (From Melamed, *et al.*, 1970.)

Apart from metal ions, paramagnetic molecules such as O_2 can effect quenching although not, perhaps, entirely as a result of their paramagnetic properties. For example, Sevchenko (1968) described how the increase in the number of nitrogen bridges in phthalocyanine causes a corresponding decrease in the phosphorescence output, because they increase the number of radiationless relaxation processes.

In inorganic systems, the most common quenching process is caused by the presence of a transition metal ion. Because the valence orbitals in the transition elements of the first series are close together, deactivation processes tend to occur non-radiatively. This is particularly true for iron in the Fe^{2+} state. It should be noted that Fe^{3+} can and does act as an activator in oxygen dominated lattices (see Butler, 1968), and Melamed *et al.* (1970) described the fluorescence of Fe^{3+} in ordered and disordered states of the lithium compound $LiAl_5O_8$ (see Fig. 26). In this case, the concentrations of Fe^{3+} were very low, and a strong red fluorescence was observed. Melamed *et al.* were able to interpret their results in terms of crystal field theory, and showed that the Fe^{3+} ions occupied the octahedral sites in the inverse spinel structure of the $LiAl_5O_8$.

Quenching can, in fact, be defined as radiationless deactivation of luminescence centres, but it should not be confused with luminescence poisoning, where continued excitation ceases to produce transitions. The mechanisms for this are obscure, but may be related to radiation or thermal damage to the specimen.

XI. Laser Action

A brief account of cathodoluminescent laser action in II–VI semiconducting compounds can be found in Chapter 10 of this volume.

XII. Field Enhancement

As with photoluminescence, cathodoluminescence can be enhanced by the application to the phosphor of an electric field. Röppischer (1968) gave an example of this when he described the voltage dependence of the field enhancement of photoluminescence in Mn activated ZnS and CdS. The effect of the field is to redistribute the excitation energy causing the charge carriers to be transferred to Mn centres. Therefore with a constant excitation intensity, there is no change in the total number of recombinations per unit time with or without the field, but there is an increase in the number of radiative recombinations.

XIII. Applications of Cathodoluminescence

Cathodoluminescence studies have been applied to a variety of materials, although to date there is no very great bibliography available. Without going into specific cases, cathodoluminescence studies are valuable for the study of defects and impurities in crystalline solids, both manufactured, and as natural minerals. In organic materials, the volume of literature is even less, but the potential reward is great. It should prove possible to distinguish between closely related organic compounds in the solid state, in particular between polymers of various kinds which are difficult to distinguish by any other means (Muir *et al.*, 1971). Cathodoluminescence studies in the SEM offer the possibility of combining analytical information with high resolution morphological information, and therefore have considerable potential for the biologist, amongst others.

ACKNOWLEDGEMENT

We wish to acknowledge the help of Dr D. B. Holt.

REFERENCES

AHMED, H. (1972). *Proc. Vth European Congr. Electron. Microscopy*, 10–11.

ALBRECHT, A. C., JOHNSON, P. M. and MCCLAIN, W. N. (1968). *Proc. Int. Conf. Luminescence* (*1966*), 405–411.

BARNETT, B. and ENGLEMAN, R. (1970). *J. Luminescence* **3,** 55–73.

BECKER, R. (1969). "Theory and Interpretation of Fluorescence and Phosphorescence". Wiley, New York.

BIRMAN, J. L. (1968). *Proc. Int. Conf. Luminescence* (*1966*), 919–968.

BOND, E. F., BERESFORD, D. and HAGGIS, G. F. (1974). *J. Microsc.* **100,** 271–282.

BRIL, A., WANMAKER, W. L. and TERVRUGT, J. W. (1968). *J. Elec. Soc.* **115,** 776–777.

BUTLER, K. H. (1968). *Proc. Int. Conf. Luminescence* (*1966*) **2,** 1313–1329.

COLBOW, K. (1966). *Phys. Rev.* **141,** 742.

CUTHBERT, J. D. (1970). *J. Luminescence* **1, 2,** 307–314.

CUTHBERT, J. D. and THOMAS, D. G. (1967). *Phys. Rev.* **154,** 763.

DAVEY, J. P. (1966). "Optique des Rayons X et Microanalyse", pp. 436–441. Hermann, Paris.

DAVOINE, F., BERNARD, R. and PINARD, P. (1960). *Proc. European Reg. Conf. on Electron Microscopy*, Delft.

DITTMANN, R., GOERSCH, H. and HAHN, D. (1968). *Proc. Int. Conf. Luminescence* (*1966*) **2,** 1958–1961.

ENGLEMAN, R. and BARNETT, B. (1970). *J. Luminescence* **3,** 37–54.

FIJNAUT, H. M. (1971). Thesis, University of Utrecht.

GEACINTOV, N., POPE, M. and KALLMANN, H. (1968). *Proc. Int. Conf. Luminescence* (*1966*), 260–264.

GONI, J. and REMOND, G. (1969). *Min. Mag.* **37,** 153–155.

HAGSTON, W. E. (1971). *J. Luminescence* **3,** 253–260.

HAYNES, J. R. (1966). *Phys. Rev. Letters* **17,** 860.

HENRY, C. H. and NASSAU, K. (1970). *J. Luminescence* **1, 2,** 299, 306.

HÖRL, E. M. (1972a). *Micron* **3,** 540–544.

HÖRL, E. M. (1972b). *Proc. Vth European Congr. Electron Microscopy*, 374–375.

HÖRL, E. M. and MÜGSCHL, E. (1972). *Proc. Vth European Congr. Electron Microscopy*, 502–503.

JONES, G. A. C., NAG, B. R. and GOPINATH, A. (1973). *Proc. VIth Ann. SEM Symposium IITRI, Chicago*, 309–316.

LE POOLE, J. B., BOK, A. B. and BOOGERD, W. J. (1968). *Geologie Mijn.* **47,** 443–450.

LEVSHIN, V. L. (1968). *Proc. Int. Conf. Luminescence (1966)*, 23–38.

LONG, J. V. P. and AGRELL, S. O. (1965). *Min. Mag.* **34,** 318–326.

LUSHCHIK, C. H. (1970). *J. Luminescence* **1, 2,** 594–609.

MELAMED, N. T., VICCARIO, P. J., ARTMAN, J. O. and DES BARROS, F. (1970). *J. Luminescence*, **1, 2,** 348–367.

MUIR, M. D. (1970). *Proc. IIIrd Ann. SEM Symposium IITRI, Chicago*, 129–136.

MUIR, M. D., GRANT, P. R., HUBBARD, G. and MUNDELL, J. (1971). *Proc. VIth Ann. SEM Symposium IITRI, Chicago*, 401–408.

MCMULLAN, M. (1952). Ph.D. Dissertation, University of Cambridge.

PASTRNAK, J. and KAREL, F. (1968). *Proc. Int. Conf. Luminescence (1966)*, 1437–1476.

PEASE, R. E. W. and HAYES, T. L. (1966). *Nature, Lond.* **210,** 1049.

RADFORD, W. E., HAGSTON, W. E. and BRYANT, F. J. (1972). *J. Luminescence* **5,** 47–56.

REMOND, G., KIMOTO, S. and OKUZUMI, H. I. (1970). *Proc. IIIrd Ann. SEM Symposium IITRI, Chicago*, 33–40.

REMOND, G., KIMOTO, S. and OKOZUMI, H. I. (1972). *Jeol. News*, **9e** (4), 24–27.

RÖPPISCHER, H. (1968). *Proc. Int. Conf. Luminescence (1966)*, **2,** 1855–1858.

SEVCHENKO, A. N. (1968). *Proc. Int. Conf. Luminescence (1966)*, 183–193.

SHAY, J. L., LEHENY, R. F., BUEHLER, E. and WERNICK, J. (1970). *J. Luminescence* **1, 2,** 851–861.

SHIONOYA, A. (1968). *Proc. Int. Conf. Luminescence (1966)*, 739–764.

SIPPEL, R. F. (1968). *Proc. Int. Conf. Luminescence (1966)* **2,** 2079–2084.

SMITH, J. V. and STENSTROM, R. C. (1965). *J. Geol. 73*, 627.

SMITH, K. C. A. (1956). Ph.D. Dissertation, University of Cambridge.

SMITH, K. C. A. (1972). *Proc. Vth Ann. SEM Symposium, IITRI, Chicago*, 1–8.

STEYN, J. B. and HOLT, D. B. (1973). *Proc. SEM 73*, Institute of Physics, London.

TEROL, S. and AGUINO, F. F. (1968). *Proc. Int. Conf. Luminescence (1966)*, **2,** 2062–2067.

THOMAS, D. G., HOPFIELD, J. J. and ANGUSTYNIAK, W. M. (1965). *Phys. Rev.* **140A,** 202.

THORNTON, P. R. (1968). "Scanning Electron Microscopy", pp. 1–350. Chapman and Hall, London.

TOLSTOY, N. A. and ABRAMOV, A. P. (1968). *Proc. Int. Conf. Luminescence (1966)*, 1403–1407.

VENEKLASEN, L. H. and SIEGEL, B. M. (1972). *J. appl. Phys.* **43,** 1600.

WEBER, G. (1960). *Biochem. J.* **75,** 335–338.

WILLIAMS, F. (1968). *Proc. Int. Conf. Luminescence (1966)*, 113–123.

WILLIAMS, P. M. and YOFFE, A. D. (1969a). *Radiation Effects* **1,** 61–64.

WILLIAMS, P. M. and YOFFE, A. D. (1969b). *Nature, Lond.* **221,** 952–953.

YOFFE, A. D., HOWLETT, K. J. and WILLIAMS, P. M. (1973). *Proc. VIth Ann. SEM Symposium IITRI, Chicago*, 301–308.

Chapter 10

Quantitative Scanning Electron Microscope Studies of Cathodoluminescence in Adamantine Semiconductors

D. B. HOLT

Department of Metallurgy, Imperial College of Science and Technology, London, England

Symbols

n = minority carrier density in excess of equilibrium cm^{-3}
D = minority carrier diffusion coefficient
g = number of excess carriers produced $cm^{-3}\ sec^{-1}$
τ = minority carrier lifetime
L = minority carrier diffusion length ($= \sqrt{D\tau}$)
g_0 = (assumed) constant minority carrier generation rate (throughout a volume of radius R)
A, A_1 = constants
B = a constant
c' = a constant
$\mathscr{I}$ = intensity of CL radiation
τ_r = lifetime for radiative recombination

z = depth in the material
ξ = a radial coordinate in the plane of the layer parallel to the surface (so that $r^2 = \xi^2 + z^2$)
α = linear absorption coefficient
G = generation factor (number of hole-electron pairs generated per incident beam electron)
I_b = electron beam current
q = charge on the electron
$\tau_1, \tau_2 \ldots$ = lifetimes for particular non-radiative recombination processes
n_0, p_0 = equilibrium densities of electrons and holes respectively
τ_n, τ_p = lifetimes for electrons and holes respectively
n_i = intrinsic carrier density
τ_i = intrinsic carrier lifetime
$\beta_{nn}, \beta_{pp}, \beta_{np}, \beta_{pn}$ = capture probabilities for Auger recombination at impurities
τ_{nr} = lifetime for all non-radiative processes taken together
k = a constant
g' = total carrier generation rate ($= GI_b/q$)
ρ = density of the material
s = surface recombination velocity
S = reduced surface recombination velocity ($= s/V_d$)
V_d = diffusion velocity ($= L/\tau$)
N' = carriers lost by surface recombination per sec
R = range (maximum penetration distance) for electrons
E_0 = energy of incident beam electrons
ζ = coefficient in assumed relation $R = c(E_0)^\zeta$
W' = reduced electron range ($= R/\rho L$)
d = thickness of assumed dead surface layer due to energy band bending
V_b = electron beam accelerating voltage
m = coefficient in empirical relation $\mathscr{I} \propto I_b^m$ for p-type GaAs
$F(z)$ = Gaussian function for distribution of carrier generation with depth
F_0 = a constant
a = a constant
D_n = minority carrier (electron) diffusion coefficient
u = mass depth ($= \rho z/R$)
k = Boltzmann's constant
T = absolute temperature
D_p = minority carrier (hole) diffusion coefficient
P = photomultiplier output current ("CL signal strength")
$Q(\nu)$ = quantum efficiency ("sensitivity") of the photomultiplier, i.e. number of electrons emitted per incident photon (< 1)
η = refractive index
c = speed of light
y = reduced depth ($= z/R_G$) in section 2c
R_G = $3{\cdot}98\ \mathrm{V}^{1{\cdot}75}\ \mu\mathrm{gm\ cm}^{-2}$
A'/A = "degeneracy factor" = 1 for non-degenerately doped material
$\mathscr{I}_{obs}$ = observed CL intensity, i.e. that emitted from the material
$\mathscr{I}_{gen}$ = generated CL intensity in the interior of the material
t = thickness
L_1 = minority carrier diffusion length in n-type GaAs
L_2 = characteristic absorption length for n-type GaAs ($= 1/\alpha_{eff}$)
α_{eff} = effective absorption coefficient i.e. the value corresponding to the wavelength of the peak transmitted through a distance x
$I(X)$ = the charge collection current due to the barrier electron

voltaic effect for a distance X between the point of impact of the beam and the p-n junction

P_p = photomultiplier current (CL signal) for p-type material

P_n = photomultiplier current (CL signal) for n-type material

η_e = external quantum efficiency of an injection luminescent p–n junction diode ($= \eta_{rad}\eta_a\eta_f$)

η_{rad} = radiative recombination efficiency of the injection region

η_a = fraction of internally generated photons that are emitted from the material

η_f = fraction of the total number of carriers passing through the device that recombine in the luminescent region

I_s = specimen current, i.e. the leakage current to earth from the electron beam through the specimen

η_N = extreme CL efficiency normalized to 1 at high beam voltage

V_0 = dead voltage, i.e. the beam voltage at which $R = L$

I. Introduction

The information obtainable from the analysis of cathodoluminescent emission spectra might be divided for practical purposes into extrinsic and intrinsic. Extrinsic features of cathodoluminescence (CL) spectra are those which depend on the presence of impurities present in the material whereas intrinsic features of CL spectra arise from the energy band structure of the material itself. For many practical purposes intrinsic CL emission peaks may be thought of as arising from electronic transitions of the type indicated in Fig. 1(a), taking place between states in or near the two energy bands on either side of the forbidden energy gap of width E_g. Extrinsic cathodoluminescence includes all the types of transition indicated schematically in Fig. 1(b). That is, the term extrinsic cathodoluminescence will be used to refer to light emitted as a result of an electron transition which starts or finishes on a localized impurity level in the forbidden energy gap or which both starts and finishes on such a level. This terminology is a logical extension of the standard usage in semiconductors. Unfortunately an almost opposite terminology is in use in the phosphor literature where certain emission bands or emission spectra from a certain material are described as "characteristic" of the particular impurity or "activator" element that gives rise to that characteristic colour cathodoluminescence. Conversely the emission that is not indicative of any impurity is described as "uncharacteristic". However, no harm will arise from this terminological babel provided that it is remembered that extrinsic cathodoluminescence is characteristic of some particular impurity or impurities while uncharacteristic cathodoluminescence is intrinsic to the semiconductor (or "phosphor") material.

The distinction between intrinsic and extrinsic is useful but in many important practical cases the simplified model of Fig. 1 needs modification. When the concentration of impurities in a semiconducting material exceeds a certain level of the order of 10^{18} cm^{-3} the individual impurity atoms become so closely spaced that the wave functions of their outermost electrons begin to overlap significantly. In consequence their energy levels broaden out into a band. If the initial level is near to a permitted band, the impurity band and the adjacent band for example,

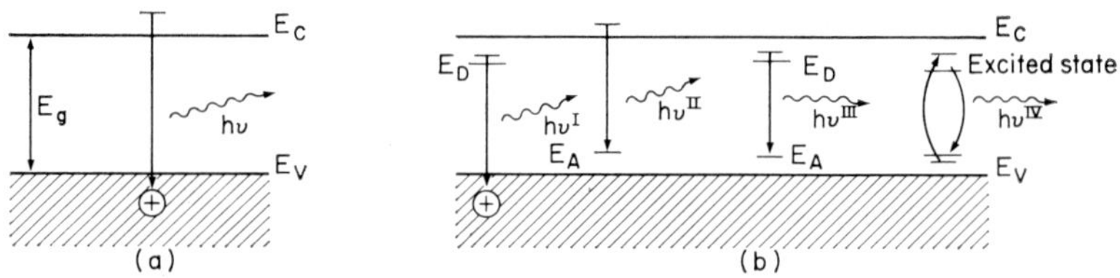

FIG. 1. Intrinsic and extrinsic cathodoluminescence. (a) Intrinsic cathodoluminescence arises from the recombination of holes and electrons produced by the electron beam bombardment, by an electron recombining from a state in the conduction band (above the level E_c) into an empty state (hole) in the filled valence band, below the energy level E_v. E_g is the width of the forbidden energy gap and is an important characteristic of the material, and the radiation with $h\nu = E_g$ is referred to as the band edge emission. (b) Extrinsic cathodoluminescence arises from electron transitions which start or finish (or both) on localized states in the forbidden energy gap. A variety of cases are shown. The localized states arise from the presence of point defects especially impurity atoms.

the conduction band will eventually overlap. In such a case the semiconductor is said to be degenerately doped. The width of the forbidden energy gap then becomes a function of the doping concentration and the wavelength and the intensity of the formerly "intrinsic" band-to-band recombination radiation become functions of the impurity concentration. This is the case with the GaAs and GaP used in lasers and light emitting diodes (LEDs).

At low temperatures the hole-electron pairs produced by the electron bombardment (see Chapter 9) tend to form bound pairs called excitons. The lowering of their total energy in consequence of the binding of the excitons by the coulombic attraction of their opposite electrostatic charges brings their energy levels closer together than the width of the forbidden energy gap. In other words excitons can exist in a series of bound states lying in the forbidden gap. Exciton recombination then produces photons of wavelengths (or energies) which are "intrinsic", i.e. essentially corresponding to the forbidden gap width of the particular structure. Exciton lines as will be reported below are found to be important in the CL emission spectra of II–VI compounds at low temperatures including II–VI CL laser emission spectra.

A second term needs to be introduced here. A "direct" energy band gap is one like that in Fig. 2(a) in which the minimum of the conduction band occurs at the same value of the wave vector **K** as the maximum of the valence band. Vertical transitions, which conserve momentum,

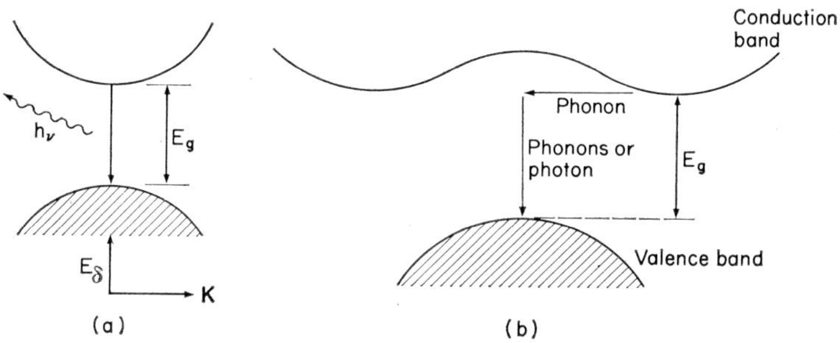

FIG. 2. Direct and indirect gap energy bands. (a) In a direct gap material vertical transitions can occur and the probability of radiative recombination is high. (b) In indirect gap materials, vertical transitions across the minimum gap are not possible. Phonons must be emitted to conserve momentum when hole-electron pairs recombine and radiative recombinations are much less likely.

are then possible between the most probably filled (minimum) states in the conduction band and the (maximum) states most likely to be unoccupied in the valence band. Consequently radiative recombination is likely between holes and electrons in such materials. However, in "indirect" gap materials as illustrated in Fig. 2(b) the conduction band minima and valence band maximum do not occur at the same place in **K** space. To conserve momentum in a material with such a band structure, phonons must be emitted when hole-electron pairs recombine. The probability of a photon also being emitted is considerably reduced in such cases as compared with direct gap materials. Cathodoluminescence of any intensity therefore in general in indirect gap materials arises from impurities or other imperfections. Conversely high efficiency emission of intrinsic cathodoluminescence will only occur in direct gap materials, e.g. CL laser action only occurs in direct gap materials. High efficiency is of course a relative term. GaP, which is an indirect gap material, can be used to make *p–n* junction diodes which will emit electroluminescence with efficiencies of up to 6%.

A third distinction that needs to be made in relation to cathodoluminescence observations is that between non-dispersive and dispersive measurements. Most measurements are made by allowing all the emitted spectrum to fall on the photocathode of a photomultiplier and recording the photomultiplier output current. This non-dispersive signal is the integral of the photomultiplier quantum efficiency times the cathodoluminescence intensity over the whole wavelength range for which the product is non-negligible. An alternative type of CL detection system passes the CL spectrum through a monochromator first before allowing the light to fall on the photomultiplier. In such a dispersive measurement the signal is the product of the photomultiplier quantum efficiency and the CL intensity for some relatively narrow range of wavelengths. This method of detection is of course essential if CL spectra are to be recorded for analysis.

Attention in this chapter will be confined to the adamantine semiconducting materials. These are the materials with "diamond-like" crystal structures. This is a somewhat elastic term referring to a large number of elements and compounds all having tetrahedrally co-ordinated crystal structures (Parthe, 1966) and related energy band structures all with a filled valence band separated from an empty conduction band by a forbidden energy gap (Mooser and Pearson, 1960; Suchet, 1965). This group of materials can be considered as beginning with the diamond-structure elements of the IVB sub-group of the periodic table: diamond (carbon), silicon and germanium. A number of closely related binary materials are produced by making 50–50 atomic percent compounds of elements lying equal distances on either side of the IVB elements. Thus the III–V compounds such as GaAs and GaP and the II–VI compounds like ZnS, CdS and ZnSe arise. By "cross-substitution" numerous compounds of three or four elements can be produced such as the II IV V_2 compounds like $ZnSnAs_2$ etc., etc. These adamantine materials include all the semiconducting materials of any importance except for the chalcogenide glasses, the IV–VI compounds (e.g. PbS) and Se. The adamantine compounds, especially the III–V and II–VI compounds include all the materials that are of importance at present in solid state opto-electronic devices and all the materials with which this chapter is concerned are either III–V or II–VI materials.

The III–V materials are the better developed and more rapidly increasing in commercial importance of the two groups. More CL SEM work has therefore been done on these materials than on any others and we will begin by reviewing the literature on the III–V compounds before outlining the work that has been done on the II–VI compounds. Finally the work which has been done relating the barrier electron voltaic effect in the conductive mode to the CL emission in an effort to assess electroluminescent *p–n* junction devices will be treated.

II. Quantitative Studies of III–V Compounds

Because GaAs is an efficient source of CL as it has a direct band gap, a large number of quantitative studies of CL in scanning electron beam instruments have been made of this material than of any other since strong signals can be obtained. At low temperatures and high excitation levels GaAs produces infrared CL with a quantum efficiency approaching 100%. The high radiative recombination probability under these circumstances is the reason that GaAs is a good laser material. Even under the circumstances in an SEM at room temperature the efficiency of CL emission in GaAs is of the order of 1% as we shall see below.

A. Dependence of the Cathodoluminescence from GaAs on Specimen Properties and on Excitation Conditions

A valuable although simplified analysis of the phenomena involved in CL observations on GaAs in scanning beam instruments was given in one of the earliest publications in the field (Kyser and Wittry, 1964). In this section we shall follow their account closely.

The first thing that has to be treated is the distribution of charge carriers around the point of impact of the electron beam. The basic mechanism of the generation of CL in GaAs is the production of hole-electron pairs by the "ionization" of electrons up from the filled valence band to the empty conduction band and the subsequent recombination of the electron and hole with emission of the energy of the pair as a photon. It is only necessary to consider the minority carriers in general, i.e. the holes in n-type material or the electrons in p-type material.

The differential (continuity) equation for the diffusion of minority carriers in a semiconducting material is

$$D\nabla^2 n - \frac{n}{\tau} + g(r) = 0 \tag{1}$$

where n is the minority carrier density cm^{-3} in excess of the equilibrium value, D is the diffusion coefficient, $g(r)$ is the number of excess carriers produced cm^{-3} sec^{-1} and $1/\tau$ is the probability of recombination sec^{-1}, i.e. τ is the minority carrier lifetime.

Assuming that the distribution is spherically symmetric and the size of the source is infinitesimal, the appropriate solution of the continuity equation is

$$n = \frac{A_1 e^{-r/L}}{r} \tag{2}$$

where L is the minority carrier diffusion length and is equal to $(D\tau)^{\frac{1}{2}}$.

For a source of finite size the simplest assumption is that the generation rate is a constant g_0, throughout a volume of radius R and zero elsewhere. The solution in this case is

$$n = \frac{A\,e^{-r/L} + B\,e^{+r/L}}{r} + \tau g_0 \tag{3}$$

By setting the flow of carriers out of a sphere of radius r_1 equal to the number of carriers generated inside this volume per second minus the number of carriers lost per second by recombination it can be shown that $A = -B$. The solution can therefore be rewritten as

$$n = \frac{C^1 \sinh(r/L)}{r} + \tau g_0 \tag{4}$$

where C^1 is a constant. By differentiating, the value of r at which n is a maximum can be shown to be $r = 0$. At this beam impact point the excess carrier concentration is $g_0\tau + C^1/L$.

Both the point source expression (2) and the uniform, spherical energy-dissipation volume solution (4) can be used to obtain expressions for the self absorption corrections and to obtain information on the dependence of the intensity of the recombination radiation (CL) on the doping of the specimen.

First we use the point source solution to treat the effect of absorption. Equation (2) can be rewritten and employed to obtain the intensity produced by radiative recombination in a layer of thickness dz at a depth z in the specimen as follows

$$\mathscr{I}(z)\,\mathrm{d}z = \frac{n}{\tau_r} = \frac{A_1}{\tau_r}\int_0^\infty \frac{\exp\left(-\sqrt{\xi^2+z^2}/L\right)}{\sqrt{\xi^2+z^2}}\,2\pi\xi\,\mathrm{d}\xi\,\mathrm{d}z \tag{5}$$

where ξ is a radial co-ordinate in the plane of the layer so that $r^2 = \xi^2 + z^2$ and τ_r is the lifetime for radiative recombination. Changing variables this can be integrated to give

$$\mathscr{I}(z)\,\mathrm{d}z = (2\pi L A_1/\tau_r)\exp(-z/L)\,\mathrm{d}z \tag{6}$$

The absorption correction is obtained from the fact that

$$\frac{\mathscr{I}_{\text{obs}}}{\mathscr{I}_{\text{generated}}} = \frac{\int_0^\infty \mathscr{I}(z)\exp(-\alpha z)\,\mathrm{d}z}{\int_0^\infty \mathscr{I}(z)\,\mathrm{d}z} = \frac{1}{1+\alpha L} \tag{7}$$

where α is the linear absorption coefficient and we have assumed

Lambert's Law:

$$\frac{\mathscr{I}(z)}{\mathscr{I}(0)} = \exp(-\alpha z) \tag{8}$$

to hold.

The distribution function $\mathscr{I}(z)$ can be rewritten in terms of the total number of excess carriers generated per second, GI_b/q, where the generation factor G is the number of hole-electron pairs generated per incident beam electron, I_b is the beam current and q is the charge on the electron. G is discussed in more detail in Chapter 9. In the present model $GI_b/q = 2\pi A_1 D$

$$\mathscr{I}(z)\,\mathrm{d}z = \frac{GI_b L \exp(-z/L)\,\mathrm{d}z}{q\tau_r D} \tag{9}$$

The actual number of photons generated per second at a depth z is then given by

$$\mathscr{I}_{\text{generated}} = \frac{GI_b L^2}{q\tau_r D} \tag{10}$$

However $L^2 = D\tau$ where $1/\tau$ is the probability per second that recombination will occur by any of the radiative and non-radiative recombination mechanisms that are operative in the specimen.

Hence

$$\mathscr{I}_{\text{gen}} = \frac{GI_b\tau}{q\tau_r} \tag{11}$$

The intensity of the recombination radiation generated will thus depend on the relative probability of radiative recombination and of recombination by all possible mechanisms. This result follows by definition of τ and τ_r but the calculation of Kyser and Wittry which we have just given shows that this conclusion is unaltered by carrier diffusion so long as the two lifetimes are constant.

In general there will be a number of different non-radiative recombination mechanisms operative. We can then write

$$\frac{1}{\tau} = \frac{1}{\tau_r} + \frac{1}{\tau_1} + \frac{1}{\tau_2} + \frac{1}{\tau_3} + \cdots \tag{12}$$

where τ is the effective lifetime of the excess minority carriers, τ_r is the radiative lifetime and τ_1, τ_2, etc., are the lifetimes for the operative non-radiative recombination processes.

Kyser and Wittry calculated the lifetimes for four recombination processes that are known to be significant in GaAs. The four sets of lifetimes are plotted in Fig. 3 against the number of free charge carriers (electrons) $no. = Nd - Na$ where Nd is the concentration of donors and Na is the concentration of acceptors—assumed negligible unless the material is known to be (partially) compensated, i.e. to contain both types of impurity. The four recombination mechanisms are listed in the caption to Fig. 3. The calculations were made on the assumptions that

(i) The injection level is low, i.e. $n \ll no$, except that in the case of Auger recombination in the host crystal $(A-H)$ a high injection level was assumed.

(ii) In the case of direct radiative recombination $(D-R)$ it was assumed that

$$\tau_n = \tau_p = 1/B(no+po+n)$$

where τ_n is the electron lifetime and τ_p is the hole lifetime with $B = 6{\cdot}3 \times 10^{11}\ \text{cm}^3\ \text{sec}^{-1}$ at 300°K.

(iii) The capture probability for multiphonon recombination at an impurity $(MP-I)$ is $10^{-10}\ \text{cm}^3\ \text{sec}^{-1}$ and an expression derived for a large impurity density applies.

(iv) Auger recombination in the host crystal $(A-H)$ gives lifetimes

$$\tau_n = \tau_p = \frac{3n_i^2}{n^2}\tau_i$$

where n_i is the intrinsic carrier density and $\tau_i = 10^{+19}$ sec at 300°K.

(v) Auger recombination at impurities $(A-I)$ involves capture probabilities of

$$\beta_{nn} = \beta_{pp} = 10^{-27}\ \text{cm}^6\ \text{sec}^{-1}$$
$$\beta_{np} = \beta_{pn} = 10^{-23}\ \text{cm}^6\ \text{sec}^{-1}$$

and the impurity density is small enough for charge storage to be neglected.

(vi) For the impurity controlled processes $(MP-I$ and $A-I)$ an impurity concentration of $10^{16}\ \text{cm}^{-3}$ and a level 0·5 eV below the middle of the band was assumed.

Kyser and Wittry remarked that in view of the assumptions involved the results of Fig. 3 should be regarded as order-of-magnitude estimates. They also pointed out that similar results would apply for p-type material (plotting the lifetimes versus p_0 the hole concentration) except that in the case of multiphonon recombination at an impurity there might be a different capture probability than that of assumption (iii) above so that curve $MP-I$ of Fig. 3 might be significantly affected. Kyser and Wittry discussed the changes in the results of Fig. 3 that would result from impurity concentrations different from the assumed value.

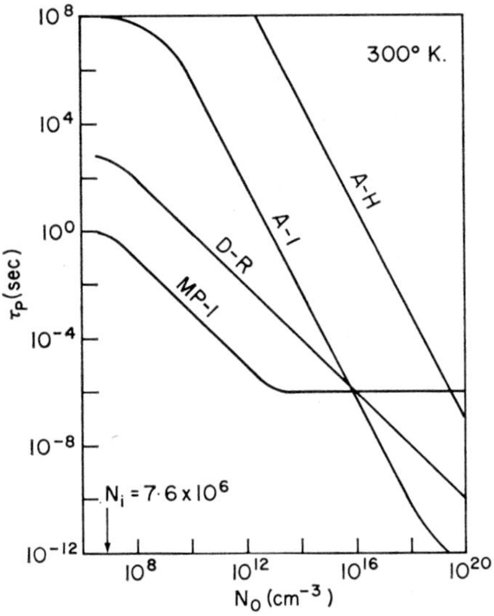

FIG. 3. Calculated minority lifetime versus electron concentration in GaAs for four different recombination processes. These are direct band-to-band radiative recombination (D–R), multiphonon emission at an impurity (MP–I). Auger recombination at an impurity (A–I) and Auger recombination in the host crystal (A–H). (After Kyser and Wittry, 1964.)

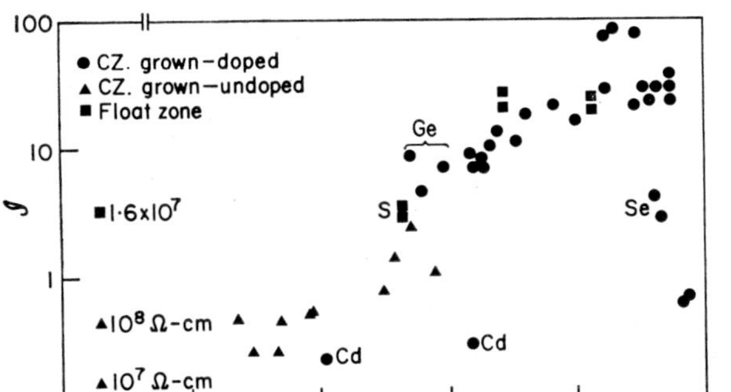

FIG. 4. Average infrared intensity *vs* the carrier concentration for GaAs at 300°K using an electron beam current of 0·1 μA and voltage 30 kV. Except where otherwise indicated all the specimens with $n_0 > 10^{16}$ cm^{-3} were Te-doped and specimens with lower values of n_0 than this were undoped. (After Kyser and Wittry, 1964.)

For future reference it can be seen that for $n0 = 10^{18}$ cm^{-3} which is the approximate value of the optimum doping level for luminescent efficiency the values obtained are approximately: $\tau_{A-I} = 10^{-10}$ sec, $\tau_{D-R} = 10^{-8}$ sec, $\tau_{MP-I} = 10^{-6}$ sec and $\tau_{A-H} = 10^{-3}$ sec. Substituting into eqn (12) shows that $\tau = 10^{-10}$ sec and substituting these values of $\tau_{D-R} = \tau_r$ and τ into eqn (11) indicates that in such material at room temperature approximately 10^{-2} of the hole-electron pairs produced by electron bombardment will produce photons, i.e. the CL quantum efficiency of 10^{18} cm^{-3} carrier concentration n-type GaAs at 300°K is expected to be about 1%.

Kyser and Wittry also gave a simplified treatment of the dependence of the CL intensity on the electron beam current as follows. As discussed above the efficiency of CL emission is given by τ/τ_r. Let the lifetime for all the non-radiative recombination processes together be τ_{nr}, then

$$\mathscr{I} = Kg^1\tau_{nr}(\tau_r+\tau_{nr})^{-1} \tag{13}$$

where $\mathscr{I}$ is the intensity in photons sec^{-1}, K is a constant with the dimensions cm^3 and g^1 is the total rate of generation of excess carriers, i.e. $g^1 = GI_b/q$.

If radiative recombination predominates then $\tau_r \ll \tau_{nr}$ and $\mathscr{I} = Kg^1$ for all values of g^1. If non-radiative transitions predominate $\tau_r \gg \tau_{nr}$ and we assume then that $n = \tau_{nr}g^1$ for all values of g^1 then from

$$\tau_n = \tau_p = 1/B(no+po+n) \tag{14}$$

which was previously used (assumption (ii)) in calculating the results of Fig. 3, we have

$$\tau_r = B^{-1}(no+po+\tau_{nr}g^1)^{-1} \tag{15}$$

Therefore at low excitation levels for which $\tau_{nr}g^1 \ll no+po$

$$\tau_r = B^{-1}(no+po)^{-1} \tag{16}$$

and

$$\mathscr{I} = KB(no+po)\tau_{nr}g^1 \tag{17}$$

At high excitation levels for which $\tau_{nr}g^1 \gg no+po$

$$\tau_r = (B\tau_{nr}g^1)^{-1} \tag{18}$$

and

$$\mathscr{I} = KB\tau_{nr}^2(g^1)^2 \tag{19}$$

Since $g^1 = GI_b/q$, i.e. $g^1 \propto I_b$ the following conclusions may be drawn from the analysis given above:

(i) If radiative recombination is dominant, the CL intensity should be proportional to the beam current and independent of the equilibrium carrier concentration $no(\mathscr{I} = Kg^1)$.

(ii) If non-radiative transitions are dominant, the CL intensity varies in different ways with the beam current in the high and the low excitation cases as follows:

(a) At low beam currents, the intensity should be proportional to *no* and to the beam current (eqn 17).

(b) At high beam currents, i.e. in the case that $n \gg no$, the intensity should be proportional to the square of the beam current but be independent of *no* provided that τ_{nr} does not depend on *no* (eqn 19).

Kyser and Wittry (1964) also reported experimental results which were in general agreement with those predicted above for the case in which non-radiative recombination is dominant and the excitation level is low. In other words, even though they used a microanalyser with relatively large beam currents they were unable to reach sufficiently high beam currents to observe the predicted dependence on the square of the current. Their measurements of relative intensity as a function of the free carrier concentration are shown in Fig. 4 and the expected linear dependence (prediction ii(a) above) can be seen to be rougly obeyed although the scatter was large. Specimens of the same carrier concentration, at the highest levels could give CL intensities that differ by up to two orders of magnitude. Variations in the intensity from point to point, to be discussed below, indicate that differences in purity and crystalline perfection may be more important than carrier concentration in determining the CL efficiency in many cases.

The results of Kyser and Wittry on intensity as a function of specimen current are shown in Fig. 5 and the approximately linear dependence (45° slope) predicted by eqn 17 can be seen to be fairly well obeyed by specimens of several different dopings. The specimens with net carrier concentrations of $1{\cdot}9 \times 10^{18}$ cm^{-3}, $2{\cdot}8 \times 10^{17}$ cm^{-3} and $4{\cdot}4 \times 10^{16}$ cm^{-3} were the samples of the highest efficiency measured in Fig. 4 and they showed a linear response with specimen current at low levels and a slightly superlinear response at high currents. This is to be expected on the treatment given above if the efficiency and the injection level are relatively low. The semi-insulating 10^8 ohm cm specimen followed a square law dependence over two orders of magnitude variation in the specimen current. This is understandable on the above theory for the case of low efficiency and a high injection level. Another semi-insulating specimen with higher efficiency had nearly the same response as that of the $4{\cdot}4 \times 10^{16}$ cm^{-3} *S*-doped sample. The specimen with a net carrier concentration of $4{\cdot}5 \times 10^{18}$ cm^{-3} was one of the lowest efficiency samples for this carrier concentration and gave only a slightly superlinear response over the entire range of specimen current. These measurements thus show that a linear response is usually associated with a relatively high efficiency for spontaneous radiative recombination as is to be expected from the theory.

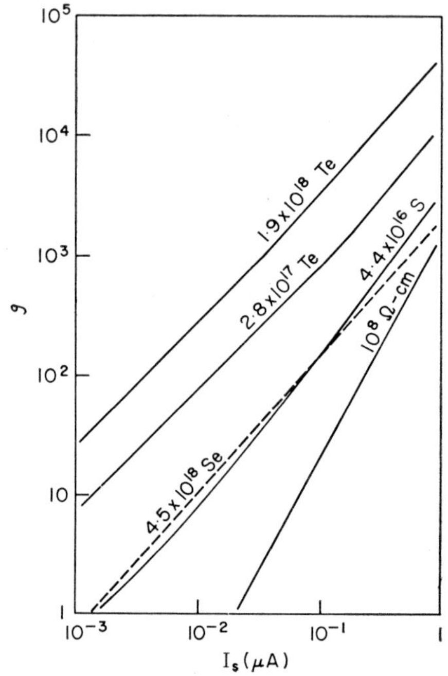

FIG. 5. Infrared intensity *vs* specimen current with an accelerating voltage of 30 kV and a static beam. 300°K. (After Kyser and Wittry, 1964.)

In the next three sections are discussed the effects of beam voltage, specimen temperature and self-absorption on the detected CL. These are all factors which have been treated in cases in which the emphasis was on the understanding of the physical properties of the material. The beam voltage is not a cause of contrast and variations in the others as possible causes of contrast have not yet been investigated. In sections II.E and II.F two sources of micrographic contrast are discussed.

B. VARIATION OF CL INTENSITY WITH ELECTRON BEAM VOLTAGE IN GaAs AND GaP

Wittry and Kyser (1966, 1967) calculated the CL intensity for *n*-type GaAs as a function of electron beam voltage as follows. Let the number of excess carriers generated per unit time at a mass depth ρZ, where ρ is the density of the material and Z the depth, be $\phi(\rho Z)\,\mathrm{d}\rho Z$. The loss of carriers by surface recombination is given using the Green's function for a point source derived by Van Roosbroeck (1955) in the form

$$n^1 = S(S+1)^{-1}\int_0^\infty \phi(\rho Z)\exp(-\rho Z/\rho L)\,\mathrm{d}\rho Z \tag{20}$$

where n^1 is the number of carriers lost per sec by surface recombination, S is the reduced surface recombination velocity $= s/V_d$ where s is the surface recombination velocity and V_d is the diffusion velocity, $V_d = L/\tau$, where L is the diffusion length and τ is the minority carrier lifetime. The net carrier concentration available for radiative recombination is thus given by

$$n = \int_0^\infty \phi(\rho Z)\,\mathrm{d}(\rho Z) - n^1 \tag{21}$$

The CL generated by recombination of the hole electron pairs will be partly absorbed in emerging from the interior of the material. However, the absorption coefficient falls with increasing wavelength. If the (dispersive) measurements are confined to wavelengths long enough so that the absorption coefficient is less than say 100 cm^{-1} and the beam penetration is small the effect of self absorption may be neglected.

At low injection levels (small n) in n-type GaAs the intensity of the recombination radiation is approximately proportional to the net carrier concentration. Thus the CL intensity can be obtained from eqn (21) by inserting a Gaussian approximation for $\phi(\rho Z)$. It was assumed that the function $\phi(\rho Z)$ does not change shape significantly with beam voltage and that it is simply scaled in depth in proportion to the maximum range R of the electrons which was assumed to be of the form

$$R = C(E_0)^\zeta \tag{22}$$

where C is a constant, E_0 is the energy of the incident electrons in keV and ζ lies somewhere between 1·5 and 1·7. It was found that the beam voltage dependence of the CL could be reconciled with the theory by postulating that there is a "deal layer" of thickness d at the surface. The Gaussian approximation to $\phi(\rho Z)$ then provided sets of curves of CL intensity against the reduced electron range $W^1 = R/\rho L$ and depending on the reduced surface velocity S, on the diffusion length L and on the relative thickness of the dead layer d/L. The dead layer was found to have a thickness consistent with the interpretation that it results from the existence of a space-charge depletion region caused by the pinning of the Fermi level by surface states. Sets of curves calculated in this way by Wittry and Kyser (1967) are shown in Fig. 6. Experimental measurements were made for beam accelerating voltages 5 to 50 kV and the resulting measurements were fitted to curves of the type shown in Fig. 6. In this way the curves giving the best agreement with

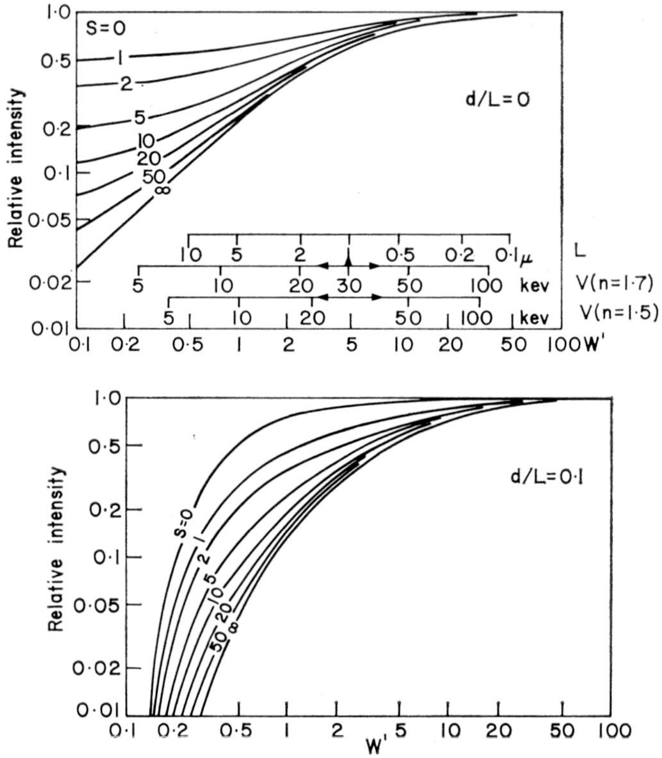

FIG. 6. Relative CL intensity for *n*-type GaAs as a function of the reduced electron range $W' = R/\rho L$, and reduced surface recombination velocity $S = s\tau/L$ fer a Gaussian approximation to $\phi(\rho Z)$, (a) assuming no dead layer and (b) assuming a dead layer of thickness $d = 0{\cdot}10$ L. (After Wittry and Kyser, 1967.)

experiment were found to give values for the diffusion length L ranging from 3·0 μm at low carrier concentrations (low doping densities) $-5{\cdot}1 \times 10^{16}$ cm^{-3} to 0·65 μm at a high concentration of 3×10^{18} cm^{-3}. The accuracy estimated for diffusion length determinations made in this way for values between 0·5 and 4 μm was $\pm 30\%$. Values for the reduced surface recombination velocity S and the reduced dead layer depth d/L were also obtained from the curve fitting. However, there is some ambiguity in the latter information since various choices of the latter parameters give nearly the same values of L for a given specimen.

Subsequently Rao-Sahib and Wittry (1969) treated the problem in the case of p-type GaAs. The theory outlined above assumed that the intensity of the cathodoluminescence (for a wavelength for which absorption in the specimen was negligible, i.e. less than say 100 cm^{-1}) was linearly proportional to the net carrier generation rate g. This relation $\mathcal{I} \propto g$ is valid for n-type GaAs over a wide range of beam voltages and currents, i.e. $\mathcal{I} \propto I_b$ for constant beam voltage V_b. In most specimens of p-type GaAs, however, the cathodoluminescence intensity varies superlinearly with specimen current, i.e. $\mathcal{I} \propto I_b^m$ where the value of m generally lies between 1 and 2 and is a function of both V_b and I_b. The theory developed for n-type GaAs could not be readily modified for $m \neq 1$ in the three-dimensional case. It was therefore found necessary to modify both the theoretical curves and the experimental technique as follows.

The distribution of carrier generation with depth for electron beam excitation was assumed to be a Gaussian function of the form

$$F(z) = F_0 \exp\left[-a^2(z-z_0)^2\right] \text{ cm}^{-3} \text{ sec}^{-1} \tag{23}$$

The steady state continuity equation in one dimensional form is

$$D_n d^2n(z)/\mathrm{d}z^2 - n(z)/\tau_n + F(z) = 0 \tag{24}$$

Where $n(z)$ is the excess carrier density, D_n the minority carrier diffusion constant and τ_n the minority carrier lifetime. The solution to this equation is required subject to the boundary condition that

$$D_n dn(z)/\mathrm{d}z/_{z=0} = sn(o) \tag{25}$$

where s is the surface recombination velocity for electrons in p-type material.

Lengthy expressions for n were given by Rao-Sahib and Wittry (1969) as solutions for these conditions both with and without the assumption that a dead layer of depth d exists. The solutions are based on the assumption that the total carrier generation is constant for constant electron beam power, i.e. for $V_b I_b$ = const. The solutions are valid for both focused and defocused electron beams provided that the excess carrier distribution in planes parallel to the surface is unimportant.

If the intensity of the CL varies with the specimen current as $\mathcal{I} \propto I_b^m$ then in terms of the steady-state carrier density, we can write

$$\mathcal{I} \simeq \int_0^\infty [n(u)]^m \,\mathrm{d}u \tag{26}$$

where $u = \rho z/R$ where ρ is the density of the material in g cm^{-3} and R is the maximum range of the electrons in g cm^{-2} given by R = const. $V_b^{1\cdot7}$.

Substituting the solutions $n(u)$ quoted by Rao-Sahib and Wittry for the case $m = 1$ the integration can be carried out analytically and eqn (26) leads to the same expression independently derived previously.

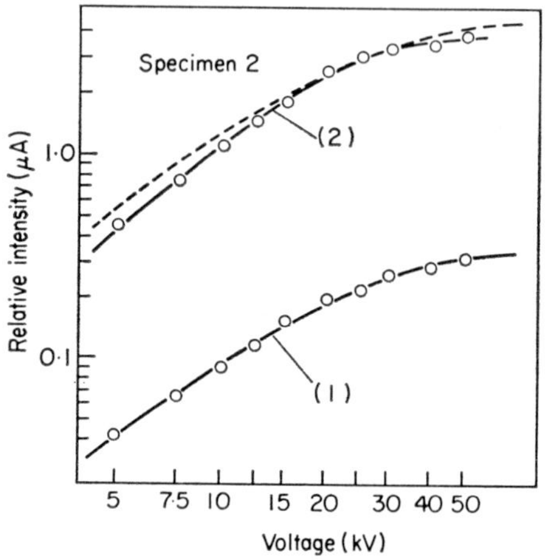

FIG. 7. Comparison of theory and experimental results for a specimen of p-type GaAs using a broad electron beam. Curve (1) is the theoretical curve for $m = 1\cdot0$, $d/L = 0\cdot033$ and $S = 10$, fitted to the experimental results at a power level of 5×10^{-3} W. Curve (2) is the theoretical curve for $m = 1\cdot2$, $d/L = 0\cdot033$ and $S = 10$ fitted to the experimental results at a power level of 50×10^{-3} W. Curve (2) is also compared with curve (1) (the dashed curve) normalized at 30 kV. (After Rao-Sahib and Wittry, 1969.)

When $m \neq 1$ the integral must be evaluated numerically. This was done for a number of different values of m and of the parameters d/L and s. It was found that no agreement could be obtained between any of these theoretical curves and experimental results obtained by using a finely focused electron beam. It was found that in fact if a focused beam was used the calculated curves were applicable only for $m = 1$ and the good agreement found previously for n-type GaAs was due to the fact that in that case m was very nearly one, and was independent of current or voltage. Since the solution for the three-dimensional case with a non-linear response is mathematically difficult, it was easiest to modify the experimental conditions to approximate to one-dimensional diffusion.

This was done by using a large diameter electron beam and using narrow monochromator slit settings to select the light from a small area on the specimen surface. (The spot size was 180 μm and the selected area 27 μm $\times$ 27 μm.) Careful alignment and adjustment was employed to obtain a uniform electron beam density over the selected area and a homogeneous area of the specimen was employed. Results obtained by this method are compared with theoretical curves in Fig. 7. It can be seen that good agreement was obtained between the experimental values for a beam power level of 5×10^{-3} W and a theoretical curve for $m = 1{\cdot}0$, and a beam power of 50×10^{-3} W and a theoretical curve for $m = 1{\cdot}2$ giving consistent values of the parameters $d/L = 0{\cdot}033$, $S = 10$, $L = 0{\cdot}74$–$0{\cdot}75$ μm.

Similarly good agreement was obtained for a number of other specimens. This work showed that the method could be used to obtain values for the minority carrier diffusion length L, the surface recombination velocity S and the dead layer thickness d in the cases of both n- and p-type GaAs. It should be emphasized that the theory for p-type material is phenomenological. It assumes that $\mathscr{I} \propto I_b^m$. The theory does not explain the non-linearity of the relationship. Rao-Sahib and Wittry (1969) reported that the values of m increased as the impurity concentration decreased and that the value varies with the depth of the excess carriers in the specimen so that the observed (averaged) value varies also with the injection levels. They reported observing values greater than 2 and less than 1 on some specimens at very high and very low injection levels.

In the case of specimens with high radiative recombination efficiencies it is possible to measure CL intensities over a wide range of beam power levels and hence for a range of different values of m. This is desirable since it reduces the range of choice of values of d/L and S for which consistent values of L are obtained by fitting the experimental results to theoretical curves.

Casey and Jayson (1971) applied the method of Wittry and Kyser (1967) to the case of GaP doped with Zn and O. The radiative recombination mechanism for efficient red emission in GaP requires the presence of a Zn acceptor and a donor oxygen atom on nearby lattice sites. When they occupy adjacent sites an electrically neutral (compensated) or isoelectronic complex results, and this acts as a deep electron trap which can bind a hole to form a bound exciton. This can then decay radiatively. At high levels of excitation, the Zn–O luminescent centres become filled with captured minority carriers. As the centres are filled, the excess minority carriers recombine non-radiatively either in the bulk or at the surface; consequently the CL intensity increases less rapidly than linearly with increasing excitation. The curves obtained by Wittry and Kyser were only applicable for $m = 1{\cdot}0$, i.e. for $\mathscr{I} \propto I_b$.

Fortunately Casey and Jayson were able to show that although the intensity had a sublinear dependence on I_b for a focused beam a linear dependence was obtained when a non-focused beam was used as shown in Fig. 8.

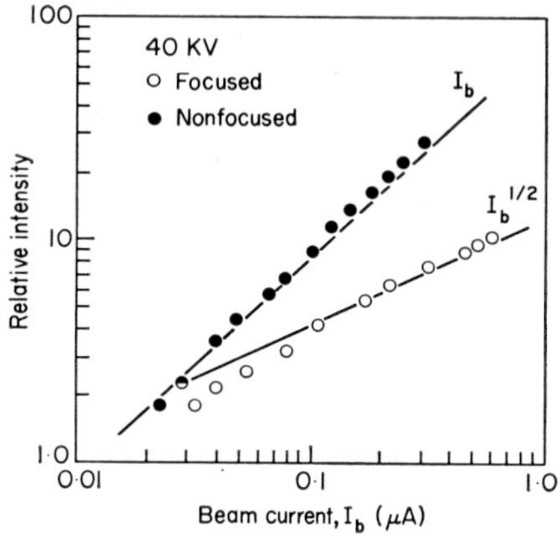

FIG. 8. The variation in the relative CL emission intensity for GaP doped with Zn and O at 40 kV for a focused beam of diameter 0·5 μm and for an unfocused beam. (After Casey and Jayson, 1971.)

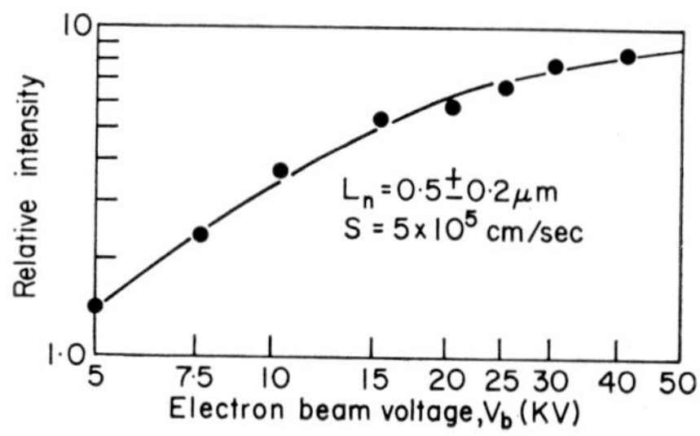

FIG. 9. Relative CL intensity *vs* electron beam voltage for a constant beam power. The data points are fitted to a theoretical curve of Wittry and Kyser (1967) for the values of the parameters shown. (After Casey and Jayson, 1971.)

The CL emission intensity was measured on a relative scale using a defocused beam and a constant beam power for beam voltages from 5 to 40 kV. These results are shown in Fig. 9 fitted to one of the curves of Wittry and Kyser (1967). Since the curves were calculated for GaAs the value of the diffusion length was multiplied by the GaAs/GaP density ratio of 1·3. Values of $Ln = 0{\cdot}5 \pm 0{\cdot}2$ μm, $S = 5 \times 10^5$ cm sec^{-1} and $d = 0{\cdot}025$ μm were obtained. These values lie in the range covered by the values previously obtained on similar material using photo-excitation.

C. TEMPERATURE DEPENDENCE OF CL INTENSITY IN DIRECT GAP MATERIALS

Jones *et al.* (1973) calculated the temperature dependence of the CL intensity in GaAs on the basis of a model for the radiative recombination of electrons and holes in Ge due to Van Roosbroeck and Shockley (1954) as follows.

The photomultiplier output current in a non-dispersive experiment in which the whole CL spectrum emitted by the specimen falls on the photocathode can be written as

$$P = \frac{8\pi\eta}{c^2}\left(\frac{A'}{A}\frac{1}{n_0}+\frac{1}{p_0}\right)\int_{\nu=0}^{\infty}\alpha(\nu)\nu^2 \exp\left(-h\nu/kT\right)Q(\nu)\,\mathrm{d}\nu \times \int_{z=0}^{\infty} p(z)\exp\left(-\alpha(\nu)z\right)\mathrm{d}z \qquad (27)$$

where P is the photomultiplier output, $p(z)$ is the excess hole concentration profile (for n-type material) $\alpha(\nu)$ is the absorption coefficient, η is the refractive index of the material, c the speed of light, A'/A is the so-called degeneracy factor, ν the frequency of the radiation and $Q(\nu)$ is the quantum efficiency of the photomultiplier.

In principle then the temperature dependence of the CL in a non-dispersive experiment can be obtained using the above equation provided that $\alpha(\nu)$, po and $p(z)$ are known as functions of the temperature throughout the range of interest. This information is not generally available, however, and so simplifying assumptions had to be made.

Jones *et al.* assumed that the refractive index is independent of temperature, that the factor $\exp\left(-\alpha(\nu)z\right)$ does not vary significantly with temperature in the range of interest and as z is relatively small for SEM beam voltages of 5 to 20 kV this term was omitted. Thus the second integral in eqn (27) becomes simply

$$\int_0^{\infty} p(z)\,\mathrm{d}z = g'\tau_p \int \psi(z)\,\mathrm{d}z \qquad (28)$$

where g' is the hole-electron pair generation rate, τ_p the total lifetime for holes given by $1/\tau_p = 1/\tau_r + 1/\tau_{nr}$ so that $g'\tau_p$ is the total number of hole-electron pairs generated by the beam that exist in the energy dissipation volume at any instant. $\psi(z)$ is the solution of the continuity equation for a particular generation function and surface recombination velocity and is a function of L and hence of T the temperature.

If the detector sensitivity $Q(\nu)$ is constant for all ν the integral with respect to ν in eqn (27) is simply proportional to $1/\tau_r$. Combining this with eqn (28) a solution for P can be obtained by assuming a particular generation function.

Jones *et al.* considered both the Gaussian function of the form of eqn (23) as used by Wittry, and a polynomial function $F(z)$ due to Everhart and Hoff (1971) and used the latter as the simpler and not too different from the Gaussian function. To obtain $p(z)$ then the diffusion equation

$$D_p \frac{d^2p(z)}{dz^2} + \frac{p(z)}{\tau_p} = -F(z) \tag{29}$$

must be solved subject to the boundary conditions that

$$D_p \frac{dp(z)}{dz} = Sp(0) \quad \text{and} \quad p(\infty) = 0 \tag{30}$$

The solution is $p(z) = Ce^{-y/L_r}$ for $y \geqslant 1$ where $y = z/R_G$, $L_r = L/R_G$ and R_G which appears in $F(z)$ is $R_G = 3{\cdot}98\ V^{1{\cdot}75}\ \mu g\ cm^{-2}$ and for $0 \leqslant y \leqslant 1$ p is given by a more complicated expression. Substituting these expressions into eqn (28) and assuming that $\mathscr{I}$ is proportional to the integral in (28), $\mathscr{I}$ as a function of the reduced range $W' = R/\rho L$ can be obtained for various values of S just as in the theory of Wittry and Kyser in section II.B. The results obtained in both ways are similar and when plotted (Jones *et al.*, Fig. 2) it can be seen that for large values of S, as are appropriate for GaAs, over a limited range of values of W' the results are approximately a constant $\times L^{-a}$ where a is a constant appearing in the expression for p which was found as the solution to eqn (29) under the boundary conditions (30).

Hence the excess minority carrier integral can be written as

$$\int_0^\infty p(z)\, dz = g'\tau_p \times \text{constant} \times L^{-a} \tag{31}$$

and substituting this into eqn (27) we can write

$$P = \text{constant}\left(\frac{A'}{An_0} + \frac{1}{p_0}\right)\frac{\tau_p}{\tau_r} L^{-a} \tag{32}$$

Jones *et al.* (1973) computed a curve of P versus temperature from expression (32) on the assumption that S was about 50, W' was in the range 0·1–4·0 and hence $a = 0{\cdot}717$, by substituting expressions for the temperature dependences of each of the terms L and $(A'/An_0 + 1/p_0)$. The temperature dependences of τ_p and τ_n are the same so that for the present purpose the temperature dependences of the lifetimes can be ignored. Their calculated curve is shown in Fig. 10 together with experimental results. The two curves have been set equal at room temperature. Agreement is good in view of the many simplifications introduced into the mathematical model, except at low temperatures where specimen degeneracy may invalidate the expression used for the term $(A'/An_0 + 1/p_0)$.

Jones *et al.* then used the CL photomultiplier output in linescans across Gunn Oscillators to obtain information on the temperature distribution across the device with various bias voltages applied.

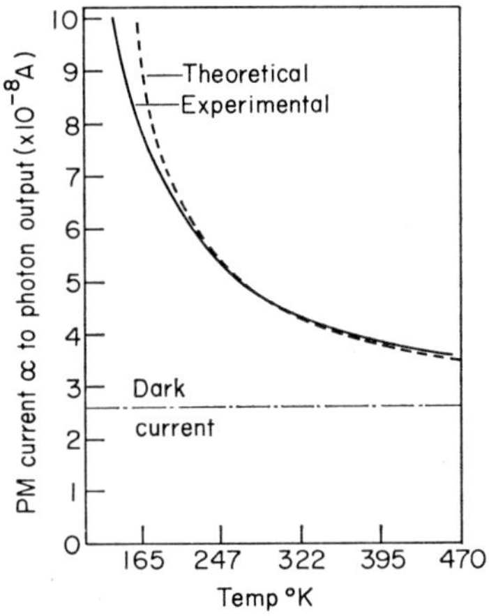

FIG. 10. Experimental and calculated temperature dependencies of the total (non-dispersively measured) CL emission from n-type GaAs with 10^{16} electrons cm^{-3}. (After Jones *et al.*, 1973.)

D. EFFECT OF SELF-ABSORPTION ON THE DETECTED CL

Equation (27) contains a term $\exp(-\alpha z)$ to take account of the effect of self-absorption on the emergent CL spectrum although Jones *et al.* (1973) then omitted it for simplicity. The correction of the observed CL intensity to take account of the absorption of the recombination radiation during its propogation through the material was also discussed by Kyser and Wittry (1964) (see section II.A, eqns (7) and (8)) for the simple model of the electron beam as introducing a point source of hole-electron pairs in the material.

Kyser and Wittry used this approach to explain the differences that they observed between the CL spectra of p- and n-type GaAs and the injection electroluminescence spectrum observed on passing a forward current through a p–n junction in GaAs as shown in Fig. 11. They pointed out that the CL from p-type GaAs had a maximum at 0·90 μm and a half-width of about 0·04 μm while the CL from n-type GaAs had a maximum at 0·89 μm and a half width of about 0·025 μm. The peak intensity for the injection luminescence occurred at a longer wave length (0·925 μm) than the peak CL from either p- or n-type material and it had a sharper cut-off at short wavelengths. These facts could be attributed to absorption. Since a large fraction of the radiation resulting from injection—which is generated all over the *p–n* junction plane normal to the surface scanned—originates at much greater depths than does the CL.

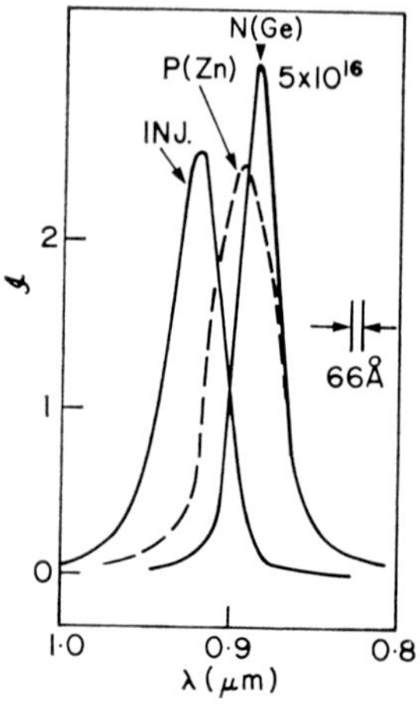

FIG. 11. The observed spectra of cathodoluminescence with a beam current of 1·0 μA and a beam voltage of 30 kV for *p*- and *n*-type GaAs compared with the electroluminescence spectrum from injection at a *p–n* junction using a current density of about 15 A cm^{-2}. (300°K). (After Kyser and Wittry, 1964.)

This argument can be expressed algebraically as follows. Let $\mathscr{I}_0$ be the (constant) intensity of the recombination radiation produced per unit depth of the junction, then the absorption correction factor for the injection luminescence can be written

$$\frac{\mathscr{I}_{obs}}{\mathscr{I}_{gen}} = \frac{\mathscr{I}_0 \int_0^t \exp(-\alpha z)\, dz}{\mathscr{I}_0 t} \tag{33}$$

where t is the thickness of the material and hence the maximum depth of any point on the *p–n* junction below the surface.

Hence

$$\frac{\mathscr{I}_{obs}}{\mathscr{I}_{gen}} = \frac{1-\exp(-\alpha t)}{\alpha t} \quad \text{(for injection)} \tag{34}$$

The correction factor for CL was derived on the point source model (eqn (7)) as

$$\frac{\mathscr{I}_{obs}}{\mathscr{I}_{gen}} = \frac{1}{1+\alpha L} \quad \text{(for CL)} \tag{35}$$

The injection luminescence spectrum is therefore altered more by absorption than are the CL spectra so long as $L \ll t$.

The uncertainties in correcting the observed spectrum in the injection case were large since low values of intensity had to be multiplied by a large correction factor. It was therefore not possible by this means to determine whether the actual spectrum generated, $\mathscr{I}_{\text{gen}}$, in the injection case was indicative of *p*-type or of *n*-type recombination radiation.

Another problem in which the effect of self-absorption has been treated is that of the self-detected CL term in the barrier electron voltaic effect observations on *p–n* junctions in GaAs scanned normal to the barrier. This was dealt with in section III.A.4 of Chapter 8. However, to recapitulate briefly the charge collection current due to the barrier electron voltaic effect in this case has the form

$$I = A \exp(-X/L_1) + B \exp(-X/L_2) \quad (\text{in } n\text{-GaAs}) \tag{36}$$

where I is the charge collection current, A and B are constants, X is the distance of the beam impact point from the *p–n* junction, L_1 is the minority carrier diffusion length. The first term is the minority carrier diffusion term and is well understood. The second term was suggested by Wittry and Kyser (1965) to be due to infrared CL photons propagating through the crystal to the junction where they were absorbed to produce a new hole-electron pair which then contributed to the charge collection current.

The value of L_2 requires to be explained. Wittry and Kyser pointed out that on the interpretation of the second term in eqn (36) as due to the self detection of the CL radiation, comparison with Lambert's Law (eqn (8)) indicates that $L_2 = \alpha^{-1}$. However the experimental value for α_{eff}, the effective infrared absorption coefficient they obtained in this way (300 cm^{-1}) was not equal to the value of α_{peak}, the absorption coefficient for the wavelength of the peak of the CL emission spectrum (1500 cm^{-1}).

Holt and Chase (1973) showed that this disagreement could be removed by taking account of the effect of self-absorption in the varying distance X in the material, which continually alters the position of the peak of the CL emission spectrum, and hence the value of α. Their method for calculating α as a function of X was as shown in Fig. 12. Beginning with the initial emission spectrum shown by the solid curve in Fig. 12(b) and applying Lambert's Law for each of the values of $h\nu$ for a given thickness X the reduction in the intensity for each frequency can be calculated. This was done for each $h\nu$ using the corresponding value of α for *n*-type GaAs from data of Hill shown in Fig. 12(a). In this way modified spectra were calculated, as in the two cases shown

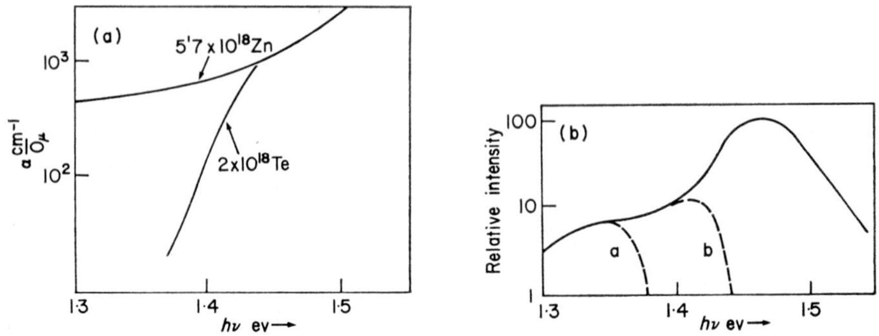

FIG. 12. (a) The optical absorption coefficient, α, of p- and n-type GaAs at 300°K. (After Hill, 1964.) (b) The cathodoluminescent emission spectrum for n-type GaAs with a free charge carrier density of $1{\cdot}75 \times 10^{18}$ cm^{13}. (After Casey and Kaiser, 1967.) The dotted curves represent the modified spectrum after absorption through (a) 1 mm and (b) 0·1 mm thicknesses of material. (After Holt and Chase, 1973.)

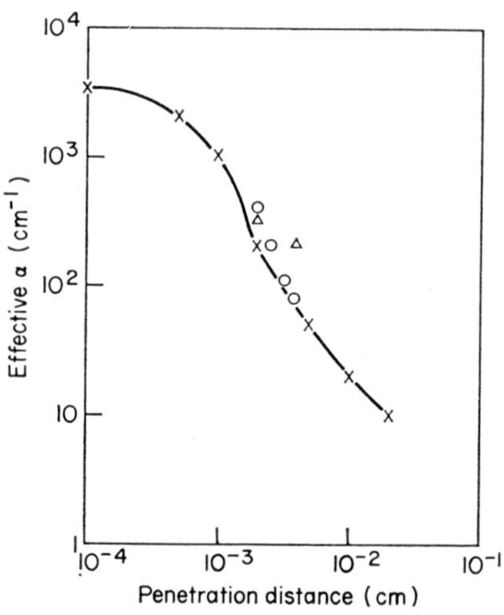

FIG. 13. The Xs mark values obtained from calculations of modified spectra like those shown by dotted curves in Fig. 12(b). The curve through these points thus represents the calculated variation in the effective self-absorption coefficient with the distance travelled by the infrared radiation through the GaAs. Circled and triangular points are values of $\alpha = 1/L_2$ obtained from analyses of charge collection current, line scan traces taken across p–n junctions in two GaAs laser diodes. (After Holt and Chase, 1973.)

by dotted curves in Fig. 12(b), for a series of values of X. Then the value of α corresponding to the peak value of $h\nu$ was taken as α_{eff} for the value of X giving the particular modified spectrum. These calculated values of α_{eff} are plotted against X in Fig. 13. Experimental values of $\alpha = 1/L_2$ obtained from charge collection current traces recorded for two GaAs laser diodes are also plotted in this figure. The agreement is well within the errors due to (1) small differences in the doping of the laser material as compared with that used in the measurements of α by Hill (1964) on which the calculations were based and (2) errors in obtaining L_2 values from the charge collection current measurements. This agreement leaves no room for doubt that this interpretation of the physical origin of the second term in eqn (36), as due to the self-detection of CL, is correct.

E. IMPURITY STRIATIONS AND SEGREGATION IN GaAs

The first form of micrographic contrast, due to variations in CL efficiency from point to point, reported in the literature is what came to be known as dot-and-halo contrast and striation contrast.

As mentioned above, GaAs has an unusually great CL efficiency. In consequence, it emits relatively intense infrared radiation so that as discussed in the preceding section the *p–n* junction charge-collection current contains a term due to the self-detection of the CL. This term predominates in this type of signal for large distances of the beam from the *p–n* junction. In the regions of *n*-type material far from the junction therefore the same contrast appears, because the two signals have the same physical cause, in both the direct (non-dispersive) CL micrographs and the indirect (self-detected CL) charge-collection micrographs. One observation that has been made in both types of micrograph is that of dot-and-halo contrast.

Dot-and-halo contrast and striation contrast was found in Te-doped GaAs by Kyser and Wittry (1964) and Wittry and Kyser (1964), and an example is shown in Fig. 14. These workers and Casey (1967) observed dot-and-halo contrast in CL micrographs and Holt and Chase (1968) observed it in charge collection micrographs and it was confirmed in each case by etching that the effect was associated with dislocations. Shaw and Thornton (1968) explained the fact that this effect is frequently seen in Te-doped GaAs, but not in the case of Se or Si doping, in terms of the relative misfits of the various impurity atoms on the GaAs atomic sites. Of these three dopants Te has the largest misfit. It is about 12% larger than the As atom, whereas the Se atom is 3% smaller than the As atom and Si is about 7% smaller than the Ga atom. The tendency to form condensed Cottrell atmospheres at edge dislocations will therefore be much larger for Te than for Se or Si.

Shaw and Thornton (1968) and Chase and Holt (1973) noted that in some Te-doped GaAs specimens dot-and-halo contrast is seen while in other specimens the dark dots are seen without surrounding bright regions. Chase and Holt explained this in terms of the dependence of

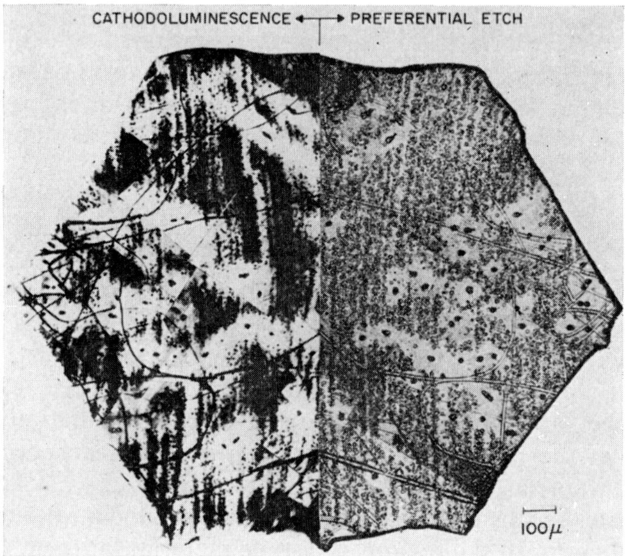

FIG. 14. Comparison of the cathodoluminescent and preferential etch features in a sample of GaAs doped with Te to a free electron concentration of $2{\cdot}2\times10^{18}$ cm^{-3}. The left half of the picture is a montage of CL SEM micrographs. The right half is an optical micrograph of the same area, laterally inverted in printing, after etching with a reagent known to reveal dislocations. Each dot-and-halo in the CL micrograph corresponds to a dislocation etch pit in the optical micrograph. The vertical striations are also visible in both micrographs. (After Casey, 1967.)

CL intensity on the Te concentration as shown in Fig. 15. If a specimen has the average Te concentration indicated by the open circle numbered 2 in Fig. 15(a), it will exhibit the corresponding CL intensity over most of its surface area. Segregation of Te atoms to a dislocation will increase the Te concentration near the line to some value such as that indicated by the right-hand solid circle (no. 3) and reduce the local CL intensity producing the dark dot on the CL micrograph. This will denude a region round the dislocation line, reducing the concentration to a lower value such as that at the left hand solid circle (no. 1). This will *increase* the CL intensity in this region and result in a bright halo. Specimens with average Te concentrations very near the peak of the curve of CL intensity *vs* impurity, the optimum doping density, will have the CL intensity reduced in both the higher concentration dislocation region (right-hand solid circle (no. 3) in Fig. 15(b)) and in the reduced concentration denuded volume outside the dislocation region (left hand solid circle (no. 1) in Fig. 15(b)). Shaw and Thornton (1968) published schematic diagrams to illustrate the detailed form of the dislocation contrast to be expected in each of the cases shown in Fig. 15 as well as in the case in which the average Te concentration is well below the optimum value as shown in Fig. 16. They reported that they had observed experimental contrast patterns of each of the four types shown in Fig. 16.

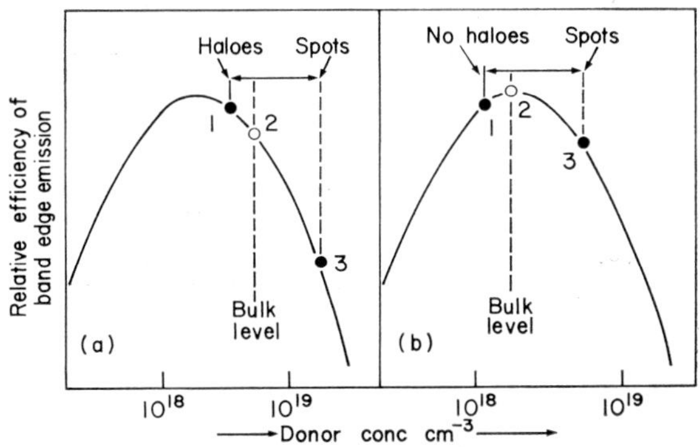

FIG. 15. Relative efficiencies for radiative recombination in Te-doped GaAs at 300°C. (After Cusano, 1964.) Compositions are indicated that could give rise to (a) the dot-and-halo contrast and to (b) dot-without-halo contrast. Points 1, 2 and 3 in each case indicate the relative efficiencies for the depleted region, the bulk material and the dislocation spot respectively. (After Holt and Chase, 1973.)

Impurity growth striations, which are bands of varying impurity concentration resulting from temperature fluctuations etc. during growth, have also been reported in Te-doped GaAs in a number of these papers (Wittry and Kyser, 1964; Casey, 1967; Shaw and Thornton, 1968). Examination of Fig. 15 shows that the contrast in Te-doped GaAs showing growth striations in CL micrographs could correspond to the bright bands having either higher or lower Te concentrations than the darker bands depending upon whether the average concentration is lower or higher respectively than the optimum level. Wittry (1966) demonstrated that in his specimens the brighter regions had the lower Te concentrations by comparing the CL results with Te concentration results obtained by using the X-ray mode. Shaw and Thornton (1968) observed cases in which the brighter striations corresponded to higher Te concentrations as well as cases in which they corresponded to lower concentrations of Te than the average. This was established by comparing CL micrographs with transmission infrared (optical) micrographs of the same area. The non-dispersive CL signal, i.e. the photomultiplier output current with the whole CL emission spectrum incident on the photocathode, was also measured for both the bright and dark areas as a function of temperature from 130° to 300°K and as a function of

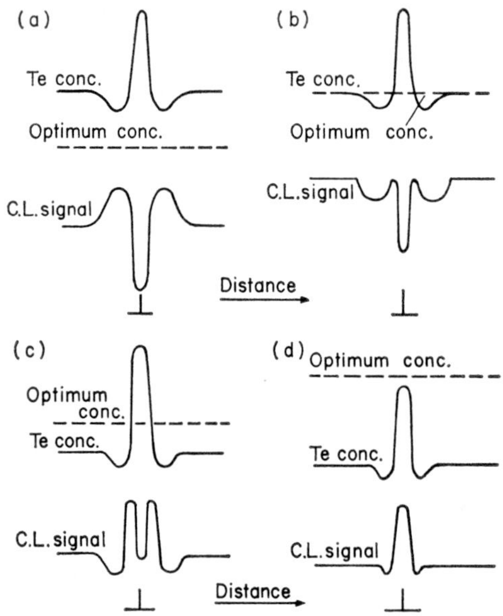

FIG. 16. Schematic diagrams of the redistribution of Te occurring at dislocations and the corresponding CL signal when (a) the optimum concentration is below the bulk concentration; (b) the optimum concentration equals the bulk level; (c) the optimum level is above the bulk level but below the peak level; and (d) the optimum level is above the peak level. (After Shaw and Thornton, 1968.)

beam current from 10^{-9} to 10^{-6} A. The observations indicated that the CL signal varied at $T^{-1\cdot64\pm0\cdot15}$ and as $I_b^{0\cdot9\pm0\cdot1}$ both in the bright regions and in the background. This linear dependence of intensity on beam current shows in terms of eqn (17) and its interpretation discussed in section IIA, that in their material non-radiative recombination was dominant and low excitation levels obtained in both types of material, bright and dark. By comparing their results with the data of Cusano (1964) they estimated the variations in concentration across the striations to lie between 10 and 40% at dopant levels of 4×10^{18} atoms cm^{-3}.

Casey and Kaiser (1967) recorded the CL emission spectrum of n-type GaAs at room temperature and showed that a number of parameters derived from the peak shown in Fig. 17 could be used to determine the doping concentration to within $\pm 15\%$ near concentrations of 5×10^{17} cm^{-3} and to within $\pm 7\%$ at 5×10^{18} cm^{-3}. The parameters that were used were the slope of the low energy tail of the CL emission peak, the energy of the intensity maximum (the peak energy) and the half width of the peak (Fig. 17). These three parameters are plotted

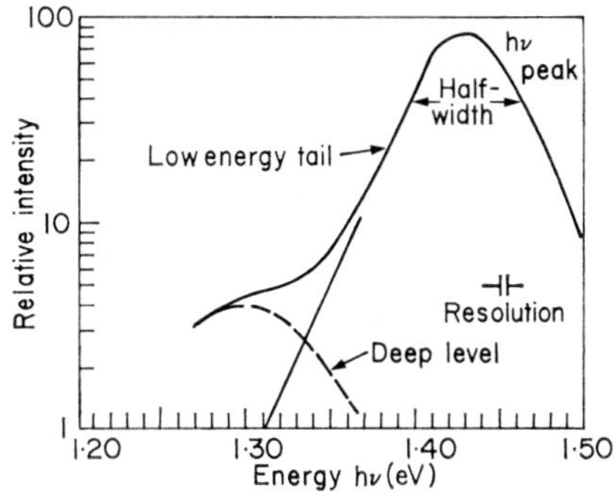

FIG. 17. Typical cathodoluminescent emission spectrum for n-type GaAs containing $1{\cdot}75 \times 10^{18}$ electrons cm^{-3}. The beam voltage was 40 kV and the beam current was 0·3 μA. (After Casey and Kaiser, 1967.)

against electron concentration (the donor concentration) in Fig. 18(a), (b) and (c). By comparing the value of one or more of these parameters obtained from the observed emission spectrum of a given sample with the appropriate curve of Fig. 18, the donor concentration can be read off. Because the luminescent intensity and efficiency of GaAs goes through a maximum (Fig. 15) near a doping concentration of 2×10^{18} much GaAs material is produced with concentrations near this value and analytical techniques intended for use in this concentration region like this method of Casey and Kaiser are therefore particularly valuable. Electrical measurements give the number of free charge carriers. The concentration of electrons in n-type material is given by the density of donor atoms minus the density of acceptor atoms. If there is a significant density of acceptors, the material is said to be (partially) compensated. Both donors and acceptors influence the CL emission spectrum so the emission peak for compensated material has a different shape from that for uncompensated material of the same carrier density shown in Fig. 17. Casey and Kaiser (1967) showed that compensated samples, when the three parameters were determined from their emission spectra as in Fig. 15 and the values were used to obtain values for the free electron concentration n from Fig. 18, gave the following results. The peak position is too low for the value of n indicated by the half-width and the low-energy tail slope is much less than that corresponding to the half-width value for n. The technique thus also provides a means to detect the presence of acceptor impurities in n-type material.

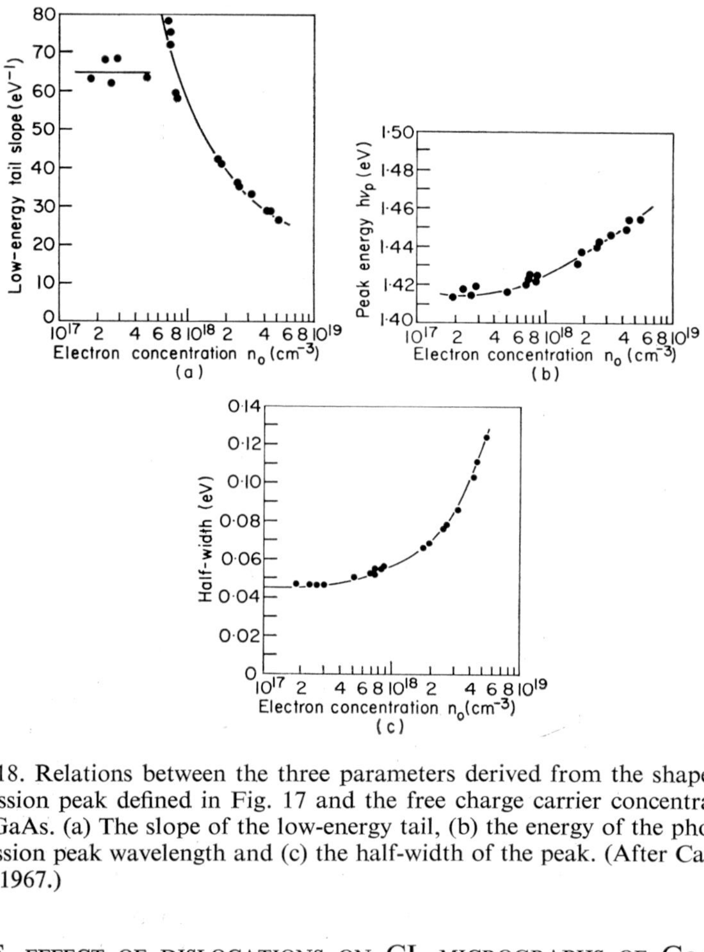

FIG. 18. Relations between the three parameters derived from the shape of the CL emission peak defined in Fig. 17 and the free charge carrier concentration in n-type GaAs. (a) The slope of the low-energy tail, (b) the energy of the photons at the emission peak wavelength and (c) the half-width of the peak. (After Casey and Kaiser, 1967.)

F. EFFECT OF DISLOCATIONS ON CL MICROGRAPHS OF GaAs

In section II.B it was found that dot-and-halo contrast was associated with dislocations. That was however an indirect effect. The variations in intensity were due to the doping concentration variations (Fig. 8) and not directly to the dislocations themselves.

In a recent paper Esquivel *et al.* (1973) reported contrast which might be due directly to dislocations.

GaAs has the sphalerite (or zincblende) crystal structure. This is non-centrosymmetric and consequently exhibits polarity in the [111] directions. A consequence is that there are two types of dislocations in sphalerite-structure materials corresponding to each type of dislocation in the diamond-structure materials. In the case of dislocations with an edge component the two types are described as α- and β-type dislocations. α-dislocations are those in which the atoms with "dangling" (unsaturated) bonds along the edge of the extra-half plane are of the lower valence element, Ga in the present case. β-dislocations have As atoms with dangling bonds at the core (Holt, 1962). The classic method for studying α- and β-dislocations is to bend bars of the material in two opposite senses one of which introduces preponderantly α-dislocations and the other β-dislocations.

The orientations used by Esquivel *et al.* are shown in Fig. 19(a) and (b). Both undoped and Te-doped (10^{17} cm^{-3}) specimens were bent in both senses to radii corresponding to about 5×10^6 cm^{-2} and 10^7 cm^{-2} each of α and β dislocations compared to an initial density of dislocations of 10^3 to 10^4 cm^{-2}. Figures 19(e) and (f) are infrared CL micrographs revealing checkered patterns (and a subsurface flaw in Fig. 19(f)) in the heavily deformed areas marked "1" in Figs 19(a) and (b). No such features of plastic deformation were visible in optical, transmission infrared micrographs (not shown) or in secondary electron emissive mode micrographs of the same areas (Figs 19(c) and (d)).

It was suggested that the black dots, approximately 3 μm in diameter, one of which is indicated by an arrow in Fig. 19(e), were related to dislocations lying on the [111] slip planes. In the case of the specimen of Fig. 19(e) a count of these dots gave a density of 5×10^6 cm^{-2} compared with a dislocation density calculated from the radius of curvature of the bend of 8×10^6 cm^{-2}.

Unfortunately GaAs is brittle at room temperature. In order to bend it plastically, therefore, it had to be heated to 700°C in argon for 15 min. Comparison of CL micrographs of bent and unbent control specimens subjected to the same heat treatment indicated that some diffusion of the impurities present had occurred. Measurements of the infrared emission peak at a photon energy of 1·41 eV (see Fig. 17) indicated that the average intensity of this peak was increased by the heat treatment in both the bent and the unbent samples. However, in the case of the bent samples the intensity from the central portions of the samples (e.g. area 1 in Figs 19(a) and (b)) which contained high densities of dislocations was 40–70% lower than the intensity from the relatively undeformed ends of the samples (e.g. area 2 in Fig. 19(b)). The effects of α and of β dislocations at calculated densities of 5×10^6 and 10^7 cm^{-2} were not distinguishable.

The results of this work are inconclusive since impurity segregation to the dislocations may be the cause of the observed contrast in this case just as it was in the case of dot-and-halo contrast. However, it is possible, as Esquivel *et al.* (1973) suggested, that this contrast is due to the dislocations themselves and in that case it could be interpreted

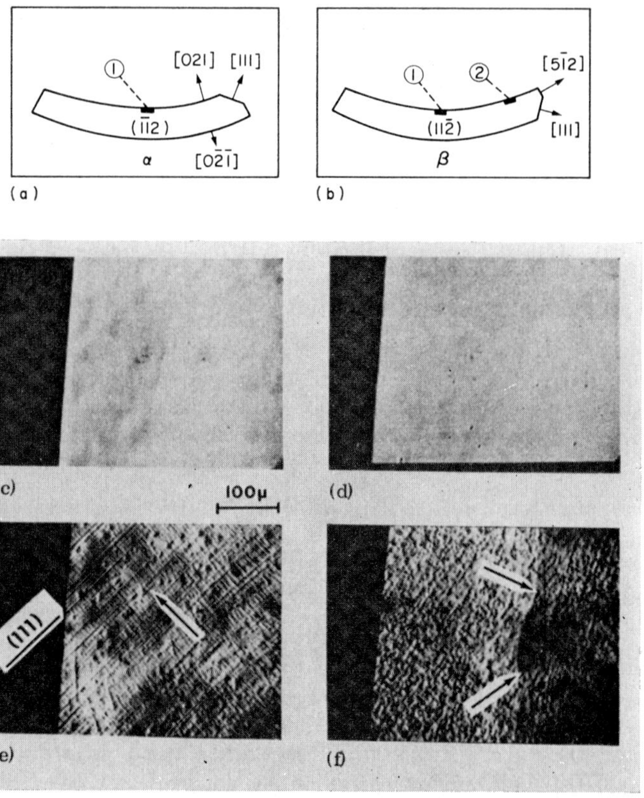

FIG. 19. (a) and (b) are schematic diagrams of the orientations of the specimens deformed in four-point bending in order to introduce an excess of α or of β dislocations in GaAs. Areas (1) and (2) are the areas examined as described in the text. (c) and (d) are secondary electron emissive mode micrographs of areas 1 on crystals bent as in (a) and (b) respectively. (e) and (f) are CL micrographs of the same areas 1 as in (c) and (d) respectively. The arrow in (e) points to black dots associated with the intersections of dislocation lines with the ($\bar{1}\bar{1}$2) surface. The trace of the (111) slip plane on the ($\bar{1}\bar{1}$2) surface is also indicated. In (f) the arrows point to a subsurface flaw. (After Esquivel *et al.*, 1973.)

as further evidence that dislocations introduce recombination or trapping levels into the forbidden energy gap of GaAs. Similar observations of dislocations or slip planes as lines of reduced signal strength were made in Si using the barrier electron voltaic effect in the conductive mode. That effect was interpreted as due to the recombination centres associated with the dangling bonds in the cores of the dislocations (see Chapter 9).

III. Quantitative Studies of II–VI Compounds

A. Analyses of Faulted Structures in II–VI Compounds

Many platelets of II–VI compounds, and especially of ZnS, grown from the vapour are found to be "striated". That is, they are crossed by grids of fine lines, visible to the naked eye, at right angles to each other. One set of fine lines runs at right angles to the *c*-axis of the crystal structure and consists of stacking faults and fine bands of material with different crystal structures: hexagonal (wurtzite), cubic (sphalerite) and more complex crystal structures called polytypes. The other set of lines called linear markings are growth steps, i.e. thickness variations arising during growth. An example of such a striated ZnS platelet is shown in Fig. 3 of Chapter 9. It can be seen that certain of the striations of both perpendicular sets act as localized sources of cathodoluminescence (Holt and Culpan, 1970). It is thought that this localized, room-temperature, visible CL is extrinsic and preliminary analyses of the emitted spectrum are consistent with Cu and Ag as the impurities segregated at the localized sources.

Williams and Yoffe (1969a) used both a monochromator-photomultiplier and a quartz light pipe-photomultiplier detector to receive ultraviolet cathodoluminescence from liquid nitrogen cooled platelet crystals of II–VI compounds. With this equipment micrographs could be produced using either the total cathodoluminescence (CL) emission or a selected CL line as video signal. The ultraviolet CL studied by Williams and Yoffe was intrinsic, i.e. it consisted of photons having energies with values near to E_g, the energy of the forbidden gap. By cooling the platelets, sufficiently sharp spectra were obtained that spectral emission lines could be observed as shown in Fig. 20(a), which is taken from the work of Williams and Yoffe (1969a) on ZnSe. The peaks marked H_1 and C_1 were identified as characteristic of excitons bound to crystal defects in hexagonal and cubic material respectively. The other peaks corresponded to exciton transitions with the emission of one or more phonons. By detecting and amplifying light of the peak

H_1 selected by the monochromator, as video signal the CL micrograph of Fig. 20(b) was obtained in which the regions of wurtzite-structure material appear bright. Figure 20(c) is the CL micrograph obtained by using peak C_1 as video signal.

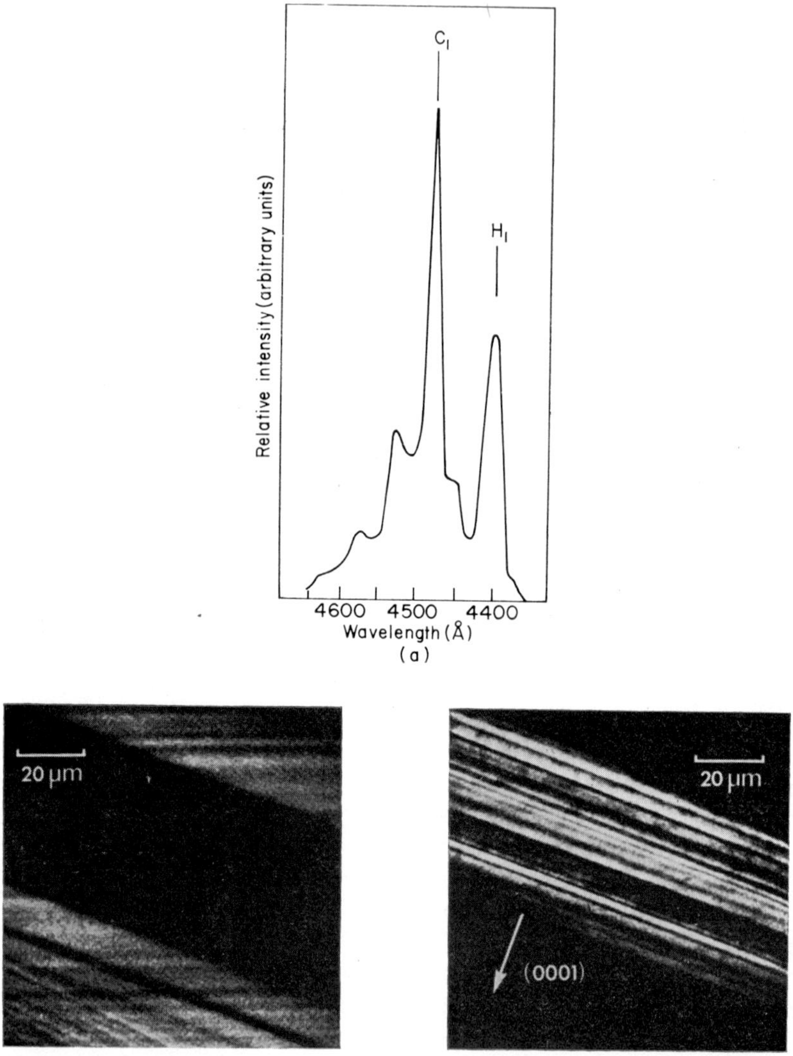

FIG. 20. (a) Emission spectrum of ultraviolet cathodoluminescence from a striated platelet of ZnSe at liquid nitrogen temperature. (b) CL micrograph using peak H_1 as video signal. (c) CL micrograph obtained by recording peak C_1. (After Williams and Yoffe, 1969a.)

In another study of the cathodoluminescence of ZnSe platelets, Williams and Yoffe (1968) found that in thin platelets both stacking faults and twin boundaries cause a local decrease in the intensity of the CL emission. The nature of the defects responsible for the dark lines observed on the CL SEM micrographs was determined by transmission electron microscopy. In the case of thicker hexagonal structure platelets of ZnSe (1 to 10 μm) the intensity of the CL was sufficient to enable a spectrum to be taken. It was found that the relative intensities of the CL from the hexagonal phase matrix of the platelet and from thin basal plane layers of cubic material varied with the beam current density. At low beam current densities the cubic layers appeared dark on a bright background while at higher current densities (10^3 A cm^{-2} as against 4×10^2 A cm^{-2}) the cubic layers appeared bright on a dark background. It was shown that this effect was associated with the threshold excitation behaviour of the cubic emission peaks.

Williams and Yoffe (1969b) reported similar studies made on hexagonal platelet crystals of ZnS. 30 keV electron bombardment of crystals held at 100°K resulted in CL emission spectra in which emission lines could be ascribed to either hexagonal or cubic phase exciton recombination. For thin platelets comparison of CL SEM micrographs with transmission electron micrographs showed that intense luminescence was associated with stacking faults or groups of stacking faults. The fact that a local increase in intensity was observed in this case whereas a local decrease near stacking faults had been observed in thin platelets of ZnSe was tentatively explained in terms of a dependence of the sign of the effect in both materials on the beam current density. In thicker platelets it was again found that the emission from faulted or cubic layers in hexagonal crystal became more intense than that from the hexagonal matrix when the beam current was increased to its maximum value. Similar results were reported again in 1971 by Williams and Yoffe together with an observation of a twin boundary as a line of reduced CL intensity in an SEM micrograph of CdS. It was suggested that in this case the effect might be due to the segregation of non-radiative recombination centres to the twin boundary, i.e. it was suggested that this was an extrinsic cathodoluminescence effect. No evidence was presented to support this hypothesis, however.

B. CATHODOLUMINESCENT LASER ACTION IN II–VI COMPOUNDS

The laser action induced by electron bombardment of films and platelet crystals of a number of II–VI compounds has been extensively studied. None of this work involved scanning and so only the references to this work are listed in Table I and the values of the various relevant parameters at laser threshold reported in the literature are listed in Table II.

Williams and Yoffe (1969a) looked for laser action in their SEM studies but did not observe it. They used a 30 keV beam, a 500 Å diameter spot and current densities from 1 to 100 A cm^{-2} with specimens held at liquid nitrogen temperature, but did not observe the onset of stimulated emission in ZnS, ZnSe or other II–VI compounds. The CL laser literature suggests several possible reasons for this. Comparison

TABLE I. Reports of Cathodoluminescent Laser Action in II–VI Compounds

Material	References
CdS	Basov *et al.* (1964), Benoit à la Guillaume *et al.* (1967); Hurwitz (1966a, 1967a); Grillot and Bancie-Grillot (1967); Kurbatov *et al.* (1967); Nicoll (1968a); Dirochka *et al.* (1969); Nicoll (1970); Bille *et al.* (1972); Benoit à la Guillaume *et al.* (1969).
CdSe	Hurwitz (1966a, 1967a); Kurbatov *et al.* (1968a,b); Filinski *et al.* (1971); Smiley *et al.* (1971)
Cd (Se : S)	Hurwitz (1966b)
CdTe	Vavilov *et al.* (1967)
ZnO	Nicoll (1966, 1968a, 1968b)
ZnS	Hurwitz (1966c)
ZnSe	Bogdankevich *et al.* (1967)
ZnTe	Vlasov *et al.* (1967); Hurwitz (1967b)

A review of laser action in II–VI compounds with special emphasis on the growth of the material was written by Reynolds (1967). Much of the literature listed here is concerned with the interpretation of the observed emission spectra above and below the laser threshold in terms of exciton recombination processes. A review of the identification of lasing transitions in II–VI compounds was given by Dean (1969, pp. 120–122).

with Table II indicates that 30 kV is rather lower than the beam voltages used in most of the CL laser literature. It is generally reported that the threshold beam current for a given set of conditions varies widely from one specimen to another. It also varies from one part of a particular crystal to another, i.e. only certain filaments within the material lase (Hurwitz, 1966a; Smiley *et al.*, 1971). The threshold current for a given voltage or the threshold voltage for a given beam current also depends strongly on the surface coatings, if any, present on the mirror surfaces at the ends of the laser cavity (Nicoll, 1970; Smiley *et al.*, 1971). Thus careful selection of material and the application of reflecting or dielectric films to the crystal surfaces might make it possible to attain stimulated emission thresholds in conventional tungsten filament gun equipped SEMs.

TABLE II. Thresholds for Cathodoluminescent Laser Action in II–VI Compounds

Compound	Threshold values				References
	Temperature	Voltage	Current	Pulses	
CdS	4·2°K	50 keV	1000 mA cm^{-2}		Hurwitz (1966b)
CdS	77°K	50 keV	1200 mA cm^{-2}		
CdSe	4·2°K	50 keV	50 mA cm^{-2}		
CdSe	77°K	50 keV	200 mA cm^{-2}		
CdSe CdS	liq. He or liq. N (no significant difference)	20 keV to 75 keV	1·5 A cm^{-2} 0·2 A cm^{-2} ~5 *X* higher	0·5 mm diam. beam 100–200 nsec pulses at 60 per sec	Hurwitz (1966a)
ZnSe	100°K rising to 150°K	45 to 150 keV	about 20 A cm^{-2}	150 nsec pulses	Bogdankevich *et al.* (1967)
ZnTe, one face Al coated	liq. N	40 keV 60 keV	13 A cm^{-2} 4 A cm^{-2}	50 nsec pulses at 60 sec^{-1}	Hurwitz (1967b)
	liq. He	40 keV 60 keV	6 A cm^{-2} 1·5 A cm^{-2}		
CdSe	10°K	60 keV	1·5–3 A cm^{-2} at threshold	4×10^{-8} sec at 300 cps	Kurbatov *et al.* (1968a,b)
CdS for filaments under bubbles in surface Al films	77°K room temp.	5 keV 9 keV	? up to 5 A cm^{-2}		Nicoll (1970)
CdSe	4·2° to 77°K	30 keV	up to 40 mA	200 nsec, 50 Hz	Filinski *et al.* (1971)
CdS	120°K	215 keV	>5 A cm^{-2}		Bille *et al.* (1972)

Dean (1969, p. 121) ascribes the greater threshold beam current densities for lower beam voltages to surface recombination effects which must be relatively greater when the electron penetration range is less. He also points out that the reported power efficiencies for the best CdS crystals at 4·2°K and at 77°K for 60 keV electrons correspond to quantum efficiencies of 100%. That is the best CdS under laser conditions has a radiative recombination transition probability of unity, just as is the case for GaAs.

The advent of higher brightness guns for SEMs (field emission and LaB_6 sources—Chapter 1) may also help in attaining laser action in the CL mode. It is not possible at present to be certain which type of electron source will give the optimum results in this field. Each type of source gives the maximum beam current for some range of spot sizes. While the CL laser literature quotes threshold current density values (see Table II) it is clear that in all this work the areas irradiated with electrons were very large (mm to cm across) compared with the beam diameters available in scanning beam instruments. It is also clear that there is bound to be a minimum beam diameter, of the order of the diameter of the low threshold laser filaments in II–VI crystals, below which laser current-density thresholds must be expected to rise. This is because the condition for the onset of light amplification by stimulated emission of radiation is that the intensity gain per unit length as the light propagates through the material must exceed the intensity losses per unit length due to absorption and diffraction. The diffraction losses will rise rapidly if the active region, i.e. the region containing a high density of hole-electron pairs constituting an "inverted population", is of a very small diameter perpendicular to the beam direction, i.e. parallel to the mirror surfaces. Use of larger spot sizes than that used by Williams and Yoffe (1969a) thus should also be beneficial although no data regarding active filament diameters are given in the literature.

Both for its intrinsic interest and as a means of investigating the material's properties or structures responsible for filamentary laser action in II–VI crystals this field would appear to warrant further investigation.

CL laser action can also be induced in a number of III–V compounds. However in these materials laser action can be produced in other ways, e.g. by *p–n* junction injection which are of greater practical value. There is therefore little or no interest in III–V CL laser emission. II–VI injection lasers cannot be made and so CL laser emission in these materials is of considerable practical interest.

IV. Quantitative Studies of Electroluminescent P–N Junctions

The technique which has been found valuable in these studies is the intercomparison of charge collection current curves and CL signal curves recorded in line scans across *p–n* junctions which meet the scanned surface at right angles (see Fig. 25, Chapter 8). The barrier electron voltaic effect produces a peak, in the conductive mode linescan, which marks the position of the electrical junction. The relation between this peak and the CL linescan curve provides information relevant to the luminescent properties of the junction under injection (electroluminescence) conditions. This work was discussed in Chapter 8 with the main emphasis on the interpretation of the barrier electron voltaic effect observations. Here the emphasis is naturally on the CL observations.

A. GaAs LASER DIODES

Wittry and Kyser (1965) reported that a local decrease in the non-dispersive CL signal occurred at *p–n* junctions in GaAs diodes as shown in Fig. 21. The coincidence of the junction and the region of reduced CL signal was confirmed by three independent methods. These were comparison of the CL trace dip with: (1) the position of the step produced by etching the surface with a reagent which preferentially attacked the *p*-type material; (2) comparing the CL signal with the barrier e.v.e. signal as in Fig. 21; and (3) the position from which infrared emission (injection luminescence, i.e. electroluminescence—EL) was emitted when the diodes were forward biased.

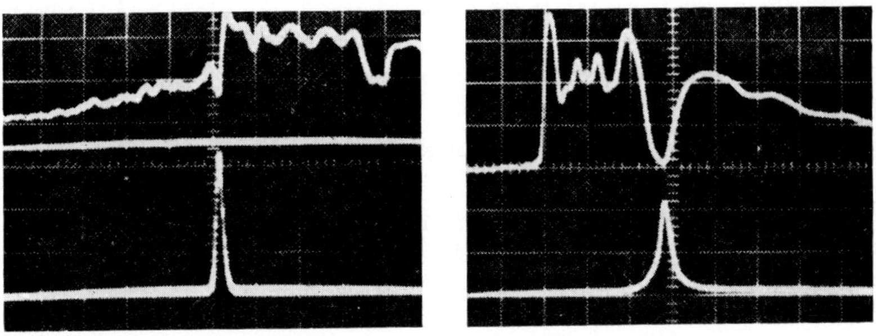

FIG. 21. Dual trace oscilloscope recordings of the non-dispersive CL signal (upper curves) and the barrier e.v.e. currents for two different diodes in (a) and (b). (After Wittry and Kyser, 1965.)

The relative strengths of the photomultiplier current with the entire CL spectrum incident, on the *p*- and the *n*-sides depended on the beam current since the CL intensity from the *n*-region increased more rapidly than that from the *p*-region as the beam current increased.

Wittry and Kyser pointed out that the dip of the CL intensity at the junction could not be due only to the lack of free charge carriers in the depletion region (see Chapter 8 for a brief introduction to this and other features of *p–n* junctions) since the probe size was approximately 1 μm whereas the width of the depletion regions in the two diodes shown in Fig. 21 were 0·07 μm and 0·53 μm. Other possible causes for the dip might be non-radiative recombination by carrier tunnelling at the junction or by multiphonon or Auger processes at impurities segregated to the junction by chemical potential effects during diffusion of zinc into the Te-doped material to produce the junctions. Wittry and Kyser argued that these seemed unlikely because the current–voltage characteristics of the diodes showed low reverse currents and the relative magnitude of the decreased intensity at the junction appeared to depend on the carrier concentration of the bulk material.

Wittry and Kyser proposed instead that the effect was due to the fact that the barrier electron voltaic effect current consists of those charge carriers accelerated across the depletion region by the built-in field of the junction. These carriers are separated from the other carrier of the hole-electron pair in which they were generated. Such carriers do not recombine and so are lost to CL radiative recombination. The CL signal should therefore have the form $1 - C'I(X)$ where $I(X)$ is the barrier electron voltaic effect current, X is the distance of the electron beam from the junction and C' is a constant of the order of unity. The barrier e.v.e. current was deduced by Wittry and Kyser on a simple model (see Chapter 8, section III.A.4) to have the form $\exp(-X/L)$ where L is the minority carrier diffusion length. That this was so on the p-side was experimentally confirmed by them and by Holt and Chase (1973) while on the n-side an additional exponential term appeared due to the self-detection of the CL as noted above. This does not of course affect the CL signal detected directly. The second term in eqn (36) can therefore be neglected for the present purpose.

Thus, on the assumption that the CL intensity is linearly dependent on the excess minority carrier density, we have

$$\mathscr{I} \propto 1 - C' \exp(-X/L) \tag{37}$$

on both sides of the junction (L the minority carrier diffusion length is of course in general different on the two sides). The intensity at the junction, i.e. for $X = 0$, will not be zero since the probe (energy dissipation volume) has a finite size and the charge collection efficiency of the junction as discussed in Chapter 9 is therefore less than unity.

It can be seen in Fig. 21 that the dips in the CL traces are approximately related to the peaks in the barrier e.v.e. traces as expression (37) predicts. No attempt has been reported to confirm quantitatively that the CL curve has this form, however. The possibility that non-uniformities in the distribution of radiative or non-radiative recombination centres near the junction may also influence the form of the CL trace (through variations in the proportionality constant in expression (37)) cannot therefore be ruled out. It will be seen in the next section that there is good reason to suppose in the case of *p–n* junctions in GaP that the impurity distribution plays an important role in determining the form of the CL signal.

B. GaP LIGHT-EMITTING DIODES

The CL and charge collection (CC) line scans recorded across *p–n* junctions in GaP have the form shown in Fig. 22. The important practical point is that of all the parameters that can be measured on either curve or relating the two curves only one (Holt *et al.*, 1973) correlates with the performance of the device, i.e. with the electroluminescent efficiency of the diode. It is therefore of some importance to understand this parameter, the "CL–CC shift", i.e. the separation of the 50% CL photomultiplier current level position from the position of the peak of the CC trace.

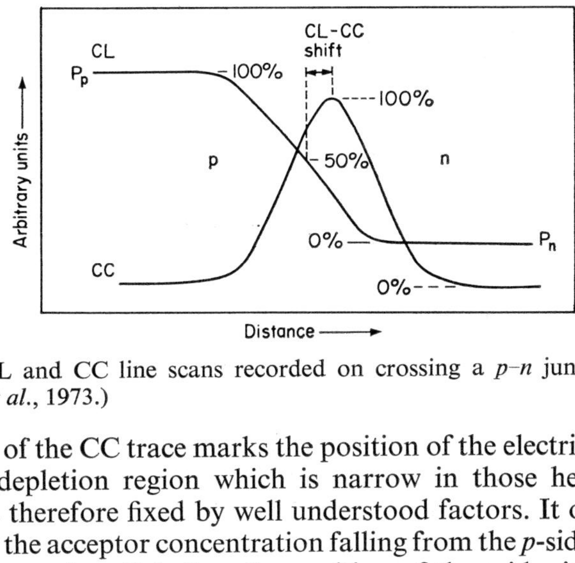

FIG. 22. CL and CC line scans recorded on crossing a *p–n* junction in GaP. (After Holt *et al.*, 1973.)

The peak of the CC trace marks the position of the electrical junction, i.e. of the depletion region which is narrow in those heavily doped diodes. It is therefore fixed by well understood factors. It occurs at the point where the acceptor concentration falling from the *p*-side, equals the donor concentration. It is thus the position of the midpoint of the CL trace that needs to be explained.

The form of the CL trace in a non-dispersive observation like that of Fig. 22 is that of a smooth fall from a high value P_p on the *p*-side, the photomultiplier output current with the beam incident far from the junction on the *p*-side, to a much lower value P_n on the *n*-side. This does not have the form $P \propto 1 - C' \exp(-X/L)$, i.e. a dip corresponding to the peak in the CC trace. If the intensity emitted on the *n*-side is supposed to be negligible the CL trace would show a fall of the form

$P \propto 1 - C' \exp(-X/L)$ on the p-side and a horizontal (zero) line on the n-side. However, the CL–CC shift would then depend only on L. It is found that this is not so. The electroluminescent efficiency of the diodes correlates with the CL–CC shift (Calverley and Wight, 1970; Holt *et al.*, 1973) but not with L. Moreover both the CL–CC shift and the electroluminescent efficiency can vary among diodes with the same value of L on the p-side. Actually this argument is not conclusive in demonstrating that the barrier electron voltaic effect is not the only factor affecting the form of the CL trace. The reason is that, as discussed in Chapter 8, the barrier electron voltaic effect peak in the CC trace in the case of GaP does not behave according to the simple theory predicting a dependence of the form $I(X) = \text{const.} \exp(-X/L)$, and this behaviour is not understood at present.

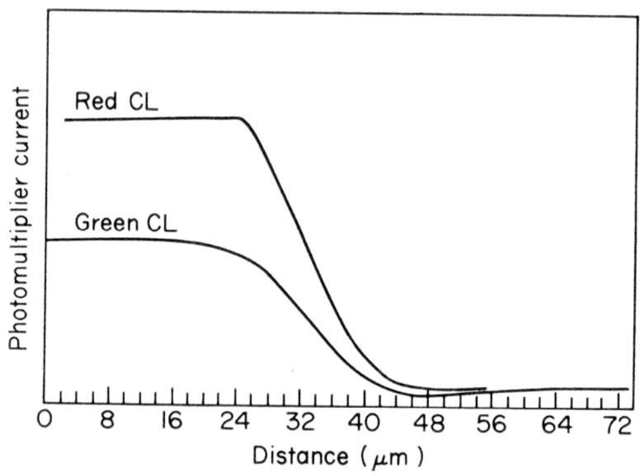

FIG. 23. The photomultiplier current recorded on line scans across a GaP p–n junction using red and green filters to select the two peaks of the CL spectrum. (After Holt *et al.*, 1973.)

However, it does seem probable that the distribution of radiative and/or non-radiative recombination centres near the p–n junction is an important factor in the CL behaviour. Experimental support for this view was obtained by Holt *et al.* and is shown in Fig. 23. Red and green filters were used to record the photomultiplier outputs due to the two peaks in the GaP CL spectrum. The red emission is associated with the Zn–O impurity pairs and the green with other impurities. It can be seen that the two curves are quite different in shape and that as the green curve reaches its maximum value much further from the junction the CL–CC shift would be found to be wavelength dependent in dispersive CL measurements.

Calverley and Wight suggested that the correlation of EL efficiency with the magnitude of the CL–CC shift could be understood on the hypothesis that the CL–CC shift was a measure of the width of an hypothetical dead layer on the *p*-side of the junction due to the presence of nonradiative recombination centres. A large CL–CC shift would then correspond to a wide region of low radiative recombination efficiency in the active region for injection electroluminescence and a low EL efficiency would result.

C. $ZnSe_yTe_{1-y}$ DIODES

Aven *et al.* (1972) used scanning electron microscopy to measure the CL efficiency of the material, to study the junctions and to determine where the injection luminescence was generated in diodes of the II–VI solid solution or "mixed crystal" material $ZnSe_yTe_{1-y}$.

According to Aven *et al.* the external quantum efficiency η_e of an injection luminescent *p–n* junction diode can be written as

$$\eta_e = \eta_{rad}\eta_a\eta_f \tag{38}$$

where η_{rad} is the radiative recombination efficiency, i.e. the ratio of the number of radiative recombinations to the total number of recombinations of the region into which minority carriers are injected when the device is forward biased, η_a is the fraction of the internally generated photons that emerge from the crystal and η_f is the fraction of the carriers passing through the device that recombine in the region of the junction in which the observed luminescence is generated. These factors are all temperature dependent. $ZnSe_yTe_{1-y}$ diodes have been made with external quantum efficiencies up to 18% at 70°K but this falls to between 0·001 and 0·01% at room temperature. A major objective of research in this field is to maximize room temperature efficiencies. The work of Aven *et al.* was designed to investigate the temperature dependence of each of the factors in eqn (38) in turn.

1. *Cathodoluminescent Efficiency of* $ZnSe_yTe_{1-y}$

Measurements were made of the efficiencies of $ZnSe_yTe_{1-y}$ crystals that were undoped or phosphorus doped to have *p*-type conductivities and then the efficiencies were measured again after Al-doping to make the crystals *n*-type. A beam voltage of 20 kV was used. The characteristics of the crystals in these two conditions are those of the materials on the two sides of the *p–n* junctions in the diodes. The compositions investigated ranged from $y = 0{\cdot}35$ to $0{\cdot}50$. The results showed that before Al diffusion the efficiency dropped by a factor of 10^2–10^3 as the temperature rose from 77°K to 300°K, whereas in the Al-doped crystals the efficiency only dropped by a factor of 2 over this temperature range. The n-type (Al-doped) crystals at 300°K had high quantum efficiencies (near 4%).

The n- and p-type materials also had significantly different dependences of CL emission intensity on beam voltage as shown in Fig. 24. Before doping the crystal showed a superlinear relation between photomultiplier output and beam voltage whereas after Al-doping the relation was roughly linear above about 4 kV. Aven *et al.* analysed these results by the method of Gergely (1960) as follows.

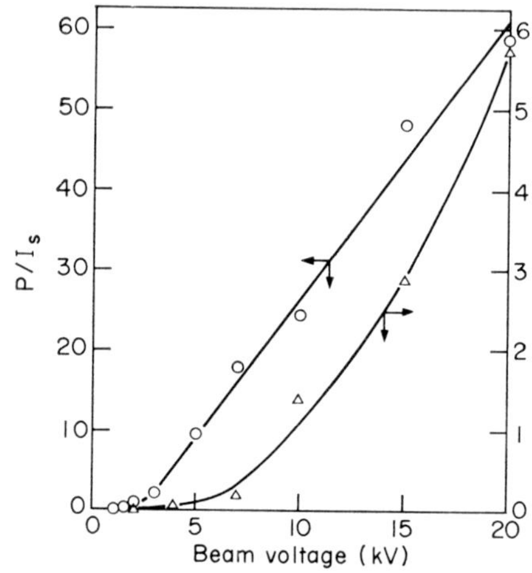

FIG. 24. Photomultiplier current for a non-dispersive observation divided by specimen current versus beam voltage for a $ZnSe_{0\cdot35}Te_{0\cdot65}$ crystal before (triangles) and after (circles) Al diffusion. (After Aven *et al.*, 1972.)

According to Gergely the range of the beam electrons in the material can be written as

$$R = AV_b^b \tag{39}$$

where A and b are constants of the material and $b = 1{\cdot}6$ for $ZnSe_yTe_{1-y}$ according to Aven *et al.* The CL efficiency normalized to 1 at high voltage can be written as

$$\eta_N = 1 - Q\frac{1 - \exp[(-V_b/V_0)^b]}{(V_b/V_0)^b} \tag{40}$$

where V_0 is a dead voltage and Q is a surface recombination loss parameter. Aven *et al.* assume $Q = 1$ corresponding to an infinite recombination rate at the surface. Therefore (40) becomes for $ZnSe_yT_{1-y}$

$$\eta_N = 1 - \frac{1 - \exp[(-V_b/V_0)^{1.6}]}{(V_b/V_0)^{1.6}} \tag{41}$$

Figure 25 shows the results of Fig. 24 fitted to conform to eqn (41). The best fit for the crystal before Al-doping was obtained by setting $V_0 = 10$ kV whereas after Al-doping the data were best fitted with $V_0 = 3{\cdot}7$ kV. Assuming that the dead voltage is the beam voltage for which the range equals the ambipolar carrier diffusion length of carriers gives values for this parameter of approximately 0·1 μm in the Al-doped crystals and 0·4 μm in the undiffused crystals. These results are only approximate but Aven *et al.* concluded that the diffusion length of carrier in $ZnSe_yTe_{1-y}$ is less than 1 μm and is considerably reduced by Al doping.

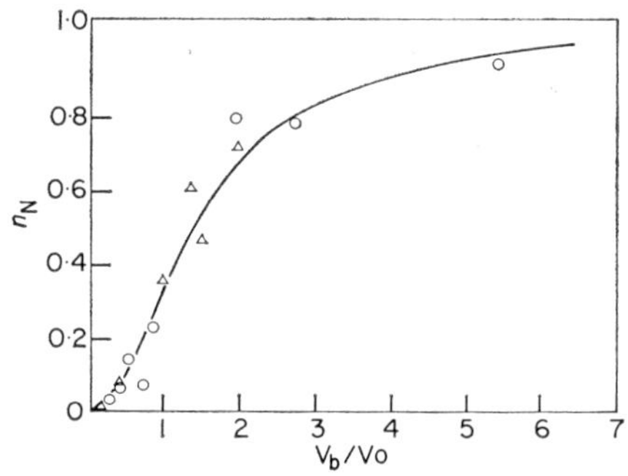

FIG. 25. The relative (normalized) CL efficiency of the same crystal as in Fig. 24 *vs* the parameter V_b/V_0. The triangles are points obtained before Al diffusion and the circles, after. The curve follows eqn (41). (After Aven *et al.*, 1972.)

2. *Effect of Self-absorption on the* CL *Spectrum*

The CL emission spectra observed for various values of X the beam to junction distance on a $ZnSe_yTe_{1-y}$ diode are shown in Fig. 26. The absorption spectra for *n*-type Al-doped and *p*-type undoped crystals of $ZnSe_{0{\cdot}5}Te_{0{\cdot}5}$ are shown in Fig. 27. The absorption in the undoped *p*-type material is due to fundamental lattice (band to band) absorption. This effect, since it overlaps the "edge" emission band (the 2·1 eV peak) can lead to a loss of efficiency and distortion of the emission spectra of injection luminescent diodes. The additional absorption in the *n*-type material is probably due to the impurity levels of the Al atoms spreading into energy bands at the high (degenerate) doping levels (10^{19} cm^{-3}) used.

The injection diodes are usually so made that the light has to traverse about 100 μm of *p*-type material before emerging. Aven *et al.* used the absorption coefficient data of Fig. 27 to calculate the effect of self absorption in attenuating and shifting the peak of the high energy band in Fig. 26. The calculated curve together with the observed diode

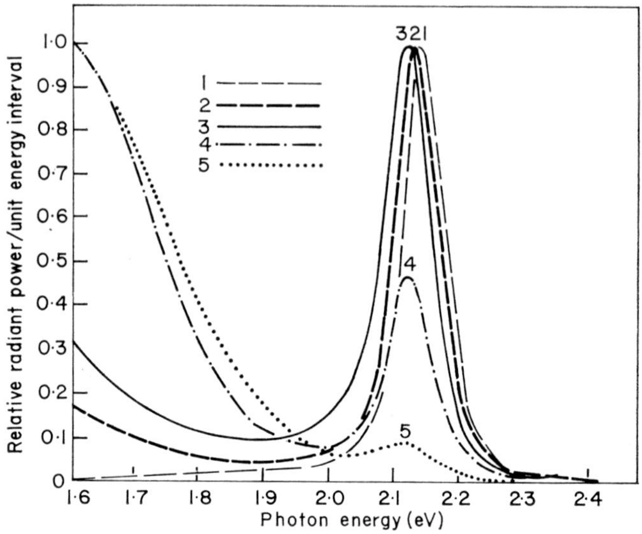

FIG. 26. CL spectra obtained from a $ZnSe_yTe_{1-y}$ diode with the electron beam incident at the following positions: (1) p-type bulk, (2) 4 μm from the junction on the p-side, (3) 4 μm from the junction on the n-side, (4) 16 μm from the junction on the n-side and (5) n-type bulk. Spectra 1 to 3 were normalized at the 2·1 eV peak, spectra 4 and 5 at the 1·6 eV peak. The positions are those of the beam impact point. The diameter of the beam excited region in the material was about 4 μm. (After Aven *et al.*, 1972.)

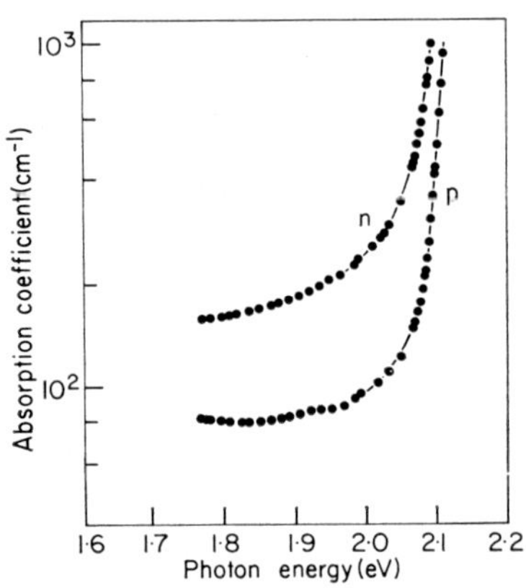

FIG. 27. Absorption spectra of undoped p-type and Al-doped n-type $ZnSe_{0·5}Te_{0·5}$. The thickness of the n-type crystal was 64 μm and that of the p-type crystal was 67 μm. (After Aven *et al.*, 1972.)

injection luminescence spectrum is shown in Fig. 28. The self-absorption reduced the area under the high-energy peak by about a factor of 10. It can be seen that the agreement on the high energy side is good, with the lattice absorption cutting off a substantial portion of the high-energy side of the emission band of Fig. 26 and shifting the emission peak from 2·13 to 2·06 eV. The agreement on the low-energy side is not as good because of overlap with the tail of the 1·6 eV band. These results show that the high energy band of the injection luminescence spectrum of the diodes with an apparent peak at 2·06 eV arises from radiative recombinations involving the same luminescence centres as the 2·13 eV CL band characteristic of p-type $ZnSe_yTe_{1-y}$. This demonstrates that the injection electroluminescence arises from recombinations occurring on the p-side of the junction.

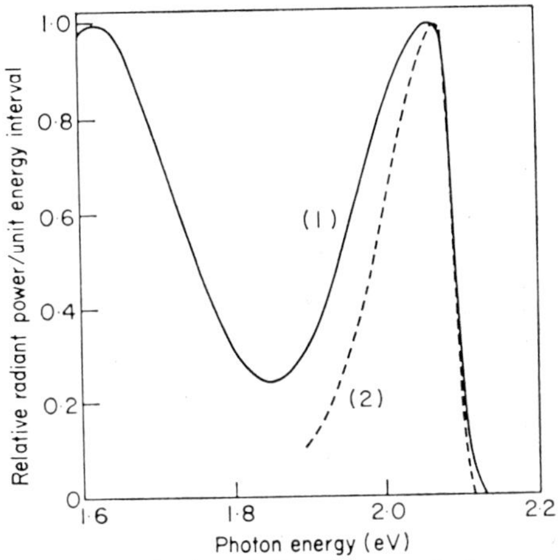

FIG. 28. Observed (1) and calculated (2) diode emission spectra. The calculated spectrum is that obtained by assuming the light generated to correspond to spectrum (3) of Fig. 26 and to pass through a 100 μm layer of p-type material with the absorption spectrum shown in Fig. 27. (After Aven *et al.*, 1972.)

Aven *et al.* also carried out voltage contrast and barrier electron voltaic effect linescan observations under reverse bias conditions as discussed in Chapter 8. They took the half width of the CC peak to be the sum of the minority carrier diffusion lengths on the two sides plus the depletion region width. The halfwidth of 1 μm less the carrier diffusion lengths of 0·1 μm and 0·4 μm obtained from V_0 by fitting the curve of Fig. 25 gives a significant depletion region width. It was concluded that a significant fraction of the total current through the diodes occurs through non-radiative recombinations in the depletion region. This conclusion was in agreement with the form of the current–voltage curve found for these diodes. Thus η_f is significantly less than 1.

η_e can now be estimated using eqn (38). η_f was taken to have an upper limit of 0·5. η_a is mainly influenced by the strong lattice absorption (Figs 27 and 28) and was assumed to have an upper limit of 10^{-1}. Taking $\eta_{rad} = 0{\cdot}04$, the value found for one particular crystal, makes

$$\eta_e < 0{\cdot}04 \times 0{\cdot}1 \times 0{\cdot}5 = 2 \times 10^{-5}$$

which is close to the average quantum efficiency over the visible spectral range found for a number of diodes that were measured.

The strong temperature dependence of diode efficiency is simply due to that of the *p*-type material since the injection luminescence is generated in the *p*-type material.

This work demonstrates that results can be obtained with the SEM that would be difficult if not impossible to obtain in any other way, and which are of the greatest importance for understanding and developing practical electroluminescent light sources.

References

Aven, M., Devine, J. Z., Bolon, R. B. and Ludwig, G. W. (1972). *J. appl. Phys.* **43,** 4136–4142.

Basov, N. G., Bogdankevitch, O. V. and Devyatkov, A. G. (1964). *Sov. Phys. Dokl.* **9,** 288.

Benoit à la Guillaume, C., Debever, J. M. and Salvan, F. (1967). *In* "II–VI Semiconducting Compounds, 1967 Internat. Conf." (D. G. Thomas, ed.), pp. 669–681. Benjamin, New York.

Benoit à la Guillaume, C., Debever, J. M. and Salvan, F. (1969). *Phys. Rev.* **177,** 567–580.

Bille, J., Bojničič-Kninski, St. V. and Ruppel, W. (1972). *Phys. Stat. sol.* **a14,** 141–145.

Bogdankevitch, O. V., Zverev, M. M., Krasilnikov, A. I. and Pechenov, A. N. (1967). *Phys. Stat. sol.* **19,** K5–K6.

Calverley, A. and Wight, D. R. (1970). *Solid St. Electron.* **13,** 382–385.

Casey, H. C. (1967). *J. electrochem. Soc.* **114,** 153–158.

Casey, H. C. and Jayson, J. S. (1971). *J. appl. Phys.* **42,** 2774–2776.

Casey, H. C. and Kaiser, R. H. (1967). *J. electrochem. Soc.* **114,** 149–153.

Chase, B. D. and Holt, D. B. (1973). *Phys. Stat. sol.* **a19,** 467–478.

Cusano, D. A. (1964). *Solid St. Commun.* **2,** 353–358.

Dean, P. J. (1969). *Appl. Solid St. Sci.* **1,** 1–151.

Dirochka, A. I., Kurbatov, L. N. and Mashchenko, V. E. (1969). *Sov. Phys. Semicond.* **2,** 1399–1401.

Esquivel, A. L., Lin, W. N. and Wittry, D. B. (1973). *Appl. Phys. Letters* **22,** 414–416.

Everhart, T. E. and Hoff, P. (1971). *J. appl. Phys.* **42,** 5837–5846.

Filinski, I., Wojtowicz–Natanson, B. and Hvam, J. M. (1971). *J. phys. Chem. Solids,* **32,** 2193–2199.

GERGELEY, G. (1960). *J. phys. Chem. Solids*, **117,** 112.
GRILLOT, E. and BANCIE-GRILLOT (1967). *J. Phys.* **28,** 898–906.
HILL, D. E. (1964). *Phys. Rev.* **133,** A866–A872.
HOLT, D. B. (1962). *J. phys. Chem. Solids*, **23,** 1353–1362.
HOLT, D. B. and CHASE, B. D. (1968). *J. Mater. Sci.* **3,** 178–182.
HOLT, D. B. and CHASE, B. D. (1973). *Phys. Stat. sol.* **a20,** 135–144.
HOLT, D. B. and CULPAN, M. (1970). *J. Mater. Sci.* **5,** 546–556.
HOLT, D. B., CHASE, B. D. and CENSLIVE, M. (1973). *Phys. Stat. sol.* **a20,** 459–467.
HURWITZ, C. E. (1966a). *Appl. Phys. Letters*, **8,** 121–124.
HURWITZ, C. E. (1966b). *Appl. Phys. Letters*, **8,** 243–245.
HURWITZ, C. E. (1966c). *Appl. Phys. Letters*, **9,** 116–118.
HURWITZ, C. E. (1967a). *In* II–VI Semiconducting Compounds 1967 Internat. Conf. (D. G. Thomas, ed.), pp. 682–693. Benjamin: New York.
HURWITZ, C. E. (1967b). *I.E.E.E. J. Quantum Electron.* **QE–3,** 333–334.
JONES, G. A. C., NAG, B. R. and GOPINATH, A. (1973). *In* 6th Ann. SEM Symp., *IITRI, Chicago*, pp. 309–316.
KURBATOV, L. N., MASHCHENKO, V. E. and MOCHALKIN, N. N. (1967). *Opt. Spectrosc.* **22,** 232–236.
KURBATOV, L. N., GROSS, E. F., MASHCHENKO, V. E., RAZBIRIN, B. S., MOCHALKIN, N. N. and BRITOV, A. D. (1968a). *Sov. Phys. Semicond.* **2,** 237–238.
KURBATOV, L. N., GROSS, E. F., MASHCHENKO, V. E., RAZBIRIN, B. S., MOCHALKIN, N. N. and BRITOV, A. D. (1968b). *Phys. Stat. sol.* **26,** K139–K142.
KYSER, D. F. and WITTRY, D. B. (1964). *In* "The Electron Microprobe" (T. D. McKinley, K. F. J. Heinrich and D. B. Wittry, eds), pp. 691–714. Wiley, New York.
MOOSER, E. and PEARSON, W. B. (1960), *Progress in Semiconductors*, **5,** 103–139.
NICOLL, F. H. (1966). *Appl. Phys. Letters*, **9,** 13–15.
NICOLL, F. H. (1968a). *I.E.E.E. J. Quantum Electron*, **QE–4,** 198.
NICOLL, F. H. (1968b). *J. appl. Phys.* **39,** 4469–4470.
NICOLL, F. H. (1970). *Appl. Phys. Letters*, **16,** 501–503.
PARTHE, E. (1966). "Crystal Chemistry of Tetrahedral Structures". Gordon and Breach, New York.
RAO-SAHIB, T. S. and WITTRY, D. B. (1969). *J. appl. Phys.* **40,** 3745–3750.
REYNOLDS, D. C. (1967). *Trans. Met. Soc. AIME*, **239,** 300–309.
SHAW, D. A. and THORNTON, P. R. (1968). *J. Mater. Sci.* **3,** 507–518.
SMILEY, V. N., TAYLOR, H. F. and LEWIS, A. L. (1971). *J. appl. Phys.* **42,** 5859–5861.
SUCHET, J. P. (1965). "Chemical Physics of Semiconductors." Van Nostrand, London.
VAN ROOSBROECK, W. and SHOCKLEY, W. (1954). *Phys. Rev.* **94,** 1558–1560.
VAVILOV, V. S., NOLLE, E. L., GOLUBEV, G. P. MASHTAKOV, V. S. and TSARAPAEVA, E. I. (1967). *Sov. Phys. Solid St.* **9,** 657–659.

VLASOV, A. N., KOZINA, G. S. and FEDOROVA, O. B. (1967). *Sov. Phys. J.E.T.P.* **25,** 283–285.
WILLIAMS, P. M. and YOFFE, A. D. (1968). *Phil. Mag.* **18,** 555–560.
WILLIAMS, P. M. and YOFFE, A. D. (1969a). *Nature, Lond.* **221,** 952–953.
WILLIAMS, P. M. and YOFFE, A. D. (1969b). *Radiation Effects,* **1,** 61–64.
WILLIAMS, P. M. and YOFFE, A. D. (1971). *Radiation Effects,* **9,** 139–144.
WITTRY, D. B. (1966). *Appl. Phys. Letters,* **8,** 142–144.
WITTRY, D. B. and KYSER, D. F. (1964). *J. appl. Phys.* **35,** 2439–2442.
WITTRY, D. B. and KYSER, D. F. (1965). *J. appl. Phys.* **36,** 1387–1389.
WITTRY, D. B. and KYSER, D. F. (1966). *J. Phys. Soc. Japan* **21,** Supplement (Proc. Internat. Conf. Phys. Semicond. Kyoto, 1966), pp. 312–316.
WITTRY, D. B. and KYSER, D. F. (1967). *J. appl. Phys.* **38,** 375–382.

PART III

X-rays in Scanning Electron Beam Instruments

Chapter 11

The X-ray Mode and Basic Microanalysis

J. A. BELK

Metallurgy Branch, Royal Military College of Science, Shrivenham, England

Symbols

E_0 = accelerating voltage.

E_c = critical excitation potential below which characteristic X-rays will not be generated

I_A = measured peak height—background of space

$I_{(A)}$ = measured peak height—background of standard

k_A = count ratio

C_A = weight fraction

Q_{AX} = correction factor relating k_A and C_A

ε_K = minimum count ratio detectable

B = background count rate

t = counting time

ε_c = minimum weight fraction detectable

d = thickness of lamellae

p = density in g cm^{-3}

I. X-RAY GENERATION

Any sufficiently energetic electron beam incident on a solid specimen will generate X-rays. The X-rays have intensities and wavelengths that are characteristic of the quantity of the individual elements of which the specimen is composed. It is very convenient to use the X-radiation that is necessarily of limited intensity and energy in scanning electron microscopy to determine the composition variations within the specimen. This may be done by incorporating some method of measuring the characteristics of the emitted X-rays and comparing these with X-rays emitted from standard samples under exactly similar excitation conditions. Unfortunately the optimum excitation conditions for X-ray emission do not coincide with those for high resolution electron images so before analysis can be attempted the excitation conditions must be corrected. Also X-ray microanalysis in a scanning electron microscope takes a considerably longer time than the production of electron images; hence, one needs to optimize the conditions initially and not waste too much time on trial and error for each specimen and instrumental set-up.

A. ACCELERATING VOLTAGE

The accelerating voltage E_c that is applied to the electron beam determines the range of X-ray wavelengths that will be excited. Each element has a critical excitation potential E_c below which its characteristic X-rays will not be generated. Values of E_c range from 0·282 kV for carbon K_α through 1·487 kV for aluminium K_α to 17·478 kV for molybdenum K_α radiation. If the L X-ray spectra are used 2·293 kV is sufficient to excite the molybdenum L_α and 13·613 kV for uranium L_α. M spectra of course have even lower critical excitation potentials.

In general it is advisable to exceed the critical excitation potential by at least a factor of two in order to get a sufficiently intense characteristic X-ray peak. As all the analysis is carried out at constant accelerating voltage a suitable value must be chosen that is sufficiently high to cover all the elements of interest. There is an upper limit to the accelerating voltage set by the instrument but it is not advisable to choose a value higher than necessary as the spatial resolution of X-rays in the specimen deteriorates as the accelerating voltage increases. This point will be covered in more detail under the heading of X-ray resolution (section III.D).

As an example it was required to analyse a specimen known to contain aluminium, copper, magnesium and zinc. The excitation potentials of the K_α radiation are respectively 1·487, 8·047, 1·254 and 8·638 keV. The compromise voltage chosen in this case was 15 kV, high enough to give sufficient intensity from the copper and zinc yet not so high that the resolution for aluminium and magnesium suffered.

The optimum wavelength range for X-ray spectrometry and detection is between 0·07 and 1·0 nm and this usually determines the choice of K, L, or M wavelengths to be detected for a given element, although there are circumstances where a particular wavelength will be chosen because it does not interfere with the X-rays generated by another element which is also present in the specimen.

B. PROBE CURRENT

Conventionally in electron probe microanalysis the term "probe current" is used to describe the current flowing down the scanning electron probe. Some of the electrons are back-scattered, contributing to the electron image, while others flow through the specimen giving rise to the specimen current. The proportion of back-scattered electrons varies with the characteristics of the specimen, hence the specimen current is only a guide to the probe current. The probe current can be measured by replacing the specimen with a Faraday cage which in effect eliminates back-scattering and so gives a true reading of the current flowing through the probe.

The value of the probe current is important as it influences the intensity of the emitted X-rays and so determines the sensitivity of the analysis and the accuracy of the results. In general a larger probe current is required for analysis than would be chosen for the production of an electron image although its actual value depends on the efficiency of the X-ray collector and analyser. With a wavelength dispersive system it may well be necessary to operate with a probe current of 10^{-7} A in order to obtain optimum analysis results, whereas an energy dispersive system might allow 10^{-10} A. The characteristics of the two types of X-ray analysers are different and will be commented on in detail below.

The ratio of the specimen current to probe current varies from 0·50 for very high atomic numbers up to 0·95 for low atomic number specimens although the exact value depends on the exact excitation conditions. At least a rough guide of this sort is useful in assessing the approximate value of the probe current. An accurate measurement of the probe current is not required before the analysis results can be determined; what is vitally important, however, is that the probe current shall remain constant during the whole time that the analysis is performed and this will typically be between 10 min and 1 hour.

II. X-RAY DETECTION

A system is required that can measure the wavelength and intensity of X-rays emitted from an effective point source over a wavelength range from about 0·07 nm to 7·0 nm, or an energy range from about 20 keV to 0·2 keV. The two major categories of equipment are wavelength dispersive and energy dispersive in nature and they offer different combinations of advantages but are in many ways complementary. A brief description of each of the methods will be given first, followed by a comparison of their relative merits as accessories to a scanning electron microscope.

A. WAVELENGTH DISPERSION

X-rays from a point source are dispersed into a spectrum by diffraction at a crystal or grating and the intensity of a given section of the spectrum is measured electronically by a counter system. The counter is usually of the gas-filled proportional type which produces electrical pulses whose voltage is proportional to the energy of the incident X-rays. Thus if two different wavelengths happen to be diffracted through the same angle because they have different orders of diffraction the resultant pulses will have a different height and so can be separated electronically. The geometry of diffractive X-ray spectrometers is such that the X-ray spectrum is produced around the circumference of a circle which also contains the X-ray source and the curved crystal. The Rowland circle on which the spectrum is formed has half the radius of curvature of the curved crystal. In most scanning electron instruments the direction of X-ray take-off is fixed and the only way to ensure that the correct geometric arrangement of components is maintained is to move the crystal along the X-ray take-off direction and allow it to rotate about an axis at right angles to this. The counter also has to be moved so that it remains on the circumference of the Rowland circle. Mechanically such an instrument is complicated and hence expensive to produce. A mechanically simpler alternative is to depart from the fully focusing design, and allow the curved crystal to rotate on an axis at right angles to the X-ray take-off direction while the counter rotates on an arm about the same axis. This semi-focusing arrangement requires a number of crystals each with different radii of curvature to give an approximation to the correct focusing geometry.

Fully focusing spectrometers have the best spectral resolution and can have better X-ray collection efficiency, but are complicated and very sensitive to X-ray source position, crystal quality and accuracy of setting. Semi-focusing spectrometers are simpler, can be more compact, but require a larger number of crystals and do not offer such high sensitivity or efficiency. Accuracy of measurement requires that the spectrometer mechanism matches its spectral resolution so that repeat settings on a peak may be made reproducibly. Hence, although fully focusing spectrometers should give better results than semi-focusing types this is not always the case. About five different crystal materials would be required to cover the full wavelength range and a semi-focusing spectrometer might require up to three crystals of some materials with different radii of curvature.

B. ENERGY DISPERSION

A proportional X-ray counter produces pulses whose heights are proportional to the energies of the X-rays giving rise to them. If the pulse height distribution coming from the counter can be analysed the relative proportions of the various X-ray energies can be determined. In order to perform this energy analysis satisfactorily the proportional counter must have adequate resolution. Gas-filled proportional counters such as are used in wavelength dispersive spectrometers have inadequate resolution, but adequate resolution is available from a solid state proportional counter. This consists of a solid state detector of silicon or germanium, treated with lithium, which is held at liquid nitrogen temperature. Its surface is protected from the contaminants in the microscope vacuum system by a beryllium window and the X-rays pass through this window, interact with the silicon and produce pulses. The pulses are amplified and fed into a multichannel pulse height analyser for analysis, storage and subsequent display. The detector must be kept at liquid nitrogen temperature to reduce the thermal noise but also to prevent the migration of the lithium. A few days at room temperature produces measurable effects in the performance of silicon detectors but germanium detectors are severely affected in less than an hour.

The multichannel pulse height analyser is the most important unit in the all electronic energy dispersive system. This can separate into a few hundred channels pulses of different heights, allow them to accumulate, and recall and display the pulse height distribution at will. Thus one has an almost unlimited number of elements available for analysis at the same time and as the detector can be placed close to the X-ray source, collection efficiency is high and the time required for analysis quite short. The multichannel analyser must be calibrated as it is not an absolute measure of energy in the way that a crystal spectrometer is of wavelength.

The spectral resolution of a solid state energy dispersive system is similar to that of a crystal spectrometer for X-rays of wavelength about 0·07 nm. At longer wavelengths, or lower energies, the crystal spectrometer improves and the solid state detector deteriorates so that at a wavelength of about 0·8 nm there is a factor of 10^2 in their relative spectral resolutions. Thus a specimen containing potassium, calcium and tin would be much more difficult to analyse by energy dispersive than by wavelength dispersive means. The speed and the large number of elements analysed simultaneously is a very obvious attraction of the energy dispersive system which in the case of a scanning electron microscope may make it the best choice if only one type of analyser can be installed. Although an energy dispersive system is not sensitive to specimen positioning it has a limited wavelength range, and requires a continuous supply of liquid nitrogen—so the two alternative systems are in many ways complementary. The crystal spectrometer is a slow way of obtaining information and although it is the obvious first choice for an electron probe microanalyser it has limitations as an accessory to a scanning electron microscope, but the final choice must depend on the type of work to be undertaken.

III. Basic Microanalysis

The X-ray mode differs in a number of fundamental ways from the conventional use of a scanning electron microscope for high resolution topographical studies. It gives quantitative information on the distribution of elements within the specimen and is able to provide this information in a variety of ways. It implies restrictions on the specimen which do not apply in microscopy. In this section an attempt will be made to describe briefly the importance of specimen shape, size and topography and to indicate which elements may be analysed, what the limit of detection is, over what distances in the specimen compositional variations can be measured, how accurate the results are and why it is necessary to correct the experimental determinations to allow for the characteristics of the specimen. These topics are all dealt with in more detail in a review on experimental electron probe microanalysis by Poole and Martin (1969).

A. Analysis Techniques

There are three basic ways in which information about the distribution of elements in a specimen may be obtained and displayed: scanning images where the distribution of an element is displayed on a micrograph; point analyses where the detailed composition of very small volumes of material are obtained; and line scans where the distribution of a series of elements along a particular line on the specimen surface is measured. All these methods have their particular field of application.

1. *Scanning Images*

X-ray images are easily appreciated as they often closely resemble a micrograph of the specimen. The scanning image of an element is produced by setting the X-ray spectrometer to accept the particular characteristic X-rays of interest and displaying each pulse as a bright spot on the face of an image tube that is scanned in synchronization with the electron probe. The image is composed of a large number of spots, the brighter parts of the image indicating the areas of highest concentration in the specimen. X-ray images can be recorded photographically by exposing for a total of about 10^5 counts. Thus if the

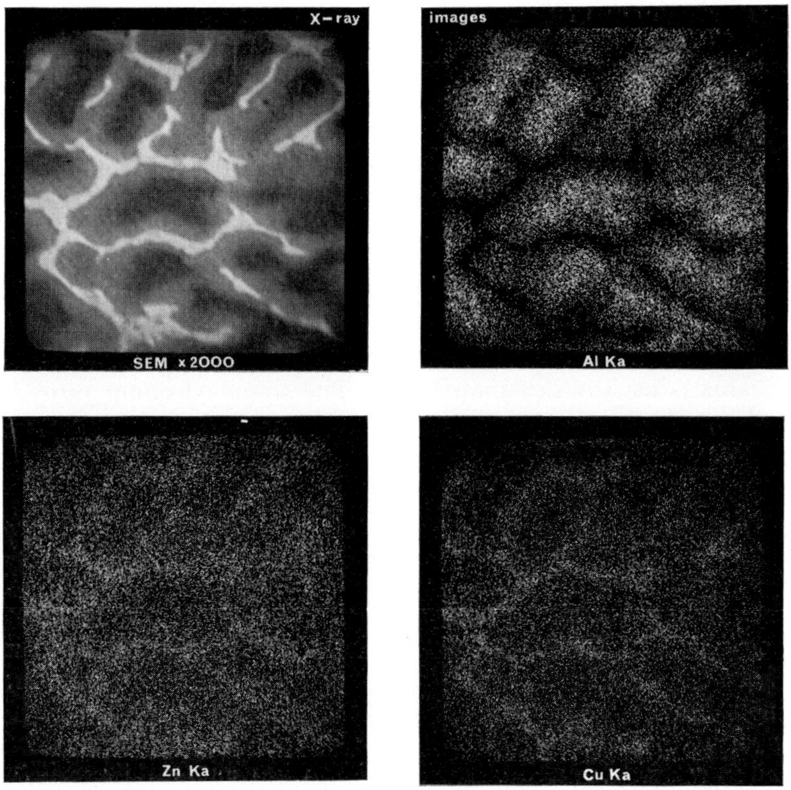

FIG. 1. Scanning electron and X-ray images on cast zinc–aluminium alloy.

concentration is low a long exposure will be required. A sequence of X-ray images for various elements is normally taken and a comparison made subsequently to estimate the presence of the various elements in the different parts of the structure. Figure 1 is a composite series of three X-ray images and the back-scattered electron image. It is quite clear into which parts of the structure the elements are segregated. The contrast of X-ray images is such that unless the concentration varies by a factor of two it is difficult to detect the intensity variation on the scanning image. The alloy used in Fig. 1 has zinc as its most abundant element, yet because the zinc concentration only varies by about a factor of two the contrast of the zinc X-ray image is not as good as that of either copper or aluminium. Because the exposure time is adjusted to give the same total number of counts it is not possible to obtain the relative concentrations of two elements by comparing their X-ray images.

2. *Point Analyses*

Point analyses offer the greatest precision of measurement of composition. Electrons diffusing into the specimen from a static probe excite X-rays in a very limited volume of material and it is the composition of this small volume that is determined in a point analysis.

The intensity of a particular characteristic X-ray line emitted by the specimen is measured by determining the peak height and subtracting the background radiation level; this is usually designated I_A. The specimen is then replaced by a standard, often a pure element, and the intensity of the same peak is measured under identical excitation conditions. This is usually designated $I(A)$. The so-called count ratio k_A is defined as:

$$k_A = \frac{I_A}{I(A)}$$

The count ratio k_A is related to the weight fraction C_A by means of a complex correction factor Q_{AX} in accordance with the equation:

$$C_A = Q_{AX} k_A$$

It can happen that Q_{AX} has a value close to unity but more often the value of Q_{AX}, which depends on the specimen composition and the detailed excitation conditions, must be calculated. The calculation of Q_{AX} will be discussed again under section E below.

Obtaining accurate point analyses is a slow process as each element has to be standardized in the manner described above and the probe current and the position and orientation of the specimen and all the standards must be constant throughout. The provision of suitable standards can pose problems. Pure elements are often satisfactory but if an oxide specimen is being analysed it is usually advisable to use oxide standards to minimize the difference in characteristics between the specimen and the standards.

3. *Line Scans*

A line scan is simply a graph of the concentration variation for a particular element against distance along a line on the specimen surface. Normally a number of line scans for different elements along the same line are superimposed to give the kind of result shown in Fig. 2. As each of the curves is reproduced directly from the recorder chart the concentration axis has a different calibration for each element. This is a convenient way of displaying semi-quantitative data and gives a better indication of the quantitative variations of composition than scanning images do.

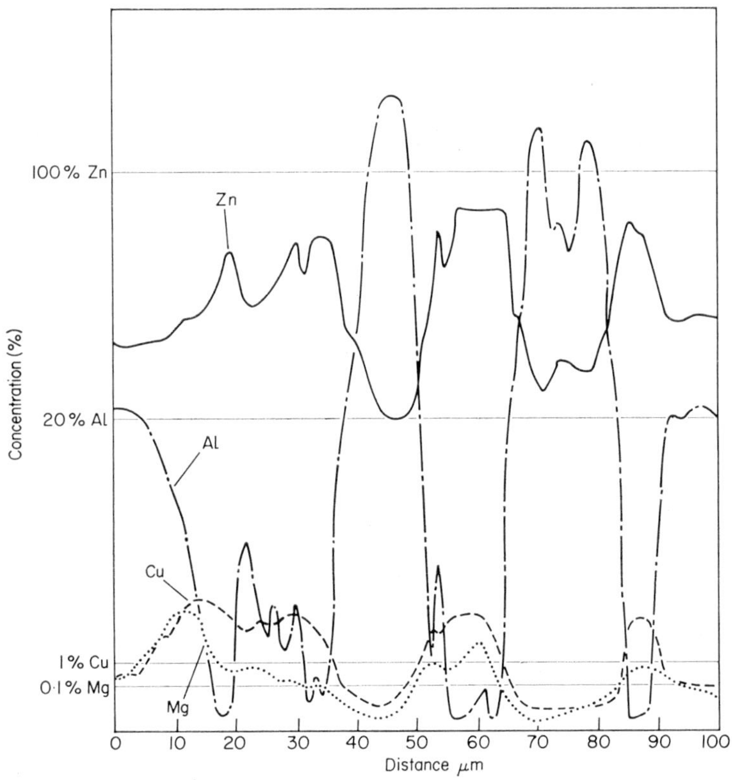

FIG. 2. Line scans of concentration *vs*. distance for the alloy shown in Fig. 1.

B. SPECIMEN REQUIREMENTS

The theory of electron probe X-ray microanalysis which is now well established assumes that the specimen and standards have flat polished surfaces, are all oriented in exactly the same way with respect to the electron beam and the X-ray spectrometer, and are thick enough to absorb all the electrons that are not back-scattered at the surface of entry. Scanning electron microscope specimens often have heavily structured surfaces, they can be rotated and tilted and moved along the axis of the objective lens, and may well not be thick enough to prevent electron penetration. For satisfactory analysis the position and orientation of the specimen and standards must be constant. A specimen with a heavily structured surface should only be used for semi-quantitative work. Thin or small specimens can only be satisfactorily analysed by determining the ratios of the concentrations of the various elements present rather than their absolute values.

C. SENSITIVITY OF DETECTION

The lowest concentration of a given element that can be detected is a measure of the sensitivity of the analysis system. In X-ray micro-analysis this depends on the element, the specimen, the instrument and the time allowed for the analysis. The problem is to decide how small an X-ray peak can be detected above a continuous background of radiation. If we assume that a peak is significant if its height is more than twice the standard error of the difference between the peak and background levels we find that the minimum count ratio detectable is:

$$\varepsilon_k = 2{\cdot}82\frac{\sqrt{B/t}}{I(A)}$$

where B = background count rate
t = counting time for both peak and background
$I(A)$ = intensity of line from pure standard.

The minimum count ratio is related to the minimum weight fraction detectable ε_c by the formula

$$\varepsilon_c - Q_{AX}\varepsilon_k$$

where Q_{AX} is the correction factor mentioned in section III.A.2 above. The above treatment assumes that the probe current and other instrumental conditions stay constant during the whole time of the analysis. This is not always so and a method for allowing for the influence of instrumental instability on the sensitivity of detection has been proposed by Long (1967). The sensitivity of detection would normally be between 0·01 and 0·2 wt %. Improved sensitivity can be obtained by increasing the accelerating voltage E_0, the probe current and the peak to background ratio of the X-ray spectrometer but the sensitivity only changes with the square root of the change of any one of these.

D. X-RAY RESOLUTION

The size of the X-ray source is normally determined by the diffusion distance of the electrons in the specimen. Reed (1966) calculated that equal thickness lamellae of spacing d could just be resolved if

$$d = 7{\cdot}7\times 10^{-2}\frac{1}{\rho}(E_0^{1.5}-E_c^{1.5})$$

where ρ = density in g cm^{-3}

E_0 and E_c are in kV

d is in μm

This criterion for qualitative resolution gives values of d for metals that are about 1 μm. This is of course much poorer than the resolution obtainable from the electron image. For quantitative analysis the structure must be three or four times coarser than for qualitative analysis if thick solid specimens are to be used. The analysis of small individual particles that are not set in a solid matrix can be done by comparative rather than absolute methods.

It occasionally happens that one emitted X-ray wavelength interacts with the atoms of another element in the specimen to produce X-rays of the second element. This X-ray fluorescence has to be allowed for in the correction factor Q_{AX} but also results in a very considerable increase in the effective X-ray source size for the second element. X-ray fluorescence occurs most markedly between elements whose atomic numbers lie between 22 and 32 and differ by 2. Thus nickel K X-rays fluoresce iron and iron X-rays fluoresce chromium. There are many other less efficient interactions whose effect can be allowed for (Reed, 1965).

E. ACCURACY OF MEASUREMENT

The ability of X-ray microanalysis to measure accurately the chemical composition of a solid sample depends on the ability to produce reproducible count ratios and the ability to calculate correction factors which can be applied to transform the count ratios into weight concentrations. The production of accurate reproducible count ratios depends on the spectrometer mechanism and stability, the electronic stability of the instrument as a whole and the experimental technique. Rarely can count ratios be quoted with an accuracy better than 2% relative and often the reproducibility from laboratory to laboratory is much worse than this.

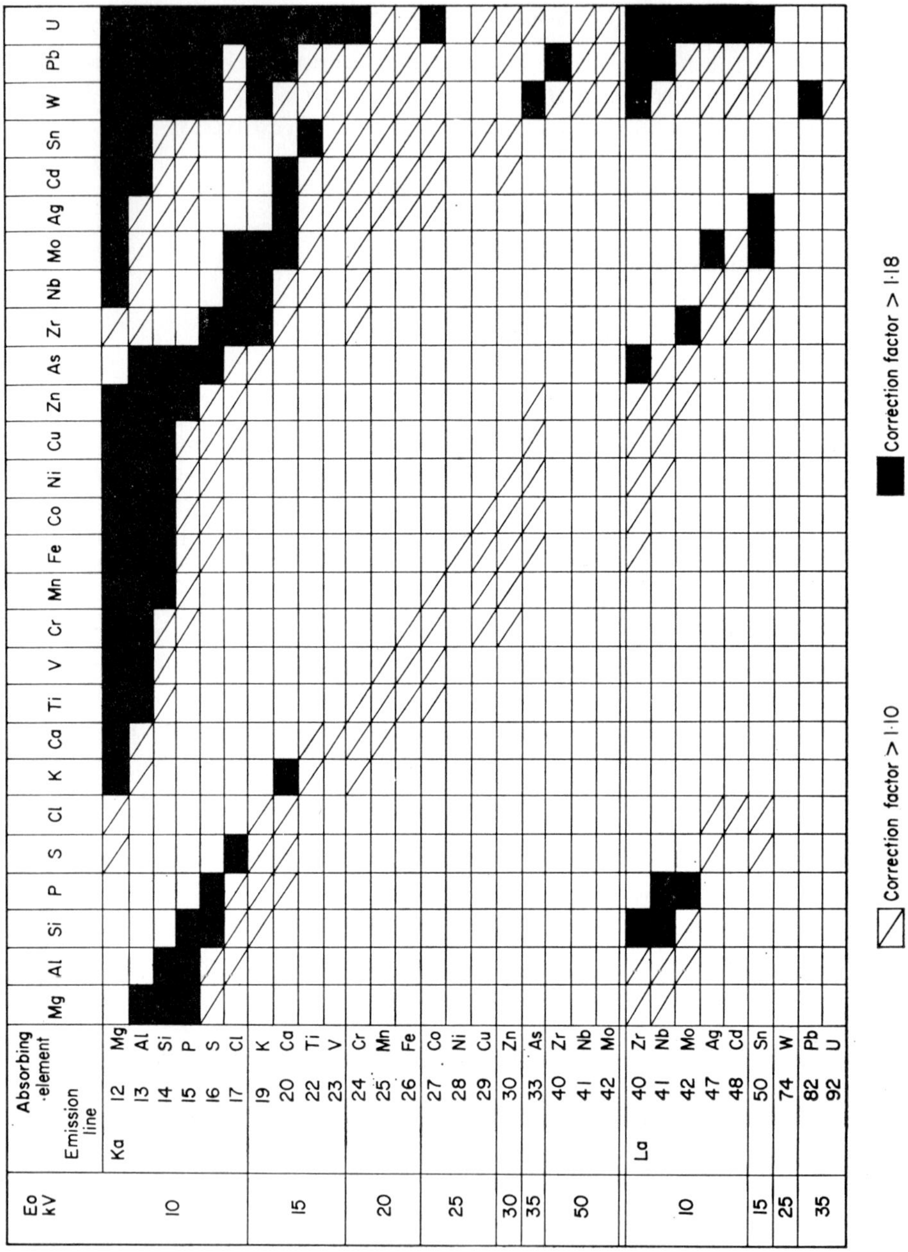

FIG. 3. Approximate values of the absorption correction for the stereoscan for a trace of emitting element in solution in the absorbing element.

Correction factors have received considerable attention, a very good review being that by Martin and Poole (1971). Corrections are required because: X-rays are absorbed to different extents by the specimen and standard; the shape and size of the X-ray source depends on the accelerating voltage, the atomic numbers of the specimen and standard and the critical excitation potential of each wavelength analysed; and X-ray fluorescence occurs.

Figure 3 is a block diagram indicating the combinations of elements where the absorption correction is appreciable. This has been drawn up for the Stereoscan and indicates the range of correction factor that would have to be applied for analysis of a small quantity of the emitting element dissolved in the absorbing element. The atomic number correction is numerically fairly small in the majority of cases and the fluorescence correction is limited to relatively few combinations of elements. Martin and Poole (1971) pointed out that correction factors are at least as good as the experimental count ratios warrant but that practically all correction procedures assume normal incidence of the electron beam on the specimen. In scanning electron microscopy this is often not the case and as there is no simple angle-of-incidence-dependent factor that can be applied, the correction procedures should be treated with some reserve.

IV. Conclusion

In the early days of electron probe microanalysis the obvious quest was for reliable accurate quantitative analysis. This has to a large extent now been achieved in microanalysers by improvements in design, operation, technique and correction methods. It is clear that using an X-ray spectrometer in conjunction with a scanning electron microscope does not allow the same single mindedness that is possible in a purpose-built microanalyser. The X-ray microanalysis results are, however, of very great assistance to the scanning electron microscope user to supplement his pictorial data. Also, for the price of an accessory X-ray spectrometer one can obtain many of the operational features of an electron probe microanalyser at a fraction of the cost. But, as in all fields, the more versatile multi-purpose instrument cannot be expected to reproduce the high quality of results that are obtainable from a specially designed single-purpose instrument. As long as this is understood and results from the X-ray mode are not endowed with attributes that they do not justify there will be very many investigations that will benefit from the extra information that is available from the X-ray mode.

References

LONG, J. V. P. (1967). *In* "Physical Methods in Determinative Mineralogy" (J. Zussman, ed.), p. 215. Academic Press, London and New York.

MARTIN, P. M. and POOLE, D. M. (1971). *Metall. Rev.* **16,** 19–46.

POOLE, D. M. and MARTIN, P. M. (1969) *Metall. Rev.* **14,** 61–84.

REED, S. J. B. (1965). *Br. J. appl. Phys.* **16,** 913–926.

REED, S. J. B. (1966). *In* "X-ray Optics and Microanalysis" (R. Castaing, P. Descamps and J. Philibert, eds), p. 339. Hermann, Paris.

Chapter 12

The Si (Li) X-ray Energy Spectrometer for X-ray Microanalysis

D. A. GEDCKE

ORTEC Inc., Oak Ridge, Tennessee, U.S.A.

Symbols

Z = Atomic Number

n = no. of electron-pairs produced by a photon

E = photon energy

ε = energy to produce an electron–hole pair

τ = RC time constant of amplifier filter

K = energy spanned by one channel of ADC

E_i = energy of centre of ith channel of ADC

C_f = capacitance of feedback capacitor

R_f = resistance of feedback resistor

V_1 = initial voltage across feedback capacitor due to photon of energy E

V_2 = voltage across feedback capacitor due to photon of energy E as a function of t

t = time

$r_i(E)dE$ = counting rate of photons with energies between E and $E+dE$

τ = also used for shaping time constant of amplifier and time constant of integrator

N = channel index in ADC

Γ = energy resolution, FWHM

Γ_{noise} = noise contribution, FWHM

F = Fano factor

R = Separability ratio

I. Introduction

The Si(Li) X-ray energy spectrometer was first applied as a tool for X-ray microanalysis on the electron beam microprobe in 1968 (Fitzgerald *et al.*, 1968). Since then, there have been substantial improvements in the energy resolution, counting rate capability, and data analysis features of the spectrometer. As a result, this instrument has become a very popular addition to electron beam microprobes and scanning electron microscopes. More recently, it has been finding its way onto a variety of transmission electron microscopes.

The utility of the Si(Li) energy spectrometer is derived chiefly from the following features:

1. Quick, simultaneous analysis of all elements with $Z \geqq 11$.
2. High detection efficiency.
3. Compactness, allowing close sample-to-detector spacing.
4. Excellent stability (does not rely on moving mechanical parts).
5. Does not employ "optical" focusing (relatively insensitive to sample position and surface roughness).
6. Visual presentation of the entire spectrum commences when data accumulation is initiated.
7. The method of data accumulation and presentation is well suited to on-line computer data processing.

Points 1 and 2 combined with 6 define the features with major impact. What makes this spectrometer most useful is that all energies (atomic elements) are detected and displayed simultaneously. It is hard to convey in words the many subtle benefits derived from this multichannel characteristic. Most obvious is the fact that the entire spectrum can be monitored as the analysis proceeds and the precision of the spectral information improves. The analysis can be stopped as soon as the significance of the data in the spectrum is seen to be adequate. Effects of changing both sample excitation conditions and points of analysis on the sample can be observed quickly, allowing the operator to "home in" efficiently on the information of real interest. It is in this area of total spectrum analysis that the Si(Li) energy spectrometer is most efficient relative to the scanning, Bragg-diffraction, wavelength spectrometer. The wavelength spectrometer is most efficient as a single-channel or single-element spectrometer. At energies below the 14 keV K-lines of strontium ($Z \leqslant 38$ in the Periodic Table), a well-designed wavelength spectrometer has superior resolving power (Fitzgerald and Gantzel, 1971; Belk, this volume, Chapter 11). This improved resolution is achieved at the expense of several orders of magnitude in detection efficiency. However, in electron beam instruments where high probe currents are feasible, the better peak-to-background ratio resulting from the improved resolution leads to lower minimum detectable limits for trace elements. The better resolution of the wavelength spectrometer also makes it useful for eliminating peak overlaps encountered in the energy spectrometer, particularly in the light element range.

High detection efficiency, coupled with compactness, makes the Si(Li) detector attractive for samples having a low X-ray yield. End cap diameters on the cryostat housing the cooled detector have been trimmed down to as low as 19 mm, making it possible to place the detector front surface within 10 mm of the sample. In this geometry the detector subtends a large solid angle with respect to the source of X-rays on the sample. Consequently, the X-ray collection and detection efficiency is many orders of magnitude above that obtainable with a wavelength spectrometer. This capability has been important in opening up the area of X-ray microanalysis in biological thin sections, where X-ray yields are low (Sutfin, 1972).

The analysing function of the Si(Li) energy spectrometer is entirely electronic. It does not depend on the accuracy of moving mechanical parts, as the wavelength spectrometer does, to obtain reproducible peak positions and detection efficiencies. Outside of the limitations set by counting statistics, which apply to both types of spectrometers, the major factor determining reproducibility in the energy spectrometer is sensitivity of the energy calibration to temperature. In modern systems, drifts in the calibration have been reduced to the point where recalibration is seldom necessary.

The wavelength spectrometer uses optical focusing, and as a result, places rather severe restrictions on the location of the source of X-rays. In the Si(Li) energy spectrometer the detection efficiency is controlled only by the solid angle and sample attenuation effects. Consequently, precise sample positioning is not as critical as with the wavelength spectrometer (Sutfin and Ogilvie, 1970). This advantage shows up in qualitative analysis where large sample areas can be scanned for X-ray maps, and moderate sample roughness does not prevent useful analysis. However, for accurate quantitative analysis, a smooth sample surface is necessary, since sample absorption and enhancement effects must be accounted for in a predictable fashion.

In the Si(Li) energy spectrometer a great deal of information is collected in a short time. Frequently an SEM or microprobe operator will find that the rate of production of information about his sample exceeds by at least an order of magnitude his ability to analyse and interpret it. Fortunately, the method by which data are accumulated in the multichannel analyser section of the energy spectrometer lends itself to efficient coupling to a small on-line computer. Increasing use of dedicated computers can be expected as the payoff in sophisticated data analysis, and information interpretation becomes more evident in this field, as it has in others.

Although the above discussion of the relative merits of the Si(Li) energy spectrometer and the wavelength spectrometer is by no means exhaustive, it is clear that the two instruments both overlap and compliment each other in performance. The choice of spectrometers will depend on many factors relating to the type of work that is anticipated on the SEM or microprobe. Where a heavy load of quantitative analysis is expected and sufficient funds are available, a microprobe with several wavelength spectrometers and a Si(Li) energy spectrometer is the optimum solution. The energy spectrometer provides rapid analysis in survey work. For precision quantitative analysis the wavelength spectrometers offer a versatile range of single-element channels. If a fair amount of conventional SEM topographic analysis is also envisioned, one of the new generation combination SEM/microprobe instruments will be advantageous. In many instances funds are limited, and quantitative analysis is a reduced part of the work load. Smaller, less expensive SEMs are usually designed with the Si(Li) detector in mind as the primary X-ray spectrometer. However, many of these offer a multiple-crystal wavelength spectrometer as an option. This combination offers an effective, low-cost solution. The Si(Li) detector is used for most of the X-ray analysis, with the Bragg-crystal spectrometer being used to resolve difficult cases of overlap or to enhance trace-element detection, particularly in the light element range. More frequently, the analyst is unable to justify the extra expense of the multi-crystal wavelength spectrometer. Although some of the flexibility of the more sophisticated installation is sacrificed, the low-cost SEM combined with a Si(Li) energy spectrometer is still quite versatile and can handle a surprising variety of problems, including quantitative analysis.

Historically, the technique of using diffraction to selectively analyse X-rays based on their wavelengths has been termed "dispersive". This label finds its basis in optics where dispersion is the process of deflecting the monochromatic components of a polychromatic beam through varying angles as a function of their wavelengths. Detectors not employing the diffraction technique are called "nondispersive". Until the advent of the Si(Li) detector, the energy resolution of nondispersive detectors was not adequate to provide an energy spectrum with usable element separation. In an attempt to emphasize the improved resolution of this new member of the nondispersive detector family, the Si(Li) energy spectrometer was termed "energy-dispersive". This use of the word "dispersive" no longer fits with the original term from optics. It serves instead to indicate that the energy spectrometer sorts detected X-rays of different energies into an energy spectrum presentation with usable atomic-element separation. A more recent trend is to refer to the two types of spectrometers on the basis of the coordinate used to identify peak positions in the spectrum. In this scheme the dispersive, Bragg-diffraction instrument is called a "wavelength spectrometer" (WS), whereas the nondispersive Si(Li) detector system is termed an "energy spectrometer" (ES). In addition to the latter terms being less awkward, they are not in conflict with the correct meanings of the words "dispersive" and "nondispersive". Furthermore, they indicate the type of spectrum being produced. The WS and ES terminology will be used in the remainder of this article.

The following sections outline the principles of operation and the important performance characteristics of the Si(Li) energy spectrometer as they relate to use on electron beam instruments.

II. The Energy Spectrometer

In this section the overall function of the spectrometer will be outlined, followed by a more detailed description of each part of the system. The spectrometer is more easily understood once the operation of the Si(Li) detector diode is clear.

A. the Si(Li) detector diode (Brown *et al.*, 1969)

Although actual Si(Li) diodes come in a variety of physical shapes due to different manufacturing processes, Fig. 1 provides a useful schematic representation of the sensitive detector volume for the purpose of understanding its function. In order to produce a solid-state ionization chamber, a cylindrical piece of *p*-type silicon is lithium compensated to increase its electrical resistivity. The lithium is drifted through the crystal from the rear under high temperature and voltage. A Schottky barrier contact is applied to the circular front face of the detector to produce a *p-i-n*-type diode. Electrical contacts are formed with thin gold coatings on the front and rear of the diode. By applying a reverse bias of about 1000 V to the diode and operating the device at liquid nitrogen temperature, most of its volume is depleted of the remaining free charge carriers, and no charge flows between the two electrodes. The entrance window for X-rays is formed by the front

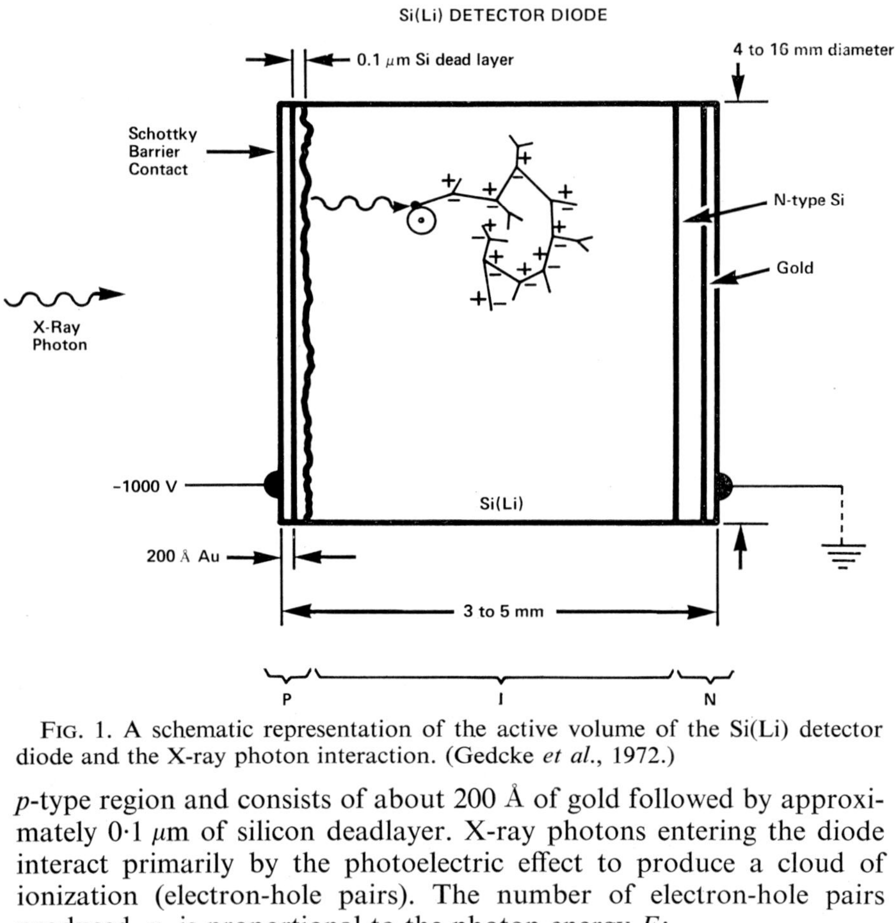

FIG. 1. A schematic representation of the active volume of the Si(Li) detector diode and the X-ray photon interaction. (Gedcke *et al.*, 1972.)

p-type region and consists of about 200 Å of gold followed by approximately 0·1 μm of silicon deadlayer. X-ray photons entering the diode interact primarily by the photoelectric effect to produce a cloud of ionization (electron-hole pairs). The number of electron-hole pairs produced, n, is proportional to the photon energy E:

$$n = E/\varepsilon, \tag{1}$$

where $\varepsilon = 3{\cdot}8$ eV for cooled silicon. Consequently, measurement of this charge should provide a measurement of the X-ray photon energy.

The charge is swept from the diode by the bias voltage. A charge-sensitive preamplifier connected to the rear detector electrode collects the charge and integrates it on a capacitor to produce a voltage pulse with an amplitude proportional to n. Consequently, the amplitude of the pulse produced by the preamplifier is proportional to the original X-ray energy E.

Detector active diameters range from 4 to 16 mm depending on the application. Larger diameter devices yield higher preamplifier noise (poorer energy resolution) owing to their higher capacitance. Therefore, the smaller diameters are more popular. Detector thicknesses range from 3 to 5 mm.

B. SYSTEM MODULES AND THEIR FUNCTIONS

In Fig. 2 the basic functions of the spectrometer are depicted. The detector diode and preamplifier front end are operated at liquid nitrogen temperature to restrict lithium mobility in the diode and minimize electronic noise. A light-tight, combination vacuum cryostat and liquid nitrogen dewar is used to provide a clean environment for the detector diode. The X-rays reach the detector through the thin beryllium window in the cryostat end cap. The detector bias voltage is supplied by a low current, low ripple, high voltage supply through smoothing filters in the preamplifier. The purpose of the preamplifier is to provide detector charge integration and cable driving capability, while adding as little noise to the signal as possible (Elad, 1971).

The amplifier serves to amplify the signal to a level that is measurable by the multichannel analyser. In addition, it contains filters which suppress both high and low frequency information to enhance the signal-to-noise ratio. The resulting bandpass of the amplifier is characterized by the classical RC time constant of the filter, τ (Fairstein and Hahn, 1965, 1966). Since the multichannel analyser measures pulse height with respect to ground potential, it is necessary to restore the baseline between amplifier output pulses to zero volts and remove the tendency of the baseline to shift downward with increasing counting rate. A well designed, asymmetric baseline restorer is able to stabilize the baseline up to a 30% duty cycle (Williams, 1968; Gedcke *et al.*, 1971; Gedcke, 1972). The duty cycle is the average fraction of the time that pulses are present above the baseline at the amplifier output. The height of the amplifier output pulse is proportional to the preamplifier output pulse height and hence is proportional to the X-ray energy.

The multichannel analyser (MCA) consists of four sections: the analog-to-digital converter (ADC), a memory, an electronic display,

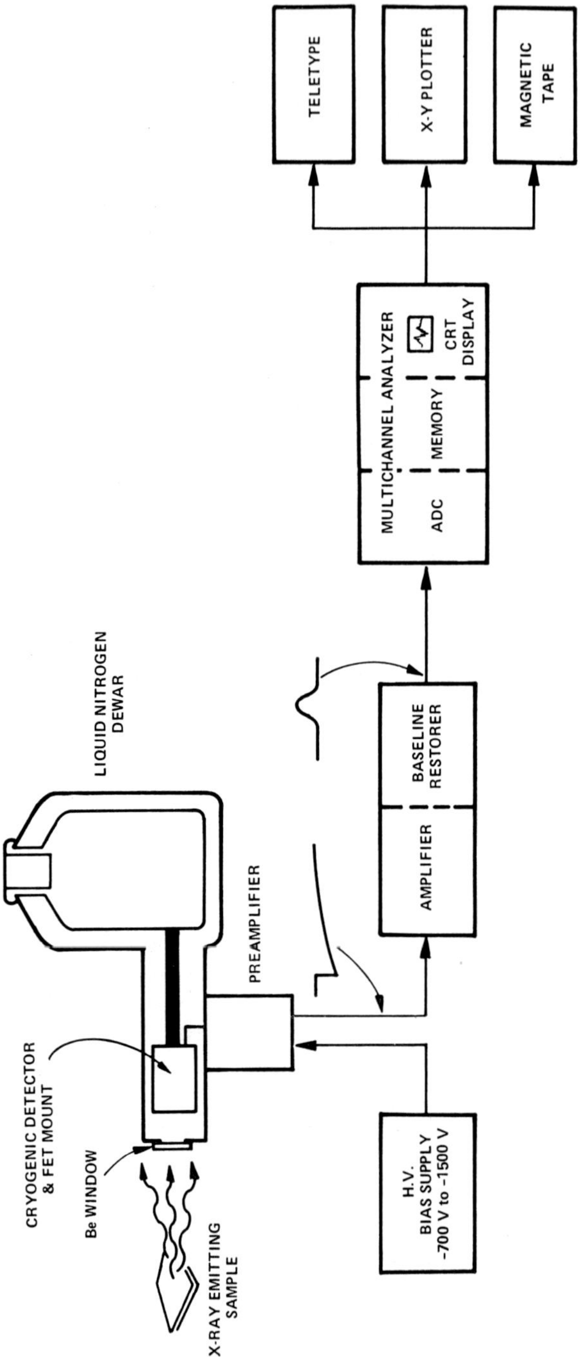

FIG. 2. Functional diagram of the spectrometer. (Gedcke *et al.*, 1972.)

and some form of hard-copy output device. In the ADC the amplifier output pulse height is measured and converted to a digital number. The precision of this measurement is governed by the number of channels in the ADC. In a 1024-channel ADC covering an energy range from 0 to 20·47 keV, each channel covers an energy range of 20 eV. The center of the *i*th channel represents the energy:

$$E_i = ki, \tag{2}$$

where k is the energy spanned by one channel. The number of times an X-ray has been measured and assigned to channel i is remembered in the *i*th channel of the memory. Functionally, the memory looks like a series of scalers (counters), each assigned to a specific channel and each having a total capacity of 10^6–1 counts. Thus, at the end of a data acquisition period, a histogram has been accumulated in the memory. The histogram is a record of the number of times X-rays were detected in each energy interval (channel). Consequently, the memory contains a spectrum of X-ray intensity as a function of energy. This spectrum is conveniently displayed on a cathode-ray tube as a series of dots, each corresponding to a channel. Usually the horizontal axis is calibrated to read as energy, and the vertical axis represents counts per channel, or intensity. Most instruments allow some sort of interaction with this display to facilitate quick interpretation of the spectrum. For hard copy this display may be photographed, or the data can be output to an X–Y plotter. If numerical analysis of the data is intended, a teletype listing of the memory content is employed. More sophisticated systems dump the data directly into a small computer or onto magnetic tape for computerized spectrum analysis.

Although not shown here, many other modular accessories are available, such as count-rate meters, energy identifiers (test pulsers), pile-up rejectors, and liquid nitrogen level monitors. Moreover, the standard NIM modular approach to electronics makes the energy spectrometer compatible with a large selection of electronic modules now popular for use on the wavelength spectrometers. A simple spectrometer can be expanded into a more versatile system at a later date simply by adding the appropriate modules.

C. THE CHARGE-SENSITIVE PREAMPLIFIER

Operation of the preamplifier is most readily explained with reference to the classical resistive feedback scheme illustrated in Fig. 3. Its characteristics are typical of the general class of "continuous feedback" preamplifiers. A negative bias voltage of about 1000 V is applied to the front contact of the detector diode through a filter which attenuates any HV supply noise. The back electrode of the detector is held rigorously at ground potential by the negative feedback supplied via the charge loop amplifier. As the charge produced in the detector by an X-ray is swept from the diode, it is forced to collect on the feedback

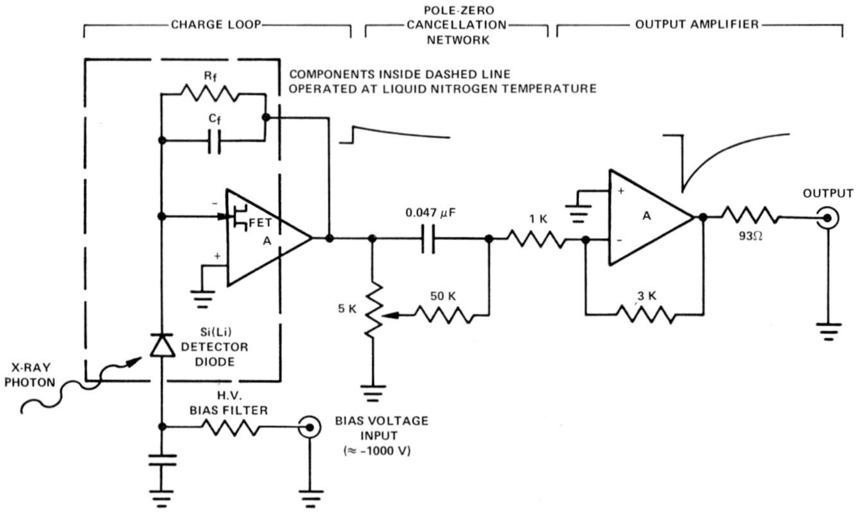

FIG. 3. The resistive feedback charge-sensitive preamplifier. (Gedcke, 1972.)

capacitor C_f because of the action of the charge loop amplifier. Consequently, a voltage is developed across C_f which appears as the leading edge step of a pulse at the charge loop output. The amplitude of this step is given by:

$$V_1 = \left[\frac{E}{\varepsilon C_f}\right] . 1{\cdot}6 . 10^{-19} * \tag{3}$$

If a means of removing the charge from C_f is not provided, then succeeding pulses will continue to charge the capacitor until the amplifier output reaches the power supply voltage and the amplifier ceases to operate. In the resistive feedback scheme the resistor R_f is used to remove the charge slowly after each pulse. As a result, the step pulse observed at the charge loop output decays exponentially as described by eqn (4):

$$V_2 = V_1 \exp(-t/R_f C_f), \tag{4}$$

for $t \geqslant 0$. This pulse shape is sketched in Fig. 3. The decay time constant $R_f C_f$ is typically in the range of 1–10 msec.

* $1{\cdot}6 \times 10^{-19}$ Coulombs = charge on the electron.

For every preamplifier there is a maximum energy rate that can be handled. The total energy rate from a detector sensing an X-ray spectrum is defined by:

$$(Er)_{\text{total}} = \int_{E=0}^{\infty} Er_i(E)\,\mathrm{d}E \tag{5}$$

where $r_i(E)\,\mathrm{d}E$ is the counting rate in the spectrum between the energies E and $E+\mathrm{d}E$. The reason for the limitation can best be described by considering a single X-ray energy E being detected at a count rate r_i. On the average, the charge produced by these X-rays in the detector must flow through resistor R_f. This average current in R_f is proportional to both the X-ray energy and the counting rate. Hence, an average voltage is developed at the charge loop output, which is proportional to the energy rate. If a sufficient increase is made in either the X-ray energy or the counting rate, the voltage at the charge loop output can be raised until the amplifier power supply voltage is reached. At this point pulses can no longer pass through the amplifier. In resistive feedback preamplifiers the maximum energy rate is determined by the value of R_f. Unfortunately, large values of R_f are required to minimize resistor noise contributions. However, energy rates in excess of 10^6 keV sec^{-1} are possible in both resistive feedback and "dynamic charge restoration" preamplifiers.

Pulsed feedback methods were developed as one means of removing noise contributions from the feedback resistor (Kandiah, 1968, 1969, 1971; Landis *et al.*, 1971). The more popular circuit is that developed by Landis *et al.* (1971). In this system the feedback resistor is removed and replaced with a pulsed, light-emitting diode, optically coupled to the field effect transistor (FET) first stage of the preamplifier. When the voltage on capacitor C_f builds up to an intolerable level, the light-emitting diode is pulsed on. The light shining on the FET generates a large leakage current across the gate to drain junction, discharging capacitor C_f. During this brief discharge period the amplifier and MCA must be gated off, since a large, negative polarity, overload pulse is generated by the discharge. The maximum energy rate of the pulsed feedback preamplifier is determined by the energy rate at which dead-time caused by resetting becomes intolerable, and ultimately by the energy rate beyond which it is impossible to completely discharge C_f. A shortcoming which is frequently encountered in the optical pulsing scheme is an energy-rate-dependent noise source (Landis *et al.*, 1971). This excess noise shows up at high energy rates and can cause the energy resolution observed on light elements to be a function of the counting rate of high energy lines. Long-lived surface states optically excited in both the Si(Li) detector diode and the FET can cause this excess noise. Continuous feedback schemes can also suffer a resolution degradation dependent on energy rate, due to noisy feedback currents (Elad, 1971, 1972; Goulding *et al.*, 1969), or excessive differential non-linearity in the charge loop. Resolution degradation due to the pre-amplifier can be confirmed by monitoring the resolution on a light element line (e.g., AlK_α), while the counting rate of a high energy line (20–40 keV) is varied. In a well-designed system the counting rate limits are normally dictated by pulse widths in the shaping amplifier.

Pole-zero cancellation is a rather fancy term derived from the mathematics of Laplace and Fourier transforms. Actually, the function of a pole-zero cancellation network is quite trivial. It is simply a scheme which replaces the long exponential decay from the charge loop with a shorter pure exponential decay. That is, the $\exp(-t/R_f C_f)$ decay is replaced by $\exp(-t/50\ \mu\text{sec})$. The net effect of this substitution is attenuation of the average charge loop offset voltage by a factor of $R_f C_f / 50\ \mu\text{sec}$. As a result, the signal can be amplified by a factor of $R_f C_f / 50\ \mu\text{sec}$ before the energy rate is again limited by supply voltages. Judicious amplifier design ensures that the energy rate limitation is determined only by the preamplifier charge loop. The output amplifier with a gain of 3 serves mainly as a coaxial cable driver.

To summarize, a well-designed preamplifier should have the following characteristics:

1. Low noise contribution.
2. A maximum energy rate compatible with the anticipated energy and count rate ranges.
3. The noise contribution must be independent of energy rate.
4. Direct current signal coupling with good gain stability as a function of temperature and energy rate.
5. Good integral linearity (pulse height proportional to energy).
6. A fast pulse rise time $\leqslant$ 100 nsec (for good performance with pile-up rejectors).

D. THE SHAPING AMPLIFIER

Figure 4 is a simplified block diagram of the shaping amplifier. It is composed of several stages of adjustable gain interconnected by high pass and low pass filters. The high pass filter is a simple CR differentiator. It serves to attenuate low frequency information where the signal-to-noise ratio is unfavorable. Usually this filter includes a pole-zero cancellation network so that the pulse produced by the filter has a pure

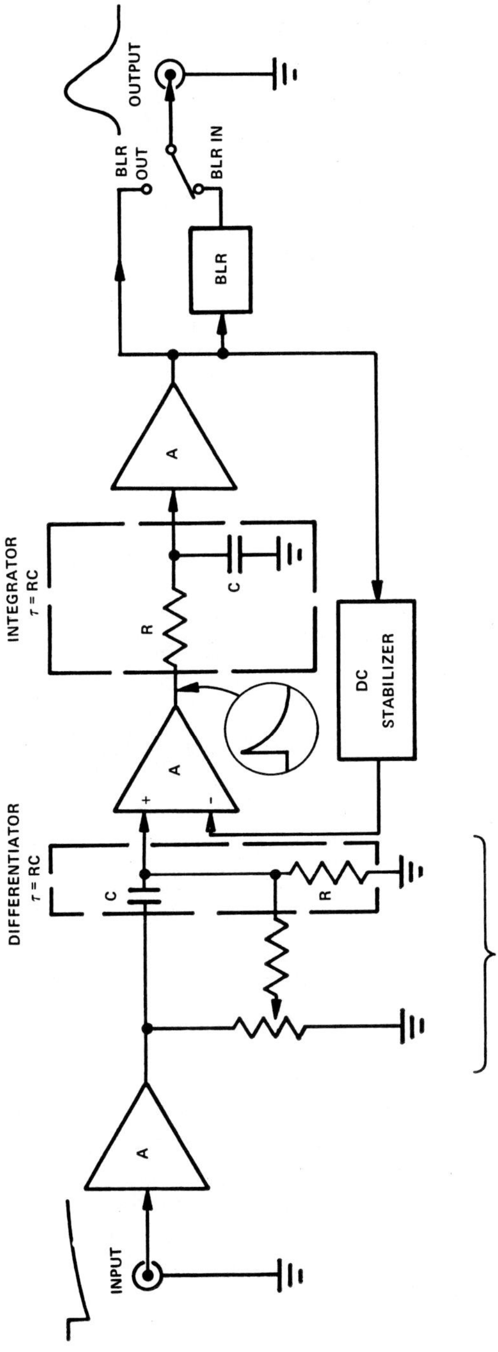

FIG. 4. A simplified diagram of the shaping amplifier and baseline restorer. (Gedcke, 1972.)

exponential decay as depicted in the circle in Fig. 4. That is, the exponential decay from the preamplifier, exp $(-t/50\ \mu\text{sec})$, is replaced with exp $(-t/\tau)$, where $\tau = RC$ is the shaping time constant selected in the amplifier. The differentiator is placed near the amplifier input so that more amplification can be added without further restricting the maximum energy rate the system can handle.

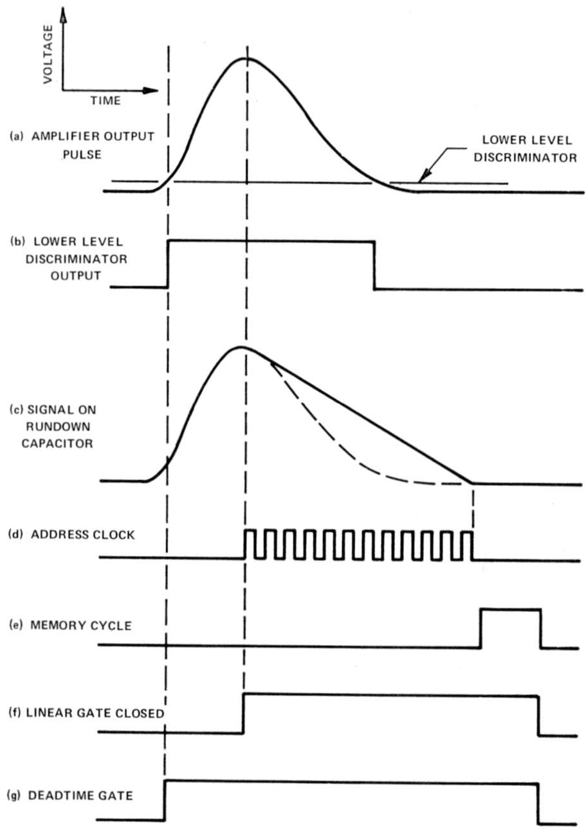

FIG. 5. Signals in the MCA during the pulse measurement process. (Gedcke, 1972.)

Although complex active filters are normally used in the low pass filter for optimum pulse shaping, the circuit has been depicted in Fig. 4 as a passive RC integrator for simplicity. This filter attenuates high frequency information where the noise contributions from the preamplifier are high. Due to the integrator the amplifier output pulse has a slow rise time, as illustrated in Figs 4 and 5. The $\tau = RC$ time constant of the integrator is normally set equal to the differentiator time constant.

Although pulse lengths are a function of the filters employed, the shapes produced by the ORTEC 716A amplifier are typical of modern, active-filter, semi-Gaussian shaping amplifiers. The total pulse width at the baseline is approximately $7{\cdot}7\tau$, while the time from baseline to peak amplitude is about $2{\cdot}2\tau$. The effect of the choice of time constant on resolution and counting rate is discussed in sections III and V.

To ensure that the pulses return quickly to baseline with no long undershoots, it is necessary to d.c. couple the entire spectrometer from detector to MCA. Long undershoots causd by a.c. coupling or inaccurate pole-zero cancellation lead to resolution degradation at medium and high counting rates. However, d.c. coupling, together with high gain in the amplifier, leads to a temperature instability of the output baseline due to input offset drifts in the first amplifier stages. Usually a d.c. stabilizer loop is added to eliminate this drift. The stabilizer inspects the amplifier output baseline and corrects for any long-term drifts. In actual fact, it is an active form of the symmetrical Robinson diode baseline restorer (Williams, 1968). Since a symmetrical restorer can hold the baseline for random pulses only up to a duty cycle of about 15%, an asymmetric baseline restorer (BLR) is added to the amplifier output (Williams, 1968). With this addition duty cycles up to 30% can be handled.* In some designs the asymmetric BLR is included in the d.c. stabilizer. It is necessary to switch off the BLR only for the purpose of adjusting the pole-zero cancellations by reference to an oscilloscope.

Coarse and fine gain controls are provided along with a baseline offset adjustment to provide full-scale energy calibration and energy zero intercept adjustment, respectively, for the spectrum viewed on the MCA. Generally, the following features are important in the amplifier (Fairstein and Hahn, 1965, 1966; Walter, 1971; Heath, 1972):

1. Sophisticated pulse shaping for optimum energy resolution and high counting rate capability.
2. Good linearity of pulse height versus energy (low integral non-linearity).
3. Good gain and baseline stability as a function of temperature and counting rate.
4. Quick, clean, overload recovery from high energy pulses beyond the range of interest.

E. THE MULTICHANNEL ANALYSER

Signals and functions occurring during the ADC pulse height measurement process are outlined in Figs 5 and 6, respectively. The process

* See note added in proof.

starts with the ADC linear gate open to receive a pulse (Figs 6(a) and 5(f)). In order for the amplifier pulse to be recognized, it must exceed the lower level discriminator threshold (Fig. 5(a)). The lower level discriminator is used to prevent noise pulses from jamming the ADC. It is usually set just above the amplifier output noise level. Once the pulse is accepted, input circuits on the ADC charge the rundown capacitor so that the signal produced on the capacitor is identical to the amplifier pulse during its rise to peak amplitude (Figs 5(c) and 6(a)).

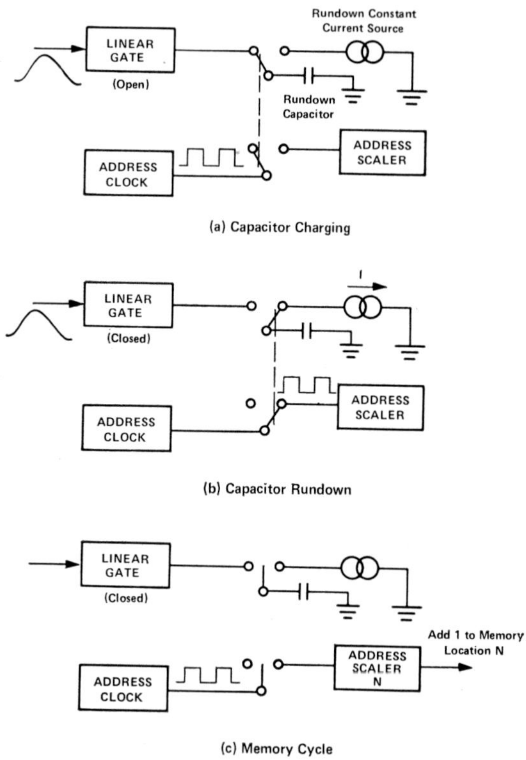

FIG. 6. Circuit functions in the ADC during pulse height measurement. (Gedcke, 1972.)

When the pulse reaches peak amplitude, a peak detection circuit senses the change in slope and simultaneously triggers several functions. The rundown capacitor is disconnected from the input, and the linear gate closes to prevent other pulses from interfering with the measurement (Figs 6(b) and 5(f)). An address clock of frequency f (typically 50 MHz) is connected to the address scaler (Figs 5(d) and 6(b)), and a constant current source begins to discharge the rundown capacitor linearly

(Fig. 5(c)). Since the discharge is linear, the time to discharge the capacitor to zero volts is proportional to the amplifier pulse height. A discriminator senses when the capacitor reaches zero volts and disconnects the address clock from the scaler (Figs 5(c), 5(d) and 6(c)). Consequently, the address scaler is left with the number of clock pulses counted, N. The number N is proportional to the amplifier pulse height and is used as a digital representation of the X-ray photon energy. During the memory cycle (Fig. 5(e)) channel N in the MCA is identified and the number of counts, m, already stored in that location is read. One count is added and the number $m+1$ is written back into memory location N. Subsequently, the linear gate waits until no signals are present at the ADC input above the lower level discriminator and then opens to accept the next pulse arriving for analysis. This process is repeated on a pulse-by-pulse basis to build up the energy spectrum as a histogram in the MCA memory.

Multichannel analysers usually contain a livetime clock to record the counting time with compensation for deadtime losses. In the simplest scheme the deadtime correction involves stopping the electronic elapsed time clock during the deadtime gate signal (line (g) of Fig. 5; Deighton, 1961). The real counting time is extended by the total system deadtime to compensate for losses. At high counting rates it becomes apparent that this simple approach does not account for losses due to pulse pile-up (Gedcke, 1972; Gedcke *et al.*, 1971; Landis *et al.*, 1972; Reed, 1972). More elaborate logic can be employed to include pulse pile-up losses in the correction scheme. Some X-ray spectrometers now include this refinement. Since the deadtime introduced by the rundown time of the ADC is inversely proportional to address clock frequency, an ADC with a clock rate $\geqq$ 50 MHz is desirable.

The following characteristics are considered to be important requirements for an MCA in the Si(Li) energy spectrometer (Williams, 1971):

1. Approximately 1024 channels in the ADC and memory with a count capacity of 10^6–1 per channel of memory.
2. Good integral linearity (channel number proportional to energy).
3. Good differential linearity (uniformity of channel widths).
4. Low deadtime.
5. DC-coupled input able to accept Gaussian pulse shapes with rise times of the order of 20 μsec.
6. An accurate and stable livetime clock which also compensates for pulse pile-up losses due to amplifier pulse shapes.
7. Good stability of the conversion gain and zero intercept with respect to temperature and count rate.

8. Provisions for easy interfacing with the SEM for distribution maps and line scans.
9. Convenient data interpretation and readout capabilities.

III. Energy Resolution

The most critical performance parameter for the Si(Li) energy spectrometer is its energy resolution. The energy resolution Γ is defined as the full width of the energy peak at one half its maximum intensity (FWHM). The factors contributing to the resolution are defined in eqn (6):

$$\Gamma = \sqrt{(\Gamma_{noise})^2 + (2{\cdot}35\sqrt{\varepsilon F E})^2} \qquad (6)$$

The second term, $2{\cdot}35\sqrt{\varepsilon FE}$, is the intrinsic Si(Li) diode contribution due to ionization statistics (not a function of amplifier shaping time constant). The Fano factor F accounts for the fact that part of the ionization process involves "branching". That is, all the events causing electron-hole pairs are not statistically independent. The Fano factor lies in the range from 0·115 to 0·135 for presently available detectors. Resolution degradation due to poor charge collection sometimes gets lumped into an inflated Fano factor, since F is usually calculated from eqn (6) when Γ and Γ_{noise} are measured. A more correct treatment would group these extra resolution contributions under a third term in eqn (6). The FWHM preamplifier noise contribution is Γ_{noise}. This component is independent of energy but is a function of amplifier shaping time constant, as shown in Fig. 7. Clearly there is an optimum time constant for minimum noise. Both shorter and longer shaping time constants yield poorer resolution. For present small-diameter detectors ($\approx$ 4 mm) the optimum time constant is around 14 μsec. Preamplifier noise is a function of detector capacitance. Larger area detectors have higher capacitance and therefore higher noise. The effect of more detector capacitance is to move the left side of the curve in Fig. 7 upwards and shift the minimum to longer time constants (Elad, 1971). For SEM and microprobe work, small area detectors (12 mm^2) with Γ_{noise} between 80 and 120 eV are generally preferred, since counting rates are usually high and light element resolution is important. Equation (6) is plotted in Fig. 8 for various values of Γ_{noise}. System resolution is normally specified on the 5·894 keV MnK_α line denoted by the dashed line in Fig. 8.

The energy difference between K-lines from adjacent elements is smallest in the light element range. Consequently, good resolution at low energies is critical. Figure 9 shows the somewhat optimistic,

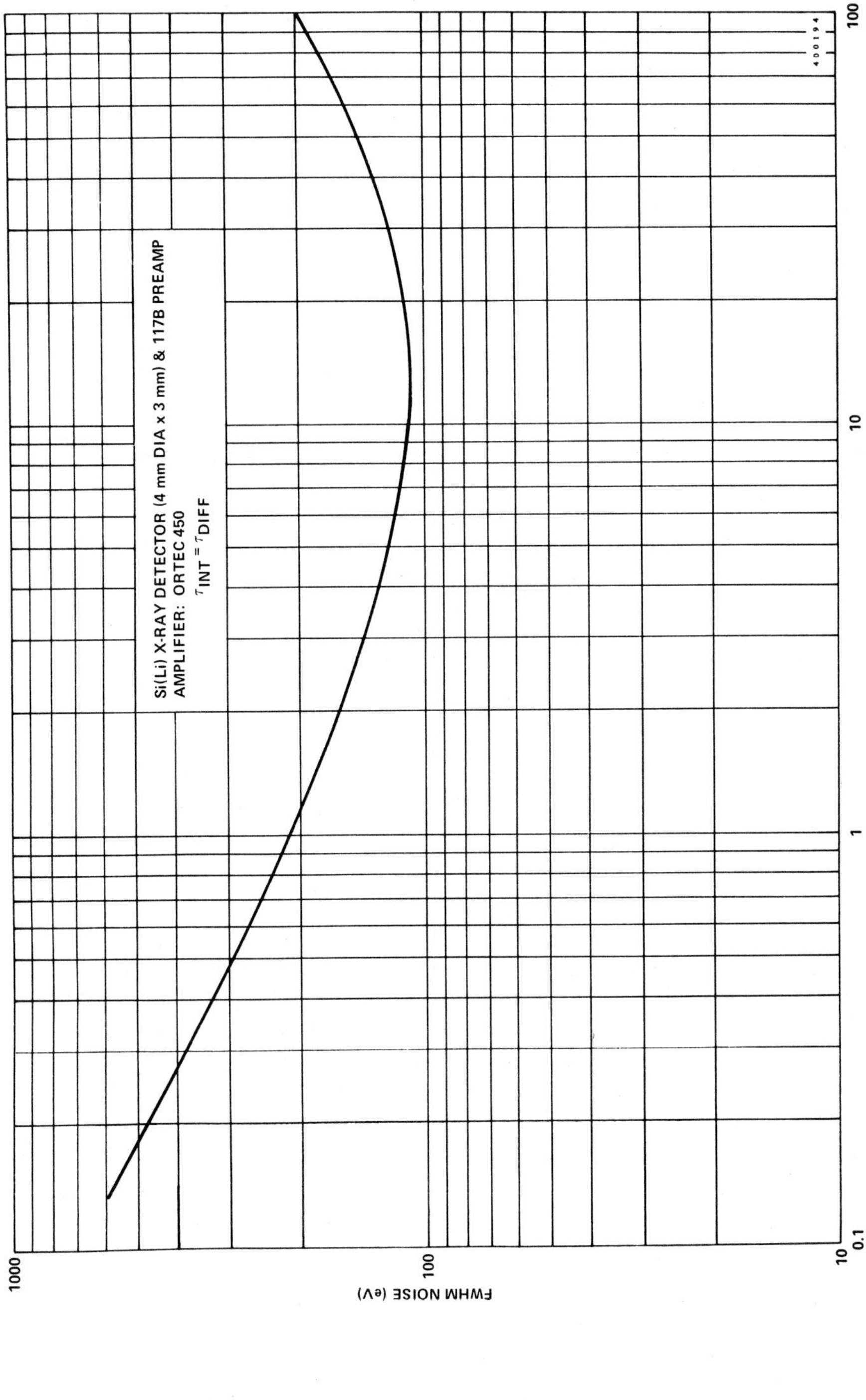

FIG. 7. Preamplifier noise, Γ_{noise}, as a function of amplifier time constant for a small detector. (Gedcke, 1972.)

theoretical separability ratios (Elad and Gedcke, 1972) of element Z and Z + 1 for three different system resolutions at 5·9 keV. A separability ratio of 2·8 for sodium with a 170 eV system implies that it will be possible to separate the NaK_α peak from the MgK_α peak only for cases where the intensity of the magnesium peak is less than 2·8 times the intensity of the sodium peak.

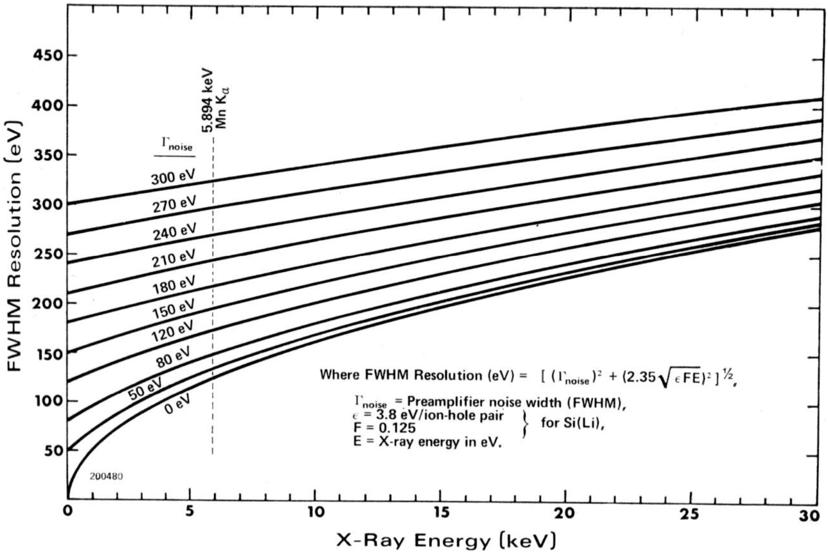

FIG. 8. Energy resolution as a function of energy and preamplifier noise. (ORTEC, 1972.)

IV. DETECTION EFFICIENCY

The theoretical full-energy detection efficiency for the Si(Li) detector diode is given in Fig. 10. At high energies the detector becomes more transparent to the X-rays, and the thicker detector provides slightly higher efficiency. At low energies, detection efficiency is dominated by the attenuation caused by the beryllium window. The calculations plotted in Fig. 10 do not include the minor attenuation due to the gold layer and silicon deadlayer on the detector front surface, nor do they include the effects of silicon escape peaks. On small detectors where the window diameter can be small enough for the beryllium to support one atmosphere of pressure differential, the thinner 0·5 mil* and 0·3 mil windows are frequently used (Elad and Gedcke, 1972). High energy electrons back-scattered from the sample can penetrate these thinner

* 1 mil (milli-inch) = 10^{-3} in.

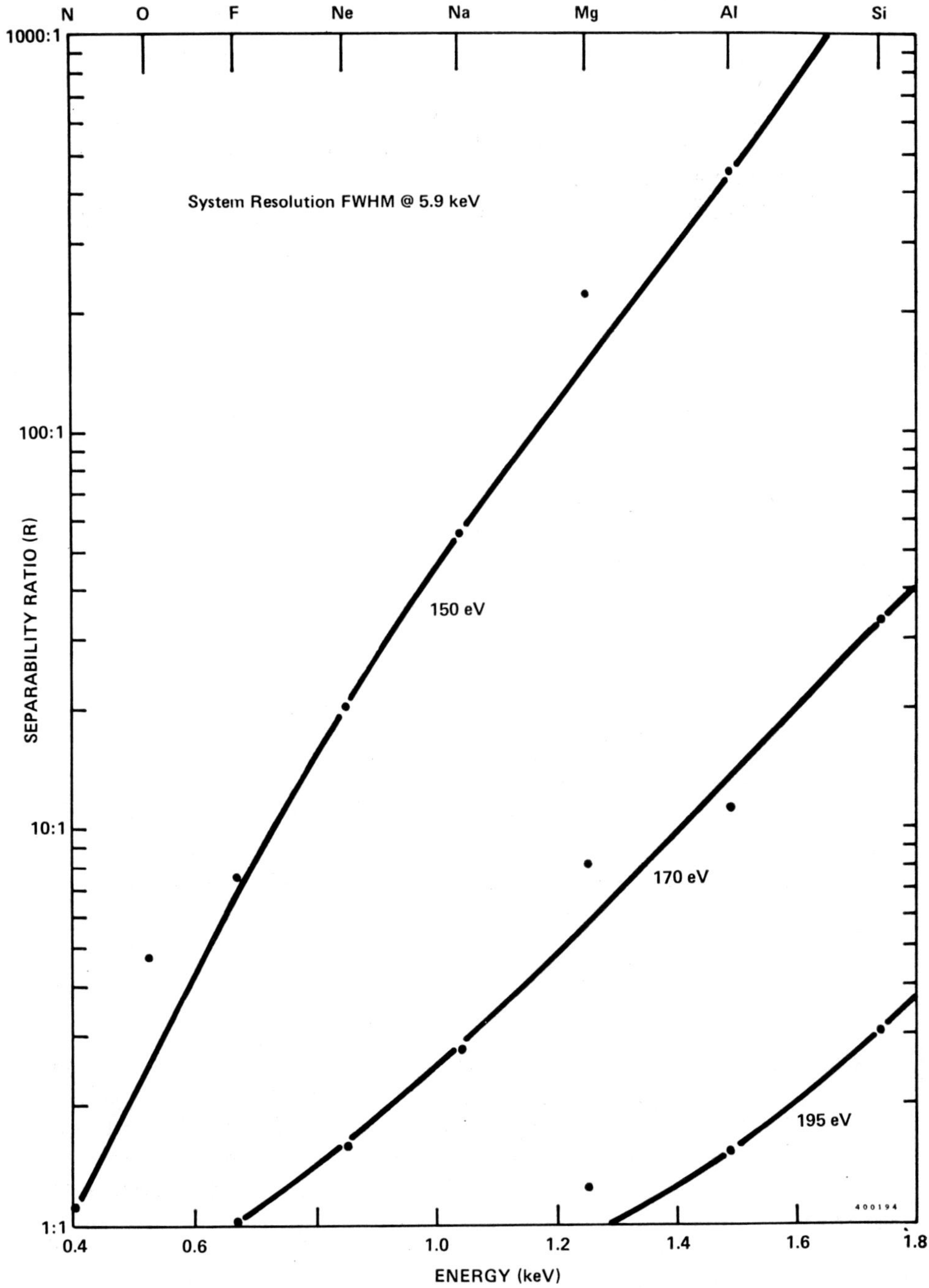

FIG. 9. Theoretical separability ratios for light elements (ratio of peak intensity of next heavier element to element shown that can be separated from it). (Elad and Gedcke, 1972.)

windows and cause noise in the spectrometer. It is sometimes necessary to restrict the SEM accelerating voltage to 30 kV with 0·3 mil windows to prevent window penetration by the electrons. Although it is possible to detect elements down to carbon with a windowless system (Jaklevic and Goulding 1971), resolution demands, and *L*- and *M*-line interferences dictate, the use of a crystal wavelength spectrometer for elements below sodium (Fitzgerald and Gantzel, 1971; Sutfin and Ogilvie, 1971).

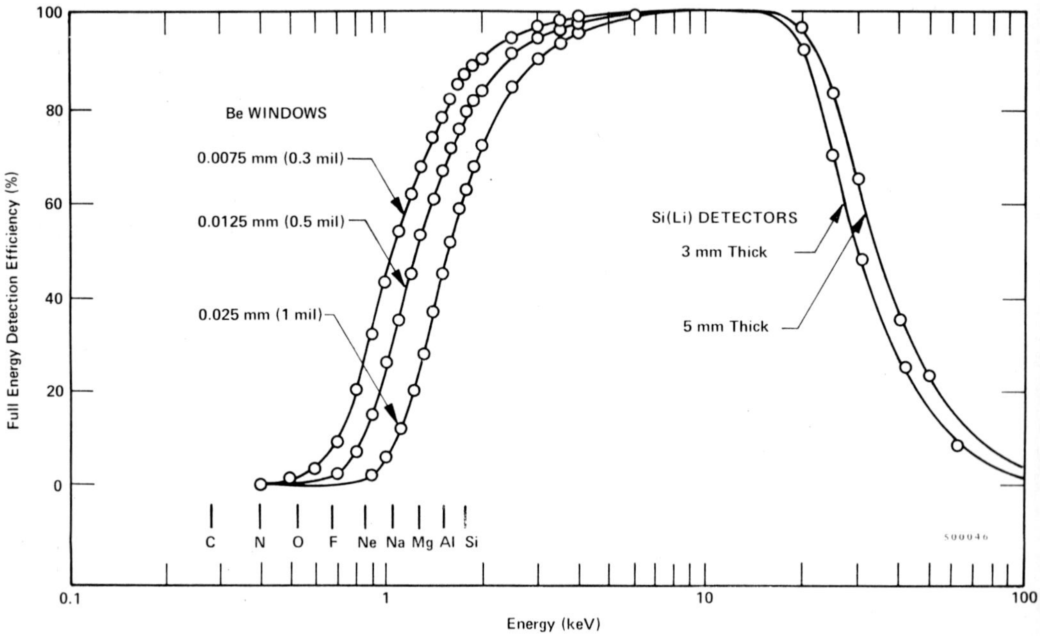

FIG. 10. Theoretical full-energy detection efficiency of the Si(Li) detector as a function of energy for various Be window and detector thicknesses. (ORTEC, 1972.)

V. High Counting Rates

A. Resolution and Baseline Stability*

Choosing the shaping amplifier time constant for optimum resolution (Fig 7) sets certain limits on the count rate capability of the spectrometer. The long shaping time constants required for good resolution are accompanied by large amplifier pulse widths. Since the maximum counting rate is inversely proportional to pulse width, the time constants providing best resolution give the poorest count rate performance. Figure 11 shows the resolution on the MnK_α line as a function of

*See note added in proof.

counting rate for three amplifier time constants (Gedcke *et al.*, 1971). In all cases the resolution is independent of counting rate up to the 20 % deadtime loss point (the point at which losses in the amplifier due to pulse pile-up are 20 %). Beyond this point the resolution degrades rapidly. The rapid degeneration is due to the baseline restorer losing control as the available baseline between pulses vanishes. This cause is

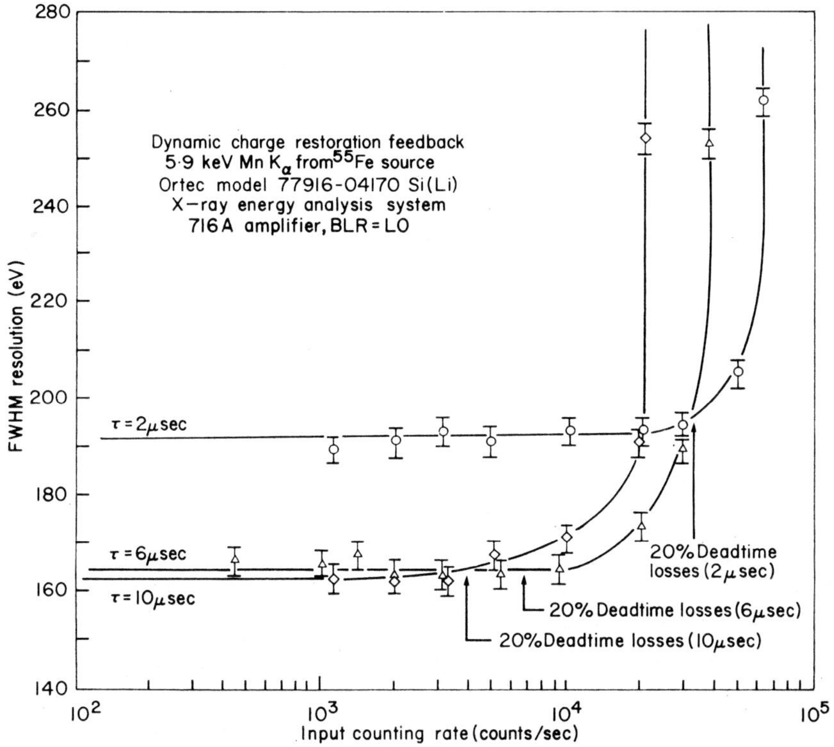

FIG. 11. Energy resolution at 5·9 keV as a function of counting rate for shaping amplifier time constants of 2, 6, and 10 μsec. (Gedcke *et al.*, 1971.)

confirmed in Fig. 12 by the downward shift of the spectrum at the same counting rate. Although Fig. 12 lists the effect as a shift in the peak centroid, it is actually the baseline that is shifting. That is, the entire energy spectrum is observed to shift downward by an amount ΔE. In practice it is wise to avoid counting rates where the baseline stability becomes questionable. This requires amplifier duty cycles definitely less than 30 % and preferably less than 20 %.* Figure 11 clearly shows that long time constants which provide the best resolution have the poorest counting rate performance.

* See note adeed in proof.

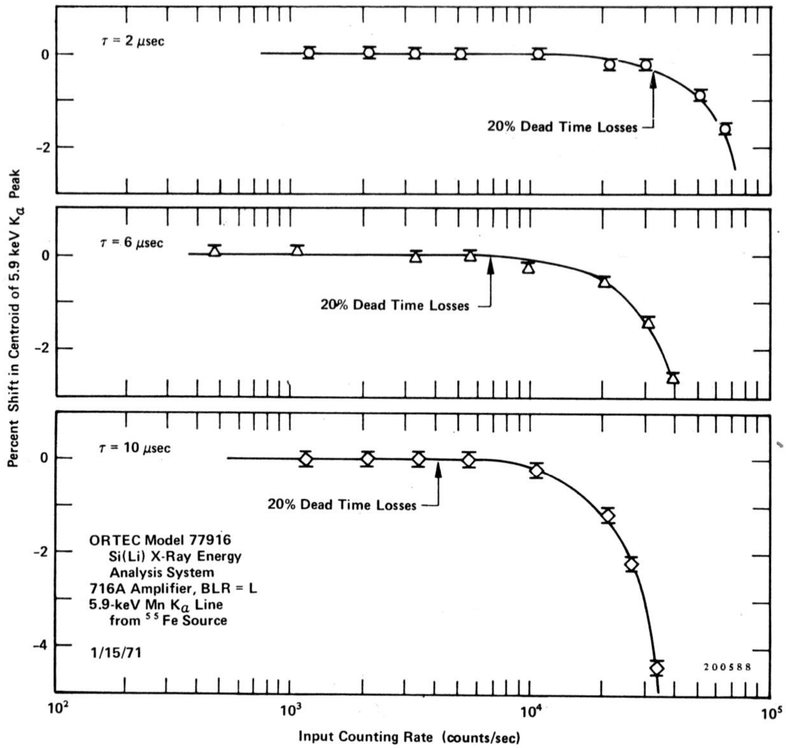

FIG. 12. Amplifier output baseline stability for $\tau = 2$, 6, and 10 μsec. (Gedcke *et al.*, 1971.)

B. SUM SPECTRA AND PILE-UP REJECTION

At amplifier duty cycles above 5%, pulse pile-up becomes important (Gedcke, 1972; Reed, 1972). Pile-up occurs at the amplifier output when a second pulse arrives during the previous pulse. Both pulses are distorted in shape as they add together. If the second pulse arrives after the peak of the first pulse, the ADC will make an undistorted measurement of the first pulse height and reject the second pulse. If the second pulse arrives between the start and the peak of the first pulse, a composite pulse height is measured. This erroneous pulse height is larger than either pulse individually and shows up as a sum spectrum, as illustrated in Fig. 13. The lower pulse heights in the sum spectrum are due to partial pulse overlaps. The full energy sum peaks arise from almost exact coincidence in pulse arrival times. The fraction of the pulses which appear in the sum spectrum is given by:

$$\frac{R_s}{r_o} = \exp[r_i T_p] - 1 \tag{7}$$

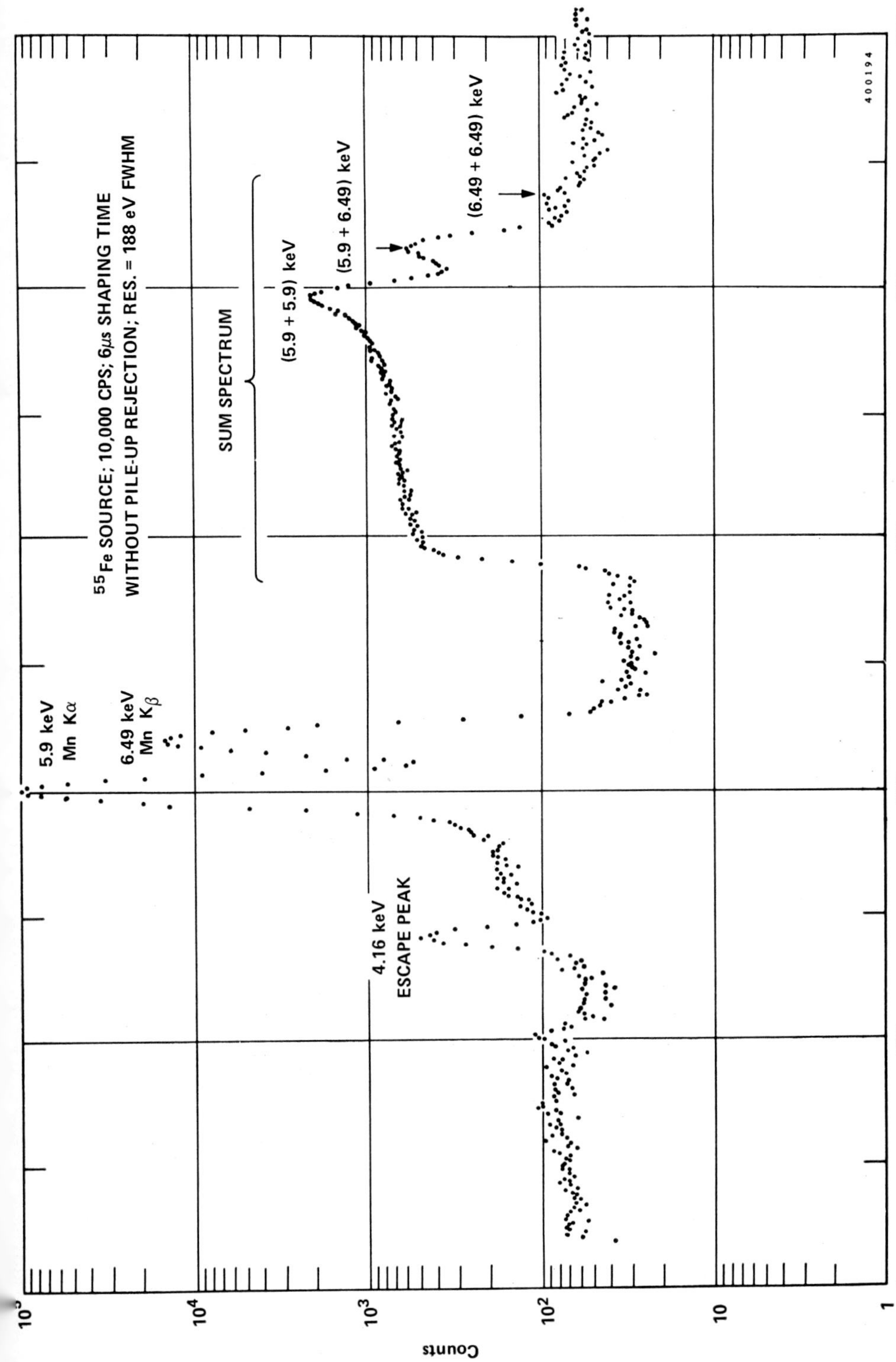

FIG. 13. A spectrum produced by an ^{55}Fe source without pile-up rejection, showing the sum spectrum (Gedcke, 1972.)

for pulses with a time from start to peak, T_p, and an input counting rate from the detector, r_i. Equation (7) shows that both long shaping constants (large T_p) and high counting rates produce intense pile-up.

One method of attenuating the pile-up spectrum is to use a pile-up rejector, as illustrated in Fig. 14. In parallel with the normal energy analysis channel, a fast amplifier and inspector channel is added. A very short shaping time constant (typically 0·1 μsec) is used in the fast amplifier, compromising energy resolution and low energy threshold in favor of short pulses. At the output of the fast amplifier it is possible to recognize two X-rays arriving as close together as 0·5 μsec. This pulse-pair resolution is short compared to the pile-up interval T_p of about 13 μsec in the slow shaping amplifier with a 6 μsec time constant.

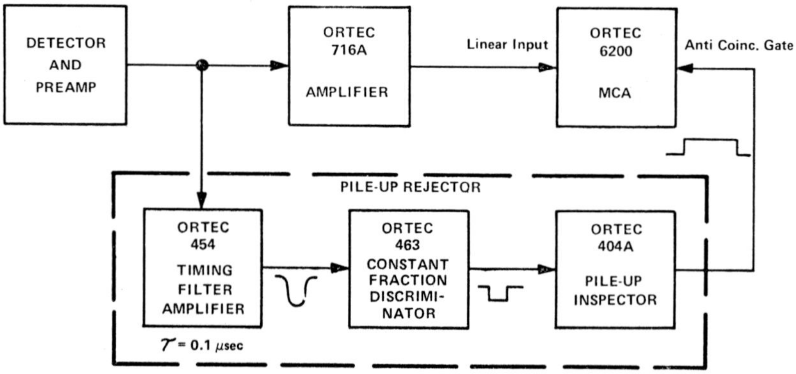

FIG. 14. A modular pile-up rejector system. (Gedcke, 1972.)

Therefore, it is possible to recognize most of the cases where two or more X-rays have arrived within the interval T_p in the slow amplifier and prevent their contribution to the spectrum. The discriminator in Fig. 14 serves as a recognition threshold, producing a short standard logic pulse for each fast amplifier pulse arriving above the noise level. The threshold is set just above noise, typically 1·4 keV for a 165 eV detector. Each pulse arriving at the inspector input causes an inspection interval of length T_{INS} to be set. This interval is chosen to be longer than the time to peak, T_p, in the slow amplifier. If a second pulse arrives during T_{INS}, a pile-up is identified, the inspect interval is extended by an amount T_{INS} from the time of pile-up, and an output inhibit pulse of length $T_{INH} = T_{INS}$ is generated. The inhibit pulse is fed to the MCA where it prevents the analysis of both X-ray pulses from the slow amplifier if the second pulse arrived before the peak of the first. Figure 15 shows the suppression of the pile-up spectrum achieved with such a system. The spectrum due to partial pulse overlaps is almost completely

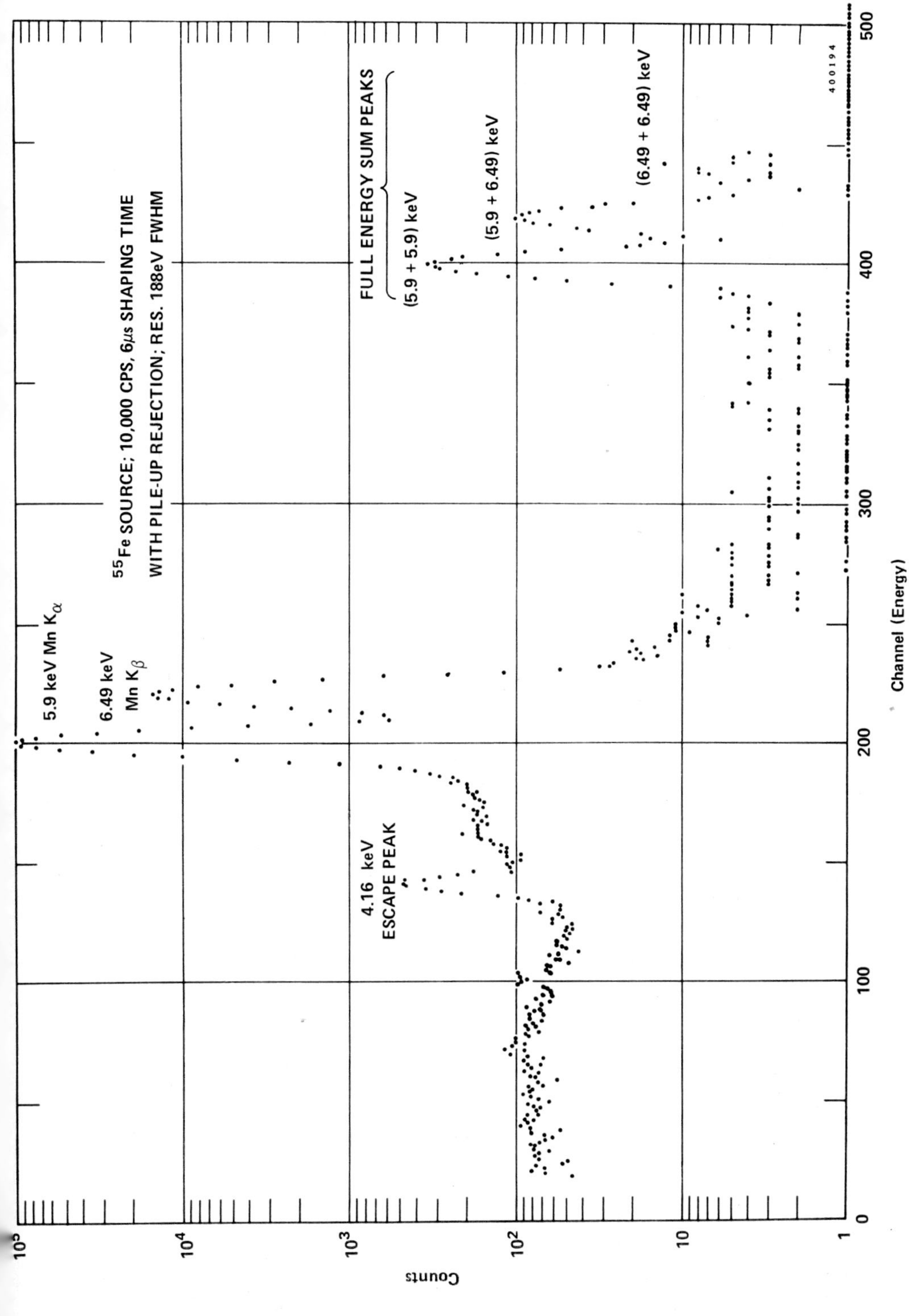

FIG. 15. A spectrum taken with pile-up rejection under the same conditions as in Fig. 13. (Gedcke, 1972.)

eliminated, while the full-energy sum peaks are attenuated by a factor of 6. Since the full-energy sum peaks result from almost exact pulse overlap (within $\approx 3\ \mu$sec), they are the most difficult to reject with the 0·5 μsec pulse-pair resolving time. In general, it is important to be aware of the full-energy sum peaks to avoid interpreting them as valid peaks in the spectrum. Although not shown in the examples here, second and third order pile-up can be significant at very high counting rates, where the probability of three and four pulses occurring at the same time becomes appreciable. With pile-up rejection, the fraction of the pulses which appear in the sum spectrum is:

$$\frac{r_s}{r_o} = \exp\left[2r_i T_F\right] - 1, \tag{8}$$

where T_F is the pulse pair resolution time of the fast amplifier. The improvement due to the rejector is given by the ratio of eqn (7) to eqn (8) and is approximately proportional to $T_P/2T_F$. Clearly, shorter fast amplifier shaping time constants would be desirable; however, Fig. 7 indicates that energy thresholds higher than 1·4 keV would result in proportion to the increased noise contribution. Reed (1972) provides an excellent discussion of this energy threshold—pulse-pair resolution compromise. The sum spectrum features due to pile-up will be most noticeable in a spectrum which contains one dominant peak intensity. In spectra with many lines of near-equal intensity, the multitude of combinations of energy which can lead to pile-up tend to wash out the sum spectra into a general background.

C. DEADTIME LOSSES

In systems counting random events obeying the interval distribution, two extreme types of deadtime can be identified: (1) extendable (paralysable) and (2) nonextendable (nonparalysable). In the extendable deadtime system each event arriving at the input initiates a new deadtime interval T, whether or not the event is lost because the system is already dead. In the nonextendable system only events arriving when the system is not already dead cause a deadtime T. Most counting systems are a combination of these two types Beaman *et al.* (1972).

Deadtime caused by the MCA is a good example of the nonextendable type. If we consider an ideal case where the amplifier pulse widths are infinitesimal, then the rundown time in the ADC and the memory cycle time of the MCA become the only system deadtimes. This deadtime is nonextendable and is described by:

$$T_{\mathrm{MCA}} = \frac{N}{f} + T_{\mathrm{MC}}, \tag{9}$$

where T_{MCA} is the total deadtime, N is the channel number into which the pulse is measured by the ADC, f is the ADC address clock frequency, and T_{MC} is the memory cycle time. In general, for the nonextendable system the relationship between input rate r_i and output rate r_0 (counted events) is (Evans, 1955):

$$r_0 = \frac{r_i}{1+r_iT} \tag{10}$$

This relationship is plotted in a dimensionless form in Fig. 16. Since only counted events cause deadtime, the fractional deadtime losses turn out to be:

$$\frac{r_i - r_0}{r_i} = r_0T \tag{11}$$

which can be read directly from the graph. In other words, if the system is allowed to count for a clock time t_c, then the number of counts actually recorded is due to the total time the system is live and able to count. This livetime t_L is given by:

$$\begin{aligned} t_L &= (1-r_0T)t_c \\ &= \frac{r_0}{r_i}t_c \\ &= \frac{1}{1+r_iT}t_c \end{aligned} \tag{12}$$

In the MCA livetime clock system, compensation for losses is achieved as follows. If we wish to record a counting time t_0 and assume no deadtime losses, then the MCA must count for a livetime:

$$t_L = t_0. \tag{13}$$

The elapsed time as measured by a realtime clock will be extended by the losses to a value:

$$\begin{aligned} t_c &= \frac{t_L}{(1-r_0T)} \\ &= t_0(1+r_iT) \end{aligned} \tag{14}$$

Using eqn (10) and (14), the total recorded counts M logged in memory against the time t_0 will be:

$$\begin{aligned} M &= r_0t_c \\ &= \frac{r_i}{1+r_iT} \cdot t_c \\ &= \frac{r_i}{1+r_iT} \cdot (1+r_iT)t_0 \\ &= r_it_0 \end{aligned} \tag{15}$$

From this it is seen that the total counts stored in memory M, divided by the selected livetime t_0, does indeed produce the true input rate r_i. In nonextendable deadtime systems, output rate increases monotonically with increasing input rate, and the only count rate limit due to deadtime is determined by the accuracy of the deadtime loss compensation scheme employed.

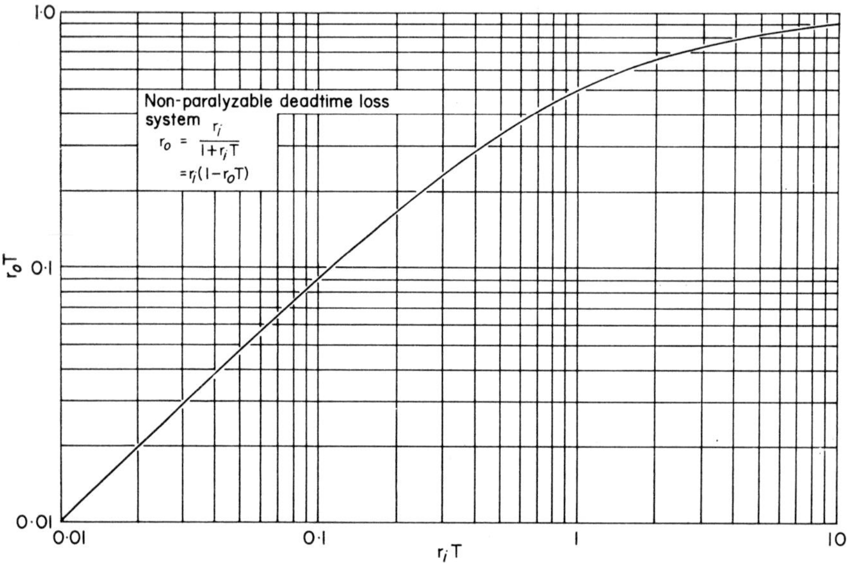

FIG. 16. Relative output rate r_0T as a function of relative input rate r_iT for a general nonextendable deadtime loss system. (Gedcke, 1972.)

With the long shaping time constants encountered in the Si(Li) energy spectrometer, amplifier pulse widths are an important source of system deadtime. In fact, pulse widths can be longer than the MCA deadtime. The duration of the amplifier output pulse above the lower level discriminator of the MCA is a clear case of extendable deadtime. A second pulse arriving during the first pulse extends the signal duration above the discriminator threshold by one pulse width. Since only one event is counted each time the discriminator threshold is exceeded, the second pulse is not counted while it contributes deadtime. The relationship between input and output counting rate for a simple extendable deadtime counting system without pulse height analysis is:

$$r_0 = r_i \exp(-r_iT), \tag{16}$$

where T is the deadtime caused by each input event. In contrast to the nonextendable system, there is an input rate which produces a maximum output rate. Increasing the input rate up to $r_i = 1/T$ increases the output rate. Beyond $r_i = 1/T$ the output rate decreases with increasing r_i. Therefore, the system is paralysable. If the lower level discriminator signal is allowed to gate off the MCA livetime clock, the MCA live timer will also correct for the deadtime losses implied by eqn (16) (Reed, 1972). However, this correction only guarantees that the total counts in the MCA memory divided by the livetime will be equal to the total count rate in the input spectrum.

In a real spectrum where pulse height analysis is important, it is highly impractical to recover the counts diverted into the sum spectrum by pile-up. Normally, the analyst only has access to the pulses that fell at their correct energies, undistorted by pile-up. To develop the formula for deadtime losses including pulse pile-up losses, it is convenient to assume an MCA with infinitesimal deadtime ($T_{MCA} = 0$). For a pulse to be recorded without pile-up distortion, its peak amplitude must be preceded by an interval containing no pulses for a time $T+T_p$, where T is the total pulse length. From the interval distribution, the number of pulses satisfying this requirement is:

$$r_0 = r_i \exp[-r_i(T+T_p)]. \tag{17}$$

Equation (17) assumes all pulses have the same shape, which is the same as assuming the discriminator threshold is set at zero energy. Consequently, the expression is only accurate for pulses whose heights are much greater than the lower level discriminator setting. The values for T and T_p are a function of amplifier design. Approximate values for the ORTEC 716A amplifier are:

$$T_p = 2{\cdot}2\tau; \tag{18}$$

$$T = 3{\cdot}5T_p = 7{\cdot}7\tau, \tag{19}$$

where τ is the amplifier shaping time constant. Equation (17) is plotted in Fig. 17 for the ORTEC 716A amplifier. The maximum output rate is achieved when $r_i(T+T_p) = 1$; however, this is far above the 30% duty cycle limit set at about $r_iT = 0{\cdot}3$ by the baseline restorer.*

Special circuits have been developed to compensate for the pile-up losses (Landis *et al.*, 1972), and some energy spectrometers include pile-up loss corrections. Figure 18 shows the results obtained with a unique circuit contained totally within the MCA on the ORTEC 6200 and 6210 analysers. The percent deadtime, uncorrected for pile-up losses, is given in Fig. 18(a) as a function of input counting rate. The amplifier time constant is 6 μsec. In Fig. 18(b) the accuracy of the pile-up compensated live timer was tested with two sources. A ^{109}Cd radioactive source was fixed in a position to produce a constant counting rate of about 300 counts sec^{-1} during the test. The total counts measured in

* See note added in proof.

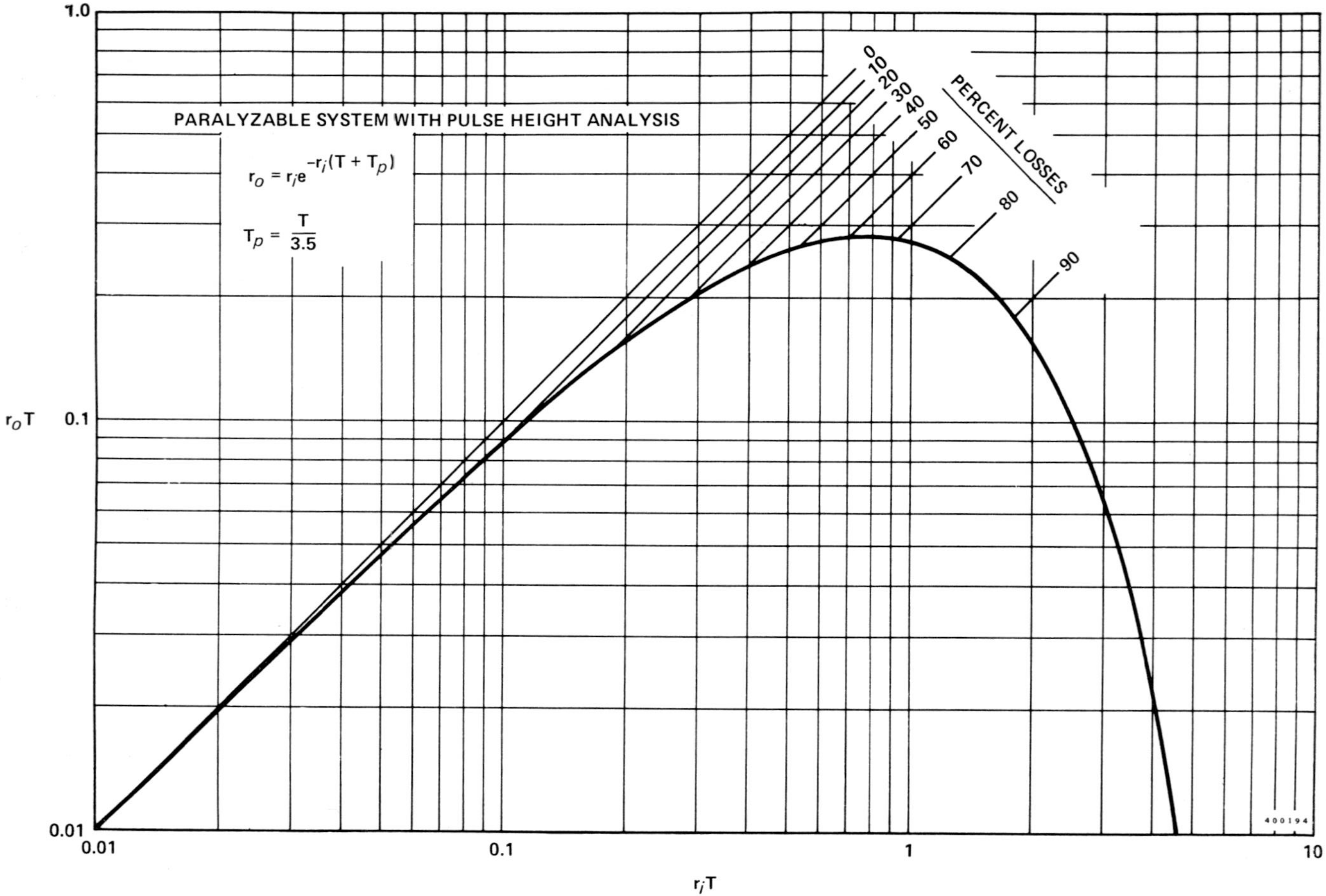

FIG. 17. Relative output rate r_oT *vs* relative input rate r_iT for an extendable deadtime loss system including pile-up losses. T_p/T is typical of the ORTEC 716A amplifier. (Gedcke, 1972.)

the Ag *K*-lines from this source, divided by the recorded livetime, were used to monitor the accuracy of the livetime correction as the counting rate of the system was varied with an intense ^{55}Fe source. In Fig. 18(b) the residual error is observed to increase slightly due to undercorrection at high count rates, followed by an overcorrection at extremely high counting rates. The residual error is within $\pm 1\%$ up to about 24,000 counts sec^{-1}. The correction is quite good even beyond the counting rate where the baseline normally becomes shaky.

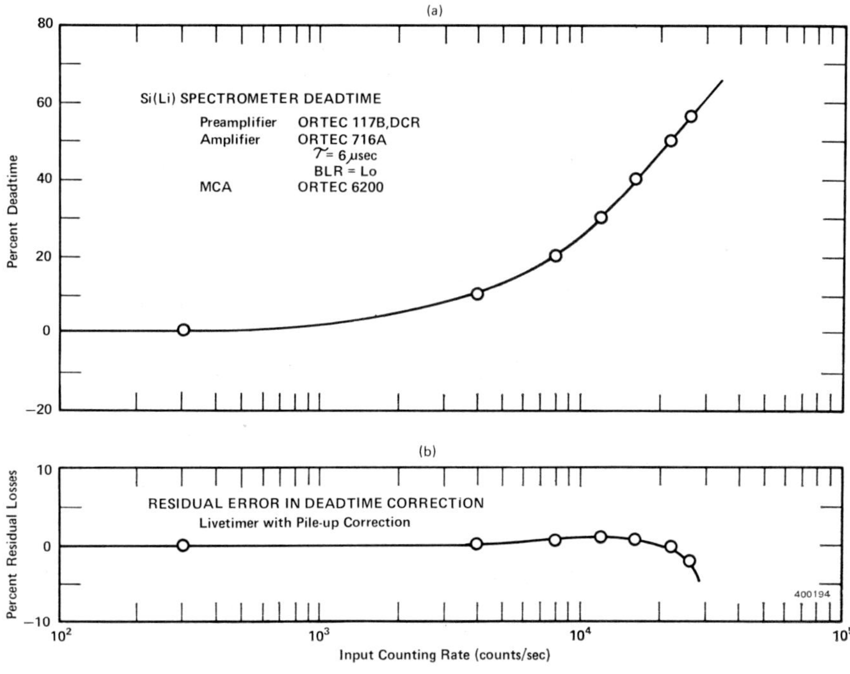

FIG. 18. The accuracy of the ORTEC pile-up compensated live timer.

Formulae for the real case where both the amplifier and the MCA contribute to the deadtime losses are quite complex. Fortunately, automatic deadtime correction schemes, as described above, can eliminate the need to calculate mathematical corrections to the data. Consequently, the analyst need only understand the sources of the losses.

VI. SPECTRUM ANOMALIES

A. ESCAPE PEAKS

Probably the most important spectral anomaly not associated with high counting rates is the silicon escape peak. When an X-ray of energy

E enters the detector, it has a high probability of ionizing the K shell of a silicon atom. As this vacancy in the shell is filled by an electron, a silicon K X-ray is emitted. In those cases where the de-exciting silicon atom is near the surface of the detector, the K X-ray has a good chance of escaping the active volume. Consequently, the energy left in the detector will be $E-1{\cdot}74$ keV, and the event will be recorded as an escape peak at the lower energy. In Fig. 15 the peak at 4·16 keV is the escape peak from the 5·9 keV MnK_α line. The ratio of the escape peak intensity at $E-1{\cdot}74$ keV to the true peak at energy E increases with decreasing energy, as the silicon K absorption edge is approached. Phosphorus is the lowest Z element capable of generating the silicon K escape peak. Reed and Ware, (1972) provide a thorough treatment of this phenomenon, including the ratio of escape to true peak intensities as a function of energy.

B. LOW ENERGY TAILING

In unravelling overlapping peaks, a deviation from pure Gaussian peak shapes can make the job more difficult. Low energy tailing is the name applied to skewing of the peak on the low energy side. This effect is due to incomplete charge collection and can occur at both high and low energies (Walter, 1971; Elad, 1972). At high energies some of the charge can be lost to trapping centers as it drifts through the detector. This is a property of the bulk material in the detector and can be minimized by using higher bias voltages, provided the detector will operate at higher voltages without generating noisy leakage currents. At very low energies a significant fraction of the ionization process takes place in the detector deadlayer. Since this is a weak field region, the electron-hole pairs recombine before they can be collected. Skewing of low energy peaks is a sensitive measure of detector deadlayer thickness (Jaklevic and Goulding, 1971).

C. THE BACKGROUND SHELF

Incomplete charge collection near the sides of the detector has also been identified as a source of spectrum distortion (Landis *et al.*, 1972; Jaklevic and Goulding, 1972). It causes the rather flat shelf of background extending from the low energy side of a peak down to zero energy, as observed in Fig. 15. In samples excited by electron beams this effect will be noticed as a slight increase in the intensity of the bremsstrahlung background below intense peaks. It represents a significant background contribution only in trace element analysis using fluorescence analysers (Landis *et al.*, 1972; Jaklevic and Goulding, 1972).

For a more detailed discussion of spectrum distortions, the publication by Walter (1971) is recommended.

VII. Application to the Microprobe and Scanning Electron Microscope

A. Detection Geometry

Owing to its compactness the Si(Li) detector allows much greater flexibility in designing the sample-to-detector geometry than is available with the wavelength spectrometers. This has proven to be a valuable asset in interfacing the detector to electron beam instruments where provisions for adding an energy spectrometer with optimum geometry were not included in the instrument's design. For qualitative and semiquantitative work on the SEM, a compromise in the detection geometry is normally tolerable. Frequently, the electron beam impinges on the sample at an angle to the surface that is not 90°, and X-ray take-off angles are compound. Although quantitative analysis is often carried out under these conditions, a more desirable situation is to have the sample surface normal to the electron beam, with an accurately defined X-ray take-off angle to the Si(Li) detector. An X-ray take-off angle greater than 20° is of course desirable. In the new generation combination microprobe/SEM instruments now becoming popular, this condition is generally satisfied. The Si(Li) detector is mounted at the same take-off angle as the two or three wavelength spectrometers to facilitate, quantitative correction calculations. When analysing metallurgical and geological specimens, especially at the higher currents required for simultaneous analysis on the wavelength spectrometers, the counting rates in the energy spectrometer are extremely high. Large sample-to-detector distances are required for these cases in order to handle the high X-ray yield. This larger distance makes take-off angle determination more accurate. For low X-ray yield samples, such as biological thin sections excited by low beam currents, it is necessary to move the detector as close to the sample as possible for efficient X-ray collection. Under these conditions large solid angles are subtended by the detector, and X-ray take-off angles become poorly defined. Quantitative corrections become less accurate for elements whose X-rays are heavily absorbed by the matrix, and the corrections must include integration over all take-off angles included in the large solid angle subtended. It becomes important to know the distance to the detector front surface and the detector active diameter rather precisely. The distance between the sample and the detector end cap often quoted by manufacturers should not be confused with the desired dimension (from sample to detector diode front surface). Retractable interfaces are available which permit the Si(Li) detector to be moved in close to the specimen for low X-ray yields, or backed off for higher yields and more precise quantitative analysis. The more satisfactory designs move the Si(Li) detector in and out while maintaining a take-off angle equal to that designed into the accompanying wavelength spectrometers.

B. MODES OF ANALYSIS

There are three modes of operation in common use: (1) energy spectrum acquisition, (2) elemental distribution mapping, and (3) single-element line scans. These modes will be illustrated with the following example.

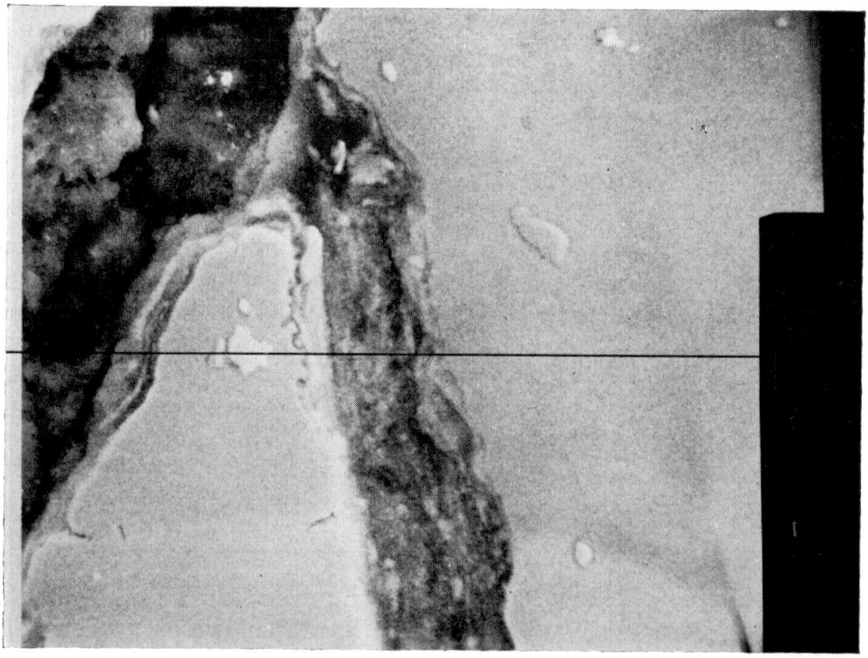

FIG. 19. Secondary electron image of the metallurgical sample from a gun barrel. The horizontal dark line marks the position of the elemental line scan taken in Fig. 24. (ORTEC, 1973.)

Figure 19 is a secondary electron micrograph of a section of a gun barrel taken on an SEM (ORTEC, 1973). The sample is imbedded in epoxy (the dark area in the upper left corner of the figure) and contains a crack suspected of collecting copper vaporized from the projectile jacket. The energy spectrum in Fig. 20 was collected as the electron beam scanned the area displayed in Fig. 19. Thus, Fig. 20 represents the elemental content of the sample averaged over a large area. Area scans with a rough sample are practical only with the Si(Li) detector. By carefully examining this spectrum with a logarithmic vertical

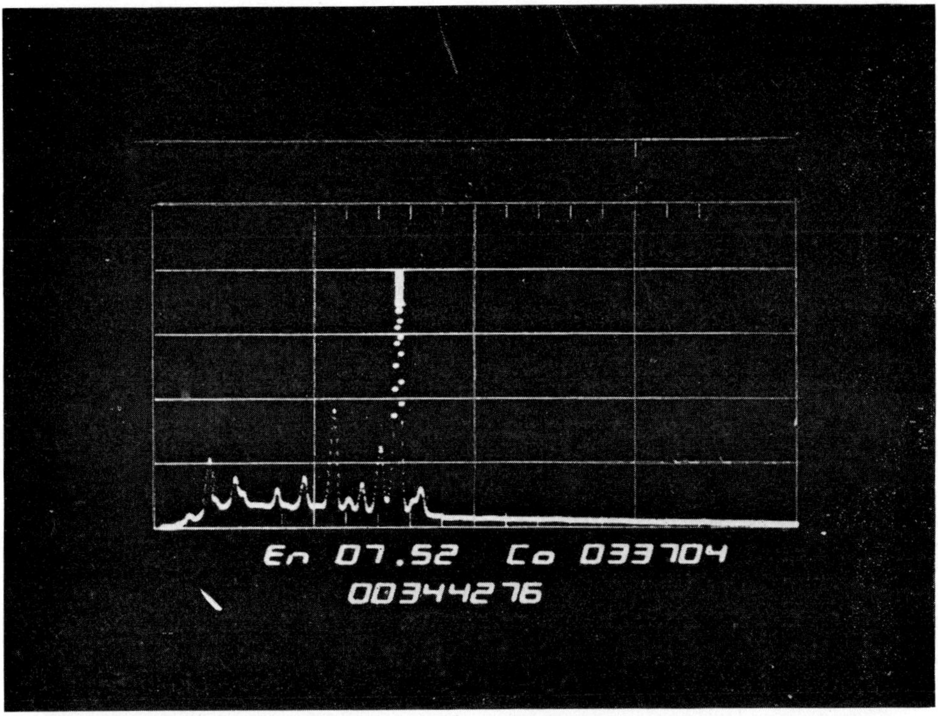

FIG. 20. The X-ray spectrum generated by scanning the area in Fig. 19. The horizontal energy axis covers the range from 0 to 20 keV. The vertical axis represents counting rate. An electronic energy marker is positioned on top of the 7·52 keV NiK_α peak. The numbers below the spectrum read the energy of the marker (En 07·52) and the counts in that channel (Co 033704). The bottom line (00344276) is the peak area contained within the region of interest (intensified channels) set on the NiK_α peak. (ORTEC, 1973.)

intensity scale, the element peaks in Table I were identified. It is also possible to position the electron beam on a fixed point in the sample for the purpose of obtaining an elemental microanalysis.

To obtain a distribution map for copper, a region of interest (ROI) was set over half the CuK_α peak on the low energy side. This selection eliminated interference from the NiK_β line. Each time an X-ray is

15

TABLE I. Lines Identified in Fig. 20

Energy (keV)	X-ray peak intensity (counts)	Element/line
0·92	1693	Cu/L_α
1·48	9949	Al/K_α
1·74	3976	Si/K_α
2·32	7192	Pb/M_α, Mo/L_α
2·60	4838	Cl/K_α
3·68	5396	Ca/K_α
4·50	7172	Ti/K_α
5·42	17586	Cr/K_α
5·89	3995	Mn/K_α
6·40	6182	Fe/K_α
6·94	11742	Co/K_α
7·52	33704	Ni/K_α
8·04	3629	Cu/K_α
8·30	5461	Ni/K_β
10·53	1089	Pb/L_α
12·63	899	Pb/L_β
17·44	424	Mo/K_α

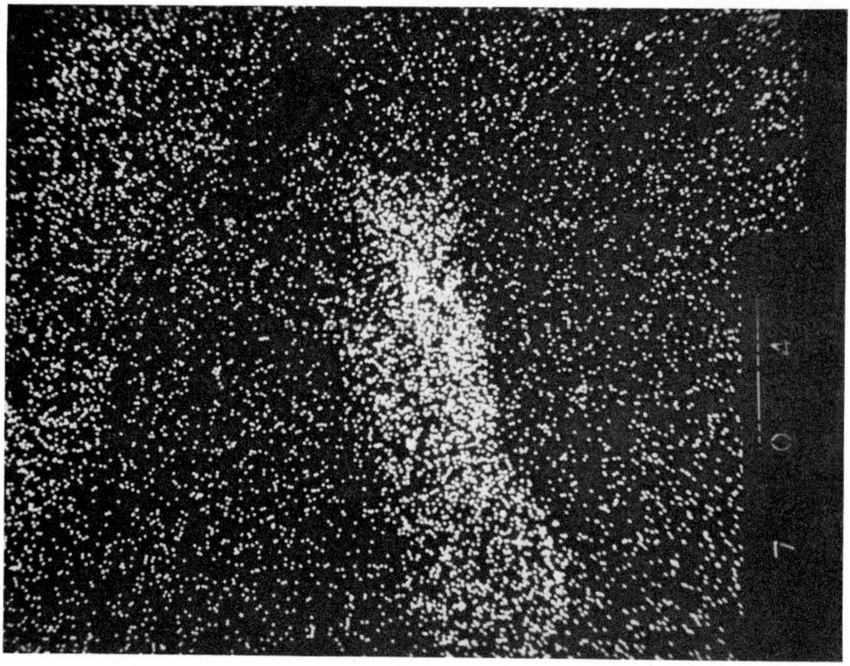

FIG. 21. The distribution map for copper. (ORTEC, 1973.)

detected within this ROI (as the beam scans the sample), the MCA sends a digital pulse to the SEM display to intensify the CRT. Figure 21 is the resulting Cu distribution map. Intense dot density indicates a high concentration of copper in the crack. Figures 22 and 23 are the corresponding maps for nickel and titanium, respectively.

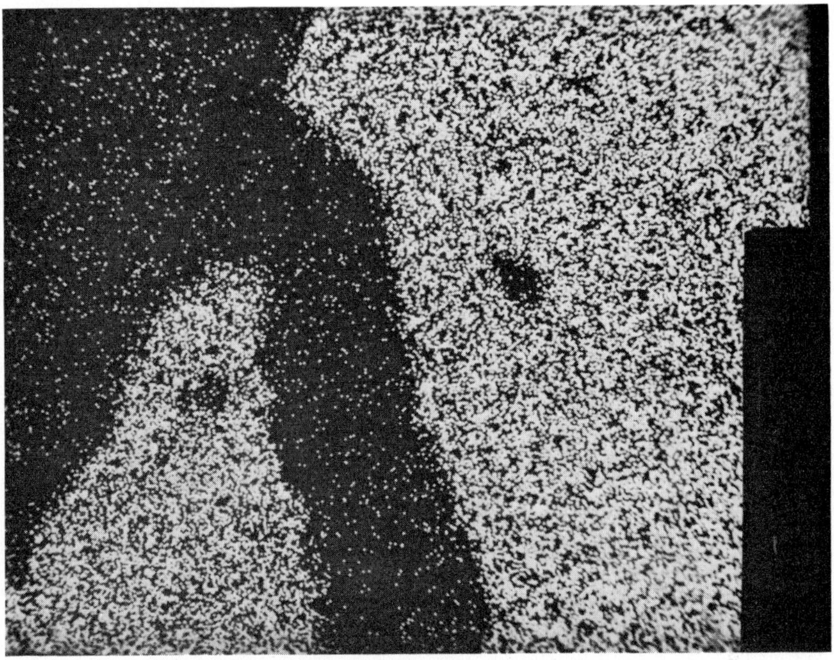

FIG. 22. Nickel distribution map. (ORTEC, 1973.)

Line scans are frequently useful as a semiquantitative tool to plot the relative intensities of elements as a function of position along a line across the sample. Two types are possible: (1) the analog line scan and (2) the digital line scan. In the analog scan the ROI output is fed to a counting rate meter to convert the digital information into an analog voltage proportional to counting rate. This ratemeter analog output is used as the Y-deflection (vertical) input to the SEM display. As the SEM scans along a predetermined line on the sample, the CRT plots a graph of element counting rate versus distance. Since the plot is made on the SEM display, it is easily related to points in the secondary electron picture. Figure 24 shows the line scans for copper and nickel taken on the gun barrel sample along the dark line in Fig. 19. The ratemeter time constant must be set long enough to provide adequate smoothing of the data. The SEM scan time must be long compared to the ratemeter time constant to avoid washing out abrupt changes in intensity over a short distance.

In the digital line scan the MCA memory is used as a series of scalers assigned to sample the counting rate in sequence as the beam scans across the sample. The result is displayed by the MCA as a spectrum of counting rate (vertical axis) versus time, or distance, across the sample (horizontal axis). The single-channel analyser window (SCA) in the ADC

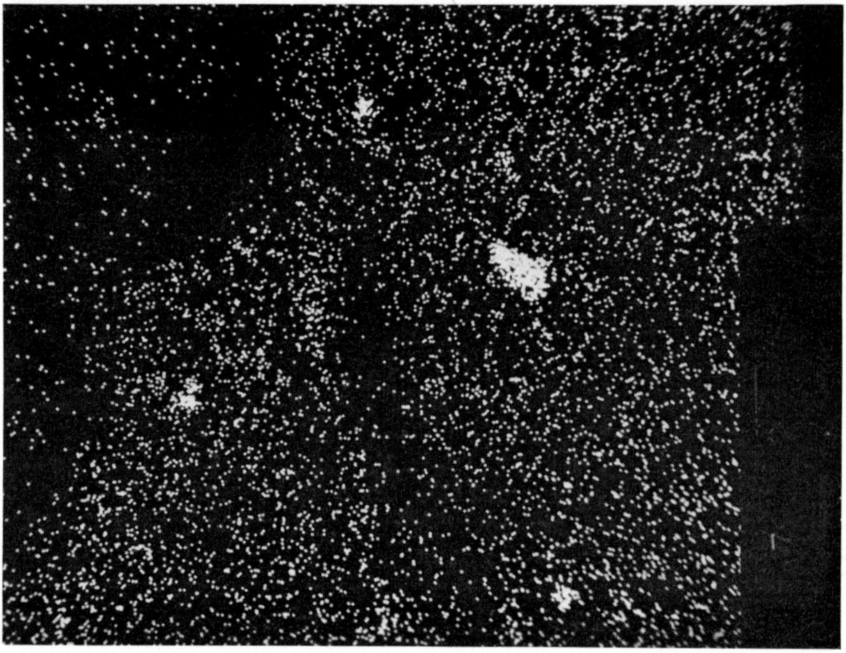

FIG. 23. Titanium distribution map. (ORTEC, 1973.)

is set across the X-ray peak of interest, and a dwell time per channel is selected on the MCA. The start of SEM sweep and the MCA scan can be synchronized either electronically or by hand. When the scan is initiated, the MCA counts the SCA output in the first channel for the preselected dwell time, and then advances to the next channel in memory. This process is repeated until the last channel in memory has been used. Distance across the sample is related to the MCA channel number through the SEM sweep time and the MCA dwell time. Consequently, the spectrum in memory represents counting rate as a function of distance across the sample. If the MCA and SEM are electronically synchronized, the line scan can be repeated several times to improve the counting statistics. Scans for two or more elements can be stored in sections of the MCA memory and compared directly on the display. Dwell times are chosen relative to the SEM sweep time to provide an adequate number of sampling points (channels) along the scan. SEM sweep speeds are usually chosen to be slow enough to produce the desired counting statistics in each channel.

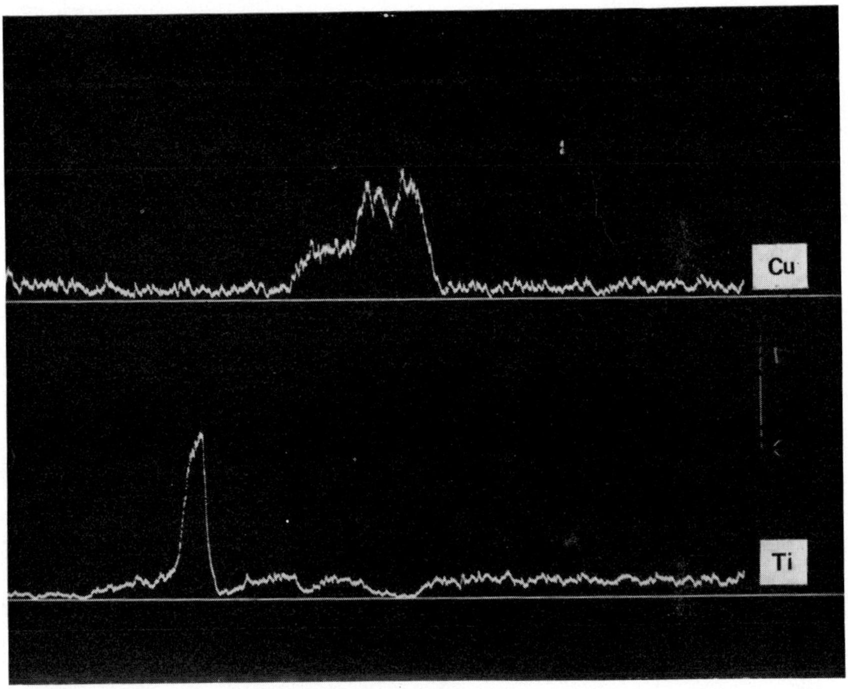

FIG. 24. Line scans for copper (upper line) and titanium (lower line) along the dark line in Fig. 19. (ORTEC, 1973.)

C. COMBINED ENERGY AND WAVELENGTH SPECTROMETERS

The multichannel pulse height analyser (MCA) is such a powerful tool for analysing spectra that it would be a shame to restrict its use to the Si(Li) detector. The ease of operation and flexibility can be significantly enhanced in the wavelength spectrometer by replacing the conventional single-channel pulse height analyser and scaler-timer combination with a multichannel analyser (Albrecht and Henson, 1972; ORTEC, 1973). Figure 25 shows an integrated X-ray analysis system

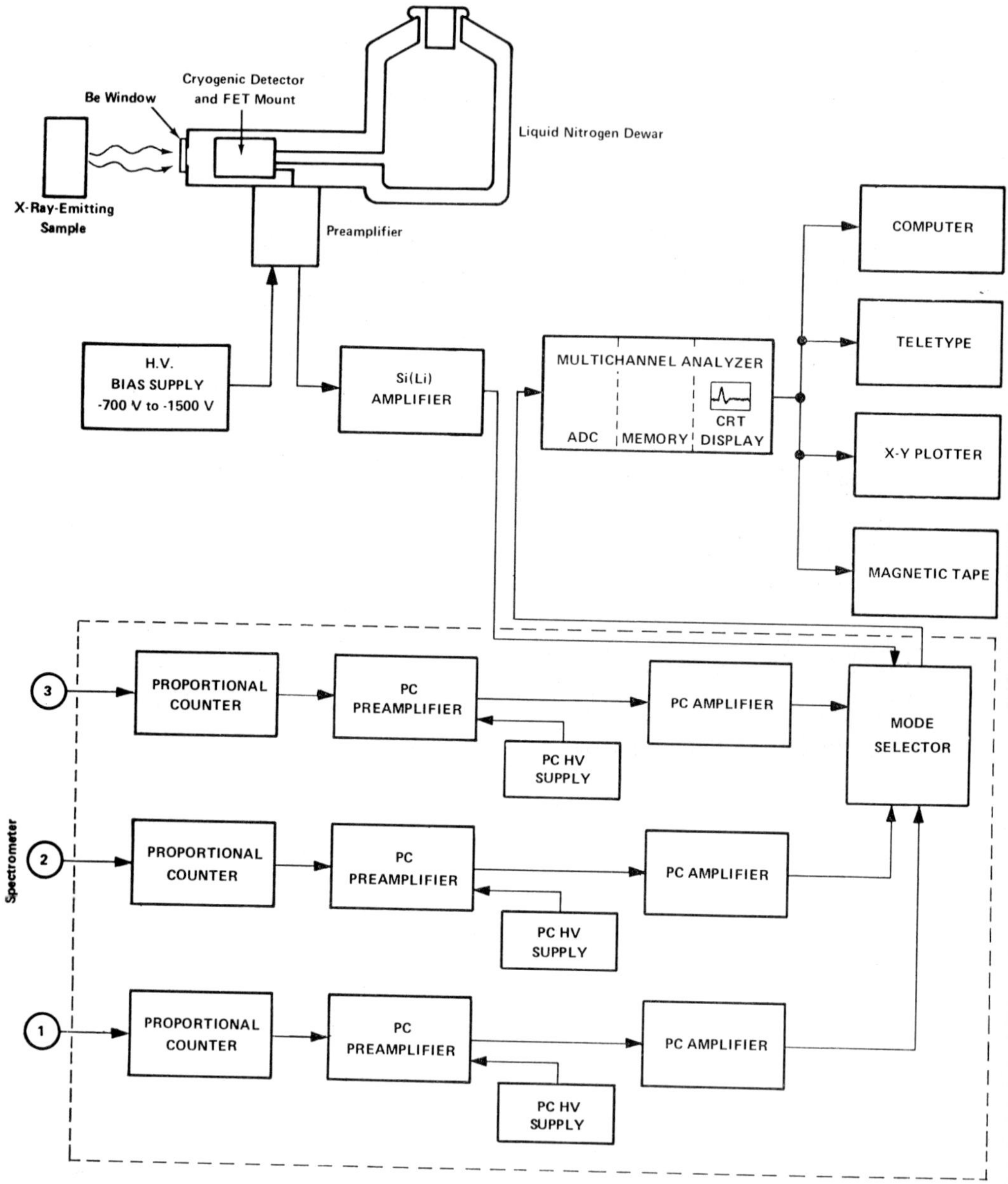

FIG. 25. The integrated X-ray analysis system. (ORTEC, 1973.)

(IX.A) where the wavelength and energy spectrometers are combined to allow the MCA to process data from any spectrometer. With the proportional counter amplifier gain adjusted so that the MCA energy scale is calibrated, a real improvement can be gained in the ease of setting a window around the line of interest. Figure 26 shows two

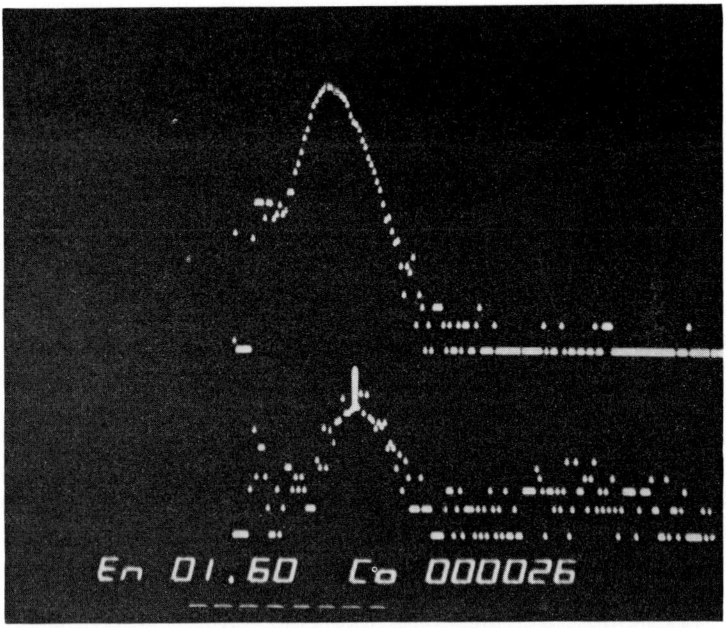

FIG. 26. Overlapped energy spectra from a wavelength spectrometer set to analyse the Mg K_α wavelength. The vertical scale (counts) is logarithmic. The horizontal scale is energy. Top spectrum: obtained from pure magnesium sample (20 sec). Bottom spectrum: bremsstrahlung background pulse height spectrum at the Mg K_α wavelength obtained from a pure aluminium sample. Peak shift is caused by a gas gain reduction at high count rates in the proportional counter. (Albrecht and Henson, 1972; ORTEC, 1973.)

proportional counter energy spectra taken with the wavelength spectrometer set to analyse the MgK_α line. The peak in the top spectrum is caused by the MgK_α line from a pure Mg sample. The peak in the bottom spectrum is due to the bremsstrahlung spectrum obtained under identical conditions with a pure aluminum sample. Note that the vertical scale is logarithmic. The peak shift on the intense peak from the pure Mg sample is due to the reduction of gas gain in the proportional counter at high counting rates. The top and bottom spectra represent the difference in counting rates encountered in alternating between a pure element standard and a trace element unknown in quantitative analysis. In using an SCA to gain better trace-element sensitivity, a narrow window would normally be set across the peak obtained with the pure element standard. With the example shown, this window position would be in error for the trace-element condition. However, with the MCA an optimum window can be selected after the spectrum has been accumulated, thus removing a major source of error, while still permitting narrow windows to be used for good peak-to-background ratios. The excellent deadtime correction system described in section V.C also works in this case to make calculated deadtime corrections completely unnecessary. The spectra in Fig. 26 were taken with the original microprobe electronics. Newer electronics employing low-noise, field effect transistor preamplifiers and sophisticated shaping amplifiers allow the use of lower proportional counter gas gains, resulting in almost complete elimination of the gain shifting problem.

The line scan mode of the MCA can be used to provide a much more usable wavelength scan than is possible with a ratemeter and strip-chart recorder. For generating digital wavelength scans, the single-channel analyser in the MCA is set to accept all pulse heights. The line scan mode is activated at the same instant as the wavelength spectrometer scan is initiated (either electronically or manually). At the end of the scan the MCA memory contains the wavelength spectrum as it would be presented on a strip-chart recorder with the following benefits:

1. Peaks cannot be slewed by too long a ratemeter time constant setting, since a digital multichannel scaling technique is used.
2. No choice of vertical scale is required prior to the scan. Both large and small peaks can be viewed after the scan by rotating the MCA vertical scale selector, or by using a logarithmic display.
3. Peak positions, heights, and areas can be read directly from the MCA display.
4. The data is readily transferred to a computer for data processing or can be plotted with several scale choices on an X–Y plotter.
5. A second scan can be added to the first to improve counting statistics.

Figures 27 and 28 compare wavelength scan spectra obtained on the same sample with an SCA/ratemeter/strip-chart recorder system versus a digital MCA scan.

No doubt, more ingenious ways to use integrated X-ray analysis systems will be devised in the future.

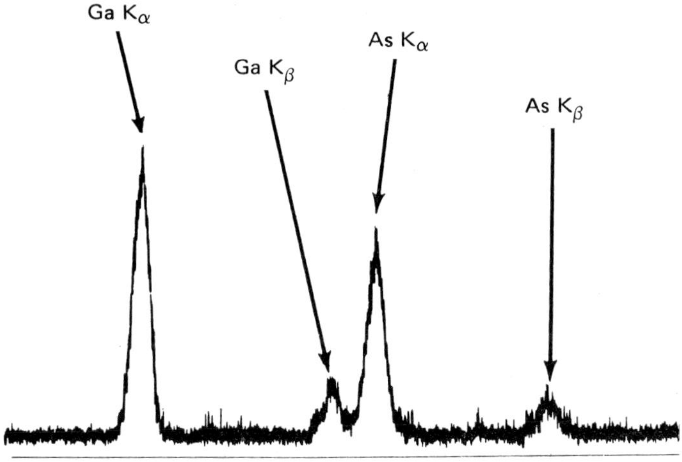

FIG. 27. A conventional wavelength scan with SCA and ratemeter. (Albrecht and Henson, 1972; ORTEC, 1973.)

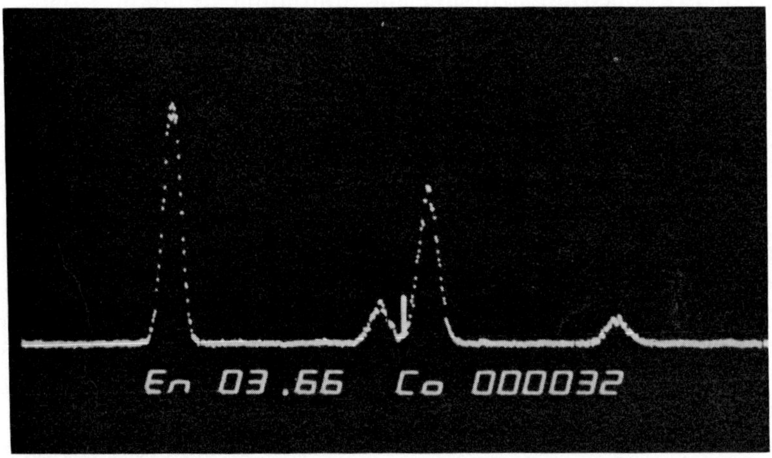

FIG. 28. A digital wavelength scan with the MCA. The spectrum source is identical to Fig. 27. (Albrecht and Henson, 1972; ORTEC, 1973.)

D. REQUIREMENTS FOR QUANTITATIVE ANALYSIS

Most of the conditions required for quantitative analysis with wavelength spectrometers also apply to the energy spectrometer. This is certainly true of sample surface conditions, take-off angle, and quantitative concentration correction calculations. Most of the differences in operating parameters have already been discussed. Outside of spectral data reduction techniques, the only major difference is the need to integrate the concentration correction formulae over the larger solid angle subtended by the Si(Li) detector.

Since the resolution below 15 keV is poorer in the Si(Li) energy spectrometer, peak overlaps and background subtraction become more important. Occasionally, peak overlaps can be avoided by choosing another line in the spectrum from the same element. Indeed a great deal of quantitative analysis has been carried out without a computer. However, in the long run the affluent analyst is bound to employ a computer to expand his data-handling capability. Large programs run on batch processing computers prove to be the most powerful. In fact, some go as far as to print out the errors in the peak intensities (Grandy and Gedcke, 1966; Tepel, 1966). On the other hand, small dedicated on-line computers provide for a better operator-instrument interaction. Although program power is restricted on a small computer, the quick answers it provides allow for efficient on-line decision-making while investigating a sample.

Too many spectrum-stripping and unfolding techniques have been developed to describe in detail here. The methods can be divided into three general groups: (1) least squares fitting, (2) Fourier analysis followed by least squares fitting, and (3) cross correlation followed by least squares fitting. In the conventional least squares fitting approach, mathematical shapes are assumed for the background and the peaks. A standard least squares fitting routine is employed to unravel overlapping peaks and background over a restricted energy region. Many programs of this type have been written. Tepel's program (Tepel, 1966; Grandy and Gedcke, 1966) is typical of this approach. Recently, Fourier analysis has been applied to spectral analysis. The most extensive development of this approach was carried out by Inouye (1964, 1969, 1971, 1972). Chiefly, it permits data smoothing and some resolution enhancement (Inouye, 1964, 1969, 1971, 1972; Colby, 1972; Vanier and Sutfin, 1972). Attempts have also been made to suppress background by Fourier filtering (Russ, 1972), although the theoretical basis for this approach has not been thoroughly established (Inouye, 1964, 1969, 1971, 1972). A danger inherent in the Fourier filtering technique is the tendency to distort the spectrum, and the method must be used with care. Of course, a least squares fitting routine is still necessary for the remaining peak overlaps.

Recently, Robertson *et al.* (1972) have proposed a simple cross-correlation technique which is practical on small computers. Significant background reduction was achieved, although the algorithm will have to be refined to allow peak overlap unravelling.

In the next few years we can expect to see considerable innovation in the data reduction technology applied to microanalysis. The enterprising microanalyst would be well advised to build an expertise in computer programming and statistical data analysis as an investment in the future.

REFERENCES

ALBRECHT, L. and HENSON, H. (1972). "Combination X-ray Energy Analysis and Wavelength Dispersive Analysis." Paper presented at Cambridge Users Meeting, London, December 1972.

BEAMAN, D. R. *et al.* (1972). *J. Phys. E, Sci. Instr.* **5,** 767.

BROWN, L. *et al.* (1969). *Semiconductor Nuclear Particle Detectors and Circuits.* National Academy of Sciences Publication 1593, Washington, D.C.

COLBY, J. W. (1972). *Proceedings of the Sixth International Conference on X-ray Optics and Microanalysis*, p. 247. University of Tokyo Press. *Proceedings of the Tutorial Session at the 7th National Conference on Electron Probe Analysis*, p. Bb–1. San Francisco.

DEIGHTON, M. O. (1961). *Nucl. Instr. Methods*, **14,** 48.

ELAD, E. (1971). Special Publication STP 485, American Society for Testing and Materials, Philadelphia, p. 57.

ELAD, E. (1972). *IEEE Trans. Nucl. Sci.* **19,** 403.

ELAD, E. and GEDCKE, D. A. (1972). *Proceedings of the 6th International Conference on X-ray Optics and Microanalysis*, p. 263. University of Tokyo Press.

EVANS, R. (1955). "The Atomic Nucleus," Chapter 28. McGraw-Hill, New York.

FAIRSTEIN, E. and HAHN, J. (1965). *Nucleonics*, **23,** 50, 56, 81.

FAIRSTEIN, E. and HAHN, J. (1966). *Nucleonics*, **24,** 54, 68.

FITZGERALD, R. and GANTZEL, P. (1971). Special Publication STP 485, p. 3. American Society for Testing and Materials, Philadelphia.

FITZGERALD, R., KEIL, K. and HEINRICH, K. F. J. (1968). *Science*, **159,** 528.

GEDCKE, D. A. (1972). *X-ray Spectrometry*, **1,** 129.

GEDCKE, D. A. and ELAD, E. (1972). *Proceedings of the 6th International Conference on X-ray Optics and Microanalysis*, p. 253. University of Tokyo Press.

GEDCKE, D. A., ELAD, E. and DYER, G. R. (1971). *Proceedings of the 6th National Conference on Electron Probe Microanalysis*, p. 5A. Electron Probe Analysis Society of America, New York.

GOULDING, F. S., WALTON, J. T. and PEHL, R. H. (1969). Lawrence Berkeley Laboratory Report UCRL-19377.

GRANDY, T. B. and GEDCKE, D. A. (1966). Instructions for using the modified Tepel peak fitting program Anneliese, Nuclear Research Centre Report, University of Alberta, Edmonton, Alberta, Canada.

HEATH, R. L. (1972). *Advances in X-ray Analysis*, **15,** 1.

INOUYE, T. (1964). *Nucl. Instr. Methods*, **30,** 224.

INOUYE, T. (1971). *Nucl. Instr. Methods*, **91,** 581.

INOUYE, T. (1972). *Nucl. Instr. Methods*, **104,** 541.

INOUYE, T., HARPER, T. and RASMUSSEN, N. C. (1969). *Nucl. Instr. Methods*, **67,** 125.

JAKLEVIC, J. M. and GOULDING, F. S. (1971). *IEEE Trans. Nucl. Sci.* **18,** 187.

JAKLEVIC, J. M. and GOULDING, F. S. (1972). *IEEE Trans. Nucl. Sci.* **19,** 384.

KANDIAH, K. (1969). "Semiconductor Nuclear Particle Detectors and Circuits" (L. Brown *et al.*, eds), p. 495. National Academy of Sciences Publication 1593, Washington, D.C.

KANDIAH, K. (1971). *Nucl. Instr. Methods*, **95,** 289.

KANDIAH, K., STIRLING, A., TROTMAN, D. L. and WHITE, G. (1968). *International Symposium on Nuclear Electronics*, vol. 1, p. 69–1. Versailles.

LANDIS, D. A., GOULDING, F. S. and JARRETT, B. V. (1972). *Nucl. Instr. Methods*, **101,** 127.

LANDIS, D. A., GOULDING, F. S., PEHL, R. H. and WALTON, J. T. (1971). *IEEE Trans. Nucl. Sci.* **18,** 115.

ORTEC (1973). X-ray Analytical Studies, XAS-11 and XAS-12 ORTEC Inc., Oak Ridge, Tennessee, U.S.A.

ORTEC (1972). 7000 Series Si(Li) X-ray Detector Operating and Service Manual, ORTEC Inc., Oak Ridge, Tennessee, U.S.A.

REED, S. J. B. (1972). *J. Phys. E, Sci. Instr.* **5,** 994, 997.

REED, S. J. B. and WARE, N. G. (1972). *J. Phys. E, Sci. Instr.* **5,** 582.

ROBERTSON, A., PRESTWICH, W. V. and KENNETT, T. J. (1972). *Nuc l.Instr. Methods*, **100,** 317.

RUSS, J. C. (1972). *Proceedings of the 7th National Conference on Electron Probe Analysis*, p. 76. San Francisco.

SUTFIN, L. V. (1972). *Proc. Vth Ann. SEM Symp. IITRI, Chicago*, p. 65.

SUTFIN, L. V. and OGILVIE, R. E. (1970). *Proc. IIIrd Ann. SEM Symp. IITRI, Chicago*, p. 17.

SUTFIN, L. V. and OGILVIE, R. E. (1971). Special Publication STP 485, p. 197. American Society for Testing and Materials, Philadelphia.

TEPEL, J. W. (1966). *Nucl. Instr. Methods*, **40,** 100.

VANIER, M. W. and SUTFIN, L. V. (1972). *Proceedings of the 7th National Conference on Electron Probe Analysis*, p. 77. San Francisco.

WALTER, F. J. (1971). Special Publication STP 485, p. 82. American Society for Testing and Materials, Philadelphia.

WILLIAMS, C. W. (1968). *IEEE Trans. Nucl. Sci.* **15,** 297, and references contained therein.

WILLIAMS, G. (1971). Special Publication STP 485, p. 125. American Society for Testing and Materials, Philadelphia.

NOTE ADDED IN PROOF: Since the original writing of this chapter significant advances have been made in high counting rate performance, allowing reliable operation up to a 40% amplifier duty cycle. See Karlovac, N. and Gedcke, D. A. (1973). *Proceedings of the 8th National Conference on Electron Probe Analysis*, p. 17. New Orleans.

Chapter 13

Role of the Divergent Beam (Kossel) X-ray Technique in Scanning Electron Microscopy

H. YAKOWITZ

Metallurgy Division, Institute for Materials Research, National Bureau of Standards, Washington, D.C., U.S.A.

Symbols

a_0, a = lattice, parameter
P_1 = principle stress
$z\omega_g$ = channelling pattern line width in radius
ξ_g = extinction distance for the reflection excited
$|\vec{g}|$ = the reciprocal lattice vector translation i.e. $1/d$
$_f$ = final i.e. after strain
$_i$ = initial i.e. before strain
σ = standard deviation
$\uparrow$ = to the power of (given by next no.)
λ = wavelength
r = radial uncertainty in measuring X_0, Y_0
Δ = small value of
δX etc = small change in X
KISS = Kossel internal stress strain analysis
S = measured lattice strain
d_f = final "d" value
d_i = initial "d" value (unstrained)
RACE = Regressive Analysis of the Conic Equation
θ = Bragg angle
$90\text{-}\theta$ = semi apex angle of diffracted cone
ϕ = $\measuredangle$ of diffracting planes w.r.t. crystal surface nearest X-ray source.
μ = linear absorption coefficient
t = thickness of specimen
E = operating voltage
E_c = critical excitation voltage of X-ray line of interest
z = X-ray source-to-film distance
ψ = $\measuredangle$ subtended by film to amount of crystal sampled in transmission case
X, Y, Z = orthogonal coordinate system where Z is the X-ray source to film direction a centre of pattern in X-Y planes origin, X_0, Y_0

I. Introduction

The divergent beam X-ray diffraction method, usually called the Kossel method after its discoverer, offers the scanning electron microscopist a means of obtaining quantitative crystallographic data from a wide variety of specimens. In particular, reliable lattice spacing data can be obtained from regions of the specimen down to about 15 μm in size. In turn, these lattice spacings can, under certain circumstances, be used to infer the deformation state of the specimen. Orientation data can also be obtained.

The role of Kossel patterns in scanning electron microscopy will, in all probability, be as a supplement to the more easily obtained channelling patterns. A successful combination of the Kossel and channelling techniques then would offer an extremely powerful means of extracting quantitative orientation and lattice spacing data from selected regions located using the familiar emissive mode of SEM operation.

The formation of the Kossel pattern, reduction of data to lattice spacings, calculation of deformation data, and an error analysis of the latter two will be presented in detail. Orientation determination will be dealt with only briefly since, in the context of scanning electron microscopy, channelling patterns often offer the best route to orientation information.

II. Formation of the Kossel Pattern

Diverging radiation propagating radially into a crystal from any microsource of X-rays generated either within the specimen or from a source outside the specimen may encounter planes of atoms lying at the Bragg angle, θ. This radiation will be diffracted along radial directions in the form of right circular cones. Each diffracting plane gives rise to a cone and the Kossel pattern is the trace of the intersection of the set of cones with a photographic film. The axis of the cone is perpendicular to the diffracting plane, and the semi-apex angle of the cone is $(90-\theta)°$.

Two modes of pattern preparation are usually used: transmission, and back-reflection. The transmission pattern can show two types of conic traces both lighter, *and* darker than background. The lighter conic traces are called deficiency conics. As the primary X-ray beam passes through the crystal, radiation is diffracted from atomic planes lying at the Bragg angle, θ. Furthermore, these planes will lie at some angle, ϕ, with respect to the crystal surface nearest the X-ray source. Now, if $\phi < \theta$, diffracted radiation will exit on the same side of the crystal as the source. Thus, on the transmission photograph, the negative of this diffracted radiation appears as the deficiency conic. Conversely, if $\phi > \theta$, then a diffraction conic containing an excess of energy with respect to the background may be observed in the transmission pattern. Thus, both light and dark conics are to be expected on a transmission pattern.

In back-reflection, the film is placed on the same side of the crystal as the source. Diffraction conics darker than background will be observed for planes where $\theta < \phi$ for non-perfect crystals. For a perfect crystal, conics or portions of conics may appear lighter than background as predicted by the dynamic theory of X-rays.

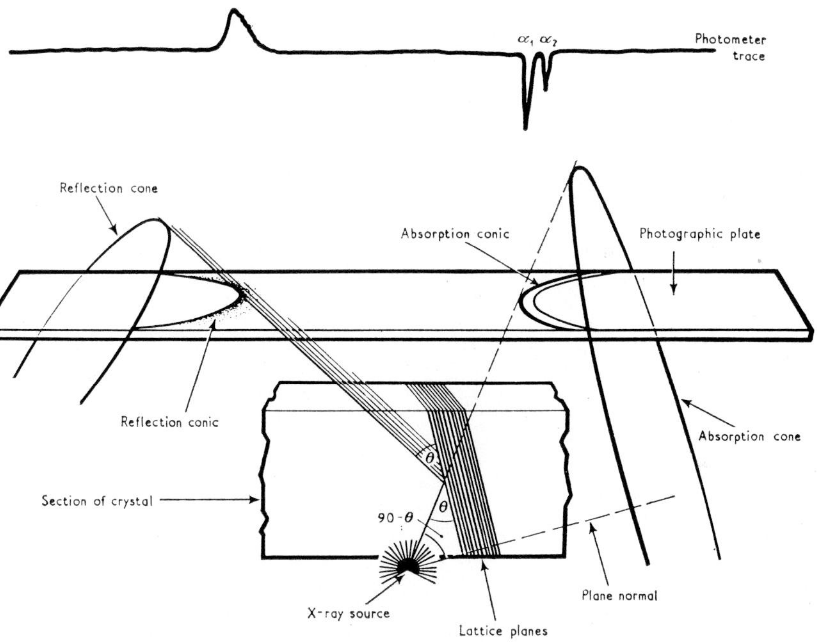

FIG. 1. Formation of Kossel conics (after Sharpe, 1965).

Figure 1, taken from Sharpe (1965), shows the geometry for the formation of both diffraction (reflection) and deficiency (absorption) conics for the case of a mosaic crystal. Note that the diffraction conic is represented as being somewhat broadened with respect to the deficiency conic. This broadening is indicated because reflection occurs from a large number of parallel planes. For each plane, there is a cone of semi-apex angle $(90-\theta)^\circ$ but all cones have vertices displayed by one interplanar distance, d. Therefore, reflection conics are generally not

FIG. 2. Typical Kossel Patterns.

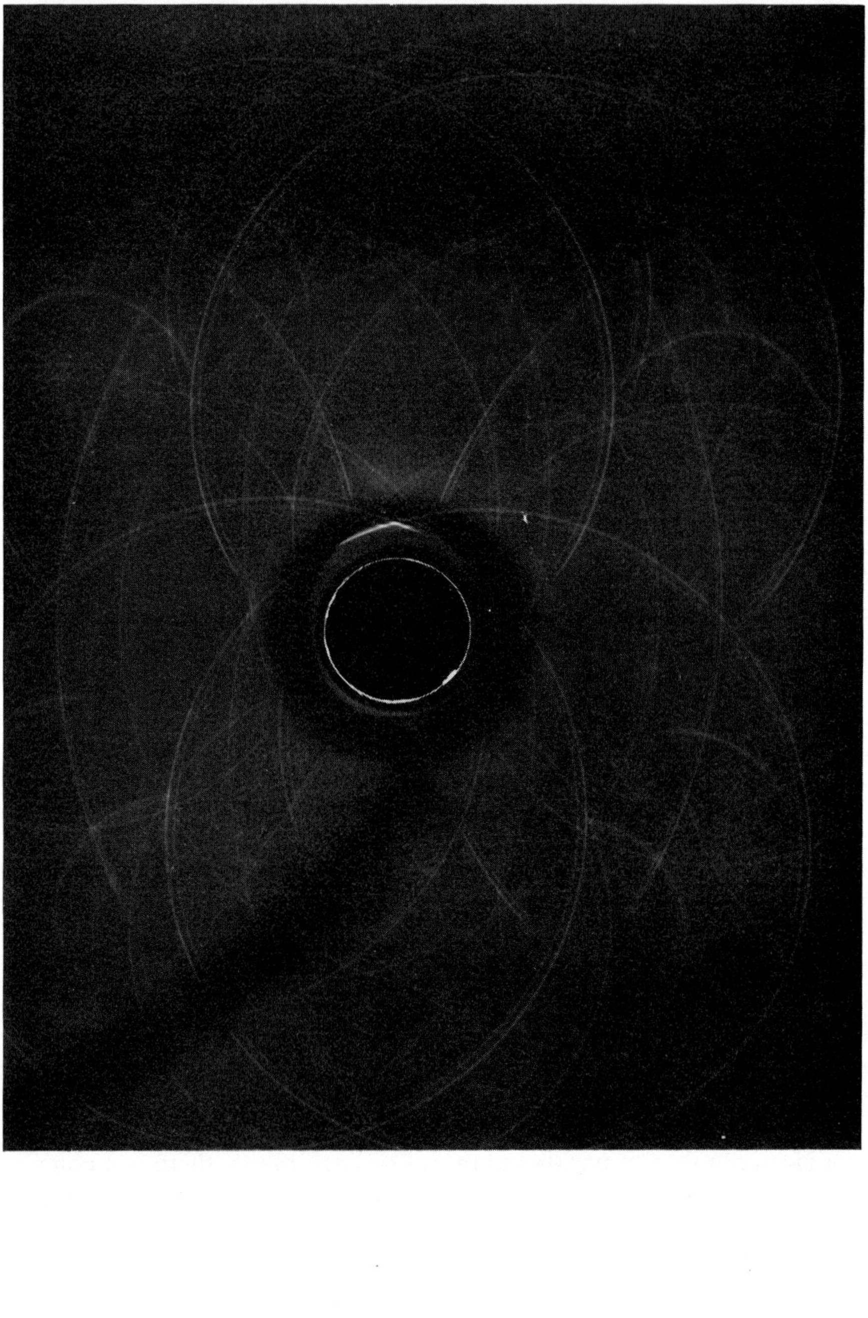

FIG. 2(a). Back-reflection Kossel pattern of single crystal ammonium di-hydrogen phosphate taken using Cu radiation at 25 kV. Note sharp lines, resolved α doublets and faint β traces (parallel to α traces). Approximate source-to-film distance 12 cm.

as sharp as deficiency conics, all of which have the X-ray source as a common vertex. However, Lonsdale (1947) pointed out that reflection conics recorded on transmission patterns are nearly as sharp as deficiency conics. Reflection conics from many such planes are significantly diminished by processes such as multiple reflection and absorption. Therefore, only reflections from planes near the exit surface are recorded on the film, thereby reducing the vertex displacement problem by decreasing the number of vertices. Figures 2(a) and 2(b) show respectively, typical back reflection and transmission Kossel patterns.

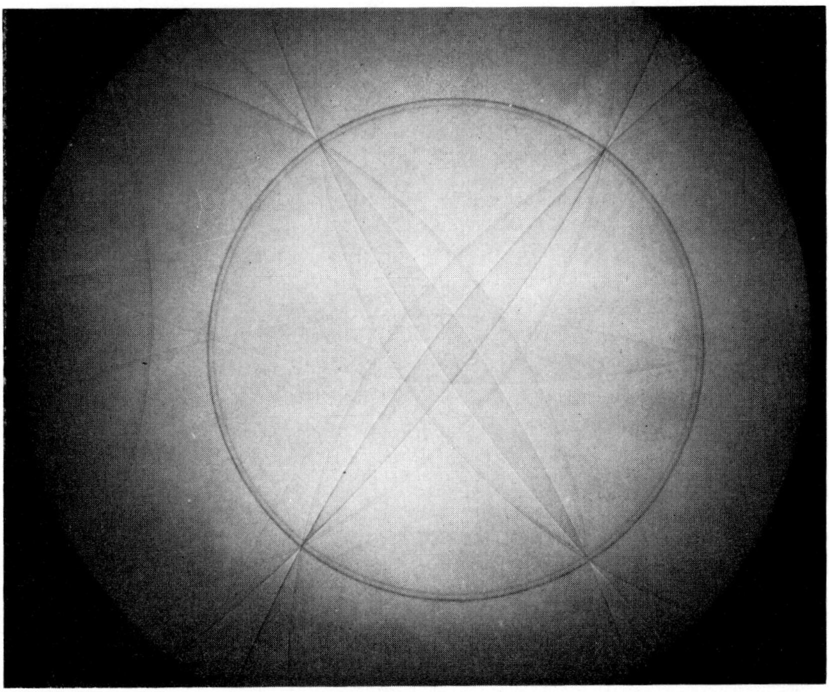

FIG. 2(b). Transmission Kossel pattern of iron–3 wt % silicon alloy using Fe radiation at 15 kV. Note dark (deficiency conic traces) and light (diffraction conic traces) lines on the print. Sample is unstrained material oriented (110) [001]. Source-to-film distance 10·5 cm.

The geometry of the pattern is governed by the natural crystal symmetry and Bragg's law. Each of the cones of interest has as a radius the angle $(90-\theta)$ and by the Bragg relation $\cos(90-\theta)^\circ = \lambda/2d$. Hence, no plane with a spacing such that $d \leqslant \lambda/2$ will give rise to an observable conic. Naturally, systematic extinctions may occur and these must be taken into account. If reflection conics appear on a transmission pattern, Lonsdale (1947) has shown that they will appear on the convex side of the corresponding absorption conic (see Fig. 2(b)). The separation between the two depends on the distance of the bottom of the crystal from the source.

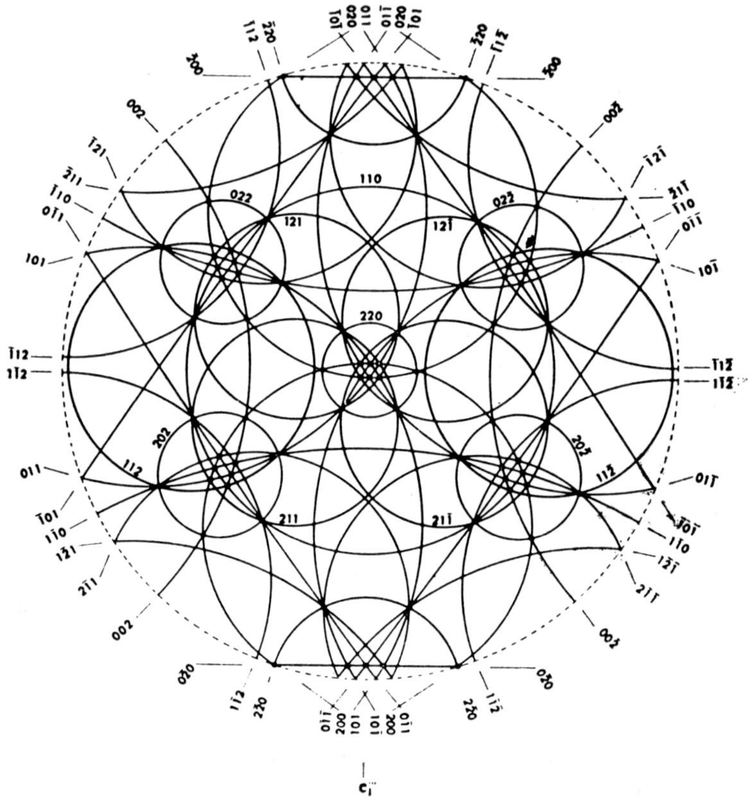

FIG. 2(c). Indexed stereographic projection of pure iron. Radiation assumed is iron Kα, and the [110] pole at the center. Note similarity between Figs (2b) and 2(c).

In a cubic lattice, two planes having Miller indices, (hkl), of say (211) and $(12\bar{1})$ respectively which lie at an angle of 60° to one another will appear as separate conic traces on the film. This separation occurs despite the fact that the Bragg angle and the d spacing are precisely the same for (211) and $(12\bar{1})$ in an unstrained cubic lattice. This separation of the individual conics from each plane of an (hkl) form is of crucial importance when the Kossel method is used to study stress–

strain configurations within the irradiated volume of crystal as described in the Appendix.

The foregoing discussion represents a simplified view of the formation and characteristics of Kossel conics. A somewhat more detailed version was given by Yakowitz (1969).

III. Relationships of the Conic Indices on a Kossel Pattern

The pattern obtained on a flat film is called a gnomonic projection of the conics. In other words, an angle between two points on the film is proportional to the arctangent of the length between the two points. This relation is precisely the same as that occurring in the films obtained from the standard Laue method used for routine crystal orientation. The net effect is a projection distortion since equal distances on the film do not usually correspond to equal angles.

Two general cases involving the indexing of Kossel conic traces exist. The first is where detailed studies are to be carried out on one or a few well characterized materials such as pure metals, simple alloys, or inorganic crystals. In such instances, approximate lattice parameters are usually known. The second case represents a true unknown about which the investigator has no prior knowledge. Both cases can be handled, but the procedures differ somewhat.

When the lattice parameters are known fairly well, a stereographic projection can be prepared where $(90-\theta) = \cos^{-1}(\lambda/2d)$ is plotted as a circle about each pole. This approximates closely to the expected pattern. Frazer and Arrhenius (1966) presented a straightforward procedure to obtain such projections by means of a computer equipped with a plotter. Figures 2(b) and 2(c) show the similarity of an actual pattern and the stereographic projection in the case of an iron alloy and iron. If extensive studies of a material are undertaken, a stereographic projection is extremely useful.

In many instances, the pattern can be indexed without the stereographic projection by tabulating $(90-\theta)^\circ$ values for each conic expected. The index of a plane is inversely proportional to the radius of curvature of the Kossel conic trace. Symmetry on the pattern is also an intuitive guide to indexing without the aid of the stereographic projection.

In the case of any pattern, including that of a complete unknown, the regressive analysis of the conic equation (RACE) method may be used to obtain the interplanar spacings or d values. Then indexing proceeds as in conventional crystallographic studies. The RACE method will be discussed in detail in the section on lattice spacing determination.

IV. Use of Transmission or Back Reflection Mode in SEM

Both modes give satisfactory data. However, the question as to which mode gives better contrast of the conics as compared to background cannot be answered unequivocally. Contrast depends to a large extent on the particular crystal under study. However, a detailed investigation of the contrast in the transmission mode has been carried out by Yakowitz (1966). For a given specimen, the best contrast will be obtained if one arranges matters so that $1{\cdot}5 \leqslant \mu t \leqslant 2{\cdot}5$ and $2{\cdot}5 \leqslant E/E_c \leqslant 3$, where μ is the linear absorption coefficient of the specimen for the radiation of interest, t is the specimen thickness, E is the operating voltage and E_c is the critical excitation voltage for the X-ray line of interest. The contrast of the pattern in either mode will also depend on the exposure of the film. Semi-empirical expressions for the proper exposure time of a Kossel pattern have been determined by Yakowitz and Vieth (1965).

The X-ray source-to-film distance, Z, plays a role as well. For back reflection, increasing Z increases the contrast since, according to James (1950), the intensity drop-off within the cones of radiation is proportional to Z^{-2}. However, the foregoing is not true for deficiency conics appearing in a transmission pattern. The contrast of such conics is governed by the crystal perfection; increasing Z may therefore reduce white line contrast by reducing the background blackening.

The amount of crystal sampled in each method is of some importance. For the transmission case, the diffracted radiation samples the full thickness of the crystal. The film subtends a total angle of ψ. In this case, the volume sampled can be approximated as a right circular cone of radius $t \cdot \tan(\psi/2)$ and height t; the volume sampled is $\pi t^3 \tan^2(\psi/2)/3$. For a typical case where ψ is 60° and t is 50 μm, this volume is $4{\cdot}37 \times 10^4$ μm^3.

For back reflection, the diverging radiation penetrates the specimen according to Beer's law and re-emerges. When the product μt becomes 1·0, the primary beam has only 36% of its former intensity. On emerging, the diffracted beam will have a maximum of 13% of the original intensity. The effective sampled volume will thus be approximated as a right cylinder of height and diameter of $1/\mu$. In the case of iron, the sampled volume is then about $4{\cdot}43 \times 10^3$ μm^3 and the cylinder has a diameter of 18 μm. This result is in line with the experimental conclusions drawn by Dingley (1971). However, serious ambiguities may result from attempts to sample grains much smaller than 15 μm, in that a contribution to the Kossel pattern observed may be made by regions beyond the grain in which the electron beam, of the SEM, may have been directed.

V. Obtaining Lattice Spacing Data

Yakowitz (1969) discussed several methods available for determining *d* values from Kossel patterns. Only one technique provides a general solution with satisfactory accuracy. This method is called the regressive analysis of the conic equation (RACE) method. The RACE method will therefore be described in detail.

The RACE method can be summarized as follows:

An orthogonal X–Y–Z coordinate system is set up where Z is the X-ray source-to-film distance and the centre of the pattern is the X–Y plane origin, X_0, Y_0. (see Fig. 3). The equation of any conic on the Kossel pattern has been given by Morris (1968) as:

$$q_1 X + q_2 Y + q_3 Z = (X^2 + Y^2 + Z^2)^{\frac{1}{2}} \sin \theta \tag{1}$$

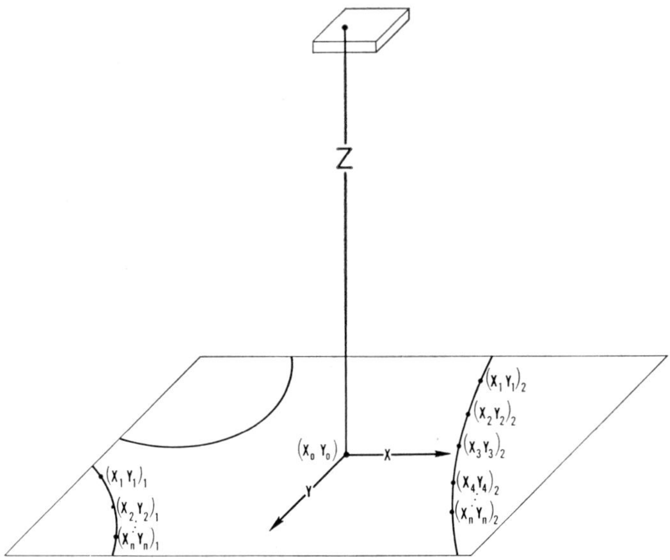

FIG. 3. Schematic indication of the experimental conditions used in the regressive analysis–conic equation (RACE) method.

Here q_1, q_2, and q_3 are the direction cosines of the plane (*hkl*) taken with respect to the X–Y–Z coordinate system. Now if:

$$U \equiv q_1 \csc\theta$$
$$V \equiv q_2 \csc\theta$$
$$W \equiv q_3 \csc\theta$$
$$M \equiv (X^2 + Y^2 + Z^2)^{\frac{1}{2}}$$

Then eqn (1) becomes

$$UX+VY+WZ = M. \tag{2}$$

In terms of the definitions above

$$q_i = \frac{U, V, W}{(U^2+V^2+W^2)^{\frac{1}{2}}}, \qquad i = 1, 2, 3 \tag{3}$$

$$\cos(90-\theta) = \sin\theta = \frac{q_1}{U} = \frac{q_2}{V} = \frac{q_3}{W} \tag{4}$$

and

$$q_1^2+q_2^2+q_3^2 = 1$$

Since $\sin\theta = \lambda/2d$, we get

$$d = 0{\cdot}5\lambda(U^2+V^2+W^2)^{\frac{1}{2}} \tag{5}$$

In principle, the d values for all conics on the film could be obtained by measuring the X–Y coordinates of only three points on each conic. In practice, several points on each conic are measured and regressive analysis is used as indicated in Fig. 3.

The values U, V and W can be obtained by the principle of least squares in which case U, V and W are obtained by minimizing the sum

$$A = \Sigma(M_i - X_iU - Y_iV - W_iZ)^2 \tag{6}$$

This process of minimization leads to a system of three equations having three unknowns. The solution of this system, called the normal equations, will be, statistically, the best estimates of U, V and W.

In this case, the normal equations for U, V, W are:

$$\begin{bmatrix} [X^2] & [XY] & z[X] \\ [XY] & [Y^2] & z[Y] \\ Z[x] & Z[Y] & nZ^2 \end{bmatrix} \begin{bmatrix} U \\ V \\ W \end{bmatrix} = \begin{bmatrix} [Xn] \\ [Yn] \\ Zn \end{bmatrix} \tag{7}$$

where $[X]$ means $\sum_i^n X_i$, etc.; n is the number of coordinate measurements for the conic. Then

$$\begin{bmatrix} U \\ V \\ W \end{bmatrix} = \begin{bmatrix} [Xn] \\ [Yn] \\ Zn \end{bmatrix} \begin{bmatrix} \Gamma^{11} & \Gamma^{12} & \Gamma^{13} \\ \Gamma^{21} & \Gamma^{22} & \Gamma^{23} \\ \Gamma^{31} & \Gamma^{32} & \Gamma^{33} \end{bmatrix} \tag{8}$$

where $\Gamma^{11} \ldots \Gamma^{33}$ represent the inverse of the first matrix in eqn (7).

The minimum value of eqn (6) is taken as an indication of the goodness of the fit of eqn (7) to obtain the d values. The standard error of the estimate is defined as

$$s_e = \left(\frac{\Sigma(M_i - X_iU - Y_iV - W_iZ)^2}{(n-3)}\right)^{\frac{1}{2}} \tag{9}$$

The standard errors associated with the estimates of U, V and W are:

$$s_U = s_e\sqrt{\Gamma^{11}} \tag{10a}$$

$$s_V = s_e\sqrt{\Gamma^{22}} \tag{10b}$$

$$s_W = s_e\sqrt{\Gamma^{23}} \tag{10c}$$

In order to obtain the standard error for the "d" spacing, the computed values of U, V and W are used to obtain $d_1, d_2 \ldots d_n$. Then, the standard deviation for d is calculated in the usual way. In practice, all of these calculations are performed by the computer. Figure 4 is the output from the appropriate program. The value given as percent strain is the ratio of the d value computed to that determined for well annealed, unstrained, material. The direction cosines q_i in eqn (1) are called L_i in the program. The input values are (hkl) indices, number of data points, i.e. XY measurements on the conic trace, diffracted wavelength, λ, Z and the pattern center, $X_0 Y_0$.

```
RESULTS FØR FE-3SI SAMPLE ZX-1

[HKL]= 2       1       1                 PCT.STRAIN= 9.44531 E-2
LAMBDA= 1.93597 A                        Z= 10.53    CM.
DATA PTS.= 11                            D= 1.16979 A

D= 1.16979 PLUS ØR MINUS SIGMA= 2.72958 E-4    A
MEAN VALUE ØF D= 1.16979 A

STD. DEV. ØF FIT FØR U,V,W= 2.78012 E-3
A SHØULD=0;MIN. A= 6.18327 E-5

U=-.578516 PLUS ØR MINUS SIGMA= 2.09551 E-3
V= .315788 PLUS ØR MINUS SIGMA= 1.03469 E-3
W= 1.01292 PLUS ØR MINUS SIGMA= 8.95227 E-5

DIRECTIØN CØSINES ARE:
L1=-.478715         L2= .261311         L3= .838182
[L1↑2+L2↑2+L3↑2]= 1.

SIN THETA= .827487               THETA= 55.8415 DEGREES
```

FIG. 4. Typical output listing for crystallographic information from Kossel patterns using the RACE method.

The statistical uncertainties are not the only source of uncertainty in the measurement of the d values. Other contributors to d value uncertainties are measurement errors in X, Y and Z, the effect of an erroneous estimation of the location of the pattern centre, $X_0 Y_0$, and the uncertainty in the wavelength, λ.

There is apparently no direct analytical method for propagation of errors due to, say, just Z in eqn (7). Therefore, values of $\Delta d/\Delta Z$ and

$\Delta d/\Delta r$, r being the radial uncertainty in locating $X_0 Y_0$ were approximated by empirical means.

The error evaluation was carried out by measuring all the conics on a pattern of Fe–3Si taken in the transmission mode. Some eleven sets of coordinate positions were found for each conic. The computer calculated d values based on $X_0 Y_0$ and Z values determined as accurately as possible. Then the Z value was altered in the computer program and the corresponding change in d recorded. The same was done for the $X_0 Y_0$ position. Thus, values of $\Delta d/\Delta Z$ and $\Delta d/\Delta r$ could be found for each conic. As in most crystallographic studies, as θ increased, the magnitude of the errors decreased.

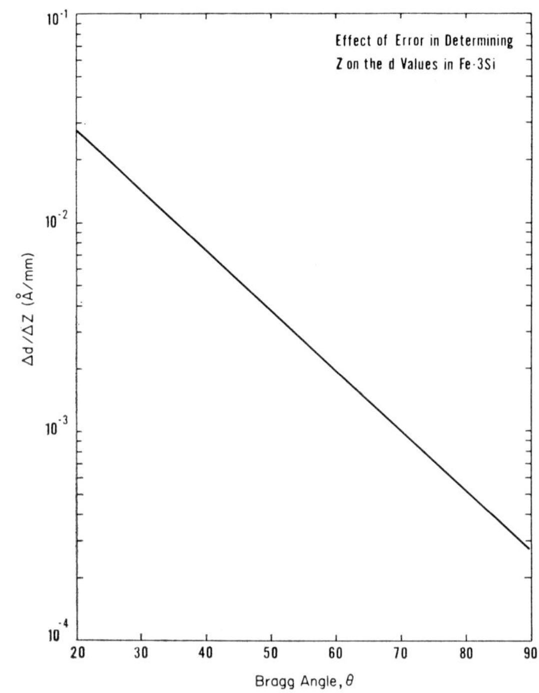

FIG. 5. Lattice spacing error due to Z uncertainty as a function of Bragg angle, θ.

The results are plotted in Figs (5) and (6). Clearly, the higher the θ value, the less the d error in the case of uncertainties in both Z and $X_0 Y_0$. Note that for a given value of θ, $\Delta d/\Delta Z$ is always considerably larger than $\Delta d/\Delta r$.

Shrier *et al.* (1966) suggested a means reducing the error in d due to uncertainties in Z. The assertion was that the difference between two

Z positions, δZ, could be measured very accurately—much more so than the absolute Z value. Experimentally, two photographic recordings of the Kossel pattern were taken, one at each Z position. Then XY coordinates of corresponding points, i.e. those from the same diffracted ray, were measured on the resulting film. The positions of correspondence were located by fixing a square grid of tungsten wire 25 μm in diameter immediately in front of the film. The double exposure showed thin breaks in the conic traces due to the presence of the wire

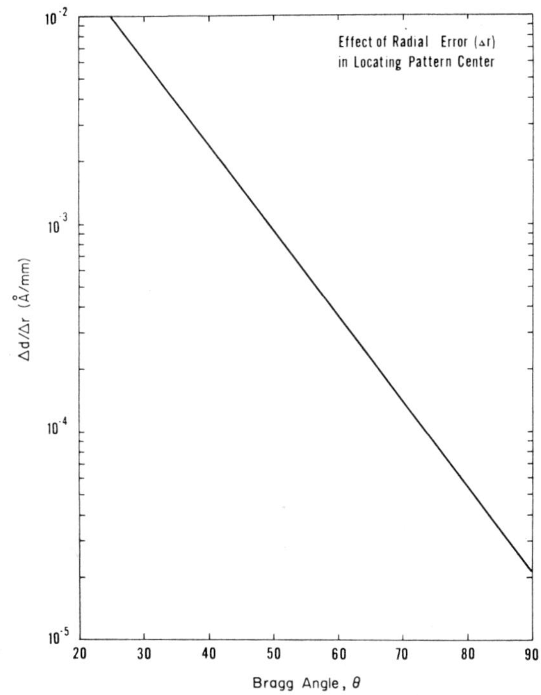

FIG. 6. Lattice spacing error due to radial uncertainty, Δr, in locating pattern centre ($X_0 Y_0$) as a function of Bragg angle, θ.

grid. The corresponding points from each trace were found by matching the breaks in the conic traces. The coordinates of the pattern centre were found by constructing lines through a number of corresponding points from different traces. The intersection of these lines is the pattern centre. If a lateral displacement occurs between the two exposures, then the lines will not interesect at a point and the exposures must be repeated.

The foregoing method seems to be worthy of consideration for use in scanning electron microscopy. The configuration of most SEM specimen chambers and the advantages of using thick specimens tend to encourage the back reflection mode for SEM work. There should be no problem in providing the wire grid before the film. Measurement of the difference in two Z positions should also be much more accurate than a single absolute determination of Z. Furthermore, the day-to-day use of the SEM almost precludes knowing an absolute value of Z without incessant remeasuring.

The relationship between U, V, and W (and hence d) to the measured coordinate differences is obtained as follows: For the first position Z_1, eqn (2) is

$$X_1U+Y_1V+Z_1W = (X_1^2+Y_1^2+Z_1^2)^{\frac{1}{2}} \tag{11a}$$

For the second value of the source-to-film distance Z_2, where $Z_1 < Z_2$, eqn (Z) is

$$X_2U+Y_2V+Z_2W = (X_2^2+Y_2^2+Z_2^2)^{\frac{1}{2}} \tag{11b}$$

Subtracting and calling $(X_1-X_2) = \delta X$, $(Y_1-Y_2) = \delta Y$, $(Z_1-Z_2) = \delta Z$ yields

$$U\delta X+V\delta Y+W\delta Z = (X_1^2+Y_1^2+Z_1^2)^{\frac{1}{2}}-(X_2^2+Y_2^2+Z_2^2)^{\frac{1}{2}} \tag{12}$$

For any conic trace, the angular dimensions on the film are proportional to $(90-\theta)^\circ$. Changing Z does not change λ, d, θ or the pole position about which each conic trace lies. Hence, altering Z_1 to Z_2 results in an equality such that:

$$\frac{X_1}{X_2} = \frac{Y_1}{Y_2} = \frac{Z_1}{Z_2} \tag{13}$$

Under these conditions, the following relationship can be proven to be an identity*:

$$(X_1^2+Y_1^2+Z_1^2)^{\frac{1}{2}}-(X_2^2+Y_2^2+Z_2^2)^{\frac{1}{2}} \equiv (\delta X^2+\delta Y^2+\delta Z^2)^{\frac{1}{2}} \tag{14}$$

Thus, for the double exposure method

$$U\delta X+V\delta Y+W\delta Z = (\delta X^2+\delta Y^2+\delta Z^2)^{\frac{1}{2}} \tag{15}$$

Under these conditions, the analogue to eqn (7) is:

$$\begin{bmatrix} [\delta X^2] & [\delta X\cdot\delta Y] & \delta Z\cdot[\delta X] \\ [\delta X\cdot\delta Y] & [\delta Y^2] & \delta Z\cdot[\delta Y] \\ \delta Z\cdot[\delta X] & \delta Z\cdot[\delta Y] & M\cdot\delta Z^2 \end{bmatrix} \begin{bmatrix} U \\ V \\ W \end{bmatrix} = \begin{bmatrix} [M\cdot\delta X] \\ [M\cdot\delta Y] \\ [M\cdot\delta Z] \end{bmatrix} \tag{16}$$

where $[\delta X] = \Sigma(X_{i1}-X_{i2})$, $[\delta Y] = \Sigma(Y_{i1}-Y_{i2})$,

* The author is indebted to S. D. Rasberry of the NBS staff for the proof.

$\delta Z = Z_1 - Z_2$ and n is the number of coordinate correspondences measured for a pair of conic traces. U, V W and M have the same meaning as before.

With a good comparator, the uncertainties in the individual X, Y measurements can be reduced to 10–20 μm. It is reasonable to assume that such errors will be randomly biased for a given conic. The effect of such errors on the computed d values is small.

Extrapolation of a line through correspondence points leads to the location of the pattern centre in the double exposure method. In a standard transmission pattern (single exposure) the $X_0 Y_0$ position can be found by allowing the electron beam to strike the photographic film or plate. This impresses a small spot at the geometric pattern centre. If the plate is in air, i.e. an X-ray window is in place, the light from the filament serves the same purpose. The $X_0 Y_0$ position can be found to within 0·1 mm in this way according to Gielen *et al.* (1965).

The possibility of the harmful effects of non-uniform film shrinkage must be considered. Gielen *et al.* (1965) found the effect to be negligible for certain X-ray films used to prepare transmission Kossel patterns. However, the problem can be removed by using glass plates for transmission studies. In the case of back reflection patterns, a hole is needed to allow the beam to pass. Presumably, it is possible to obtain glass plates with the appropriate emulsion laid on after drilling of the hole. Normally, a hole is punched through a film. Jellinek (1949) reported that punching of a hole produces large non-uniform film shrinkage near the hole. In a more recent study, Ellis and Weissmann (1964) found that a special Cronar base single emulsion film satisfied the requirements for isotropic film shrinkage even when used for back-reflection studies.

Temperature effects due to electron beam heating have been discussed by Morris (1963) and Yakowitz (1969). In SEM work, if the specimen current is kept at 10 nA or less, and the beam diameter at 0·1 μm or more, the temperature rise directly under the beam, i.e. in the electron range, for a conductor should not be more than a few degrees centigrade. The volume of material from which the Kossel pattern is obtained will experience considerably less heating.

VI. Crystal Orientation from Kossel Patterns

Orientation procedures utilizing Kossel patterns are entirely analogous to Laue procedures. The major advantage of the Kossel technique is that a microsource of X-rays is employed, thus enabling twin grains, matrix–precipitate relations, grain misorientation, and polycrystalline materials to be examined directly. Either the transmission or back-reflection mode is useful. The choice is largely a matter of convenience.

The orientation of a crystal is adequately described by the stereographic projection of the poles of the diffracting planes. Thus if a stereographic projection for the material of interest is available, the Kossel pattern of the material can be directly compared to the projection in order to determine the relative orientation of the crystal (see Fig. 2). This is possible since when a Kossel pattern is prepared on a film parallel to some crystal face (as is the usual case), the face will intersect the film centre. A reference direction is established on the film by fiducial marks. Thus, the indices of the face and a particular crystallographic direction within it can be found.

A number of methods to obtain orientation information have been offered. Most of these methods have been described by Yakowitz (1969). The RACE method provides a rapid accurate means to obtain orientation information. As shown in eqn (3), the RACE method gives the direction cosines of any (*hkl*) plane with respect to the X–Y–Z coordinate system of Fig. 3. The angle α between any two planes (called 1 and 2 respectively) can be obtained from:

$$\cos\alpha = q_{11}q_{12}+q_{21}q_{22}+q_{31}q_{32} \tag{17}$$

Hence, the orientation of the crystal in space and the angles between any two (*hkl*) planes is determined, as is the orientation of the crystal. Note that the double exposure method can also be used to obtain the q_{ij} values needed in eqn (17) through the use of eqns (16) and (3).

The orientation accuracy can be found by varying the Z and X_0Y_0 positions in the RACE calculation of several planes in an unstrained cubic crystal. In the cubic case, angles between planes are known exactly from geometrical considerations. Hence, variations are due to measurement uncertainties. Designating an error in X_0Y_0 as r—such an error is a radial error—it has been found that $\Delta\alpha/\Delta Z \simeq 0{\cdot}2°/\text{mm}$ and $\Delta\alpha/\Delta r \simeq 0{\cdot}5°/\text{mm}$. Thus, orientation can be carried out to within at least $\frac{1}{2}°$ without great difficulty.

Parenthetically, a check on α values in studies involving the cubic system offers a good internal consistency check on the d values. Departure from expected angles between planes is a sure indication of experimental error.

In SEM work, orientation determination is often more easily carried out by means of channelling pattern information. Orientation accuracy of the two methods, Kossel and channelling pattern, is comparable. The Kossel method, however, can be used where contamination or surface strains have degraded the channelling pattern.

VII. Kossel Internal Stress–Strain Analysis

The Kossel internal stress–strain (KISS) analysis yields for the sampled volume:

(a) Average Cauchy strains
(b) Principal strains and their axes
(c) Principal stresses and their axes
(d) Maximum shear strain in any (*hkl*) plane
(e) Total stored elastic energy
(f) Shear stored elastic energy

The basic information used as input to the KISS method is the change in lattice spacing for each conic after straining with respect to some reference state.

$$S = \frac{d_f - d_i}{d_i} \tag{18}$$

where S is the measured lattice strain, d_f the final value of d and d_i the initial (unstrained) value of d. The lattice strains need not come from an external straining device but may be caused by ordering, grain boundary precipitates or any other source of lattice perturbation as shown by Imura *et al.* (1962). The required measurements for d are obtained by using the RACE method.

A basic premise of the strain analysis is that the Kossel technique separates all planes of a family yielding a separate conic for each. Thus, d values can be obtained for each such plane separately. Separate strain values for each conic enable the investigator to avoid the complex line averaging required in other X-ray strain-analyses.

As Weissmann and his co-workers have stated, only the beginnings of the straining process can be observed by the means described. When plastic strain of about 1·5% occurs, line broadening, splitting and kinking may take place, thus making precise S values impossible to obtain. In this case, Fourier methods can be applied to the conic profiles in order to augment the strain analysis as shown by Slade (1963). In addition, the analysis is only valid for the elastic portion of the strain configuration. No information can be obtained concerning plastic strain components.

The details of the KISS method were given in part by Yakowitz (1969) and more extensively in a subsequent publication (Yakowitz, 1973). However, for the sake of completeness, a discussion of the stress–strain relations and the reliability analysis of the KISS method is repeated in the Appendix to this chapter.

The conclusions reached concerning experiment design and reliability of the KISS method can be summarized as follows:

(1) All conic traces in the Bragg angle range $30° \leqslant \theta \leqslant 80°$ appearing on the pattern should be used in order to provide adequate sampling for the KISS method.

(2) Depending on which method, single exposure or double exposure, was used to prepare the Kossel pattern, some eighteen X, Y or δX, δY values are sufficient to reduce uncertainties in measured d values to a 95% confidence interval width of $\pm 0{\cdot}05\%$.

(3) Under these conditions, the width of a 95% confidence interval for a computed parameter such as principal strain, derived using the measured strain in the KISS method is about $\pm 30\%$ of the computed parameter.

In a further study, Yakowitz (1973) reported that the KISS method can record the presence of very small strains but that their magnitude cannot be measured with high reliability. However, whether such small strains are present or absent can be ascertained.

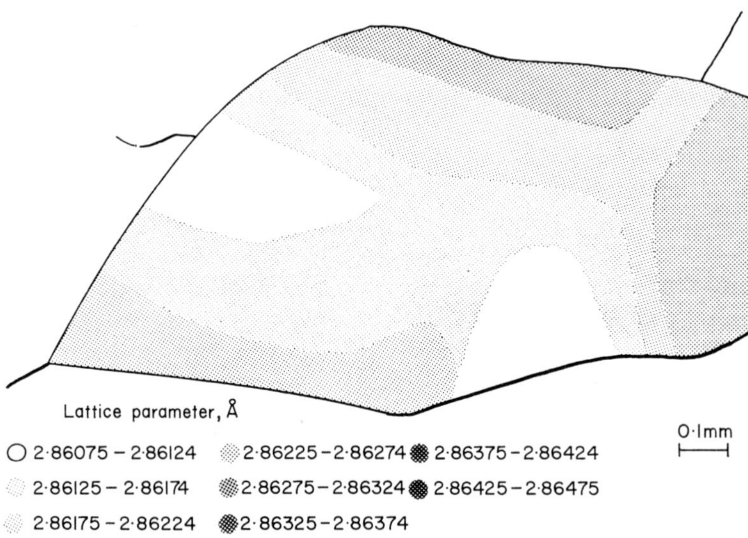

FIG. 7. Variation of lattice parameter, a_0, as a function of position in individual grain of Fe–3Si alloy oriented (110) [001].

There are two drawbacks to the extensive use of the KISS method for routine studies. The requirement for eighteen coordinate measurements per conic trace leads to a large number of measurements per Kossel pattern. This requirement can be relaxed at the expense of the accuracy of the determination of the results. The second drawback is

FIG. 8. KISS results in annealed Fe–3Si grain. Results shown as contours on individual grain.

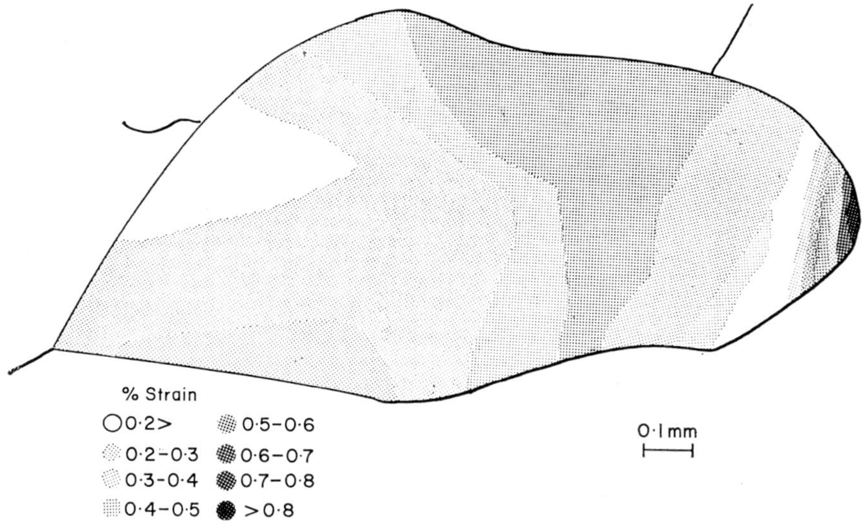

FIG. 8(a). Maximum principal strain, S_1, in %.

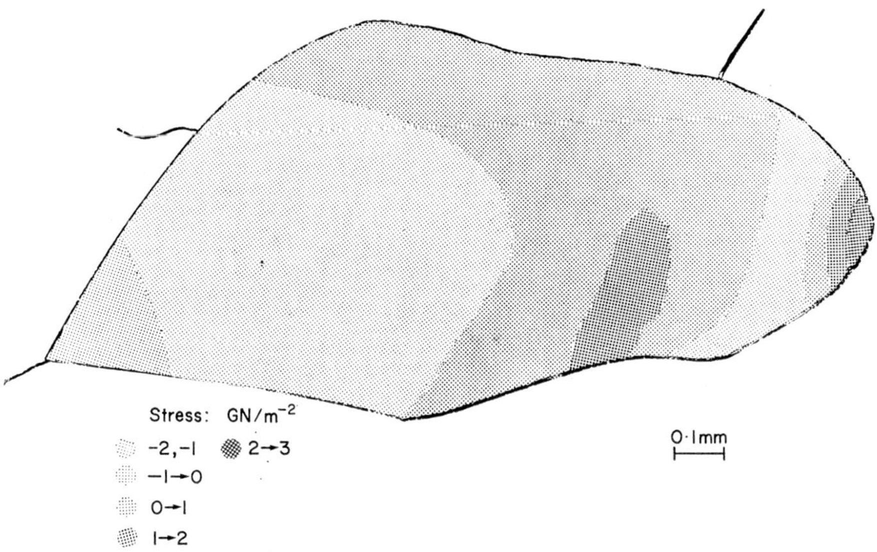

FIG. 8(b). Maximum principal stress, P_1, in GN m^{-2}.

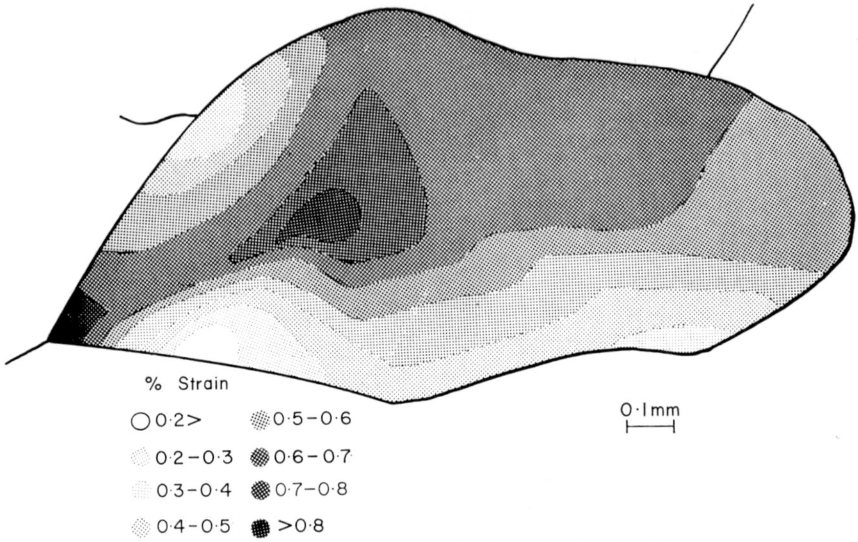

FIG. 9(a). Maximum principal strain, S_1, in %.

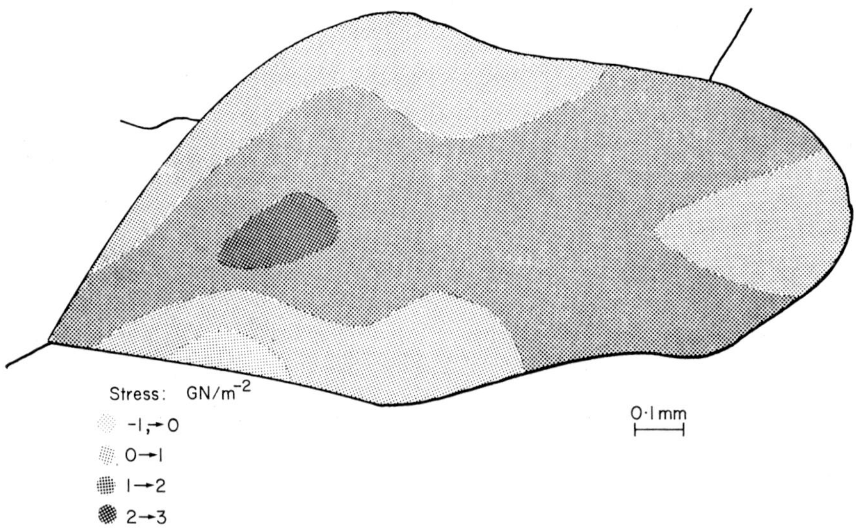

FIG. 9(b). Maximum principal stress, P_1, in GN m^{-2}. KISS results in Fe–3Si grain tensioned along long axis of grain to a level of about 400 N m^{-2}. Results shown as contours on individual grain.

that a full description of the stress-strain configuration at a single position at one stress level requires more than 100 numbers. Not only is this unwieldy, but worse, a clear picture of the situation is difficult to grasp from the numbers alone. However, in many cases, a full description is not necessary; the principal stresses and principal strains are sufficient for many purposes.

In a study of Fe–3Si alloy reported by Yakowitz (1973), the KISS method was used to evaluate the stress–strain configuration in an individual grain of Goss oriented (110) [001] material. Results were obtained from many points within the grain and synthesized into an area map of the grain. Figure 7 shows the lattice parameter a_0 for the grain in the well-annealed, presumably unstrained state. Clearly a varies within the grain. Figures 8 and 9 show the maximum principal strain, S_1, (Fig. 8(a) and 9 (b)) and maximum principal stress, P_1 (Fig. 8(a) and 9(b)) for the annealed and lightly tensioned condition—nominally 400 N m^{-2}.

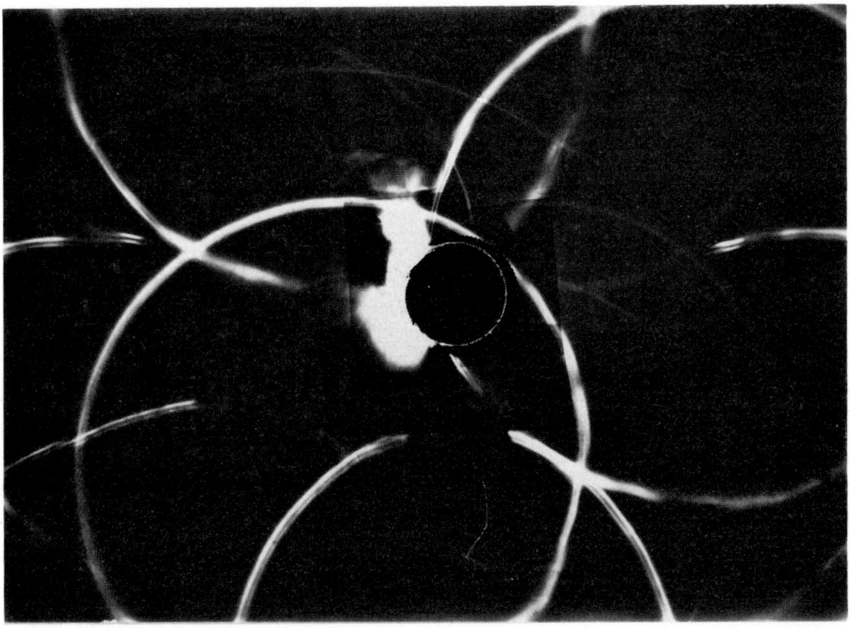

FIG. 10. Back reflection Kossel pattern of copper crystal using Cu radiation at 25 kV. Crystal is highly strained. Note broken conic traces and loss of α doublet resolution in some places. Note X-ray shadowgraph of paper holding film.

The maps in Figs 8 and 9 reveal the existence and extent of a variable strain distribution within the grain. The so-called unstrained state represented by Fig. 8 is seen to be, in fact, a strained state. These strains result from the secondary recrystallization procedure necessary to produce the (110) [001] grain texture. This texture is desired in order to minimize electrical power loss when Fe–3Si is used in transformer cores. Tension applied along the long axis of the grain causes rearrangement of the variable strain distribution as seen in Fig. (9). The metallurgical implications have been discussed by Yakowitz (1973).

When the specimen has undergone severe plastic deformation, the Kossel pattern can be expected to suffer loss of α doublet resolution and even kinking of the conic traces. Figure 10 is a back reflection pattern illustrating these features. The sample was a highly compressed copper crystal; and the X-ray source was Cu radiation. In such a case, the loss of the α doublet resolution precludes satisfactory determination of *d* values since the wavelength is unknown. However, conic traces due to Cu–K_β radiation could be used to infer *d* values since the wavelength spread is far less than that of the α radiation. The traces due to β radiation appear as singlets parallel to traces from α radiation. Furthermore, inspection of Fig. 10 indicates inhomogeneous strain within the sampled volume since partial α doublet resolution appears at certain regions of conic traces and not at others. Finally, Fourier analysis of the line broadening, as suggested by Slade (1963), could be used to infer further information concerning the stress–strain state of the crystal.

VIII. Comments on Relating Channelling Patterns to Kossel Patterns

The preparation of channelling patterns in the SEM is normally far easier and less time consuming than is the preparation of Kossel patterns. However, there appears to be no simple, sure way to obtain accurate *d* values from channelling patterns at the present time. The use of one or a few Kossel patterns to provide a basis for the inference of *d* values from channelling patterns for a given specimen may, therefore, be attractive in some instances.

Schulson (1971) has suggested the following relation for channelling patterns:

$$2\omega_g = 2/(\xi_g|\vec{g}|) \tag{19}$$

where

$2\omega_g$ is the channelling pattern line width in radians

ξ_g is the extinction distance for the reflection excited

$|\vec{g}|$ is the reciprocal lattice vector translation, i.e. $1/d$

However, ξ_g values are only poorly known. This situation has been outlined by Hirsch *et al.* (1965).

If a Kossel pattern and a channelling pattern can be prepared simultaneously in the SEM, then often, accurate d values can be obtained from the Kossel pattern using the methods already outlined. If the corresponding channelling pattern can be indexed, then $2\omega_g$ can be related to d adequately for the study of the specimen at hand. In other words, an empirical set of ξ_g values can be determined for the reflections of interest. Then channelling patterns can be used to infer d values fairly accurately for the specimen without further recourse to Kossel patterns.

Using eqn (19), and recognizing that the value of ξ_g does not change appreciably with strain, the channelling pattern analogue to eqn (18) becomes:

$$S = \frac{(2\omega_g)_f}{(2\omega_g)_i} - 1 \tag{20}$$

Then all of the discussion pertinent to the KISS method given in the Appendix applies to the channelling pattern as well. Again, a few Kossel patterns could serve to show the way to obtaining the stress–strain configuration by extensive use of channelling patterns for a given specimen. Furthermore, conditions leading to a degraded channelling pattern may still be capable of producing a usable Kossel pattern.

All of the foregoing discussion presumes that the channelling patterns and Kossel patterns are prepared under identical operating conditions of the SEM. Furthermore, the angular divergence of the electron beam in the SEM must be known in order to measure accurately the $2\omega_g$ values as shown by Schulson (1971).

IX. Brief Discussion of Experimental Technique and Equipment

The back reflection Kossel mode is that most likely to be used given the current configuration of SEM columns available. The insertion of the film with a grid in case the double exposure technique as contemplated should not be difficult. The film itself can be placed in a heavily aluminized mylar envelope to preclude light and direct backscattered electrons from causing fogging. Furthermore, the mylar will not cause an X-ray shadowgraph to be superimposed on the Kossel pattern (see Fig. 10, for which the film was in wrinkled black paper covered with aluminium foil).

In order to prevent premature exposure of the Kossel pattern, the SEM can be adjusted to operate at 1 or 2 kV in order to find the area of interest on the specimen. Then, the voltage and current can be increased to make the actual exposure. Exposure times are fairly short; Yakowitz and Vieth (1965) have discussed the estimation of exposure times. There is no reason why a channelling pattern could not be prepared simultaneously.

In many present SEM designs, the maximum value of the source-to-film distance Z can only be a few cm. This circumstance will reduce the contrast obtainable as will excess blackening of the film near the hole through which the electron beam passes. It is desirable to have Z more than 7 or 8 cm if possible. However, as Dingley (1971) has shown, entirely usable Kossel patterns can be obtained at the shorter distances. The blackening of the central region of the film can be reduced by an appropriate collimator made of a material such as beryllium (Be). The electron beam striking a Be tube will not produce much excess back-scatter, nor will bremstrahllung from the tube be excessive. An X-ray filter near the center of the film may also help in this regard.

If transmission Kossel patterns are desired, apparently some modification of most commercial SEM specimen handling devices would be necessary. Alternatively, a separate module for transmission patterns could be constructed. However, using an SEM for *extensive* Kossel studies in the transmission mode would seem to be inefficient.

The number of possible specimen–film arrangements for preparing Kossel patterns in the SEM makes it redundant to discuss detailed Kossel camera designs. However, if the double exposure method is to be used, the film holder must be rigidly fixed in space. Furthermore, the vertical motion on the SEM specimen control can be calibrated with gauge blocks so that δZ is accurately known.

X. Conclusions

The Kossel method offers the scanning electron microscopist a means of obtaining lattice spacings from appropriate specimens not otherwise available in the SEM. Typically, the back reflection Kossel method will sample a volume idealized by a right circular cylinder of height and diameter between 15 and 25 μm. Lattice spacings, orientation information and elastic deformation data can be inferred from the resulting Kossel pattern.

The precision of an interplanar spacing d, depends on the Bragg angle, θ. Precision values of $d \pm 5 \times 10^{-4} d$ can be obtained for most cases. In the case of deformation data, the reliability of a given calculated quantity will be about $\pm 30\%$ of the quantity.

In certain instances, the Kossel method can be used as a benchmark to obtain data allowing one to use the more easily prepared channelling patterns to obtain d values and deformation data. Further experimental work is needed to demonstrate fully the foregoing conclusion.

For most Kossel pattern preparation involving the SEM, the double-exposure back reflection method would seem to be most useful. Properly applied, this method should yield precise d values suitable for use in crystallographic evaluation of unknown phases or for use in the KISS method of deformation determination.

Appendix

The Kossel Internal Stress–Strain Method (KISS)

This appendix describes the physical and mathematical basis for determining the strain configuration of a crystal by means of the Kossel method (see text, section VII). In the following analysis, the crystal will be assumed to be cubic. The analysis can be carried out for non-cubic crystals but the mathematics are far more cumbersome. Furthermore, when the reliability of the results of the KISS method is considered, virtually nothing can be gained by assuming that a cubic crystal subjected to a strain no longer stays cubic.

In the cubic system, Imura *et al.* (1962) showed that it is convenient to choose a point within the irradiated volume of the crystal as a reference. For the cubic lattice, three mutually perpendicular axes corresponding to the crystallographic directions [100], [010] and [001] can be made to pass through the reference point. In this case, the components of the vector $\vec{H} = (hkl)$ are the direction numbers for the family of planes called H. In general, if j is a unit vector, the normal strain component in the j-direction is:

$$\varepsilon_j = jTj' \tag{A1}$$

where j is the transpose of j' and T is the strain tensor given by

$$T = \begin{pmatrix} \varepsilon_{11} & 0{\cdot}5\varepsilon_{12} & 0{\cdot}5\varepsilon_{13} \\ 0{\cdot}5\varepsilon_{21} & \varepsilon_{22} & 0{\cdot}5\varepsilon_{23} \\ 0{\cdot}5\varepsilon_{31} & 0{\cdot}5\varepsilon_{32} & \varepsilon_{33} \end{pmatrix} \tag{A2}$$

The ε_{ij} are the Cauchy strain components. (For the cubic system, $\varepsilon_{ji} = \varepsilon_{ij}$.)

The relation between the components of the strain tensor in terms of the Miller indices of a plane and the lattice strain, S, associated with that plane is

$$|\vec{H}|^2 S = \vec{H}' T \vec{H} \tag{A3}$$

where $\vec{H}' = (hkl)$ is the row matrix of the Miller indices h, k, l $\vec{H}$ is the column matrix of hkl, S is the lattice strain in the plane with Miller indices (hkl) and T is defined by eqn (A2).

Explicitly, eqn (A3) after the matrix multiplication becomes:

$$(h^2+k^2+l^2)S_d = h^2\varepsilon_{11}+k^2\varepsilon_{22}+l^2\varepsilon_{33}+kl\varepsilon_{32}+lh\varepsilon_{13}+hk\varepsilon_{12} \tag{A4}$$

In order to solve this equation for the six strain values, at least six independent measurements of S values are needed. Independent means that the six sets of hkl numbers lead to a matrix of rank six. There is apparently no simple way to decide whether a particular set of hkl values is suitable without constructing the matrix and testing to see that its determinant does not vanish, according to Imura *et al.* (1962).

With an appropriate set of hkl values a unique strain tensor can be computed. But, as with the determination of the d values by the RACE method, setting up the normal equations for the strain tensor yields the best estimates of $\varepsilon_{11}, \varepsilon_{22}, \varepsilon_{33}, \varepsilon_{13}, \varepsilon_{23}, \varepsilon_{12}$. Calling the best estimates $\langle\varepsilon_{11}\rangle, \langle\varepsilon_{22}\rangle \ldots \langle\varepsilon_{12}\rangle$, the six normal equations are given by Imura *et al.* (1962) as:

$$\begin{pmatrix} [h^2S(h^2+k^2+l^2)] \\ [k^2S(h^2+k^2+l^2)] \\ [l^2S(h^2+k^2+l^2)] \\ [klS(h^2+k^2+l^2)] \\ [lhS(h^2+k^2+l^2)] \\ [hkS(h^2+k^2+l^2)] \end{pmatrix} \begin{pmatrix} [h^4] & [h^2k^2] & [h^2l^2] & [h^2kl] & [h^3l] & [h^3k] \\ [h^2k^2] & [k^4] & [k^2l^2] & [k^3l] & [hk^2l] & [hk^3] \\ [k^2l^2] & [k^2l^2] & [l^4] & [kl^3] & [hl^3] & [hkl^2] \\ [k^2kl] & [k^3l] & [kl^3] & [k^2l^2] & [hkl^2] & [hk^2l] \\ [h^3l] & [hk^2l] & [hl^3] & [hkl^2] & [h^2l^2] & [h^2kl] \\ [h^3k] & [hk^3] & [hkl^2] & [hk^2l] & [h^2kl] & [h^2k^2] \end{pmatrix} \begin{pmatrix} \varepsilon_{11} \\ \varepsilon_{22} \\ \varepsilon_{33} \\ \varepsilon_{23} \\ \varepsilon_{31} \\ \varepsilon_{12} \end{pmatrix}$$

where $[hk\ldots] = \sum_i^N h_ik_i \ldots$ etc.

There exists for all points in a deformed continuous material a set of three directions in space such that linear elements placed along these directions in the undeformed condition will be subject to strains which have maximum or minimum values with respect to strains of linear elements in all other directions. These three maxima are called the principal strains, $S_1 \geqslant S_2 \geqslant S_3$, and the directions on which they lie after deformation are called the principal strain axes. The principal strain axes are orthogonal to each other in the undeformed and in the fully deformed state of the material.

In order to obtain the principal strains and axes, the following relation can be used:

$$\text{Det}\,(T - S_iI) = 0 \tag{A6}$$

where S corresponds to the three principal strains and I, the invariant of eqn (A6) has the components:

$$\begin{aligned} I_1 &= S_1 + S_2 + S_3, \\ I_2 &= S_2S_3 + S_1S_3 + S_1S_2, \\ I_3 &= S_1S_2S_3 \end{aligned}$$

In explicit form, the determinant of eqn (A6) is written from its characteristic matrix as:

$$\begin{vmatrix} \langle\varepsilon_{11}\rangle - S_i & 0{\cdot}5\langle\varepsilon_{12}\rangle & 0{\cdot}5\langle\varepsilon_{31}\rangle \\ 0{\cdot}5\langle\varepsilon_{12}\rangle & \langle\varepsilon_{22}\rangle - S_i & 0{\cdot}5\langle\varepsilon_{23}\rangle \\ 0{\cdot}5\langle\varepsilon_{31}\rangle & 0{\cdot}5\langle\varepsilon_{23}\rangle & \langle\varepsilon_{33}\rangle - S_i \end{vmatrix} = 0 \tag{A7}$$

This determinant reduces to a cubic equation in S_i; the roots of which are the latent roots of eqn (A7). All of the latent roots of eqn (A7) are real, as shown by Imura *et al.* (1962). A positive principal strain value represents a tensile strain while a compressive strain results in a negative sign. In practice, eqn (A7) is solved for its roots by machine computation, and the principal strains $S_1 \geqslant S_2 \geqslant S_3$ are obtained. For the principal strain axes the relation is

$$\begin{pmatrix} \langle\varepsilon_{11}\rangle - S_i & 0{\cdot}5\langle\varepsilon_{12}\rangle & 0{\cdot}5\langle\varepsilon_{31}\rangle \\ 0{\cdot}5\langle\varepsilon_{12}\rangle & \langle\varepsilon_{22}\rangle - S_i & 0{\cdot}5\langle\varepsilon_{23}\rangle \\ 0{\cdot}5\langle\varepsilon_{31}\rangle & 0{\cdot}5\langle\varepsilon_{23}\rangle & \langle\varepsilon_{33}\rangle - S_i \\ i = 1,2,3 & & \end{pmatrix} \begin{pmatrix} l_{1i} \\ l_{2i} \\ l_{3\mathrm{I}} \end{pmatrix} = 0 \qquad \text{(A8)}$$

The eigenvalues of the l eigenvector giving the direction cosines of the principal axes determined from eqn (B8) may be written as the orthogonal matrix:

$$\begin{pmatrix} l_{11} & l_{21} & l_{31} \\ l_{12} & l_{22} & l_{32} \\ l_{13} & l_{23} & l_{33} \end{pmatrix} = l \qquad \text{(A9)}$$

The row vectors in the l matrix determine the principal axes for the strains. Explicitly, l_{11}, l_{21}, l_{31}, are the direction cosines of the principal axis associated with principal strain, S_1, while l_{12}, l_{22}, l_{32} and l_{13}, l_{23}, l_{33} are the direction cosines of the principal axes associated with principal strains S_2 and S_3 respectively.

There is a uniqueness problem associated with the solution of eqn (A8). The Cauchy strains as determined by the statistical algorithm described are the best *estimates* available. The principal strains are derived from these estimates. What happens in the solution of eqn (A8) is that the nine equations never quite give "l" values that make the solution exactly to zero. Instead, a typical value is 0·0007. The set of l values yielding a solution to eqn (A8) closest to zero is taken to be the best estimate of the "l" eigenvector. The computer tests all possible combinations of signed solutions. The result chosen meets the "closest to zero" criterion described. The range in the solutions for the eigenvector gives an idea as to the reliability of the location of the principal strain axes.

The maximum shear strain in any (hkl) plane is of importance since it may be used to quantitatively evaluate the anisotropy of strain in each hkl plane of a family. Furthermore, the exact planes on which strain accommodation occurs may be able to be found from comparisons of the maximum shear strain values according to Slade *et al.* (1965).

The principal strain axes will be taken to be the reference axes. Within the (hkl) plane of interest, there will be some direction where the shear strain attains its maximum magnitude. Let a unit vector in this direction be called $\vec{k}$ and let a unit vector perpendicular to $\vec{k}$ be called $\vec{\jmath}$. Now $\vec{k}$ and $\vec{\jmath}$ are column matrices whose transposes will be called $\vec{k}'$ and $\vec{\jmath}'$ respectively. These vectors are three-vectors. In the principal axis system (the principal frame), the strain tensor, T, reduces to the diagonal matrix:

$$\begin{pmatrix} S_1 & 0 & 0 \\ 0 & S_2 & 0 \\ 0 & 0 & S_3 \end{pmatrix} = T\,(\text{principal frame}) \tag{A10}$$

The shear component associated with $\vec{k}, \vec{\jmath}$ is

$$\varepsilon_{\vec{k}\vec{\jmath}} = 2\vec{k}'T\vec{\jmath} \tag{A11}$$

The factor of 2 comes from the mechanics of the frame transformation of Eqn (A10) to the $\vec{k}$, $\vec{\jmath}$, $(\vec{k}\times\vec{\jmath})$, frame.

The problem involves finding (1) the magnitude of the maximum shear strain in the kj plane along $\vec{k}$ called $(e_{kj})_{\max}$, (2) the magnitude of the shear strain along $\vec{\jmath}$, i.e. perpendiclar to the maximum shear strain in the (hkl) plane and (3) to transform from the $\vec{k}$, $\vec{\jmath}$, $(\vec{k}\times\vec{\jmath})$ frame back to the principal frame in terms of the (hkl) values of interest. The final result gives the maximum shear strain in any (hkl) plane and the direction cosines of the $\vec{k}$, $\vec{\jmath}$ $(\vec{k}\times\vec{\jmath})$ system in terms of the hkl values and the direction cosines of the already determined principal strain axes. The solution to this problem was first presented by Slade *et al.* (1965). A derivation was given by Yakowitz (1973).

In any (hkl) plane, the shear strain is zero normal to a line through some point in the plane. The direction of maximum shear strain in the plane lies at 90° to this line.

On the $\vec{k}$, $\vec{\jmath}$, $(\vec{k}\times\vec{\jmath})$ frame then $\varepsilon_{(kxj)} = 0$. The maximum shear strain occurs in the kj plane and is given by

$$(\varepsilon_{kj})_{\max} = 2[(S_1-S_2)^2k_1^2k_2^2+(S_2-S_3)^2k_2^2k_3^2+(S_3-S_1)^2k_3^2k_1^2]^{\frac{1}{2}} \tag{A12}$$

Note that the kj plane lies in the (hkl) plane of interest. Furthermore, the value of "k_i" in eqn (A12) is needed in terms of (hkl) and the already determined principal strain axes positions. The direction cosines of the hkl axes are given by:

$$h_n = \frac{h}{h^2+k^2+l^2}; \qquad k_n = \frac{k}{h^2+k^2+l^2}; \qquad l_n = \frac{l}{h^2+k^2+l^2} \tag{A13}$$

The required transformation is then given by:

$$k_i = (h_n k_n l_n)\begin{pmatrix} l_{1i} \\ l_{2i} \\ l_{3i} \end{pmatrix} \tag{A14}$$

Thus, the maximum shear strain in any (hkl) plane can be evaluated. With these results, the entire shear strain effect can be followed as a function of strain and/or position in an individual grain of a material.

The next step in the evaluation of the stress–strain configuration in the elastic region is the determination of the stresses in the cubic frame, and the principal stresses and their axes. In the elastic region, the generalized form of Hooke's law relates the stresses to the strains:

$$\sigma_{CUB} = CE \tag{A15}$$

which in explicit form for the cubic case is

$$\begin{pmatrix} \sigma_{11} \\ \sigma_{22} \\ \sigma_{33} \\ \sigma_{23} \\ \sigma_{31} \\ \sigma_{12} \end{pmatrix} = \begin{pmatrix} C_{11} & C_{12} & C_{12} & 0 & 0 & 0 \\ C_{12} & C_{11} & C_{12} & 0 & 0 & 0 \\ C_{12} & C_{12} & C_{11} & 0 & 0 & 0 \\ 0 & 0 & 0 & C_{44} & 0 & 0 \\ 0 & 0 & 0 & 0 & C_{44} & 0 \\ 0 & 0 & 0 & 0 & 0 & C_{44} \end{pmatrix} \begin{pmatrix} \langle\varepsilon_{11}\rangle \\ \langle\varepsilon_{22}\rangle \\ \langle\varepsilon_{33}\rangle \\ \langle\varepsilon_{23}\rangle \\ \langle\varepsilon_{31}\rangle \\ \langle\varepsilon_{12}\rangle \end{pmatrix} \tag{A16}$$

The C values are the elastic constants in the cubic case and the $\langle\varepsilon\rangle$ values are those determined using eqn (A5). Thus, the cubic frame stresses are obtained. The principal stresses, $P_1 \geqslant P_2 \geqslant P_3$ are determined by solving for the latent roots of the stress-analogue of eqn (A7):

$$\begin{pmatrix} \sigma_{11}-P & \sigma_{12} & \sigma_{31} \\ \sigma_{12} & \sigma_{22}-P & \sigma_{23} \\ \sigma_{31} & \sigma_{23} & \sigma_{33}-P \end{pmatrix} = 0 \tag{A17}$$

The solution gives P_1, P_2 and P_3. The principal stress axes can be found in a manner entirely analogous to that outlined for determining the principal strain axes. Again, tensile and compressive stresses are indicated by the sign of the P value.

The stresses with reference to any frame can be determined following the method outlined by Slade *et al.* (1965). These authors showed that a transformation matrix could be deduced in terms of the direction cosines of the axes of the new frame and the results of eqn (A8). Thus, the stresses in any frame can be found by this transformation.

For example, suppose that the stresses are required with respect to the principal strain axes. In other words, what stress configuration is necessary to give the observed three principal strains? Following the treatment of Slade *et al.* (1965) the required transformation matrix $T_{m\sigma}$, is given by:

$$T_{m\sigma} = \begin{pmatrix} l_{11}^2 & l_{21}^2 & l_{31}^2 & 2l_{21}l_{31} & 2l_{31}l_{11} & 2l_{11}l_{21} \\ l_{12}^2 & l_{22}^2 & l_{32}^2 & 2l_{22}l_{32} & 2l_{32}l_{12} & 2l_{12}l_{22} \\ l_{13}^2 & l_{23}^2 & l_{33}^2 & 2l_{23}l_{33} & 2l_{23}l_{13} & 2l_{13}l_{23} \\ l_{12}l_{13} & l_{22}l_{33} & l_{32}l_{33} & (l_{22}l_{33}+l_{32}l_{23}) & (l_{32}l_{13}+l_{12}l_{33}) & (l_{12}l_{23}+l_{22}l_{13}) \\ l_{13}l_{11} & l_{23}l_{21} & l_{33}l_{31} & (l_{23}l_{31}+l_{33}l_{21}) & (l_{33}l_{11}+l_{13}l_{31}) & (l_{13}l_{21}+l_{23}l_{11}) \\ l_{11}l_{12} & l_{21}l_{22} & l_{31}l_{32} & (l_{21}l_{32}+l_{31}l_{22}) & (l_{31}l_{12}+l_{21}l_{32}) & (l_{11}l_{22}+l_{21}l_{12}) \end{pmatrix} \tag{A18}$$

The required result is:

$$\sigma_{PRIN} = T_{m\sigma}\sigma_{CUB} \tag{A19}$$

where σ_{CUB} is obtained from eqn (A16). In this way, the stresses with respect to any frame can be found provided the axes of the frame can be written in explicit form.

The elastic energy stored in the crystal, W, is equivalent to the work done in straining the crystal according to Slade *et al.* (1965). Specifically:

$$W = \int \sigma' \delta \varepsilon = \int \varepsilon' C \delta \varepsilon = \tfrac{1}{2} \varepsilon' C \varepsilon \tag{A20}$$

since $\varepsilon'\varepsilon = \varepsilon^2$.

In explicit terms:

$$2W = (\varepsilon_{11}\varepsilon_{22}\varepsilon_{33}\varepsilon_{23}\varepsilon_{31}\varepsilon_{12}) \begin{pmatrix} C_{11} & C_{12} & C_{12} & 0 & 0 & 0 \\ C_{12} & C_{11} & C_{12} & 0 & 0 & 0 \\ C_{12} & C_{12} & C_{11} & 0 & 0 & 0 \\ 0 & 0 & 0 & C_{44} & 0 & 0 \\ 0 & 0 & 0 & 0 & C_{44} & 0 \\ 0 & 0 & 0 & 0 & 0 & C_{44} \end{pmatrix} \begin{pmatrix} \varepsilon_{11} \\ \varepsilon_{22} \\ \varepsilon_{33} \\ \varepsilon_{23} \\ \varepsilon_{31} \\ \varepsilon_{12} \end{pmatrix} \tag{A21}$$

The final expression is:

$$W = \tfrac{1}{2}C_{11}(\varepsilon_{11}^2 + \varepsilon_{22}^2\varepsilon_{33}^2) + C_{12}(\varepsilon_{22}\varepsilon_{33} + \varepsilon_{11}\varepsilon_{33} + \varepsilon_{11}\varepsilon_{22}) + \tfrac{1}{2}C_{44}(\varepsilon_{23}^2 + \varepsilon_{31}^2 + \varepsilon_{12}^2) \tag{A22}$$

This result is for the cubic case and is referred to cubic axes. For W to be a reliable estimate of the stored elastic energy in the irradiated region, the region must essentially be free of lattice point defects which result in random strains. Sharp Kossel lines indicate that random strains are small since such strains cause line broadening.

To sum up, the Kossel method yields lattice strains for each (hkl) plane of a family. From these can be calculated the Cauchy strains, the principal strains and their axes, the maximum shear strain in any (hkl) plane, the cubic stresses, the principal stresses and their axes, and the total stored elastic energy. This information completely describes the stress–strain configuration of the crystal in the irradiated volume. The reliability of these computed stress–strain results must now be considered.

The standard error, s_d, of the d values can be obtained from eqn (10); the computed values of U, V and W are used to obtain $d_1, d_2 \ldots d_n$ and the standard error calculated in the usual fashion. The determination of the d values and s_d is based on repetitive measurements. The lattice strains, S are computed in turn from the d values and so s_d will propagate as an uncertainty throughout the KISS computations. It is interesting to calculate the number of repetitive measurements necessary to obtain a suitably narrow 95% confidence interval width, $w_{95\%}$, for a given d value. While certain specialized techniques can give relative uncertainties in a d value of about 0·002%, as shown by MacKay (1966), as a more usual value is 0·01 to 0·05%. The 0·05% value can be arbitrarily taken as the maximum acceptable width of the 95% confidence interval for a d value to be used in the stress–strain analysis.

For Fe–3Si, the [110] conics have a Bragg angle, θ, of about 30° when the diffracted wavelength, λ, is Fe–Kα_1 ($\lambda = 1{\cdot}936$ Å). The value of s_e for the [110] conics was found to be 0·0016 (11 determinations). Allowing for error contributions from sources, such as the X, Y measurements and the $X_0 Y_0$ position, the value if s_d is equal to about 0·002; d is 2·024 Å. In order to reduce the width of the 95% confidence interval to 0·05%, enough determinations to reduce the interval to 0·001 Å are needed. By going to a table of "t" values and applying the usual relation

$$W_{95\%} = \frac{(s_d)(t_{n-3})}{\sqrt{n}} = 0{\cdot}001 \tag{A23}$$

Where t_{n-3} is Students t limited by 3 constraints (the U, V, W equations) and n the number of individual d measurements, n is equal to 18. Therefore, even for a conic having θ equal to 30°, some 18(X, Y) coordinate measurements are sufficient to reduce the uncertainty in d to about 0·05%. Other conics, with higher θ values will have smaller ranges of uncertainty. Thus, 18 coordinate measurements are sufficient for all conics in the Fe–3Si case and in general for conics having $\theta \geqslant 30°$ to meet the $d \pm w_{95\%}$ criterion. The usual probable error often reported is $d \pm w_{50\%}$. For 18 determinations of the {110} conic d spacing, $w_{50\%} \simeq 0{\cdot}0004$ Å. Transferred to percentages $d_{110} \pm w_{50\%} = 0{\cdot}02\%$.

The error in the lattice strain, S, is the next quantity to be determined. Assuming a Gaussian error distribution and starting with eqn (18), it is a straightforward matter to arrive at the relation

$$\Delta S = \sqrt{(d_i \Delta d_f)^2 + (d_f \Delta d_i)^2 / d_i^2} \tag{A24}$$

Here $\Delta d_f = w_{f,95\%}$ and $\Delta d_i = w_{i,95\%}$. Now in the elastic or low strain region where the KISS method is valid, d_f will be nearly equal to d_i, and Δd_f will be nearly equal to Δd_i. So as a working approximation

$$\Delta S \simeq \sqrt{2}\, \Delta d_i / d_i \simeq \sqrt{2}\, \Delta d_f / d_f \tag{A25}$$

But as outlined above $(\Delta d_f / d_f)_{\max} = 0{\cdot}05\%$. Hence, for a maximum width of 95% confidence interval $S \pm w_{95\%} = 0{\cdot}07\%$. The probable error in this case will be $S \pm w_{50\%} = 0{\cdot}03\%$. It is worth repeating that this is the case of a conic with a Bragg angle, θ, of 30°. For high θ conics, it is possible to reduce the value of $w_{95\%}$ for S to the neighbourhood of $\pm 0{\cdot}015\%$.

The next step is to determine the uncertainties of the Cauchy strains. The Cauchy strains are computed by a least squares regressive analysis in a manner exactly analogous to the computation of the U, V, W and d values as outlined on p. 460. By analogy to eqn (9), for the case of the strains:

$$s_e = \Sigma[(h_i^2 + k_i^2 + l_i^2) S_{d_i} - l_i^2 \langle \varepsilon_{11} \rangle - k_i^2 \langle \varepsilon_{22} \rangle - l_i^2 \langle \varepsilon_{33} \rangle - k_i l_i \langle \varepsilon_{32} \rangle - l_i h_i \langle \varepsilon_{13} \rangle - h_i k_i \langle \varepsilon_{12} \rangle]^2 / (n-6) \tag{A26}$$

Here n is the total number of (hkl) planes for which there have been obtained S_d values. Then by analogy to eqn (10):

$$s'_{\langle\varepsilon_{11}\rangle} = s_e\sqrt{\Gamma^{11}}$$
$$s'_{\langle\varepsilon_{22}\rangle} = s_e\sqrt{\Gamma^{22}} \qquad \text{and so on to}$$
$$s'_{\langle\varepsilon_{12}\rangle} = s_e\sqrt{\Gamma^{66}}$$

where the Γ values come from the inversion of the second matrix of eqn (A5).

Thus, the standard error is obtained for each of the computed Cauchy strains. These values of $s'_{\langle\varepsilon_{ij}\rangle}$ will decrease as the number of (hkl) planes, n, used to obtain the Cauchy strains increases j increasing n reduces s_e in eqn (A26). For the Γ^{ii} values to decrease, a large spread in the Bragg angles of the (hkl) planes used is necessary. In other words, one must make an adequate sampling of all the planes. There are potentially large errors in S if low θ conics are used. On the other hand, line broadening is usually associated with conics having $\theta \geqslant 80°$. Therefore, a suitable range of (hkl) values is one yielding a range of $30° < \theta < 80°$. Variation of a few degrees at either end will probably not lead to serious problems.

The effect of lattice strain errors on computed $\langle\varepsilon_{ij}\rangle$ values must be combined with the $s'_{\langle\varepsilon_{ij}\rangle}$ values in order to obtain a final estimate of the uncertainties in the Cauchy strains. First, the effect of a systematic error in the S values may be considered. To observe the error propagation, a test case was run through the computer. The Cauchy strains and the $s'_{\langle\varepsilon_{ij}\rangle}$ values were calculated from a published set of typical S data from a tungsten sample strained to about 0·2% by Ellis and Weissmann (1967). Some 20 planes were used in the test. The range of θ was $40{\cdot}2° \leqslant \theta \leqslant 65{\cdot}8°$. Then the S values were varied by set amounts of $+0{\cdot}0005$, $-0{\cdot}0005$, $-0{\cdot}001$. The values of $\langle\varepsilon_{ij}\rangle$, $i \neq j$ did not change while the values of $\langle\varepsilon_{ij}\rangle$, $i = j$ varied by exactly the amount of the systematic error. None of the $s'_{\langle\varepsilon_{ij}\rangle}$ values changed.

It is unlikely that a systematic error in the S values of significant magnitude will occur. The nature of the measurements and the computation of S tend to minimize this possibility. However, random errors of individual S values will almost certainly occur. The magnitude of these errors as a function of θ has already been discussed. In order to test the effect of the random errors, the following procedure was used: (For purposes of this test, a constant increment of $\Delta S = 0{\cdot}003$ representing an average value was used.) The computer was made to list tables of random digits; the sign and the order of the ΔS values was determined by these random digits. Thus, from one table even digits meant a subtraction and vice versa while another table determined the order of the (hkl) plane to which these signs were to be applied. Some four sets of random data suffering uncertainties of S were synthesized in this way. Next the $\langle\varepsilon_{ij}\rangle$ and $s'_{\langle\varepsilon_{ij}\rangle}$ values were computed for each of these four sets. The $s'_{\langle\varepsilon_{ij}\rangle}$ did not vary significantly from their previous values. All of the $\langle\varepsilon_{ij}\rangle$ values changed. The results are shown in Table I. As a test for randomness, all of the new $\langle\varepsilon_{ij}\rangle$ values were averaged. Agreement between this mean value and the original $\langle\varepsilon_{ij}\rangle$ value was taken to indicate randomness.

TABLE I. Reliability of Cauchy Strains Using Data from 20 planes of a Tungsten Crystal Compressed 0·2%

Cauchy strain (values in %)	No assumed error	s'_{ε} (eqn A-26)	RND 1[a]	RND 2[a]	RND 3[a]	RND 4[a]	Mean of RND (1–4)	Range of RND (1–4)	s_{RND}	$s\langle\varepsilon_{ij}\rangle = [(s'\langle\varepsilon\rangle)^2 + s^2_{RND}]^{\frac{1}{2}}$	$(t_{14})s\langle\varepsilon_{ij}\rangle/\sqrt{20}$ $(= 0·48s_{\varepsilon_{ij}})$	Probable error $(= 0·155s\langle\varepsilon_{ij}\rangle)$
ε_{11}	0·077	0·059	0·096	0·050	0·073	0·090	0·077	0·046	0·023	0·063	0·030	0·010
ε_{22}	−0·137	0·074	−0·154	−0·116	−0·120	−0·166	−0·139	0·050	0·025	0·078	0·037	0·012
ε_{33}	−0·053	0·054	−0·042	−0·065	−0·060	−0·037	−0·051	0·028	0·014	0·056	0·027	0·009
ε_{23}	−0·105	0·098	−0·092	−0·121	−0·068	−0·086	−0·092	0·053	0·027	0·102	0·041	0·016
ε_{31}	0·044	0·073	0·046	0·037	0·053	0·044	0·045	0·016	0·008	0·073	0·035	0·011
ε_{12}	0·0026	0·105	0·015	0·057	−0·016	0·057	0·028	0·073	0·037	0·111	0·053	0·017

[a] All strain, S, values incremented by $\pm 0·03\%$. A random numbers scheme was used to assign positive or negative error values in the best of $20S$ values. Range of $|S|$ is 0·001 to 0·189%. Range of θ: $65·8° \geqslant \theta \geqslant 40·2°$. Thus, ΔS of 0·03% represents a reasonable value as indicated by eqn (A25).

Table I shows that the agreement is very good except for $\langle \varepsilon_{12} \rangle$. The range of $\langle \varepsilon_{ij} \rangle$ values for the four cases and the standard error in $\langle \varepsilon_{ij} \rangle$ was calculated. This final standard error, s'_{RND}, and the $s'_{\langle \varepsilon_{ij} \rangle}$ can be combined to give:

$$s_{\langle \varepsilon_{ij} \rangle} = \sqrt{(s'_{\langle \varepsilon_{ij} \rangle})^2 + (s_{RND})^2} \tag{A27}$$

Table I shows that $s_{\langle \varepsilon_{ij} \rangle}$ is about 1·05 $s'_{\langle \varepsilon_{ij} \rangle}$. Hence, the effect of uncertainties in S may be taken into account by asserting that

$$s_{\langle \varepsilon_{ij} \rangle} \simeq 1{\cdot}1\, s'_{\langle \varepsilon_{ij} \rangle} \tag{A28}$$

Equation (A28) represents a statement of the standard error of the calculated Cauchy strains which takes into account uncertainties from all contributing sources.

The width of the 95% confidence interval for the Cauchy strains is of interest. This is obtained from:

$$W_{\varepsilon,95\%} = \frac{t^{95\%}_{(n-6)} s_{\langle \varepsilon_{ij} \rangle}}{\sqrt{n}} \tag{A29}$$

In the case of Table I, some 20 (hkl) planes were used. Hence,

$$W_{\varepsilon,95\%} = \frac{(2{\cdot}145)}{20} s_{\langle \varepsilon_{ij} \rangle} = 0{\cdot}48\, s_{\langle \varepsilon_{ij} \rangle} \tag{A29a}$$

The probable error, $w_{\varepsilon,50\%}$, is $0{\cdot}155 s_{\langle \varepsilon_{ij} \rangle}$. The probable error is the interval including 50% of all future observations. For all of the work discussed here, the probable error is roughly one-third of the 95% confidence interval width.

The effect of uncertainties in the Cauchy strains on the computed principal strains will now be considered. By randomly assigning the $w_{\varepsilon,95\%}$ error on the Cauchy strains, six sets of principal strains were obtained from the data in Table I. All random sets lay within the range of principal strains obtained when all six Cauchy strain uncertainties were either $+w_{\varepsilon,95\%}$ or $-w_{\varepsilon,95\%}$. The mean of all six sets was equal to the principal strains computed from the original calculated Cauchy strains. Results are shown in Table II.

TABLE II. Reliability of Principal Strains in a Tungsten Crystal Strained 0·2%

	$\langle \varepsilon_{ij} \rangle \pm 0$	$\langle \varepsilon_{ij} \rangle + w^{95}_{\langle \varepsilon \rangle}$	$\langle \varepsilon_{ij} \rangle - w^{95}_{\langle \varepsilon \rangle}$
S_1	−0·031	−0·039	−0·019
S_2	0·081	0·116	0·052
S_3	−0·163	−0·223	−0·113

	$\langle \varepsilon_{ij} \rangle + w^{50}_{\langle \varepsilon \rangle}$	$\langle \varepsilon_{ij} \rangle - w^{50}_{\langle \varepsilon \rangle}$	Laced $w^{95}_{\langle \varepsilon \rangle}$ + − + − + −	Laced $w^{95}_{\langle \varepsilon \rangle}$ − + − + − +
S_1	−0·034	−0·028	−0·018	−0·045
S_2	0·092	0·070	0·069	0·093
S_3	−0·182	−0·145	−0·178	−0·148

Based on these results, it is reasonable to assert that a 95% confidence interval for the principal strains can be gotten by calculating three sets of principal strain values. First with the calculated $\langle \varepsilon_{ij} \rangle$ values, then in turn with $\langle \varepsilon_{ij} \rangle + w_{\varepsilon,95\%}$ and $\langle \varepsilon_{ij} \rangle - w_{\varepsilon,95\%}$. The resultant range should be a close estimate of the width of the 95% confidence interval for each of the three principal strains.

For the case used as an example, Table II shows that the 95% confidence interval is about ± 25 to 30% of the principal strain value for all three principal strains. These values appear to be typical for lightly strained Fe–3Si as well as for the tungsten test case. There is no reason to believe that they will be much different for other materials. Hence, it is concluded that for lightly strained (elastic region or residual strains), the Kossel method, carefully applied, will generally give principal strain values having 95% confidence interval widths of $\pm 30\%$ or probable errors of about $\pm 10\%$ relative to the computed principal strains.

With these results, the reliability of the stress calculations (eqn A16), (A17) can be obtained by substituting the principal strain limits into the appropriate relation. The elastic constants are assumed to be well known. In this way, the reliability of the entire stress–strain analysis is established.

As an overall generalization, all of the calculated results of the stress–strain analysis obtained by the Kossel method can be expected to have an approximate probable error of about $\pm 10\%$ relative to the calculated value. It is difficult to see how this error can be reduced significantly in the future in view of the stringent requirements for using many (*hkl*) planes having $30° < \theta < 80°$, and the difficulty in significantly improving on the regressive analysis methods used to calculate the stress–strain parameters.

REFERENCES

DINGLEY, D. J. (1971). *Proc. 25th Anniversary Meeting EMAG* (W. C. Nixon, ed.), p. 206. Institute of Physics, London.

ELLIS, T. and WEISSMANN, S. (1964–67). *U.S. Air Force Materials Lab. Technical Documentary Report* ML-TDR-64-220, Parts I-IV.

FRAZER, J. Z. and ARRHENIUS, G. (1966). *In* "Optique des Rayons X et Microanalyse" (R. Castaing, P. Deschamps and J. Philibert, eds), p. 516. Hermann, Paris.

GIELEN, P., YAKOWITZ, H., GANOW, D. and OGILVIE, R. E. (1965). *J. appl. Phys.* **36,** 773.

HIRSCH, P. B., HOWIE, A., NICHOLSON, R. B., PASHLEY, D. W. and WHELAN, M. J. (1965). "Electron Microscopy of Thin Crystals," pp. 100–103. Butterworths, London.

IMURA, T., WEISSMANN, S. and SLADE, J. J. (1962). *Acta crystallogr.* **8,** 786.

JAMES, R. W. (1950). *In* "The Crystalline State" (L. Bragg, ed.), vol. II. Bell, London.

JELLINEK, M. H. (1949). *Rev. scient. Instrum.* **20,** 368.

LONSDALE, K. (1947). *Phil. Trans. R. Soc. Lond.* **A240,** 219.

MACKAY, K. J. H. (1966). *In* "Optique des Rayons X et Microanalyse" (R. Castaing, P. Deschamps and J. Philibert, eds), p. 544. Hermann, Paris.
MORRIS, W. G. (1963). M.Sc. Thesis. Massachusetts Institute of Technology.
MORRIS, W. G. (1968). *J. appl. Phys.* **39,** 1813.
SCHULSON, E. M. (1971). *Phys. St. Sol.* (b) **46,** 95.
SHARPE, R. S. (1965). *Appl. Mater. Res.* **4,** 74.
SHRIER, A., KALMAN, Z. H. and WEISSMANN, S. (1966). U.S. Government Research Report AD 631 179 (available from U.S. Dept. Commerce).
SLADE, J. J. (1963). *Conf. Intern. Union Crystall., 6th Rome; Acta crystallogr. Abs.* **108**.
SLADE, J. J., WEISSMANN, S., NAKAJIMA, K. and HIRABAYASHI, J. M. (1965). *J. appl. Phys.* **35,** 3373.
YAKOWITZ, H. (1966). *J. appl. Phys.* **37,** 4455.
YAKOWITZ, H. (1969). *In* "Electron Probe Microanalysis" (A. J. Tousimis and L. L. Marton, eds), p. 361. Supplement VI to Advances in Electronics and Electron Physics, Academic Press, New York.
YAKOWITZ, H. (1973). NBS Monograph Series, Monograph 130, 75 pp.
YAKOWITZ, H. and VIETH, D. L. (1965). *J. Res. natn. Bur. Stand.* **C69,** 213.

Chapter 14

Application of the Kossel X-ray Back Reflection Technique in the Scanning Electron Microscope

D. J. DINGLEY and J. W. STEEDS

H. H. Wills Physics Laboratory, University of Bristol, England

Symbols

E_c = Absorption edge
$lp = \mu$ = photo-electron absorption length
ξ_p = primary extraction length
Δ_0 = Pendellösung length
θ_B = Bragg angle for reflector from g
g = crystal plane
I_c = intensity of characteristic radiation
k = constant
E = accelerating voltage
v_c = cell volume
d_g = spacing of reflecting planes
p = polarization factor
σ polarization $\perp$ reflecting planes
π polarization $\parallel$ reflecting planes
$\varepsilon_0 = 8{\cdot}85 \times 10^{-12}$ farad/metre
m = electron mass
c = velocity of light
Γ_e = radius of electron
z = depth
a = lattice parameter

I. Introduction

The application of Kossel X-ray diffraction techniques in electron microscopy is becoming more widely practised.

In this chapter we review some of the physical principles underlying the generation and observation of back reflection Kossel patterns. We give details of suitable cameras and photographic techniques, and describe a number of experiments which have been performed with a Cambridge Scientific Instruments Stereoscan. Finally we make comparison of the Kossel technique with that using back reflection (or Coates) patterns.

II. Physical Principles

A. Self-generated Kossel patterns

We first need to establish the conditions which must be satisfied in order to obtain self-generated Kossel cones. The pseudo-Kossel technique, using X-rays generated by an external source of different wavelength, will be discussed later.

The most obvious condition is imposed by Bragg's Law itself. As the elements in a crystal structure become lighter the hardest K shell X-rays must always permit a solution such that $\sin\theta < 1$. Practical conditions in fact place a more restrictive condition on θ. To illustrate this point Table I gives the Bragg angles for self-generated X-rays for a variety of elements, from Ti to Ag. Of course, it may be possible to extend the range downwards to lower atomic numbers in the case of crystals with larger repeat distances, but it is doubtful if there is much to be gained in this way. The X-rays of light elements being soft are very easily absorbed, and this provides a second limitation, both self absorption and absorption by surface films on the specimen or filter causing serious reduction of the Kossel intensity. Some photo-electric absorption lengths (lp) are included in Table I. A limitation also exists on trying to extend the range upwards beyond Zn or Ge. In our case it is the design of the microscope itself which intervenes. It is well known (see, for example, Compton and Allison, 1935) that the intensity of characteristic radiation from a thick target may be approximately represented by $I_c = k(E-E_c)^f$. Where k is a constant, E is the accelerating voltage of the electrons striking the target, E_c is the appropriate ionization potential and it has been found experimentally that $f \sim 1{\cdot}65$. On the other hand the integrated intensity of white radiation is approximately proportional to E^2. Thus, in order to maximize the signal-to-noise ratio we require

$$\frac{\mathrm{d}}{\mathrm{d}E}\frac{(E-E_c)^f}{E^2} = 0$$

or

$$E = 2E_c/(2-f) \sim 5{\cdot}8E_c.$$

Current X-ray practice indicates a value of about half this is acceptable. The values of E_c for K_α excitations are shown in Table I and it is clear that for our maximum accelerating voltage of 30 keV an unfavourable

TABLE I. Bragg angles

Element	keV E_c	μm l_p	g	θ_B	μm ξ_p	μm $\frac{1}{2}\Delta_0$
Al	1·56	11	—	—	—	—
Si	1·84	13	—	—	—	—
Ti	4·96	23	$1\bar{1}00$	32°41′	0·48	1·87
			0002	36°5′	0·27	0·92
			$1\bar{1}01$	37°48′	0·33	1·06
V	5·46	18	110	35°52′	0·40	1·57
			200	56°24′	0·78	1·26
			211	—	—	—
Cr	5·99	17	110	34°16′	0·39	1·43
			200	52°39′	0·68	1·23
			211	77°25′	1·15	0·63
Fe	7·11	17	110	28°29′	0·35	1·60
			200	42°27′	0·60	1·63
			211	55°54′	0·86	1·44
Co	7·71	18	0002	26°21′	0·16	0·80
			$10\bar{1}1$	27°58′	0·20	0·98
			$11\bar{2}0$	45°38′	0·34	0·82
			$11\bar{2}2$	57°14′	0·46	0·73
Ni	8·33	20	111	24°5′	0.34	1·86
			200	28°6′	0·42	1·93
			220	41°55′	0·73	2·02
Cu	8·98	21	111	21°43′	0·26	1·61
			200	25°16′	0·32	1·67
			220	34°58′	0·56	1·98
Zn	9·66	29	0002	16°55′	0·14	1·15
			$1\bar{1}00$	18°11′	0·31	2·34
			$1\bar{1}01$	27°15′	0·57	2·27
Mo	2·87	1·35	110	59°53′	0·25	0·70
			200	—	—	—
Ag			111	30°0′	0·45	1·93
	3·81	1·8	200	50°16′	0·55	1·16
			220	—	—	—

K_α, Al–Zn; L_α, Mo, Ag.
E_c = Absorption edge, l_p = photoelectric absorption length.
ξ_p = primary extinction length, Δ_0 = Pendellösung length.

situation occurs for $Z > 32$. On the other hand L_α radiation is much too soft in this region and it is not until elements such as Mo and Ag that L shell radiation may be used. These conclusions have been verified experimentally by Ullrich (1967).

The second point to explore is the probable depth of X-ray sources below the bombarded crystal surface. Monte-Carlo calculations

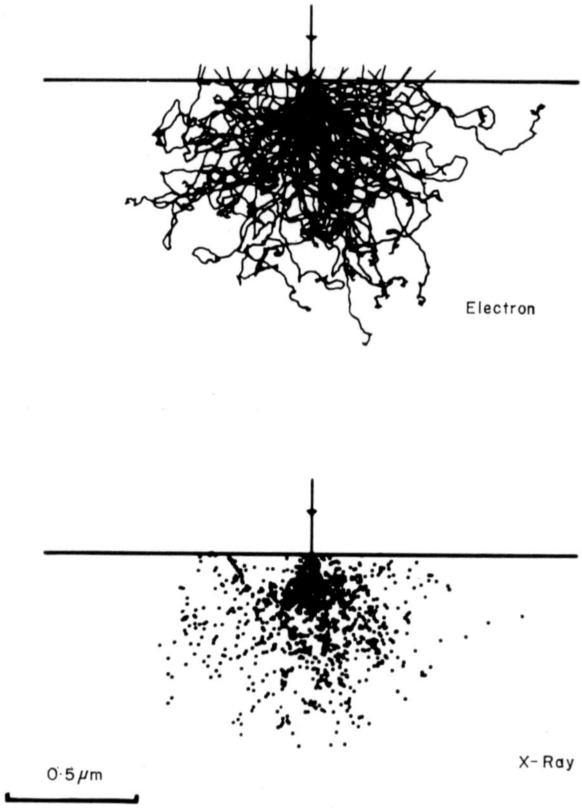

FIG. 1. Computer plot of electron trajectories and X-ray production in a copper target at 20 kV. A Monte-Carlo technique is used to simulate the tracks of 100 electrons entering the target (some of which are back-scattered) and to generate the resulting X-ray distribution. (Courtesy Duncumb, 1971.)

appropriate to the case of fine grain poly-crystalline material were performed by Duncumb (1971) and some results are shown in Fig. 1. In single crystals, bombarded along directions of high symmetry, the source distribution may be expected to be rather less concentrated. Electron crystal waves will be set up which are concentrated either on the lattice planes, or mid-way between them. The on-plane waves will generate X-rays nearer, and the between-plane waves further, from the crystal surface than in the Monte-Carlo case. In any event it is clear that the source will lie at a depth of approximately 0·5 μm.

Next we have to examine the characteristic distances associated with primary extinction (ξ_p) and Pendellösung (Δ_0), as we expect them to play an important role in controlling the intensity of the Kossel cones. Primary extinction occurs in the Bragg case (Fig. 2(a)) if the crystal is not too imperfect. Formally, it is the distance below the surface of a semi-infinite crystal in which the incident intensity drops to 1/2·718,

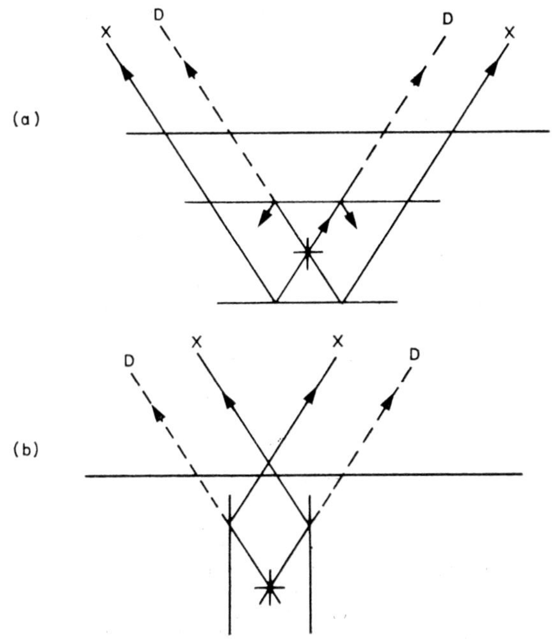

FIG. 2. Generation of back reflection Kossel cones of characteristic radiation (X = excess, D = deficiency) (a) in the Bragg case, (b) in the Laue case.

when the crystal is set for total reflection in the symmetric case. The primary extinction length is a useful parameter if the crystal has regions of almost undisturbed lattice with linear dimensions comparable with or greater than ξ_p. The Pendellösung distance measures, in the symmetric Laue case (Fig. 2(b)), the wavelength of depth oscillations of either the direct or diffracted intensity, and it is therefore similar to the extinction distance in electron diffraction. The definitions of ξ_p and Δ_0, respectively for the symmetric Bragg and Laue cases are

$$\xi_p = \frac{V_c}{\pi d_g |F_g| \, |\rho| r_e}$$

$$\Delta_0 = \frac{\pi V_c \cot\theta}{2 d_g |F_g| \, |\rho| r_e}$$

and

$$\frac{\Delta_0}{\xi_p} = \frac{\pi^2}{2} \cot\theta$$

where V_c is the volume of the cell for which the structure factor, F_g, is evaluated; d_g is the spacing of the reflecting planes; ρ is the polarization factor, $\Delta = 1$ (σ polarization) or $\cos 2\theta$ (π polarization); $r_e = e^2/4\pi\varepsilon_0 mc^2$ the classical electron radius ($2{\cdot}82 \times 10^{-13}$ cm).

It is important to notice that the ratio of ξ_p to Δ_0 can exceed an order of magnitude, and that it depends on the Bragg angle. This difference arises from three separate sources. First, as customarily defined (James, 1948; Batterman and Cole, 1964), the primary extinction length is quoted for a spherical source, introducing a factor $\pi/4$ to account for the angular average (this is clearly an artificial distinction). Secondly the attention due to primary extinction is written

$$I = I_0 \exp(-z/\xi p)$$

whereas the depth oscillations have the form

$$I = I_0 \cos \frac{2\pi\zeta}{\Delta_0}$$

This accounts for a further factor of 2π. Finally ξ_p is measured perpendicular while Δ_0 is measured parallel to the reflecting planes. This accounts for the $\cot\theta$ factor. Values of ξ_p and Δ_0 are included in Table I. In the region of interest ξ_p is dependent on the reflection operating but to a first approximation Δ_0 is not.

We are now equipped to examine the origin of Kossel cones. Let us imagine a spherical source situated at a depth of about $0{\cdot}5\ \mu$m below the crystal surface. Since the angular range over which total reflection occurs in the symmetric Bragg case is $\Delta\theta = 2/\Delta_0 g$ (Batterman and Cole, 1964) the total intensity scattered from a single crystal or from a large enough crystallite will be $\propto(\Delta\theta)^2/4\pi$. Since $\Delta\theta$ is only a few minutes of arc this intensity will be much too small to be detectable. In the symmetric Laue case the diffracted intensity from a particular set of planes arising from a spherical source is also approximately proportional to

$(1/\Delta_0 g)^2$ (Kato and Lang, 1959). It is therefore clear that Kossel lines, generated in the symmetric Laue case (Fig. 2(b)), will not be detectable owing to the short path to the surface. Although in the Bragg case, the intensity of the Kossel cones from highly perfect crystals will be too weak to be visible, if the crystal is somewhat imperfect, with good regions larger than the primary extinction length, then appreciable intensities can be built up. The situation is depicted schematically in Fig. 3. Slightly misoriented regions, each reflecting intensity $(\Delta\theta)^2/4\pi$,

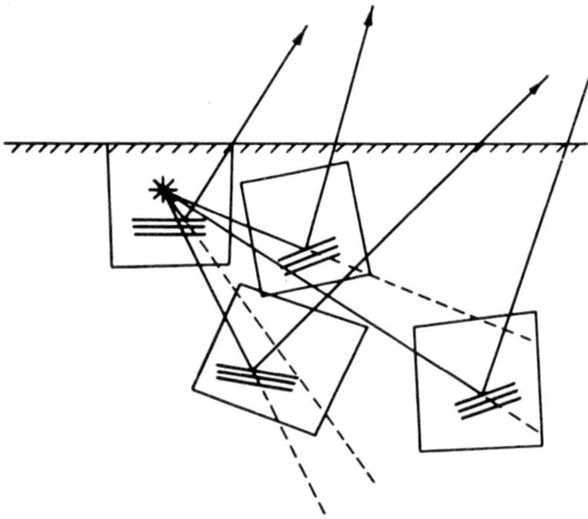

FIG. 3. Diagrammatic representation of the way in which slightly misoriented regions of a crystal lead to the production of strong Kossel cones.

contribute to the total up to a distance from the surface of approximately $lp/2 \sin\theta$. As the number of these contributions increases the Kossel cones become stronger. We now argue that the optimum size of the good regions is approximately the primary extinction length. When the good regions are smaller than this the intensity reflected from each is considerably reduced, leading to a loss in integrated intensity. The form of variation of Kossel cone intensity with crystal perfection may therefore be represented schematically by the form shown in Fig. 4. We may also use the idealized model of Fig. 3 to predict the angular spread in the Kossel cones as a function of crystal perfection. When the good regions are large there are only a few separate contributions and narrow angular spreads to the cones. However, as the good regions become smaller, the number of components in the Kossel cones increases progressively, as does the angular width.

Another interesting question is the relative strength of the various Kossel cones. In the case of good regions whose dimensions exceed ξ_p, we may say that the intensity of the Kossel cones is proportional to

$[|F_g|/g\Delta_0]$. That is, structure factors being equal, the low index cones will be strongest (Δ_0 approximately constant). Additional factors come in for less perfect crystals. Consider for example the case of a continuously deformed iron single crystal. If we start with a highly perfect single crystal, deformation will introduce misoriented regions whose mean spacing decreases progressively. Kossel lines, initially invisible, will appear weak at first and narrow, becoming stronger and broader until the mean separation of the good regions becomes close to the primary extinction length. When this condition is reached the strength of the reflection is reduced leading to a smaller contribution from each good region and weaker Kossel lines. The first reflections to be affected are those with the longest ξ_p, that is, the high index reflections (Table I).

We finally turn to the question of whether deficiency or excess Kossel cones will occur (Fig. 2). In back reflection the path of deficiency cones through the crystal is very short. They are therefore weak and normally invisible.

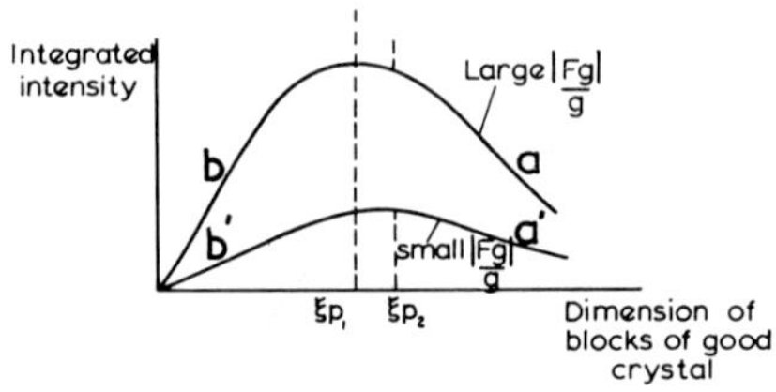

FIG. 4. Schematic graph of the variation of intensity of the Kossel cones with size of misoriented blocks.

B. BACK-REFLECTION PSEUDO-KOSSEL PATTERNS

Pseudo-Kossel cones are generated by an external source of X-rays. They may be the only available means of producing Kossel lines in the case of low atomic number materials and they are also valuable for materials in the atomic number range $32 < Z < 40$. The X-rays may be produced in either an evaporated film on the specimen surface or else in a physically separated film. Film thicknesses of the order of 0·5 μm are required and the film must clearly be in a fine grained polycrystalline form. One advantage of this technique is that by suitable choice of film material it is possible to vary the penetration of the X-rays into the crystal of interest. However, there are two important disadvantages. In the first place 0·5 μm of coating may greatly impair the topographic information obtained in scanning microscopy. Further, in back reflection work, the number of Kossel cones is markedly reduced. The reason for this can be seen in Fig. 5. Only downwardly directed X-rays enter the crystal under investigation. Important paths such as *A* pass only through the surface film. On account of these limitations it is important to work with films which are just thick enough to generate sufficient X-rays.

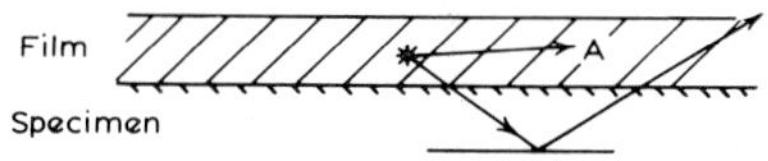

FIG. 5. Production of pseudo-Kossel cones by an evaporated film on the surface of the specimen. Only downward directed X-rays can contribute.

III. KOSSEL CAMERAS FOR THE SEM

The Kossel X-ray diffraction pattern cannot yet be imaged on the display screen of the SEM so that recording is necessarily by means of a photographic emulsion placed close to the specimen and within the vacuum system of the microscope. In practice, for examination of bulk samples, a back reflection technique must be used with the film or plate positioned vertically above the specimen and immediately below the final condenser lens of the microscope. One has to choose between film and plate. On the one hand it is easier to cut the necessary access hole for the electron beam in the case of film, while on the other hard glass-backed emulsions do not suffer from shrinkage problems on development. For the most accurate work glass, therefore, is preferred.

Transmission patterns can be obtained from thin samples and then the recording film is placed beneath the specimen and a central hole is unnecessary. However, a special specimen holder and specimen stage are required which, although not difficult to construct, increase the amount of microscope adaption.* A third possible method of recording the patterns is to surround the specimen with film wound as a cylinder whose axis is parallel to the incident electron beam. This technique looks promising and has been used in a specially built instrument (Heise, 1962), though new methods are required to interpret the patterns obtained with this geometry.

Three back reflection cameras are described below. They were built for use in the Cambridge "Stereoscan" microscopes, Marks 1,2 and S4, and although they cannot be fitted directly to other microscopes it seems likely that minor modifications would make them suitable for more general use. Two of the design criteria adopted have been that

* A specimen stage suitable for transmission work has been constructed and will be the subject of a short paper to be published in the near future.

the cameras should not interfere with the normal operation of the microscope and that resolution should be unimpaired. These features have been successfully incorporated in all three designs.

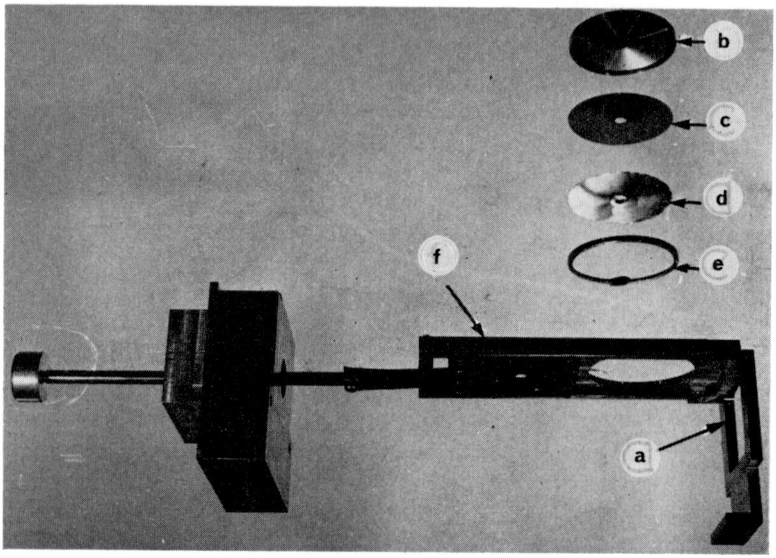

FIG. 6(a). Exploded view of components of single plate Kossel camera.

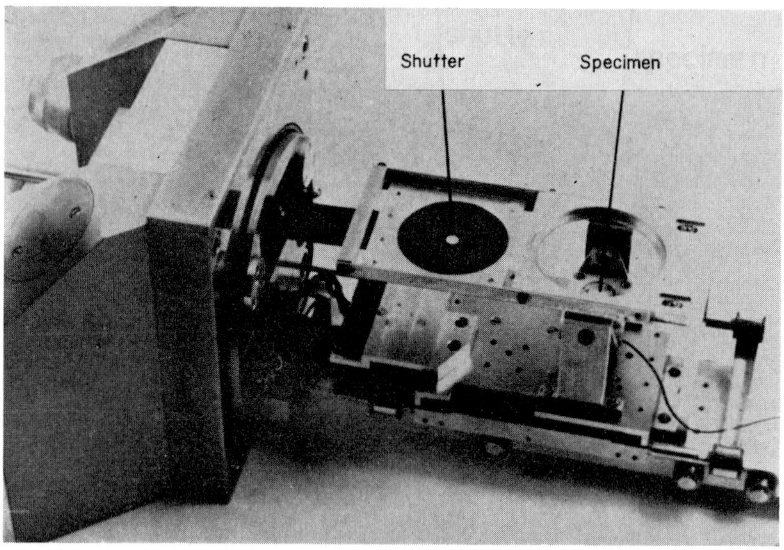

FIG. 6(b). View of assembled camera.

A. SINGLE PLATE CAMERA

This is the first design to prove successful and as described is suitable for use with the Cambridge universal specimen stage. As with the other two cameras to be described, it consists basically of a support stand, film holder, filter and shutter. These units for the single plate camera are shown in exploded view in Fig. 6(a) and assembled in Fig. 6(b). The support stand, *A*, consists of an aluminium bracket recessed to locate the film holder assembly beneath the final condenser lens of the microscope. It is mounted on the base plate of the specimen stage and is thus independent of the *X*, *Y* and *Z* shift controls so allowing the specimen to be moved independently of the camera. The design allows for minimum obstruction of scattered electrons from the specimen so that the normal collector system can be used with little degradation of the signal when viewing the sample in the normal scanning mode. The film holder, B, consists of an aluminium disc 4 mm thick by 50 mm in diameter, and is recessed to accept the film, C, and filter, D. These are in turn held in position using a circular clip, E. The filter acts to reduce the amount of the Bremsstrahlung fogging, as a sink for back-scattered electrons and as a light-tight cover for the film. The film holder, film and filter have central hole through which is threaded a short aluminium tube to maintain concentricity and to prevent charging of the inside edge of the film. The shutter F is a rectangular copper sheet 0·5 mm thick mounted in grooves in the support stand. It is positioned beneath the film holder during normal operation of the microscope and is withdrawn by sliding it along the grooves when recording the Kossel pattern. A push–pull rod acting through a neoprene seal serves as external control for the shutter.

Changing the film after exposure requires that air be admitted to the specimen chamber, the specimen stage withdrawn and a new film holder assembly fitted. The assembly is light tight so that black-out conditions are not necessary during these operations. It is a simple procedure and takes between 2 and 5 min depending on the microscope. The films used are pre-dessicated to decrease pumping time. Experience has shown that the Cambridge stage is sufficiently stable that the specimen does not shift by more than 5 μm during the change.

B. TWO-PLATE CAMERA

The first plate camera shown in Fig. 7 is completely independent of the specimen stage used, being mounted on the 5 in. diameter side entry port of the microscope. It imposes little restriction on working distance, being designed specifically for use with a liquid helium stage which clears the final lens by only 4 mm. The film holder support assembly, part A, consists of a 5 mm thick copper sheet 50 mm wide by 190 mm long with the long edges bent over to form grooves along which the film holder can slide. It is fixed to the standard side entry blanking plate, and projects into the microscope just beneath the roof of the specimen chamber. The film holder, B, is simply another copper

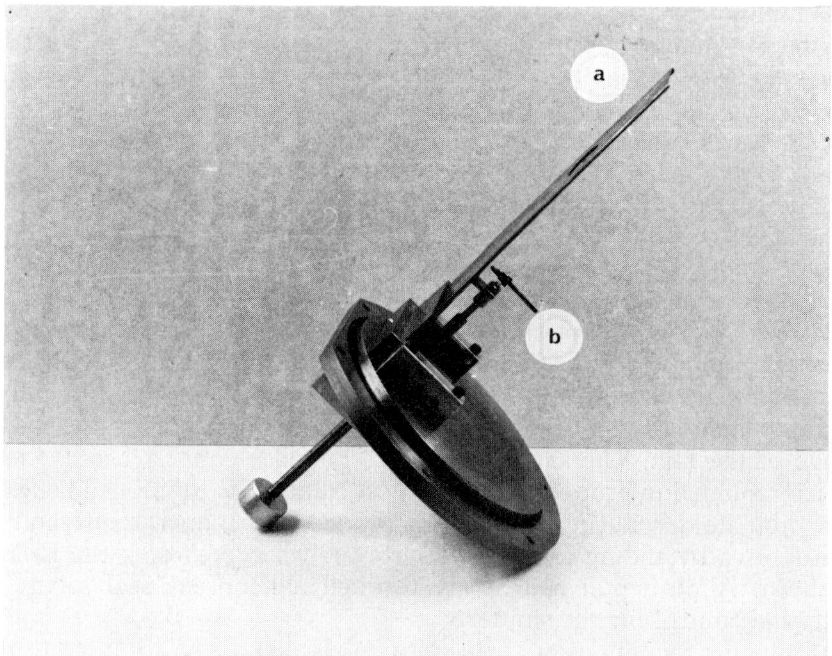

FIG. 7. View of two plate camera.

sheet 100 mm long which slides along the grooves in the support assembly. The two films are glued, using double sided Scotch tape, to the underside of this sheet beneath two holes drilled along its centre line. During normal scanning operation the film holder is withdrawn to one side of the specimen chamber and to record a Kossel pattern the holder is pushed into the centre of the microscope to locate one of the films concentrically beneath the final condenser lens. The holder is withdrawn after exposure. The second film is exposed by pushing the film holder further into the microscope so that the second film lies above the specimen. No moving shutter is necessary in this design, fogging of the plate being prevented by attaching a thick copper sheet to the underside of the support stand so that it lies directly below the film holder when in its withdrawn position. The filter material is taped directly across the film and fixed to the film holder.

As in the previous case it is necessary to admit air into the specimen chamber when changing films but this now has to be carried out under safelight conditions as the film holder is not light-tight. Unsticking the film from the film holder has presented no difficulties except some minor film cracking. Although not the best technique for really accurate work it has been found acceptable for routine studies. The chief advantage of the camera is that it is compact and projects a minimum distance beneath the final condenser lens of the microscope.

C. TWELVE-PLATE CAMERA

For general use it is desirable that many exposures can be made in relatively quick succession for example, during *in situ* tensile straining of specimens in the microscope. A multiple plate camera was therefore constructed and in order to minimize operational problems during plate changing and positioning in the microscope, was automated using digital electronics. The last version of this camera stores twelve plates

FIG. 8. View of multiple camera mounted on S4 Stereoscan SEM.

in the microscope and is shown in Fig. 8. It is completely independent of the specimen stage and as in the previous case is mounted on the 5 in. diameter side entry port of the microscope and occupies a minimum of space within the specimen chamber.

The film holder support assembly has been reduced to two parallel rods which project from the side entry port across the specimen chamber and lie 4 mm beneath the final condenser lens. One of the rods is screw threaded and acts as a lead screw for the film holder; the other is polished and acts as a guide rod. The film holder itself is a rectangular frame recessed to accept a 65 × 65 mm plate and is fixed to the lead screw and guide rod. It is driven from its standby position at the side of the specimen chamber to the centre of the microscope when a Kossel pattern is to be recorded. The drive mechanism is a stepping motor.

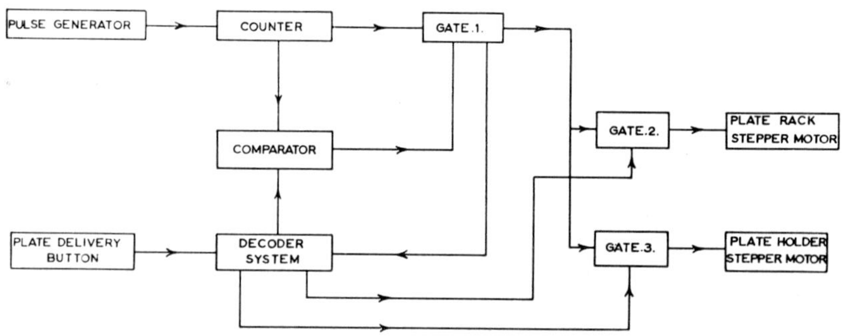

FIG. 9. Schematic of digital electronics system for multiplate Kossel camera. The decoder system determines which motor is to run and for how many pulses, and is controlled by pulses from gate 1 and plate delivery button.

The twelve plates are stacked ready for use outside the specimen chamber in an aluminium casing attached to the 5 in. side entry port and integral with the microscope column. The lead screw and guide rod extend into the aluminium case allowing the plate holder to be driven beneath the stack to deliver and pick-up plates. Movement of the plate rack is also controlled by a stepping motor and both motors are fed with pulses from a control unit alongside the microscope. A scheme of the control unit is shown in Fig. 9.

No shutter is required as the plates are stored remote from the X-ray and electron source. The filter is attached beneath the rectangular frame of the plate holder. Finally, though the camera is designed for use with glass plates, film can be used if this is first mounted on glass using one of the standard fixatives such as Kodak photoflat.

IV. Recording Techniques

The procedure for recording Kossel patterns is straightforward but requires some practice before repeatable success can be achieved. The technique adopted is as follows. The accelerating voltage is first set to 2·4–2·9 times the X-ray characteristic voltage to ensure a sufficiently large ratio of characteristic to background radiation. The specimen current signal is increased to a value of 50×10^{-9} A by reducing the first and second condenser lens currents; so providing sufficient X-ray intensity to reduce the exposure time to 1–2 min. The precise exposure time will depend on filter thickness, type of specimen, film, and specimen to film distance; and can at present only be found by experience. The specimen is next imaged at a suitable magnification to permit the probe to be positioned precisely at the point of interest. In the case of the single plate camera the shutter is now withdrawn from beneath the specimen so beginning the exposure. In the cases of the two- and twelve-plate cameras the plate is moved into position above the specimen to begin the exposure. The operating mode is altered to "point" and the probe so formed located on the area of interest. When the exposure is complete the shutter is closed or the plate withdrawn as the case may be.

There are minor variations on this technique—two of which are important. Firstly, if the size of the diffracting volume is not critical, e.g. in determining the crystal orientation of large grains, then (provided the grain diameter is substantially larger than the raster size) it is unnecessary to change to the point mode of operation. Secondly, it is possible to hold the plate in the expose position for short periods without significant fogging occurring, provided the intensity of the incident beam is reduced by increasing the first and second condenser lens currents. This is useful, for example, when a double exposure is required and the specimen has to be manipulated between exposures.

The type of emulsion used depends on circumstances and is selected by making a compromise between resolution and exposure time. Some typically useful emulsions are listed in Table II. The high beam current required for recording the Kossel pattern, 50×10^{-9} A, is 500 times greater than that used during normal high resolution operation of the microscope. This means the beam diameter is greatly increased and is of the order of 1 μm. The exact size will depend on accelerating voltage, filament height, bias control setting, lens currents, specimen working distance and size of final aperture. It is desirable to maintain the beam diameter as small as possible, especially when examining fine grained material, as it is a contributing factor to the Kossel line width. In the case where the beam diameter is comparable with the grain size, in addition to the broadening there is a loss of intensity and the lines may become indistinct. It is then necessary to reduce the beam diameter by increasing the lens currents and compensating for the decreased beam current with larger exposure times.

TABLE II. List of Photographic Emulsions suitable for Recording Kossel X-ray Photographs

Emulsion	Base	Speed	Developer	Time
Kodax Microtex	Estar	Slow	DX-80	4–12 min
			D19-b	5–15 min
Kodax Crystallex	Estar	Medium	DX-80	4–18 min
			D19-b	5–10 min
Kodax Industrex D	Estar	Fast	DX-80	4 min
			D19-b	5 min
Kodax Kodivex	Estar	Very Fast	DX-80	4 min
			D19-b	5 min

The specimen to film distance determines the amount of Kossel pattern recorded on the photographic plate. In the cameras described above, this distance could be as large as 5 cm but it was normally set at 2 cm, the higher precision obtained at the larger working distances being forfeited for increased capture angle.

V. EXPERIMENTAL OBSERVATIONS

A. KOSSEL PATTERNS FROM COARSE GRAINED MATERIAL

Two examples are shown, the first, Fig. 10(a), was obtained from polycrystalline cobalt using an iron filter and the second, Fig. 10(b), was obtained from pure iron using the same filter. In both cases the grains were greater than 50 μm in diameter and the specimen to film distance was 20 mm. A second pattern from pure iron is shown in Fig. 11 and was obtained using a manganese filter and a specimen to film distance of 30 mm. The absorption edge of the manganese, 6·54 keV filters out the Fe K_β radiation, 7·06 keV, but transmits the iron K_α doublet. The result is a sharper set of Kossel lines and increased resolution as can be seen by comparison of Figs 10 and 11 where in the latter the 112K_α doublet is easily resolved whilst in the former it remains unresolved. The mottled appearance of the background in the pattern obtained using the manganese filter is due to residual cold work in the filter material.

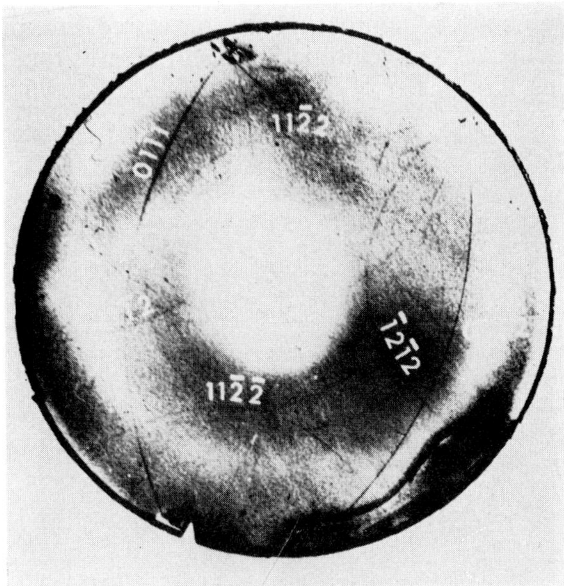

FIG. 10(a). Kossel pattern from polycrystalline cobalt.

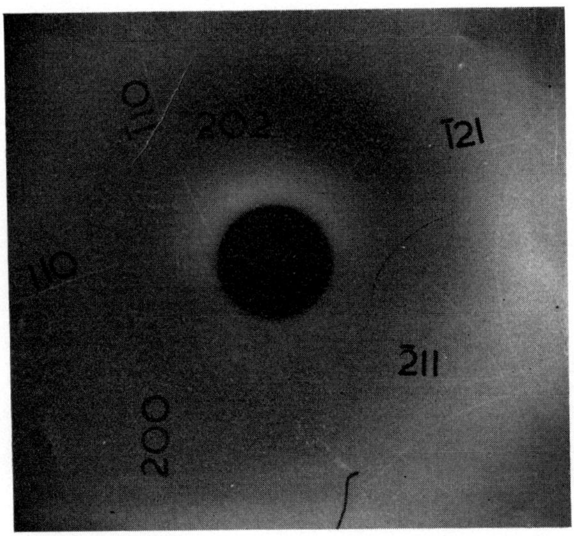

FIG. 10(b). Kossel pattern from polycrystalline iron. Specimen to film distance 20 mm, iron filter 25 μm thick.

A measure of the accuracy in determining the orientation from such patterns is attainable by comparing the measured interplanar angles with theory. Using a 30 cm Wulff net the accuracy was found to be $\pm\frac{1}{2}°$.

Such patterns have been obtained from grains down to 10 μm in

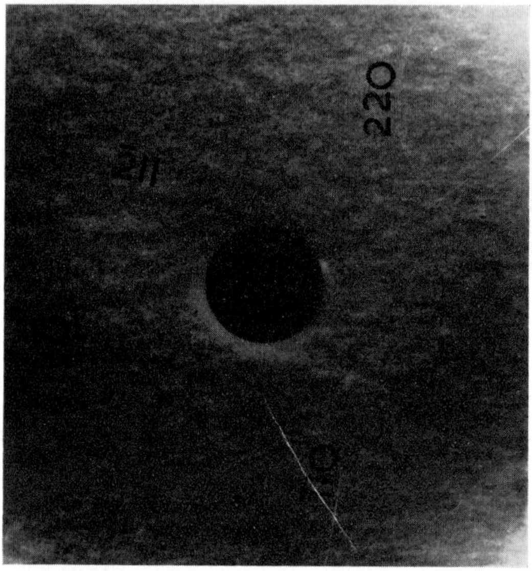

FIG. 11. Part of Kossel pattern from polycrystalline iron showing increased resolution using manganese filter.

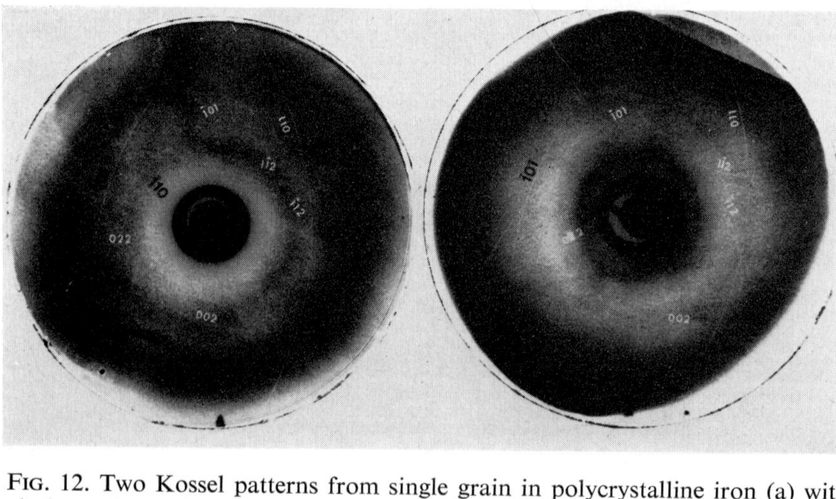

FIG. 12. Two Kossel patterns from single grain in polycrystalline iron (a) with probe located 5 μm from grain boundary, (b) with probe 15 μm from grain boundary,

diameter. However, the patterns obtained from the smaller grain sizes contain additional lines due to X-ray diffraction in neighbouring grains, as would be expected from Table I. To demonstrate this quantitatively a series of patterns was obtained from a single large grain in a polycrystalline sample, each pattern being recorded with the probe located successively at a smaller distance from the grain boundary. The result is shown in Fig. 12. Those lines indexed in white originate from the target grain; those in black from the adjacent grain. It was observed that lines from the adjacent grain are first detected when the probe is 25 μm from the grain boundary and that the low order lines are seen first. The number of lines detected increases with decreasing distance of the probe from the boundary until at a distance of 1 μm equal numbers of lines from both grains are recorded.

B. KOSSEL LINES FROM FINE-GRAINED MATERIAL

Kossel patterns have been obtained from grains as small as 2 μm diameter but as indicated above the patterns are complicated as they contain lines from neighbouring grains. An example of such a pattern is shown in Fig. 13(a). The white dot in the micrograph Fig. 13(b) locates the position of the probe during exposure of the Kossel pattern. To establish which of the lines in the pattern originate from the target grain, separate patterns were recorded with the probe located successively in each of the surrounding grains and a process of elimination followed. The lines originating from the same grain could easily be distinguished as they were simply related by the interplanar angles. In addition some of the neighbouring grains were large so that when the probe was located in one of them the Kossel lines from that grain were considerably more intense than those from the neighbouring grains permitting their point of origin to be identified. In this way it was shown that the 110 lines marked *E*, *F*, *G* came from the 2 μm grain.

Two further interesting features were noticed in the patterns from small grains. Firstly, that high order lines were never observed and secondly that the lines originating from grains neighbouring the target grain were often discontinuous. The former point is consistent with the theory discussed earlier, in as much as the primary extinction length being longer for high order reflections, then a greater volume of crystal is required before a sufficiently intense reflection to be detected can build up. The reason for the second point can be understood from a consideration of Fig. 14(a). The X-ray source in the target grain is at A, distance x from the grain boundary. The diffraction cone emanating from the second grain is labelled BCO and can be considered to be generated from an imaginary source located at A′. In this case the whole of the cone can be generated. If the diffraction plane is more steeply inclined to the surface, however, as shown in Fig. 14(b), part of the diffraction cone cannot be generated because that part of the plane necessary to generate it lies above the surface of the specimen. As the distance of the source from the boundary is reduced so less of the cone is lost in this way, until when the probe is on the boundary the entire

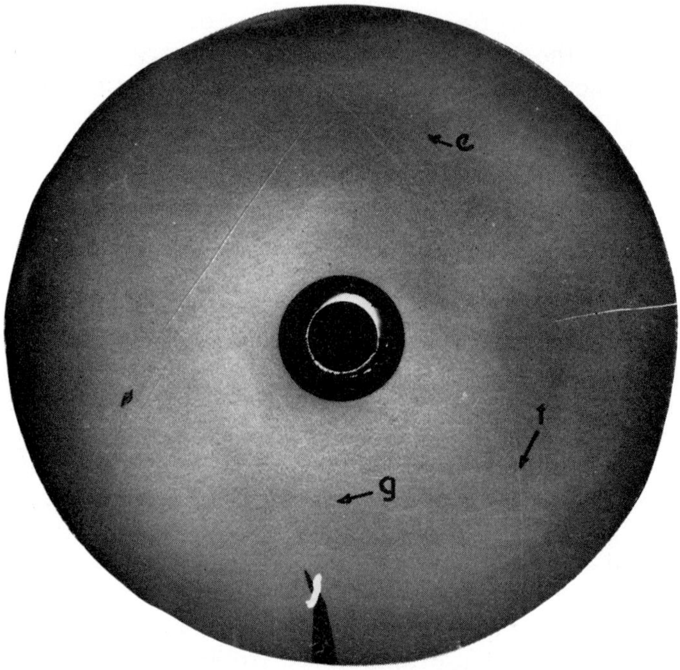

FIG. 13(a). Kossel pattern from 2 μm grain in polycrystalline iron.

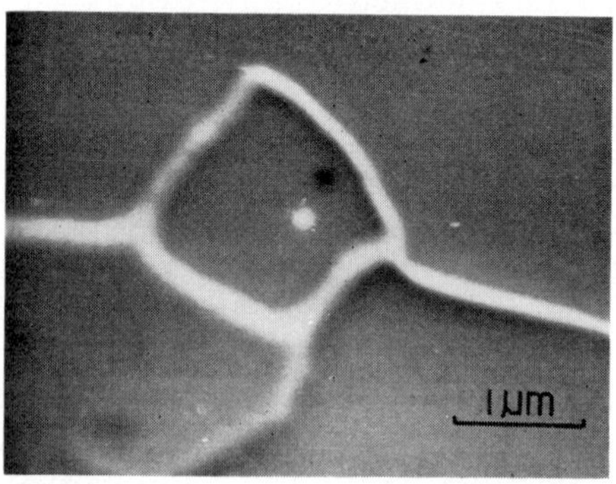

FIG. 13(b). Micrograph of area from which Kossel pattern was obtained.

cone is again generated. The limiting condition determining when part of the cone can be lost is that the diffracting plane must lie at an inclination to the surface greater than the Bragg angle. An extension of this argument leads to the conclusion that only continuous cones can be produced by diffraction in the target grain no matter how small the grain size.

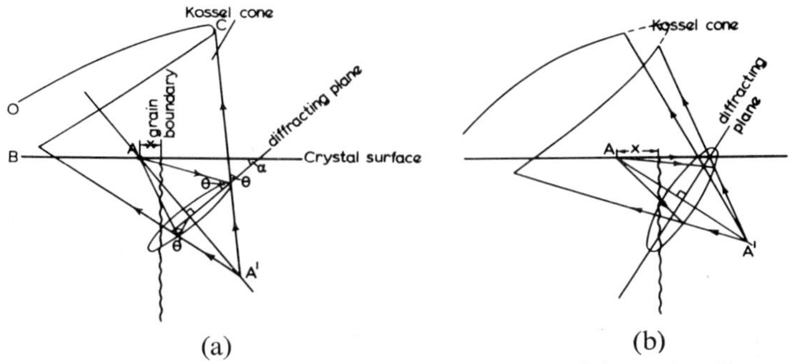

FIG. 14(a). Diffraction conditions where source is at A in grain adjacent to diffracting grain and inclination of plane is less than Bragg angle. Entire part of cone BCO can be generated.

FIG. 14(b). Same diffraction conditions as in (a) but diffraction plane inclined at an angle greater than Bragg angle. Dotted part of cone cannot be generated.

C. KOSSEL PATTERNS FROM STRAINED METALS

The inherent precision of the Kossel technique for measuring crystallographic data has encouraged several workers to use it in measurement of internal strains (Ellis *et al.*, 1964; Harris, 1970; see also Chapter 13 by Yakowitz in this volume). Following these authors the technique has been adapted for use in the scanning microscope in conjunction with *in situ* tensile straining. Two series of experiments have so far been carried out. The first concentrated on establishing the general effects of elastic and plastic strain on the Kossel lines themselves. In the second, experiments were performed to measure internal elastic strain under conditions of known applied strain. The results from some of these studies are shown in Figs 15–18. In both cases the material investigated was pure iron. The qualitative studies involved use of a miniature tensometer which had full load-recording facilities and the quantitative studies involved both this tensometer and a precision three point bending instrument.

A load elongation diagram obtained from one of the tests using the tensometer is shown in Fig. 15. At each of the points marked 1–8 the straining was arrested and a Kossel pattern recorded from a selected area of the specimen. The load was maintained during the exposure.

The Kossel patterns shown in Fig. 16 are those obtained at the points 1, 4 and 7. It is seen that prior to loading the lines are extremely sharp and faint: a result expected from strain free material. At the yield point (Fig. 16(b)) the lines are more distinct though there is little evidence of line broadening. This increased clarity is due to very small tilts, less

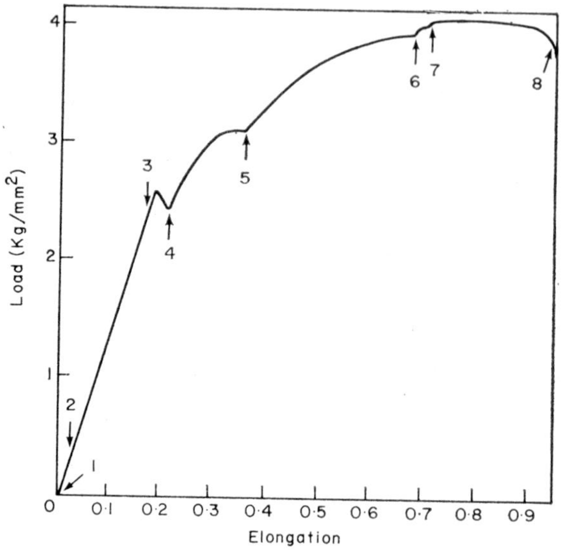

FIG. 15. Load elongation diagram for iron specimen tested in SEM.

than 10^{-4} rad, in the neighbourhood of the probe caused by the presence of dislocations. Towards fracture (Fig. 16(c)) the lines show considerable broadening due to the increased break-up of the structure in the locality of the probe. This process continues until fracture, with the additional observation that the broadened lines split to reveal sets of parallel lines. This splitting is due to the build-up of dislocation tangles and the establishment of a dislocation cell structure in the crystal, the angle between the cells being as great as 1 or 2 degrees and the dimensions of the cells of the order of 1 μm.

In Fig. 17 the Kossel patterns obtained from areas approaching a fracture surface are shown together with micrographs showing the areas from which they were taken. All were recorded immediately after fracture. The micrographs reveal increased dislocation slip damage as the fracture surface is approached and this is reflected by substantial line broadening in the Kossel patterns. At the fracture surface itself (Fig. 17(d)) the extensive line broadening and splitting makes them indistinct, though indexing is still possible and a rough orientation determination can be made.

For the unstrained material, in addition to the sharpness of the Kossel lines it is noticed that the low order 110 lines alone are visible, and again in the most highly deformed material the high order lines are less distinct than the low order lines. This is consistent with theory discussed earlier the results of which are depicted schematically in Fig. 4. In the unstrained state the blocks of undistorted material are considerably

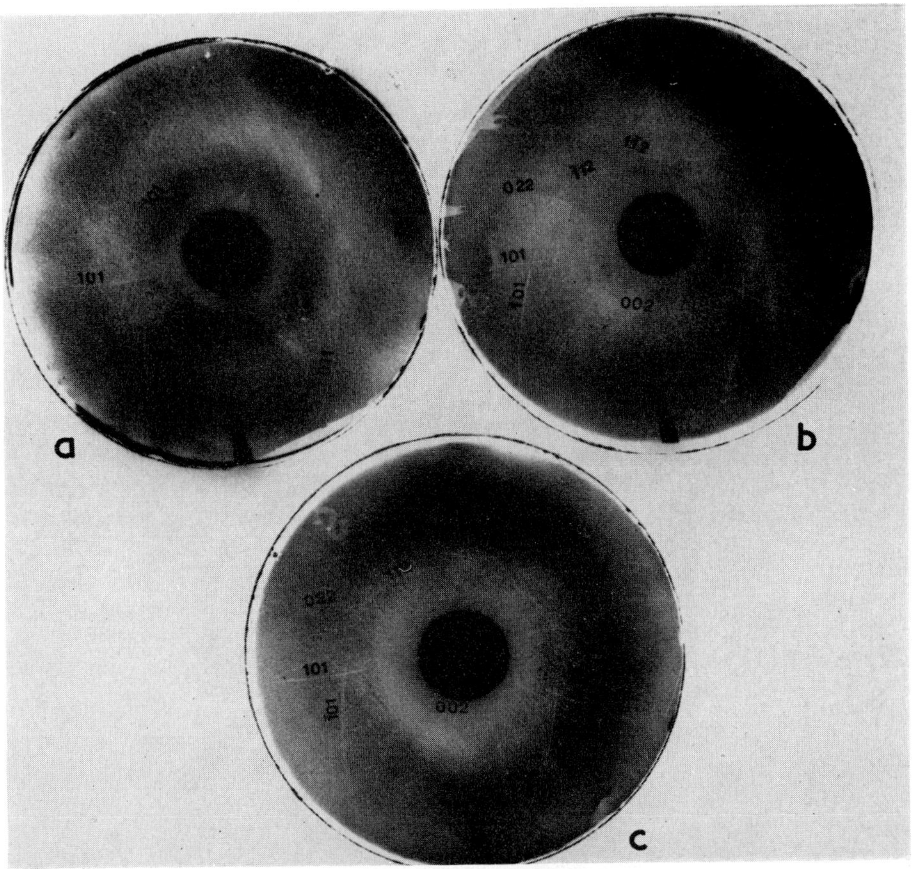

FIG. 16. Kossel patterns obtained from iron after different amounts of deformation. (a) 0 strain, (b) 2% strain, (c) 7% strain.

larger than the primary extinction distance and in the highly deformed state the converse is true. Thus the respective intensities for these conditions will be given by the points A, A′ and B, B′ in Fig. 4, where the unprimed points are for low order reflections and the primed points which fall at lower intensities, for a higher order reflection.

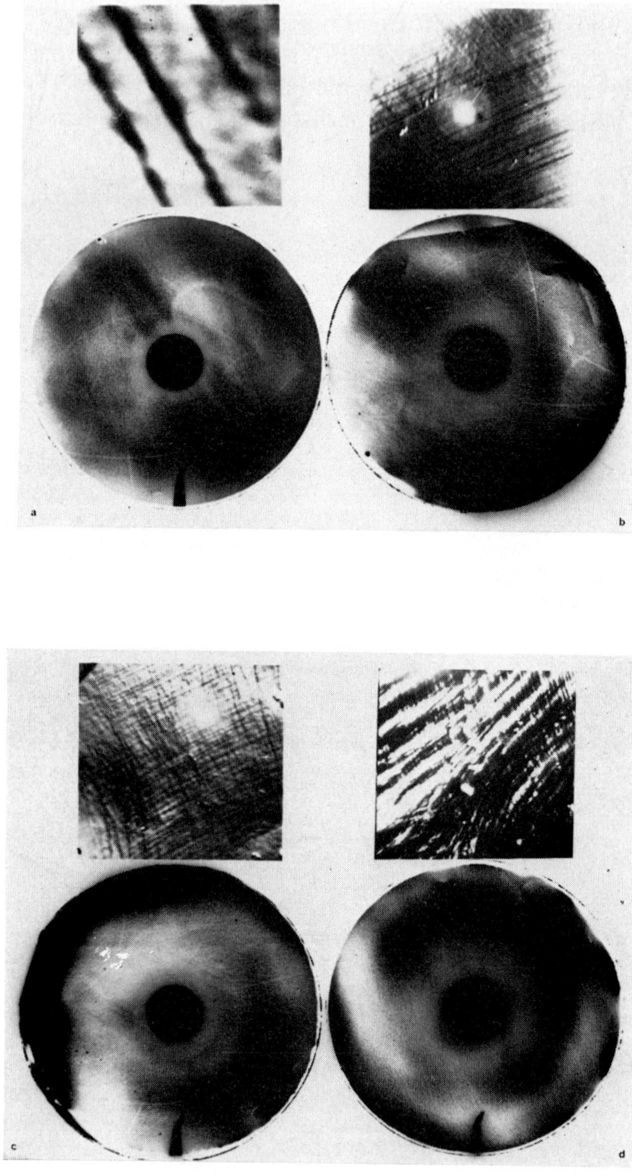

FIG. 17. Kossel patterns from zones approaching fracture surface of iron specimen together with micrograph showing slip damage at each point.

In measurement of internal strain two approaches are possible. In the first, the relative positions of the intersections, or near intersections, of two or more Kossel curves are determined and with a knowledge of the crystal symmetry the lattice parameter can be found. The value $\Delta a/a$ gives a measure of the internal strain. In the second, individual Kossel curves are studied and the lattice spacings of the planes giving rise to each of the curves determined. This permits the entire strain configuration of the crystal to be established. In current experiments in the scanning microscope the second approach has been adopted.

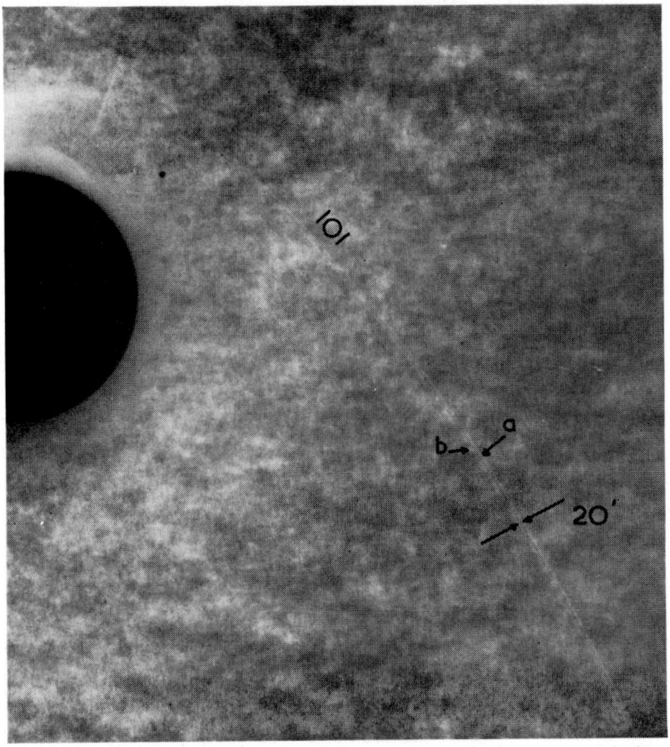

FIG. 18. Double exposure Kossel picture of iron specimen showing position of lines before and after straining. b, before straining; a, after.

The initial experiments have involved use of the double exposure technique mentioned above. That is, a Kossel pattern was first recorded from a selected area of an unstrained specimen, the specimen then strained a calculated amount in the bend test machine and a second pattern recorded from precisely the same area and on the same photographic plate. The photographic plate was not moved between exposures and the probe shifts were unaltered.

A typical result is reproduced in Fig. 18 which shows part of the Kossel pattern recorded from an iron specimen strained 0·0006 fractional strain. The displacement of the 101 line shown is 20′ of arc corresponding to an internal strain of 0·0107 fractional strain and a stress concentration of 16 times. Such differences between applied and measured strain were frequently observed and have so far been put down to stress concentrations in the sample. Control experiments were conducted to eliminate possible errors arising for example from specimen rotation during straining between exposures, change in specimen height, shift of photographic plate, etc., but in all cases the errors so introduced were small. However, further investigations are being carried out in this respect.

In the most recent experiments attempts have been made to measure internal strain by computer analysis of conic sections recorded in a single exposure of a strained specimen. Disappointingly, the results have so far been unsatisfactory, mainly because the amount of any one conic section recorded on the photographic plate has been insufficient for accurate curve fitting. This problem can be resolved by reducing the specimen to film distance but then the quality of the pictures deteriorates because of excessive contrast between the edge and the centre of the film. Filters of appropriate thickness profile are thus needed and this, together with variations in exposure procedure, is being looked into. In general, the principle of strain measurement using the computer analysis is promising, as has been shown by test runs using data calculated from known conic sections. The difficulty seems to lie in experimental methods.

D. EXPERIMENTS AT LIQUID HELIUM TEMPERATURES

In order to carry out *in situ* studies of phase changes and magnetic effects at low temperatures a liquid helium stage has been constructed and the two-plate Kossel camera described above designed to work in conjunction with it. Because of the relatively poor vacuum in the scanning microscope and the high cryogenic pumping into the liquid helium cooled surfaces, careful shielding of the specimen was required (Cort and Steeds, 1972a; Griffiths *et al.*, 1972). Thus, most of the specimen is enclosed by a liquid nitrogen cooled chamber (R) and only a small hole is left in the top surface to permit access of the electron beam (see Fig. 19). This top surface can be very thin and serves also as the X-ray filter and electron sink for the Kossel camera. The specimen is attached to an inner reservoir (B) filled with liquid helium. Since electrons from the specimen cannot escape to the conventional scintillator and light pipe arrangement, two narrow collector plates (P) are inserted in the cryo-pumped region. By applying a low external magnetic field parallel to the beam direction high collection efficiency can be achieved, and this, together with signal subtraction, goes a long way towards off-setting the loss of noise-free amplification. Since the plates are narrow they do not appreciably mask the Kossel lines.

Kossel lines have been obtained with this stage from polycrystalline V_3Si at helium temperatures using a titanium filter of 0·0125 mm thickness.

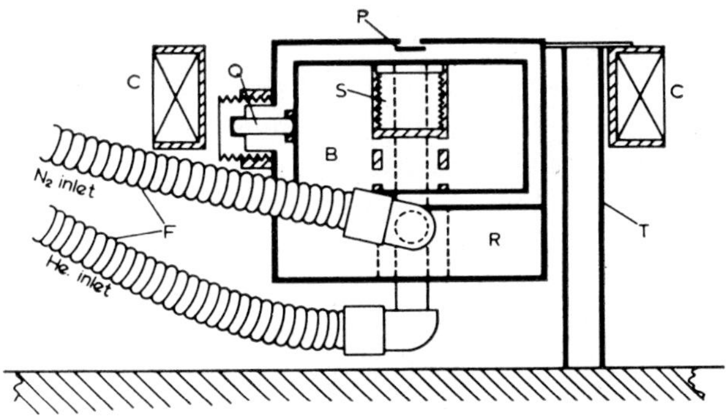

FIG. 19. Schematic of cold stage for use with Kossel camera.

VI. Comparison of Back Reflection Kossel Lines and Coates Patterns

In several respects the two techniques are complementary. For example, in studying a process of deformation, Coates patterns* yield information about the state of the specimen surface while Kossel lines give information from the interior. The major attraction of Kossel patterns is their general robustness, that is their relative insensitivity to the perfection or cleanliness of a specimen. They also score if accuracy is the prime consideration. On the other hand, Coates patterns will always be preferable if information is required from small volumes (at least in principle) or from surface layers. Let us explore the points in detail.

A. State of Surface

X-ray reflections, coming typically from depths of several microns within the crystal, are not greatly affected by the state of the surface. Thus it is not necessary to produce high polished surfaces or even specially clean surfaces. On the other hand Coates patterns depend

* Coates patterns here ≡ electron channelling patterns in Chapter 6.

critically on a well polished and clean surface condition. In some cases the sensitivity to surface condition is an asset but since one of the great advantages of the scanning microscope over transmission microscopes is its ability to cope with all states and conditions of specimens Kossel lines would seem to be the more naturally adapted technique for such an instrument.

B. PERFECTION OF SPECIMEN

Once again there is a great difference between the two techniques. Kossel lines cannot be observed from highly perfect single crystals, Coates patterns are at their best under such conditions. On the other hand Kossel lines are remarkably persistent in states of heavy plastic deformation and can even be obtained from fracture surfaces. Thus, if one is mainly concerned with semiconducting materials, Coates patterns are clearly indicated while steels and copper alloys are clear candidates for Kossel patterns.

C. VOLUME OF SAMPLED MATERIAL

Under this heading it is necessary to distinguish lateral dimensions and depth. Although Coates patterns come from close to the surface the lateral extent of the volume studied is a function of microscope development. In principle, information can be obtained from regions of only a few thousand Ångströms diameter but standard attachments to current microscopes increase this to 10 μm or more. In the case of Kossel lines it all depends on the orientation of the reflecting planes relative to the surface. Cones may arise from X-rays propagating parallel to the surface at 0·5 μm depth or else from depths of down to 10 μm. However, for fine-grained polycrystalline work it is necessary to use Coates patterns since the smallest grain size that Kossel lines can cope with is approximately 10 μm. In the case of an isolated small grain in the presence of larger grains Kossel lines may be obtained from a 2 μm grain.

D. ORIENTATION

For the purpose of determining the orientation of a crystal relative to its surface or neighbourhood there is not much to choose between the techniques. It is slightly harder to index Kossel patterns due to the more complicated geometry but either technique will rapidly give 2–3° accuracy. For controlled rotation of a specimen to a specific orientation

Coates patterns are much to be preferred owing to the speed with which they are formed (see below). The use of channel plates may help the Kossel lines in this respect but they equally facilitate the use of Coates patterns.

E. CONVENIENCE

At the present time the choice in this respect is less clear-cut. It is very easy to build a simple Kossel camera and put it into operation, while small crystal work by Coates patterns may call for modification of the scanning microscope. However, a typical exposure time for Kossel line work is three minutes and unless a more elaborate camera has been constructed, the exposed film can only be extracted by admitting air to the specimen chamber. Further a carefully chosen filter must be selected for each different X-ray wavelength. Ultimately Coates patterns must prove more convenient.

F. ACCURACY AND PRECISION

Kossel lines are capable of measuring lattice parameters to much higher accuracy and precision than Coates patterns. This is because Kossel lines are very sharp and intersect in a variety of information bearing figures while Coates patterns are subject to uncertainty in the accelerating voltage and distort on the face of the cathode ray tube. As Kossel line width is a function of the state of perfection of the crystal, accuracy will depend on the circumstances. However, 0·1 % accuracy is relatively easy to obtain and it may be possible to achieve precisions an order of magnitude better than this in favourable cases. Coates patterns are at least an order of magnitude less accurate than this.

G. MATERIALS LIMITATIONS

Kossel patterns cannot be obtained from light elements except by the use of inconvenient pseudo-Kossel techniques.

On the other hand, since X-rays are not deflected by magnetic or electric fields, Kossel lines are well suited to the study of ferro-electric or ferromagnetic materials. In demonstration of this point some work has been performed on poly-crystalline cobalt relating magnetic contrast to the orientation of the grains (Cort and Steeds, 1972b).

VII. FINAL NOTE

At the time of writing this article the authors have received the preprint of an article by Venables and Harland (1973). It is a modification of the Coates pattern technique called by the authors electron back-scattering patterns (E.B.S.P.). The important step forward is that the scanning microscope is used in the "point" mode, change of electron direction coming from thermal scattering processes in the crystal. Very high specimen currents have been necessary so far (5×10^{-7} A) but it seems that this difficulty can be overcome by the use of a channel plate and that it will then be possible to obtain crystallographic information from small crystals. This most attractive possibility makes the new technique of considerable interest.

Acknowledgements

We wish to thank Philip Lowe, Susan Biggin and Sam Worthington for allowing us to use unpublished micrographs, and also the Fulmer Research Institute and the National Physical Laboratory for the provision of specimens.

References

Batterman, B. W. and Cole, H. (1964). *Rev. mod. Phys.* **36,** 681.

Compton, A. H. and Allison, S. K. (1935). "X-rays in Theory and Experiment." Van Nostrand, Princeton.

Cort, D. M. and Steeds, J. W. (1972a). *Proc. Vth European Congr. Electron Microscopy*, p. 376. I.O.P.

Cort, D. M. and Steeds, J. W. (1972b). *Phys. Stat. sol.* **a10,** 215.

Duncumb, P. (1971). "Electron Microscopy and Analysis," 25th Anniversary of the Electron Microscopy and Analysis Group, p. 132. Institute of Physics, London and Bristol.

Ellis, T., Nanni, L. F., Shrier, A., Weissmann, S., Padawar, G. E. and Hosokawa, N. (1964). *J. appl. Phys.* **35,** 3364.

Griffiths, B. W., Mills, C. A. and Venables, J. A. (1972). *EMCON*, p. 373.

Harris, N. (1970). Ph.D. Thesis, University of Salford.

Heise, B. H. (1962). *J. appl. Phys.* **33,** 697.

James, R. W. (1948). "The Optical Principles of the Diffraction of X-Rays." Bell, London.

Kato, N. and Lang, A. R. (1959). *Acta crystallogr.* **12,** 787.

Ullrich, H. (1967). *Phys. Stat. sol.* **20,** K115.

Venables, J. A. and Harland, C. J. (1973). (submitted to *Phil. Mag.*).

Chapter 15

Quantitative Automatic Point Analysis with the Electron-Probe Analyser

J. V. P. LONG and B. JEFFERIES*

Department of Mineralogy and Petrology, University of Cambridge, England

Symbols

Z = generating factor
A = absorption factor
F = fluorescence factor

X, Y, Z — coordinate axes

I. Introduction

The automatic operation of instruments and machines aims, in general, to bring about greater ease and speed of operation and reliability. It also produces unforeseen pitfalls, and often, in retrospect, a realization that an adequate increase in efficiency could have been achieved by simple improvements in, and minor additions to, existing manual techniques. This paper attempts to assess qualitatively the value of different degrees of automation in electron-probe analysis and to describe some experience of the development and operation of one particular system.

In the present context, electron-probe analysis is restricted to mean the accurate quantitative determination of elements at single selected points on the surface of the specimen; searching techniques, scanning and modal analysis are specifically excluded, although of course these

* Present Address: *Department of Geology, University of Edinburgh.*

can be subjected to automatic control techniques as in the work of Jones and Gavrilovic (1970), Gibbard *et al.* (1971), Gibbard (this volume, Chapter 4), Jones (this volume, Chapter 16) and in the earlier work by Melford and Whittington (1965). The use of the rapidly developing technique of energy-dispersive analysis with solid-state X-ray detectors replacing the conventional X-ray spectrometer will considerably modify the design requirements of automatic systems and

TABLE I

	"Ideal" specification		"Relaxed" specification	
	Absolute	Approx. resolution	Absolute	Approx. resolution
STAGE				
X and Y movements	$\pm 1\ \mu m$ in, say 5 cm	1 in 5×10^4	$\pm 5\ \mu m$	1 in 10^4
Z movement	$\pm 2.5\ \mu m$ in 250 μm	1 in 10^2	—	—
SPECTROMETER				
Fully focusing	± 5 sec θ in 60°	1 in 4×10^4	—	
Semi-focusing	± 10–20 sec in 60°	1 in 1–2×10^4	—	
Temperature of crystals:				
Fully focusing		$\pm 0{\cdot}2$°C	—	
Semi-focusing		$\pm 0{\cdot}5$°C	—	
Gas density in flow proportional counter		$\pm 0{\cdot}3$%	—	
ELECTRON BEAM				
Probe position	$\pm 0{\cdot}5\ \mu m$		$\pm 2\ \mu m$	
Probe current control and/or measurement		± 1 in 10^3		± 1 in 500

reduce mechanical complexity. This technique will have much in its favour in scanning microscopes where facilities for operation of crystal spectrometers are generally more restricted. Practical automatic systems so far operating, however, have been designed with crystal spectrometers and these form the basis of the present discussion.

In the automatic control of a conventional electron-probe microanalyser, mechanical drives are required on the X, Y and Z movements of the specimen stage, on the angular setting of the spectrometer and, in instruments with less than three spectrometers, on the crystal changing mechanism which selects alternative crystals. In addition, for each element determined, individual settings of the threshold and window of the pulse-height analyzer, and ideally also of the proportional-counter anode voltage, are necessary. In a system designed to run for extended periods, some means of probe-current monitoring and/or control is clearly required; the most satisfactory solution here may well be the use of a lanthanum hexaboride emitter in the electron gun, which under the comparatively low-brightness conditions of the conventional probe electron-optical system should provide exceptional stability and very long life (Ahmed, 1971).

It is instructive to work back from the probe instrument to the control system, bearing in mind the objective of increasing the productivity and accuracy of the instrument and making less exacting demands on the skill of the operator. Listed in Table I are the specifications for a typical system with both ideal figures and "relaxed" figures which experience suggests will be much easier to attain in practice.

II. Mechanical Considerations

It is worth emphasizing that an essential feature of a servo-operated system is that the driven components shall take up exactly the positions indicated by the sensing devices which are often not mounted directly upon them. A further requirement is that this performance shall be maintained despite considerably increased wear rates produced both by greater usage and by higher accelerations than encountered in manual operation. Drastic redesign of the stage and spectrometer mechanisms to remove backlash may be essential before manual instruments can be successfully converted to automatic operation. It also follows that devices must be installed to protect against overrun and against malfunction of vacuum systems, X-ray windows, power supplies and, not least, the operator.

III. Servo Systems

Several types of servo drives of the required resolution are obtainable in both analogue and digital forms. In a fairly simple semi-automatic system where storage requirements are limited, analogue servos using either a.c. or d.c. drives are easily set up and relatively simple as compared to a digital system. A more complex controlling concept requires that many stage and spectrometer settings be stored within the programming part of the system and here the digital servo-loop offers a considerable advantage.

Digital servos can be of two main types: absolute closed-loop in which a given shaft position is designated by an invariant number, and incremental in which the shaft is positioned relative to a pre-determined datum point. Incremental servos can be of the open or closed-loop types, the former characterized by the use of stepping motors to move the driven member a specific incremental distance represented by the number of pulses applied to drive the motor. The closed-loop system, uses an incremental encoder, often of the optical "Moiré Fringe" type, to generate a digital signal relating to the shaft position. As each increment is indistinguishable from the next, an "up–down counter" must be used to track the position of the encoder.

High-resolution absolute encoders tend to be rather large but have the advantage of indicating the true shaft position even in the presence of severe electrical noise; moreover, no resetting is necessary on starting. An additional benefit is the possibility of "time-sharing" the error-signal generating logic among many encoders, thus leading to substantial savings in hardware if several servo loops are employed. Both types of incremental servos of even modest size are capable of extremely high resolution if reduction gearing is employed, and singly, are logically simpler than comparable absolute systems. They cannot, however, be time-shared and overall there is probably little to choose in logical complexity between any of the digital systems when the complete control loop incorporates three or more drives.

As an appreciable fraction of the machine running time for multi-point, multi-element analysis is occupied by servo movements, the servo setting time becomes an item of some importance, although it must be stressed that the highest setting speed is often determined by the maximum acceptable wear rate for the driven mechanism. Both absolute and incremental servos employing optical encoders and small a.c. or d.c. motors can attain extremely high slew rates and short settling times. Stepping motors are somewhat slower and absolute systems utilizing brush encoders, slowest of all.

IV. The Control System

Given that the various moving parts of the electron probe can be driven by devices accepting signals either in digital code or as analogue voltages and that the required electrical settings can be similarly derived, it remains to unite the separate units into a system for operating the instrument.

Because a considerable fraction of the time taken in manual operation is occupied in resetting the element parameters for the spectrometers, a very basic system as illustrated in Fig. 1 can increase the working speed of the instrument by a factor of at least 2 and at the same time reduce the probability of operator error. In this arrangement the parameters associated with each element are stored and selected manually for each point on the specimen. A system of this sort was used for some

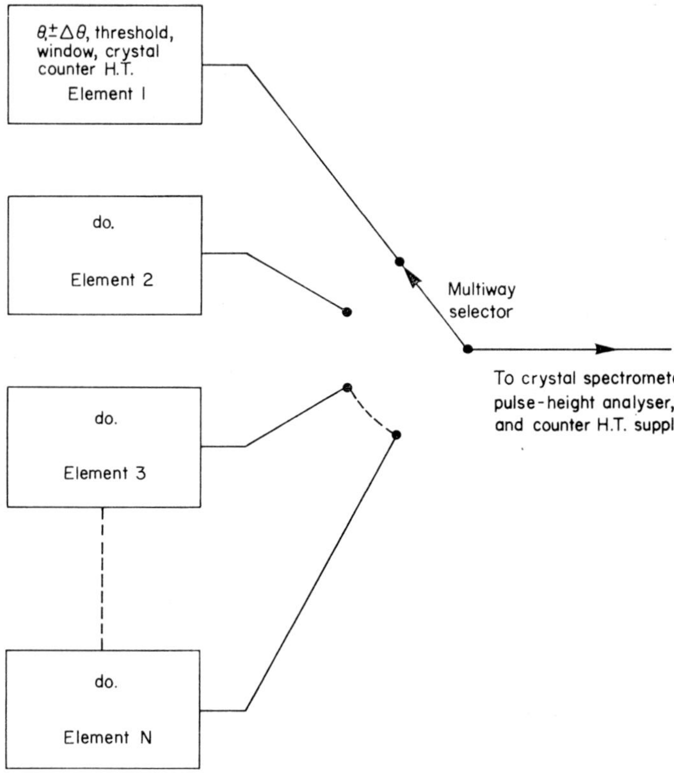

FIG. 1. Block diagram of semi-automatic system for X-ray spectrometer control.

years in the authors' laboratory and was the forerunner of the more sophisticated equipment developed subsequently. The original system used a d.c. servo drive for the spectrometer with the Bragg-angle settings being stored on helical potentiometers. A 20-turn precision helical potentiometer, driven from the θ shaft was used to produce a voltage proportional to the shaft position. A differential d.c. amplifier generated an error signal for driving a small d.c. motor geared to the θ shaft. An improved version of this spectrometer control was later used in the EMMA instrument described by Cooke and Openshaw (1969).

A semi-automatic system can easily be derived from the above unit by employing a sequencing or stepping device as the selector switch. If multi-element secondary standards are used during normal analysis the standardization procedure becomes similar to the semi-automatic sequences used for multi-element analysis on unknown points. It is assumed that a separate calibration procedure would have been carried out prior to analysis by measuring the secondary standards against primary ones.

Similar analogue servo techniques can equally be applied to driving the stage although a reasonable upper limit to the number of stored coordinates would be about ten since each requires three potentiometers. On the assumption that a complete analysis for ten elements at one point takes about 15 min, such a system would provide for automatic programmed operation for periods up to about 2 hours.

V. Systems with Program Options

The control system becomes much more complicated when provision is made for storing unlimited numbers of settings on punched tape or in a computer core and when programmable options are introduced into the operating sequences.

The configuration of the equipment is now as represented in Fig. 2. The controller can be a suitably interfaced computer in which the necessary sequence of operations is determined by means of a stored program which executes the requests of the operator, transmitted via the teletype keyboard in the case of very short runs, for example, an analysis at one point only, or from a punched tape in the case of longer runs. The stored program may be written so that it is interactive and guides the operator in the preparation of the control tape by inviting him to enter appropriate information such as element names, coordinates and code words designating particular operations or groups of operations at the correct time. The completed instruction tape, when returned to the teletype reader, initiates the required sequences and in the basic configuration prepares a second punched tape now containing, in addition, the intensity data from the nucleonic counting system.

Depending upon the complexity of the program written for the computer, operations such as automatic setting of the spectrometer on named X-ray lines and selection of the appropriate threshold and window settings can be carried out without assistance from the operator.

If the dotted connection in Fig. 2 between the counting equipment and the computer is completed, and the latter programmed accordingly, a complete correction of the measured intensities leading to a printed output of the computed element concentrations is also possible. Furthermore, by completing the path "A", the measured quantities may be used together with predetermined conditions to influence the subsequent sequence of operations. Computer-controlled equipment has been designed by Canberra Instruments Ltd. and by the Materials Analysis Company Inc. The recently announced Applied Research Laboratories Instrument (Model SEMQ), equipment being developed by Dr. P. Suddaby and Mr. K. O'Hara (personal communication) and the systems described by Colby (1972) and Lifshin (1972) are also of this type.

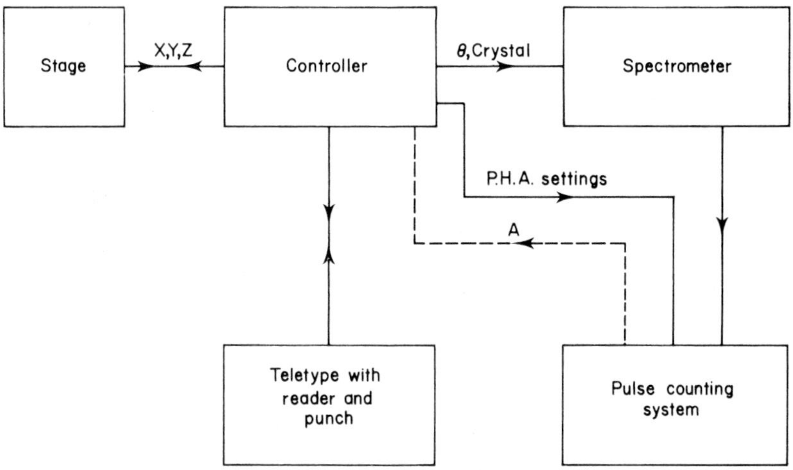

FIG. 2. Block diagram of fully automatic control system for electron probe control.

The alternative to a computer for providing overall control of the analytical system is the use of "hard-wired" logic circuitry which is already programmed to initiate a variety of different operations. These operations, identified by single characters can then be combined into a controlling sequence by the operator who punches them on to the control tape. Such a system is, by definition, less versatile than the computer based arrangement, in that except by adding more circuitry, the number of basic operations cannot be increased.

Hard-wired controllers, however, possess several compensating features: the instructions are extremely simple and the probability of error in preparation of the control tape is correspondingly reduced; no "software" or programming effort is required; they may provide an easier transition from fully-manual to fully-automatic operation so that, for example, it is easy to use the instrument in the semi-automatic mode shown in Fig. 1 for a single analysis. The use of a hard-wired controller does not, of course, preclude the addition of a computer for treatment of data, and from the point of view of ease of maintenance there may well be advantage in subdividing the functions of the different parts of the system in this way.

VI. An Actual System

The equipment developed here is of the "hard-wired" type and has been briefly described by Jefferies and Long (1971). The essential features of the electron-probe in this system are listed in Table II and the whole equipment is shown in Fig. 3. The main control console, illustrated in Fig. 4 shows the arrangement of the twelve element boards, each containing the setting of Bragg angle (a), positive and negative angular offsets for measurement of background, pulse-height analyser window (d) and threshold "potentiometers" (e), crystal selector (b) and counter voltage selector (c). Each element may be associated with any one of the twelve standards on the stage by setting the rotary switches (f) immediately below the element board. The coordinates of the twelve standards are permanently stored but the operator has individual control over the two middle digits of X and Y for each position and full control of Z by means of the twelve sets of three thumb-wheel switches (h). Once the operator has selected clear areas on the appropriate standard and brought them to a standard height by operation of the Z focusing control, these settings are retained for the duration of the run. To minimize contamination effects on the standard during subsequent measurements, a saw-tooth deflection signal may be applied to one set of scanning coils so as to produce a line 50–100 μm long on the surface of the standard in a direction parallel to the axis of rotation of the spectrometer crystal.

Elements can be grouped into six different combinations of up to twelve individuals by setting the switches in the six rows A–F (g in Fig. 3). In the analysis of minerals within a rock slice for example, one group can be used for ferromagnesians, another for feldspars, etc.

Coordinates of points on the specimen are recorded during preparation of a control tape by depressing a "load coordinates" switch after setting the point to be analysed coincident with the cross-wires of the optical microscope and bringing the surface into focus in reflected light. In this way errors in reading and transposing coordinates are avoided. The control tape is then used to operate the electron probe automatically in the "run" mode.

TABLE II

ELECTRON PROBE	
Working voltage:	25 kV max
Take-off angle:	40°
Electron beam:	diameter 1 μm
Probe current:	long term stability ± 2 μm
	precision of measurement ± 1 in 10^3
STAGE	
X and Y movements:	Travel 25×40 mm Repeatability ± 3–4 μm
Position encoders:	Harrison type 4/UG/LTB
Z movement:	Travel 250 μm Repeatability ± 2.5 μm
Position encoder:	Harrison type 2/ST/100
SPECTROMETER	
Type:	semi-focusing with four alternative crystals on automatic changer
Bragg angle:	Range 10°–60°(θ) Repeatability ± 10 sec(θ)
Position encoder:	Harrison type 12/ST/1000 with additional turns counter
PROPORTIONAL COUNTER	
Flow type with 2·5 cm absorption path;	
Gas density control providing constant density to $\pm 0{\cdot}3\%$	
CONTROL SYSTEM	
Max. number of elements:	12
No. of stored standard positions:	12
No. of possible element combinations:	6 (letters A-F)
Electron beam positioning control:	up to 64 points in matrix of variable form and size, e.g. 8×8, 4×16, etc.

A list of the control symbols is given in Table III. In addition to the fully automatic "run" mode of operation, a three-position switch allows operation in a "semi-auto" mode in which the stage is controlled manually but the spectrometer stepped automatically through a

FIG. 3. General view of fully automated electron probe analyser in the Department of Mineralogy and Petrology, University of Cambridge.

FIG. 4. Control console for automatic electron probe with hard-wired controller showing element stores, switches for element grouping etc.

selected sequence of elements at a single point on the unknown and on the appropriate multi-element standards. A third position allows for fully manual control and program preparation.

TABLE III

A,B,C,D,E,F—Element routines consisting of combinations of up to 12 elements, e.g. A = Mg, Al, Si, Ti, Fe.
S—Measure intensities on standards of elements in selected routine at $\theta-\Delta\theta$, θ(peak) and $\theta+\Delta\theta$.
H—The same but with $\theta+\Delta\theta$ background omitted.
U—The same but with $\theta-\Delta\theta$ background omitted.
V—Peak only.
O,P,Q,R—The corresponding instructions for intensity measurements in the unknown.
+ inserted after O,P,Q or R—Duplicate count sequence on each element.
++ up to ++++++ used in the same way = Repetition up to 64 times.
G inserted after O, P, Q or R—alters interpretation of the + symbols and causes the probe to be shifted electrically by increments and in a pattern (of up to 64 points for ++++++) determined by preset control.

VII. DATA PROCESSING

As noted earlier, on-line data processing within a controlling general purpose computer is possible and provided a correction procedure with "α" corrections—for example that devised by Bence and Albee (1968) is used—can be carried out within an instrument of modest size. Full *Z*, *A*, *F* (generation factor, absorption, fluorescence)-type corrections generally require larger core size and in particular backing store in the form of a disc.

In the equipment described above utilizing a hard-wired controller, the measured intensities, together with coordinate information and control symbols appear on an output tape and in parallel on the output writer. This tape is then used as input for a program on an IBM 360/44 computer which produces both lists of apparent concentrations and in parallel, a set of input cards for the final correction by a *Z*, *A*, *F* program on the same computer. It has so far been our experience that this apparently inefficient method of producing a hard-copy output at several stages has greatly simplified the process of locating and removing errors which could otherwise only have been carried out by the very costly method of generating linked programs with extensive diagnostic messages. The production of fully corrected analyses on line, already achieved in some systems, of course still remains the ultimate objective for our equipment.

In order to avoid placing unnecessary restrictions on the operator with regard to the sequence of measuring unknowns and standards and the frequency at which standard runs should be inserted, it is necessary for the tape-reading program to accept essentially unformatted data. Furthermore, if the usual procedure of using the mean of standard measurements before and after sets of unknowns of unspecified size is employed, the data relevant to the unknowns must be stored until the second standardization is available. This consideration applies equally to our tape input and to the use of an on-line computational facility and requires again that considerable storage capacity be available.

VIII. Results and Discussion

Some 500 partial and complete analyses have been made so far with the equipment described and have provided, in addition to preliminary running experience, some indications of accuracy and reliability and also of speed relative to manual techniques.

Figures 5(a) and (b) show a comparison between the determinations of iron and silicon in a set of twelve lunar glasses previously analysed on the Geoscan. Taking the Geoscan data as correct, the relative standard deviations for the two sets of analyses are $\pm 1{\cdot}7\%$ and $\pm 1{\cdot}5\%$ respectively. Additional evidence from measurements of spectrometer reproducibility suggests that, subject to an inherently poorer sensitivity, due to the lower peak/background performance of the semifocusing spectrometer, the accuracy attainable is at least equal to that of conventional instruments.

It is difficult to devise an objective basis for comparing operating speed. If, however, the rate of production of complete multi-element analyses by a single operator in charge of a manually controlled two-spectrometer instrument and responsible also for data processing, but not sample preparation, is taken as 1, then it is our present estimate that the corresponding output of the automated instrument is greater than 5.

The extent to which it is worthwhile to instal automatic control facilities in given circumstances clearly depends upon the extent to which the instrument is already employed and on whether increases of productivity of between 2 and 5 times can be justified on scientific and economic grounds. In the case of our own instrument the scientific objective is to produce a much better estimate of the variability of composition within minerals and so add significance to determinations which are frequently at present based on a very limited number of

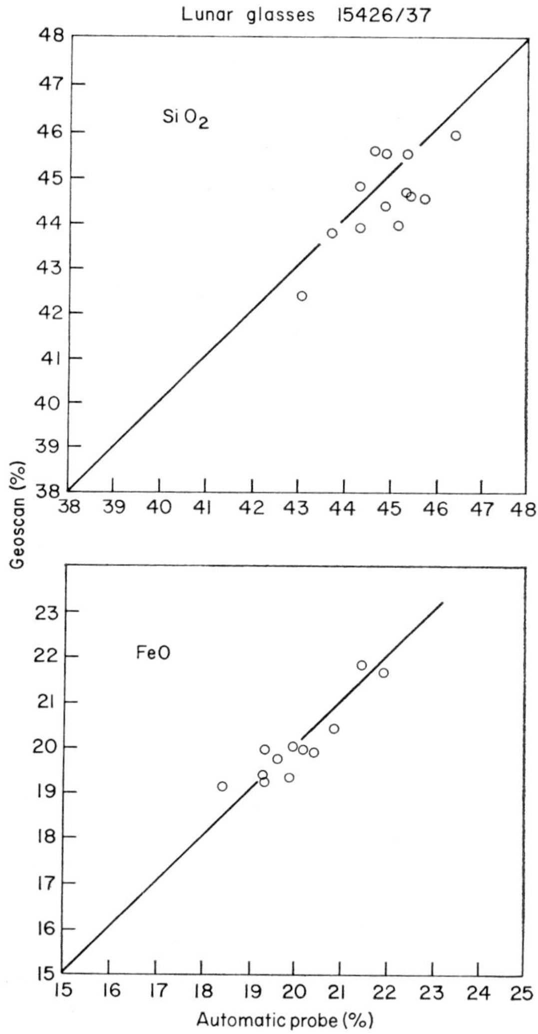

FIG. 5. Comparison of analyses for silicon and iron in lunar glasses obtained with Cambridge Scientific Instruments Geoscan and with automated instrument.

duplicate determinations. Economically, the degree of automation requires very careful evaluation: as noted earlier, simple automation of spectrometer operation could produce a factor of 2 for a capital outlay of perhaps the order of 10% of the cost of a complete probe instrument. The necessary additions are discrete devices and should not add appreciably to servicing costs. A similar independently operated system added to the stage may involve a further 10% capital cost but owing to the now extended running periods could imply improvements in probe-current monitoring or stabilizing facilities. Complete automation as undertaken in our own instrument, particularly if applied to systems with more than one spectrometer probably has a commercial value in the region of 50% of the cost of an electron-probe instrument and in addition, a very significant maintenance and servicing cost at least comparable with the salary of an operator.

Acknowledgements

We should like to express our thanks for the invaluable work carried out by Mrs E. M. Jones on the development of the computer program for tape processing, and also to Mr A. R. Arnold who carried out the measurements on the lunar glasses. We also acknowledge a research grant from the Natural Environment Research Council which has financed our work on automation.

References

Ahmed, H. (1971). *Proc. 25th Anniversary Meeting, Electron Microscopy and Microanalysis Group, Institute of Physics Conference Series*, **10,** 30–33. Institute of Physics, London.

Bence, A. E. and Albee, A. L. (1968). *J. Geol.* **76,** 382–403.

Colby, J. W. (1972). *Proc. 7th National Conference Electron Probe Society of America.* Paper 67A.

Cooke, C. J. and Openshaw, I. K. (1969). *Proc. 4th National Conference Electron Probe Society of America.* Paper 64.

Gibbard, D. W., Crawley, J. A. and Cowham, M. J. (1971). *Proc. 25th Anniversary Meeting, Electron Microscopy and Microanalysis Group, Institute of Physics Conference Series*, **10,** 158–160. Institute of Physics, London.

Jefferies, B. and Long, J. V. P. (1971). *Proc. 25th Anniversary Meeting, Electron Microscopy and Microanalysis Group, Institute of Physics Conference Series*, **10,** 150–153. Institute of Physics, London.

Jones, M. P. and Gavrilovic, J. (1970). *Rudy Ore*, **18,** 189–197.

Lifshin, E. (1972). *Proc. 7th National Conference Electron Probe Society of America.* Paper 66A.

Melford, D. A. and Whittington, K. R. (1965). "Optique des Rayons X et Microanalyse," p. 497. Hermann, Paris.

Chapter 16

Automatic Stereological Analysis by Electron Probe X-ray Microanalyser

M. P. JONES

Department of Mineral Technology, Imperial College, London, England

Symbols

N = total number of intercepts made during the traverse

L = intercept length across each feature

ΣL = sum of the intercept lengths across all the features

ΣL^2 = sum of the squares of the interceptors across each feature

ΣL^3 = sum of the cubes of the intercepts

ΣL^4 = sum of the fourth powers of the intercepts

L_T = total length of traverse

V_v = volumetric proportion of feature

D = size of maximum possible intercept on a grain

$u_i^n(D)$ = nth moment of grain size (by volume)

K_n = shape factor for the nth moment of volumetric grain size (for spheres $K_n = \frac{n+3}{3}$

$u'_{(n+1)}(L)$ = $(n+1)$ moment of intercept size distribution

$u'_1(L)$ = 1st moment of intercept size distribution

D_{oc} = size of grains as measured by screening or other convenient method

K_{cn} = shape factor relating the nth moment of the intercept distribution to the size distribution (by volume) measured by screening or other convenient method

I. Introduction

The geometrical parameters of multi-component metallurgical and mineralogical materials have seldom been accurately measured because of the difficulties of collecting the necessary data in a reasonable time. These parameters include the volumetric proportions of the various phases, the sizes and shapes of the individual grains and the spatial interrelationships between various phases or features.

Many attempts have been made to measure these parameters by using instruments in which the phase discrimination capability was based entirely on optical contrast. Unfortunately this contrast is often inadequate for the accurate delineation of mineral phases (Jones and Gavrilovic, 1970). It is possible, however, to obtain greatly improved phase discrimination by using combinations of the various signals that are produced in an electron probe X-ray microanalyser. In addition, the operation of the microanalyser can now be largely automated.

II. Stereological Considerations

The geometrical parameters of metallurgical or mineralogical specimens are almost always measured from thin sections or polished specimen surfaces, irrespective of the device that is employed to provide the data. The observations that are made can either be planar, as in the Quantimet, or linear, as in the traversing photometer and microanalyser (Figs 1 and 2).

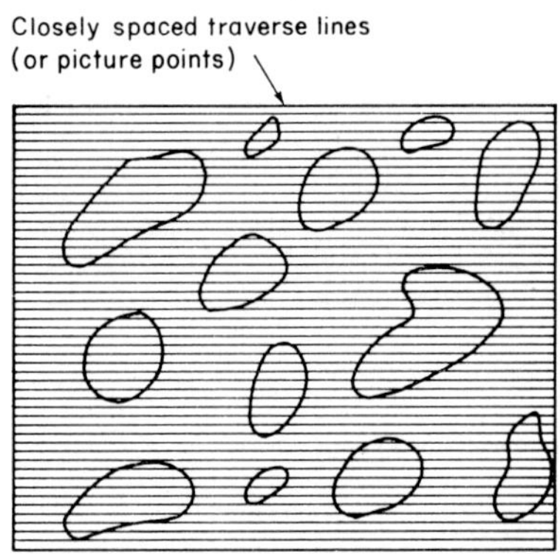

Fig. 1. Provision of planar data from closely spaced traverses or overlapping picture points: feature size is taken as the maximum intercept made horizontally across the feature.

In order that the maximum information can be obtained from such linear (one-dimensional) or planar (two-dimensional) observations they must be transformed into three-dimensional data which can then be applied to the total volume of the specimen that is being examined. Such a transformation can be achieved by:

1. Stereoscopic methods
2. Serial sectioning (see Fig. 3), or
3. By the application of stereological techniques.

Stereoscopy is the biaxial viewing of a feature in a transparent medium (e.g. aerial photography, where the transparent medium is air) but, unfortunately, stereoscopy cannot be applied to features that are embedded in an opaque matrix like the components of most rocks or metal specimens.

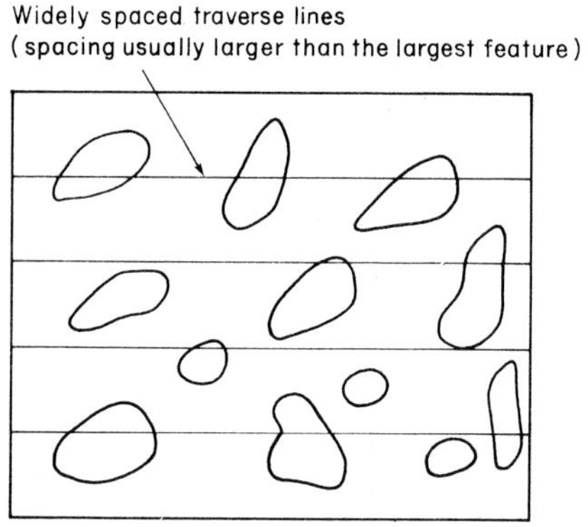

FIG. 2. Provision of linear data from widely spaced traverse lines: feature size is taken as the length of the intercept across that feature by the traverse line.

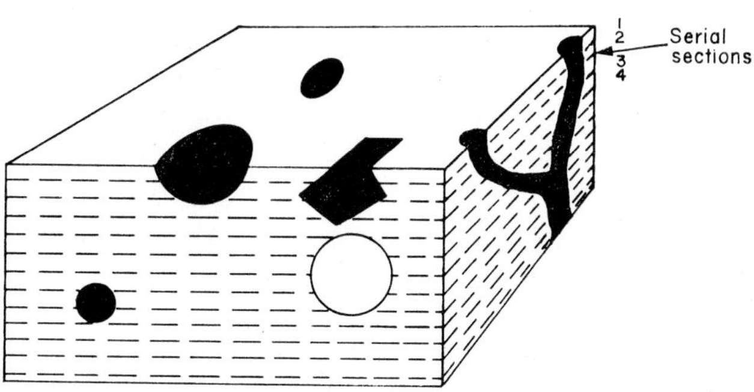

FIG. 3. Serial sectioning: the whole block is sectioned parallel to the upper face and the information obtained from each section is used to obtain a composite picture.

Serial sectioning is very tedious and is, therefore, only used for the examination of disjointed, but continuous, features such as veins or pores, or for the study of specimens in which only a strictly limited number of features occur and to which statistical methods of analysis cannot be applied. Stereology is a word coined about ten years ago (Elias, 1967) to describe the statistical relationships that exist between quantities that are measured from two-dimensional planes and the features that these quantities represent in three-dimensional space: in other words, stereology involves the derivation of information about the whole volume of a specimen from the planar or linear examination of a plane surface of that specimen.

The volumetric parameter that is most readily determined from an examination of a plane surface is the volume fraction of a component in the solid specimen. In this instance (and provided the examined area is sufficiently large) no explicit stereological transformation is necessary since lineal, areal and volumetric fractions are numerically equivalent.

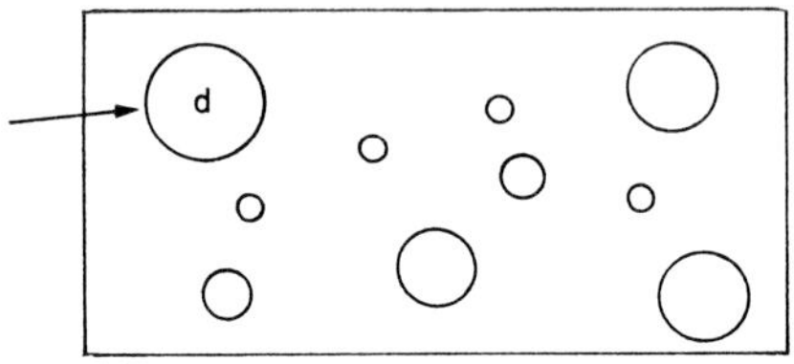

FIG. 4. Effect of sectioning uniformly sized spheres dispersed at random in space. The true diameter is $\geqslant d$ but the observed diameters range from this to zero.

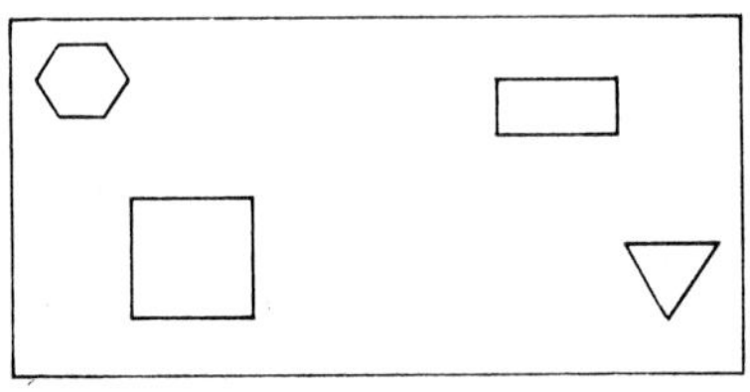

FIG. 5. Effect of sectioning uniformly sized cubes dispersed at random in space. This gives rise to squares, rectangles, triangles and hexagons of various sizes.

For other geometric parameters, such as feature size, feature shape, the contiguity of two features, the connectivity of parts of a single feature etc., there are no simple relationships between the observations that can be made on a two-dimensional surface and the three-dimensional reality. Consequently, the results of lineal or areal observations will be misleading unless the correct stereological transformation of the data is carried out. Such a transformation can either be carried out subjectively by the experimenter or it can be objectively applied by the use of stereological formulae. The skilled experimenter often tries to make intuitive stereological transformations of limited and often imprecise information but great care and experience are required to avoid misconceptions in the interpretation of size, shape or contiguity (see Figs 4-6).

Most of the mathematical stereological transformations, on the other hand, demand high precision in the raw measurement data and this precision can only be achieved within a reasonable period by the use of automatic data collection systems such as the image analysing computer or the electron probe X-ray microanalyser that is described below.

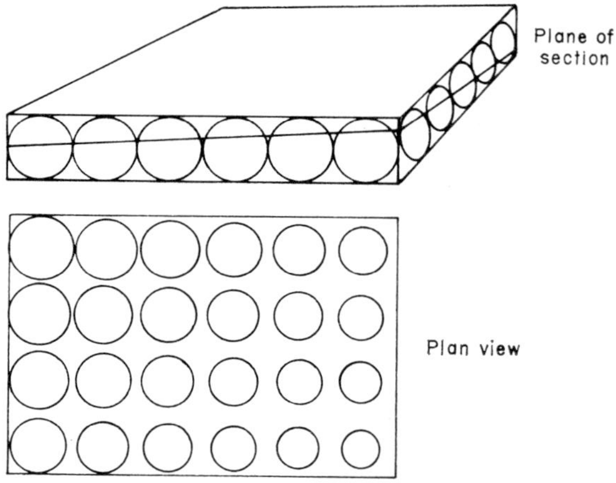

FIG. 6. Effect of randomly sectioning contiguous, uniformly sized spheres.

III. MEASURING SYSTEMS

Any system that is used to measure geometrical parameters must include the following components:

1. A suitably prepared specimen,
2. A means of illuminating that specimen,
3. A device (or devices) for collecting the signals that are produced by the illuminated specimen and,
4. A method of moving the specimen in relation to a datum point or line.

The spatial discrimination of such a system depends either on the effective size of the detector or on the size of the illuminating beam.

There are a number of measuring methods available, for example:

1. *Specimen plane devices*: where the specimen is moved past a stationary illuminator and a stationary signal detector. The specimen movement can be in the form of discrete steps (as on manual microscopes that are equipped with a ratchet-operated, mechanical stage or on the automatic photometer and microanalyser units) or the movement can be continuous as on an analogue-controlled microanalyser (see Fig. 7).
2. *Image plane devices*: where the detector is moved over a stationary, uniformly illuminated specimen; for example, television camera units such as the Quantimet where the detector is, in effect,

moved over an image of the specimen on a vidicon or plumbicon television camera (see Fig. 8).

3. *Source plane devices*: where the illuminator is moved over a stationary specimen and detector; for example, the scanning electron microscope and the scanning microanalyser (see Fig. 9).

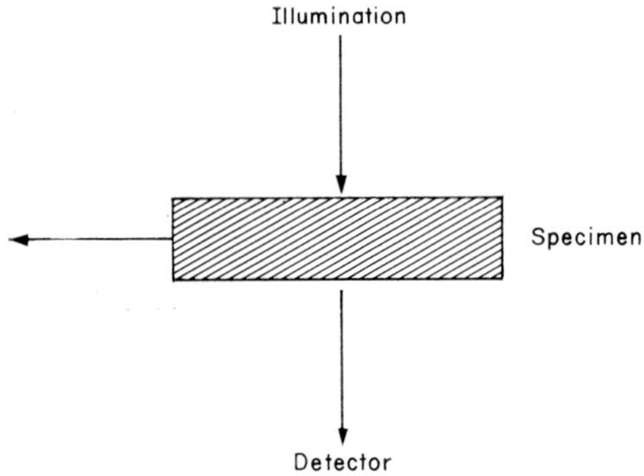

FIG. 7. Specimen plane device. Specimen moves, illuminator and detector stationary.

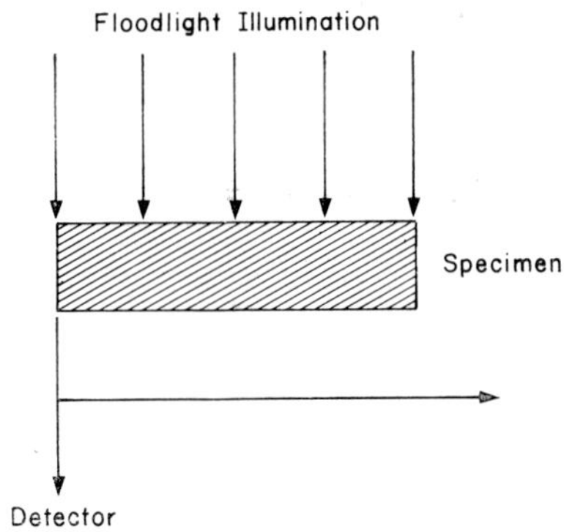

FIG. 8. Image plane device. Detector moves, specimen and illuminator stationary.

A number of combinations and variations of the above three systems are possible for specialized uses; for example, the illuminator may be pulsed whilst the specimen is moved at a continuous, uniform rate, or the specimen may be moved continuously or in discrete steps whilst the illuminator is continuously moved at right angles across the specimen—as in the Geoscan searching technique for the location of rare phases (Jones and Gavrilovic, 1968) (Fig. 10). In addition, it is possible to examine a small portion of a stationary specimen (either by moving the detector or the illuminator) and then automatically move to a new, pre-selected area where the examination is continued (e.g. Quantimet

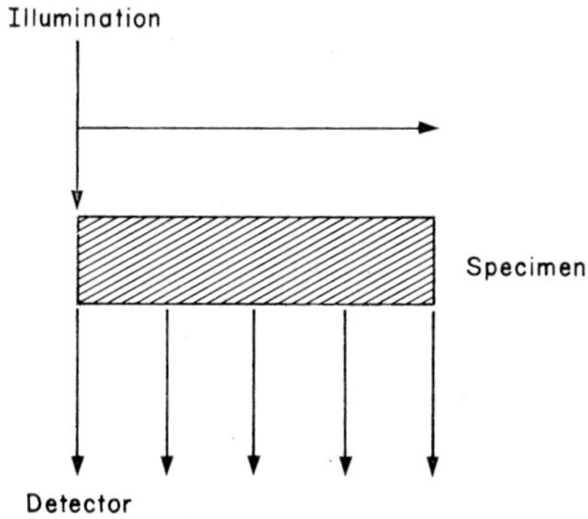

FIG. 9. Source plane device. Illuminator moves, specimen and detector stationary.

720, Fig. 11). Any of the detailed measurements made by the above methods may, if necessary, be confined to pre-selected features within a specimen. For instance, the specimen can be moved rapidly along a pre-determined path until a desired feature is encountered; measurements would then be taken at the much lower speed required for the necessary discrimination or precision. The specimen would then again be rapidly moved until the next suitable feature is located and the measuring procedure repeated, e.g. a "dynamic" traversing procedure on the microanalyser (see Fig. 12).

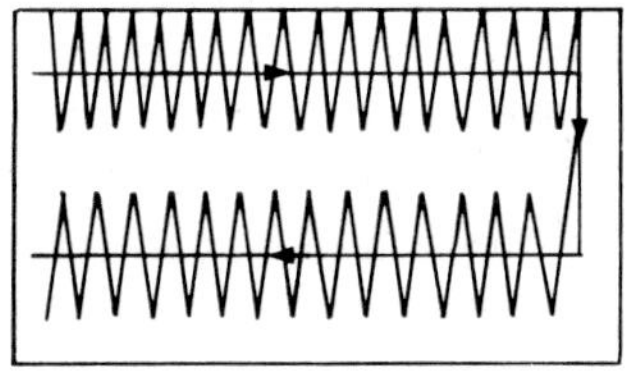

FIG. 10. Searching technique. The specimen moves and the illuminator moves at right angles.

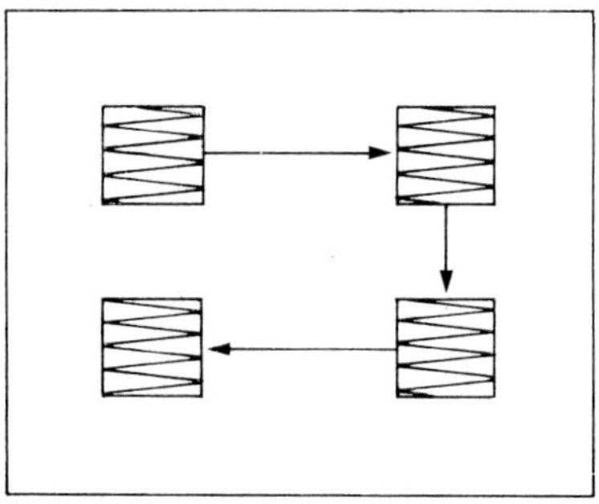

FIG. 11. Television scanning. Small area of the stationary specimen is examined; specimen is then moved to a new position and a new area is examined.

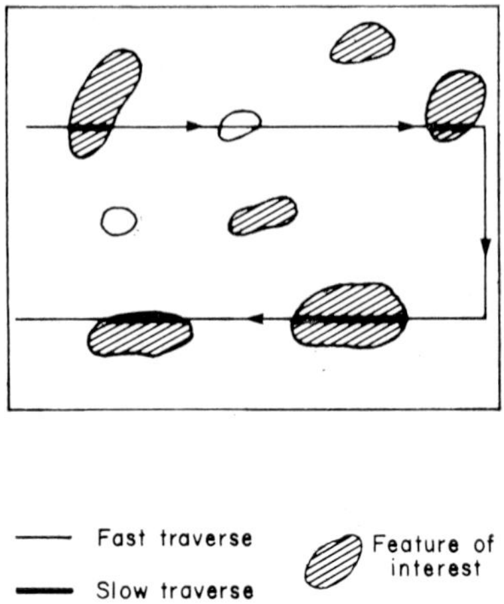

FIG. 12. Dynamic traversing.

IV. ELECTRON PROBE INSTRUMENTATION

The potential phase discrimination capability of the electron probe X-ray microanalyser is much greater than that of optical techniques since the microanalyser can employ any one, or any combination, of the various signals that are produced by electron–solid interaction (for example, characteristic X-rays, primary or secondary electrons, specimen current, cathodoluminescence, etc.). These signals are derived from a small volume of specimen and the spatial discrimination of the instrument is good.

Early attempts to collect geometrical data from the electron probe microanalyser used the instrument as a source plane device (Melford and Whittington, 1966). The maximum specimen area that can be examined as a single field by this method is about 2×2 mm but, owing to the defocusing effect on the X-ray spectrometers, the usual area that can be examined is less than 1 mm^2. Such a small area can seldom provide a representative sample of a metallurgical, mineralogical or ceramic specimen and therefore it is usually necessary to examine large numbers of these fields. Consequently, it can take a long time to produce representative information by this method. However, despite this long time, the spatial interrelationships of the data points from the various areas that are examined are not usually recorded and a great deal of potentially useful geometrical data may be lost.

The microanalyser, when used to collect stereological data, is more usually employed as a specimen plane device. In early attempts at the Department of Mining and Mineral Technology, Royal School of Mines, London, to produce an automatic measuring system the specimen movement was controlled by the analogue servomotors that were fitted on the instrument by the manufacturers. The *time* taken to make linear intercepts across selected phases was measured: the intercept times were then categorized (as equivalent intercept *lengths*) and the result of a traverse across a specimen was eventually drawn as a size/frequency histogram by the chart recorder provided in the equipment (Jones and Gavrilovic, 1970) (see Fig. 13). All additional assessments,

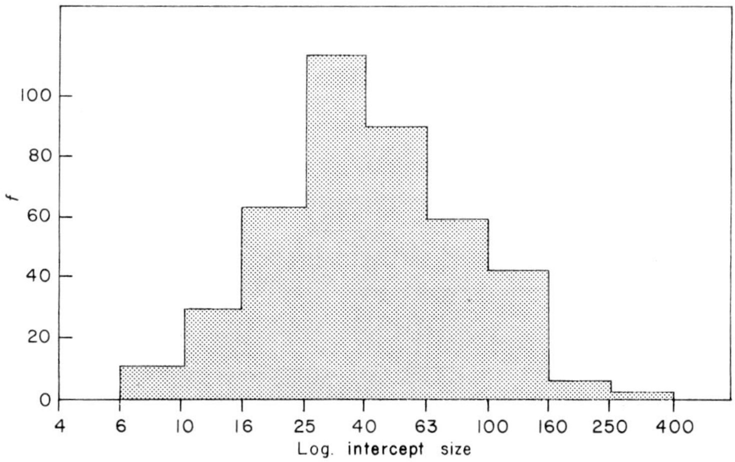

FIG. 13. Size/frequency histogram of intercept lengths produced by an analogue controlled Geoscan and drawn (originally) on the chart recorder of the instrument.

such as stereological transformations, were carried out on external calculators. The discrimination of the selected phase was based entirely on the uncorrected, ratemeter outputs from the X-ray spectrometers and these outputs could be combined at pre-set levels either in coincidence or in anti-coincidence by a small "hard-wired" logic unit. Under analogue control the specimen movement facility was not sufficiently accurate since it was almost impossible to equalize the specimen speeds in the "plus" and the "minus" directions owing to lack of symmetry within the control circuits. Greatly improved precision of specimen movement and of intercept measurement has now been achieved by controlling the operation of the Geoscan by a Minic computer of 8K, 8 bit bytes of core memory.

The automatic system is shown schematically in Fig. 14. The signals from the two X-ray spectrometers—after pulse shaping and amplification—are passed into single channel analysers and these digital signals accumulate within the computer interface for fixed periods (in a manner that is equivalent to the phase integrator unit described by Heinrich (1964)). The signals from the two spectrometers can be compared with any pre-set signal levels (windows) and used, either individually or together, to discriminate a mineralogical or metallurgical feature. At present only one mineral (or metallurgical phase) can be distinguished during a single traverse across a specimen but it would not be difficult to provide software programs that would enable up to six phases to be discriminated and measured during a single traverse.

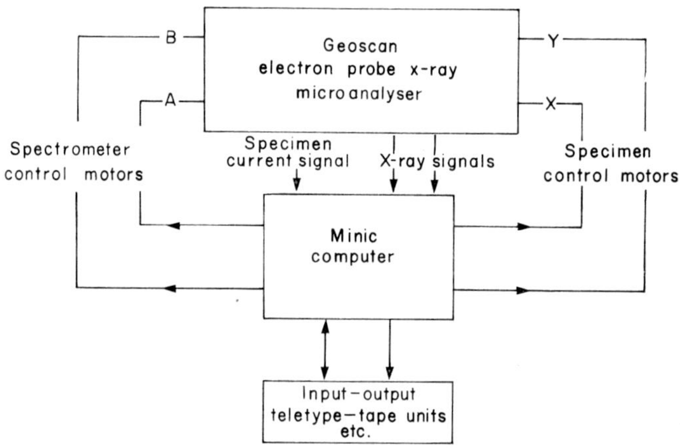

FIG. 14. Schematic diagram of Geoscan-Minic system.

The specimen is moved by two stepper motors which are controlled by the computer. These motors are coupled through electrically-operated clutches to the manual specimen movement controls on the Geoscan. The specimen can be moved in both the "X" and the "Y" directions in steps of a minimum size of 1 μm. No automatic control of the specimen position in the "Z" direction is incorporated but adequate electron beam focusing conditions over the whole specimen surface can be obtained by good specimen preparation techniques. Any point on a specimen (measuring up to $8 \times 4{\cdot}5$ cm) can be re-positioned by the stepper motors to within about 2–3 μm of the original position.

The maximum allowable stepping speed is about 450 steps sec^{-1}; in this way all specimen movement is within the "pull-in" rate of the available stepper motors. Ordinarily, however, the X-ray counting statistics do not justify a traverse speed in excess of 100 steps sec^{-1}. By using beam currents of up to 0·5 μA it is possible to obtain count rates on some minerals of 15,000 sec^{-1}; i.e. a mean value of about 150 counts per 10 m sec (or per step of traverse at 100 steps sec^{-1}). Since the production of X-rays is a random process the counts that are derived per unit time from any point on the specimen will vary in a statistical manner. The computer has been programmed to accept within its pre-set signal levels only those count rate values that occur within three standard deviations of a pre-set mean value. For example, when the mean count rate is expected to be 100 per 10 m sec period then any signal that falls within 30 counts of that value would be accepted, i.e. $100 \pm (3 \times \sqrt{100})$, from 70 to 130 counts. There is 0·99 probability that the signal received from the correct phase would fall within these limits.

TABLE I. Maximum Number of Phases that can be Differentiated on the Basis of one X-ray Signal During a 10 msec Analysing Period (Maximum Count Rate $\simeq$ 15000 sec^{-1})

Phase no.	Approximate composition of phase: Percent of an element	Counts per 10 msec			
		Range	Mean	σ	3σ
1	100	113–193	153	13	40
2	55	58–112	85	9	27
3	25	21–57	39	6	18
4	7	2–20	11	3	9
Background	0	0–1			

The background or noise level on the microanalyser is commonly only 0–1 count per 10 m sec period and, in consequence, it is possible to discriminate up to a maximum of four suitably different phases on the basis of a single X-ray signal during this period (see Table I). When more than one signal are used simultaneously then the phase discrimination capability of the instrument during a 10 m sec counting period is greatly enhanced: e.g. combinations of two such X-ray signals can discriminate up to 16 phases and the chance that a two-component phase would be misidentified is 1 : 100^2, i.e. 1 in 10,000.

By using a "dynamic" scanning procedure (see p. 537) it is possible to extend the counting period on selected phases without extending the total time taken for the examination and the phase discrimination capability of the instrument can then be greatly improved, e.g. Table II.

The X-ray spectrometer positions on the Geoscan are changed by stepper motors that are similar in all respects to the ones that drive the specimen stage. These spectrometer motors control the positions of the original spectrometer command servomotors and can reposition the spectrometers to within a few minutes of arc.

TABLE II. Maximum Number of Phases that can be Differentiated on the Basis of a Single X-ray Signal During a 100 msec Period (Maximum Count Rate ≏ 15000 sec^{-1})

Phase no.	Approximate composition of phase: Percent of an element	Counts per 100 msec			
		Range	Mean	σ	3σ
1	100	1380–1620	1500	40	120
2	83	1145–1355	1250	35	105
3	70	949–1141	1045	32	96
4	53	763–937	850	29	87
5	45	597–753	675	26	78
6	35	451–589	520	23	69
7	25	320–440	380	20	60
8	17	210–310	260	16	50
9	10	113–193	153	13	40
10	6	58–112	85	9	27
11	2·5	21–57	39	6	18
Background	0	0–10	4	2	6

The specimen current signal from the microanalyser is integrated for a fixed time period (usually 10 msec) and the results are digitized by an analogue to digital converter interfaced to the computer. This signal has a very fast response time but its phase discrimination potential has not been widely investigated in mineralogy (see Fig. 15). However, the specimen current signal can be used to delineate rapidly such topographical features as grain boundaries and this signal is used to delineate such boundaries in a "dynamic" scanning program that is being produced by the author.

The microanalyser shows a certain degree of electrical and thermal instability, or drift, with time but much of this drift can be taken into account by the computer program. For example, numerous Faraday cages are fitted at the extremities of the mechanical traverse positions and these cages are periodically used to measure the electron beam current. The results obtained are not used to maintain a steady current but to adjust the test levels that are employed for decision-making on the X-ray and specimen current data. When the measured beam current varies significantly from the original value the unit is automatically shut down.

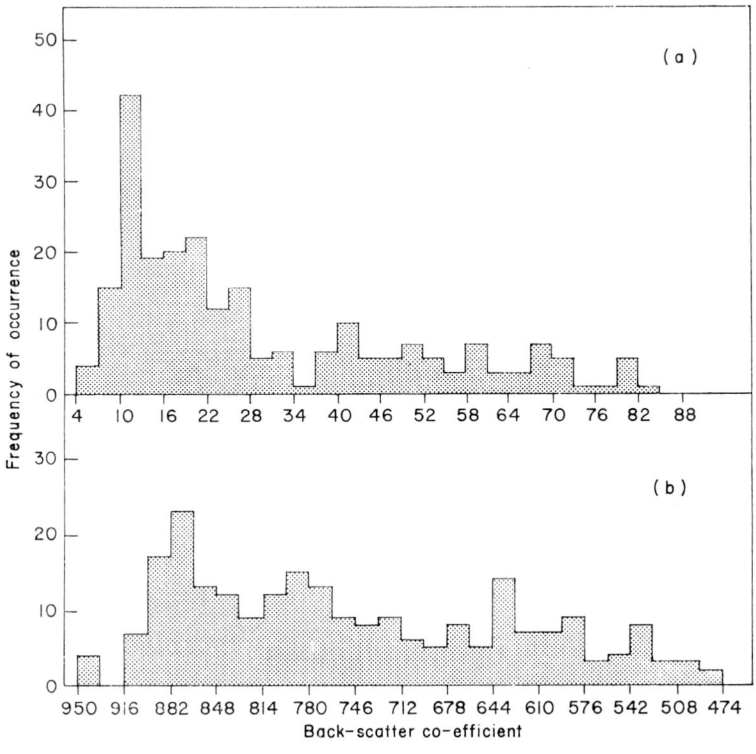

FIG. 15. (From *Micron* **2** (1970)). (a) The mean atomic numbers of the more commonly occurring mineral species; (b) the ratio of I_s/I_b for the more commonly occurring mineral species ($\times 10^3$). (The values for a total of 235 minerals are given. These minerals are those found in "The Identification of Mineral Grains", by M. P. Jones and M. G. Fleming (1965).)

The existing control programs have been written in assembly language because they occupy less computer memory and have shorter execution times than if they had been written in a higher level language. The execution times of the various sections of the existing computer programs conveniently match the time requirements of the X-ray statistics, i.e. about 10 msec. The time requirements are, of course, much longer in the microanalyser than in those instruments that use optical or electronic signals for phase discrimination. The signal that is available in optical or in electron–optical equipment can be used to provide an almost instantaneous areal analysis of a specimen. It is not possible to obtain a sufficient number of X-ray quanta from the microanalyser to provide such an area scanning facility in real-time: instead, the author has used a linear method of analysis whereby traverses 1 m long (with a spatial resolution of about 1 μm) are completed in about 2·5 hours.

If necessary, the computer program can incorporate the degree of accuracy that is required of a point analysis and the computer can then decide on the traversing speed that is required to achieve that accuracy (i.e. dynamic scanning). Scratches and polishing pits on a specimen can make a particular feature appear smaller than it really is (Fig. 16). Such an effect is countered by a software "filter" that delays the categorization of the size of a particle for a variable number of specimen movement steps: if the same phase reappears before the selected number of steps have been taken then the intercept length that is finally obtained includes the polishing artifacts within a single measurement of the phase.

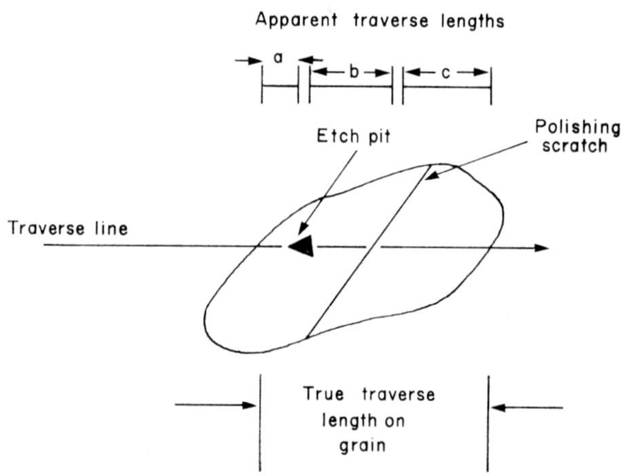

FIG. 16. Effect of pits and polishing scratches on the apparent intercept size measured during a traverse.

At present it is possible to load up to six specimens of 1 in. diameter into the Geoscan at any one time and these can be examined sequentially. In geology and in mineral technology it is often necessary to locate rare phases which occur in volumetric concentrations of only a few parts per million. If a small electronic scan in the "Y" direction is imposed upon the mechanical movement in the "X" direction (see Fig. 10) it is possible to cover a large area of specimen during a traverse. When the X-ray or the specimen current signal intensity exceeds a pre-set value then the specimen movement is momentarily stopped and the presence of the required phase can be checked by a comparatively prolonged count. When the identity of a required grain is confirmed its location is recorded on paper tape or on a teletypewriter and the grain can be re-examined later.

V. Stereological Transformation of Linear Intercept Data

The automated microanalyser permits the rapid and precise measurement of the linear intercepts that are made by moving a point across a plane section of grains. It is, however, a difficult problem to reconstruct the size distribution of those grains in space from the linear data that is produced. Successful reconstructions are possible with spherical particles (Underwood, 1967), but truly spherical particles are only infrequently seen in metallurgical or mineralogical materials. A more generalized method of reconstructing volumetric grain size distributions from linear intercepts has recently been proposed by Barbery (1972) and this method is applicable to linear measurements made on any irregular, convex grains. Transformation equations can, if necessary, be incorporated in the computer programs and the size distributions (by volume) of a specified phase can be produced directly by the computer.

The stereological transformations demand the following data:

N = total number of intercepts made during the traverse
L = intercept length across each feature
ΣL = sum of the intercept lengths across all the features
ΣL^2 = sum of the squares of the intercepts across each feature
ΣL^3 = sum of the cubes of the intercepts
ΣL^4 = sum of the fourth powers of the intercepts
L_T = total length of traverse

From these data the following parameters can be calculated:

$\Sigma L/L_T$ = linear proportion of feature in the specimen $\equiv$ volumetric proportion of feature = V_v
$\Sigma L/N$ = mean particle intercept = 1st moment of the intercept size distribution (Note: all moments are taken from the origin)
$\Sigma L^2/N$ = 2nd moment of intercept distribution (with reference to the origin)

$\Sigma L^3/N$ = 3rd moment of intercept distribution (if the 3rd moment about the mean is zero then the size distribution is symmetrical and is possibly Gaussian)

$\Sigma L^4/N$ = 4th moment of the intercept distribution

If it can be assumed that all the grains have a constant shape then the moments of the grain size distribution (by volume) can be derived from the following equation:

$$u'_n(D) = \frac{K_n(u'_{(n+1)}(L))}{u'_1(L)}$$

where

D = size of maximum possible intercept on a grain (for example: for a cube with a distance between parallel faces of "a" then $D = a\sqrt{3}$)

$u'_n(D)$ = nth moment of grain size (by volume)

K_n = shape factor for the nth moment of volumetric grain size $\left(\text{for spheres } K_n = \frac{n+3}{3}\right)$

$u'_{(n+1)}(L)$ = $(n+1)$ moment of intercept size distribution

$u'_1(L)$ = 1st moment of intercept size distribution.

In this way the first three moments of the grain size distribution (by volume) can be determined and, by assuming either a generalized log normal distribution of the grains or that the grain size distribution follows an incomplete β function, then the relevant volumetric size distribution curve can be constructed.

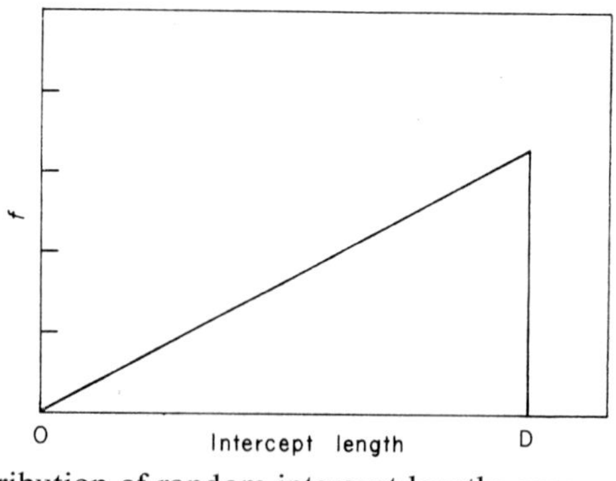

FIG. 17. Size distribution of random intercept lengths across mono-sized spheres dispersed at random in space. D = diameter of spheres.

When the shape factors of irregular, convex grains are not known then the automated Geoscan can itself be used to determine them. For example, if a selection of "mono-sized" grains (produced by careful, close screening) is mounted in a matrix and sectioned, then the plane surface will show grains having a variety of shapes and sizes. If a random traverse is made across such a plane and the intercepts across the grains are measured then it is possible to relate the size distribution of the intercepts to the grain shape and in this way to derive the necessary shape factors. Thus, for example, monosized spheres would produce an intercept size distribution similar to that shown in Fig. 17 and monosized cubes would produce a distribution similar to that shown in Fig. 18. Irregular convex shapes also produce characteristic intercept

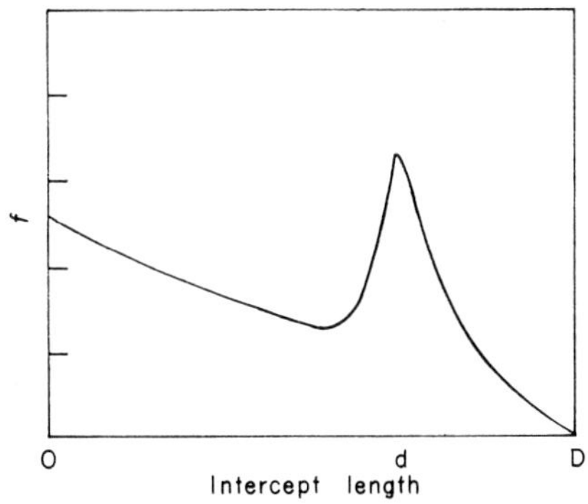

FIG. 18. Size distribution of intercept lengths across mono-sized cubes dispersed at random in space. $d =$ cube face; $D =$ cube diagonal.

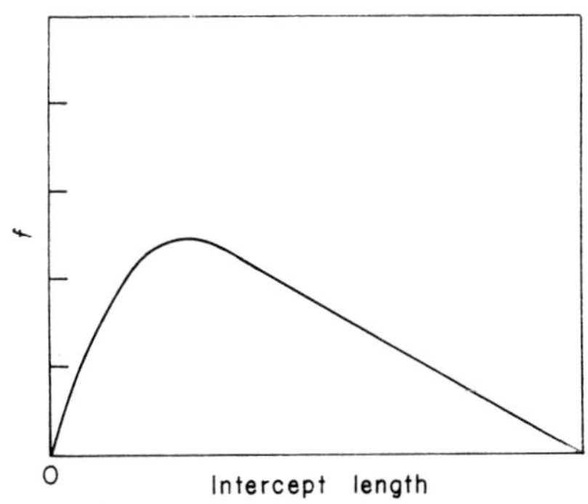

FIG. 19. Size distribution of intercept lengths across a convex irregular grain. The moments of this distribution can be used to calculate the shape factor for that grain.

distributions (Fig. 19) and the moments of these distributions are used to calculate the shape factors for stereological transformations (1972). Thus,

$$K_{cn} = \frac{D^n_{oc}\mu'_n(L)}{\mu'_{n+1}(L)}$$

where

D_{oc} = size of grains as measured by screening or other convenient method

K_{cn} = shape factor relating the nth moment of the intercept distribution to the size distribution (by volume) measured by screening or other convenient method.

From the data that are collected during a traverse it is, therefore, possible to calculate the size or shape distributions of specified phases. In addition, the collected data also allows the specific area of the specified phase, and also its liberation characteristics, to be calculated.

VI. Future Developments

Computer programs are being developed that can discriminate and measure up to six phases simultaneously by using the coincidence or anti-coincidence of two spectrometer signals. For example, it would be possible by using the copper K_α and iron K_α signals to recognize, with confidence, the following mineral phases within a single specimen:

Copper oxides: by the presence of a high copper signal and no iron signal,
Copper silicates: by low copper and no iron,
Iron oxide: by no copper and high iron,
Iron silicate: by no copper and low iron,
Copper–iron sulphide: by low copper and low iron,
Iron sulphide: by no copper and medium iron.

Such groups of minerals frequently occur together in mineralogical materials but, at present, they can only be detected and measured by using sequential traverses across a specimen. It is the author's long-term intention to use energy dispersion of the X-rays to supplement the existing wavelength dispersion system and in this way it may be possible to detect and to measure all the phases that occur in a specimen during a single traverse. The statistical and the timing problems that are involved in the use of the non-dispersive system have been studied and do not appear to be insurmountable.

References

Barbery, G. (1972). *In* "Symposium on Interscience Applications in Mineral Science and Technology, Kosice, 1972". *Mineralia Slovaca*, **5,** 1973, 345–350.

ELIAS, H. (ed.) (1967). *In* "Stereology", *Proc. 2nd Int. Congr. Stereology, Chicago*. Springer-Verlag, New York.

HEINRICH, K. F. J. (1964). "Advances in X-ray Analysis," vol. 7, pp. 384–394. Plenum Press, New York.

JONES, M. P. and GAVRILOVIC, J. (1968). *Trans. Instn Min. Metall.* **77B,** 137–138.

JONES, M. P. and GAVRILOVIC, J. (1970). *Rudy*, **18,** 189–197.

MELFORD, D. A. and WHITTINGTON, K. R. (1966). *In* "Optique des Rayons X et Microanalyse", Hermann, Paris.

UNDERWOOD, E. E. (1967). *In* "Stereology", *Proc. 2nd Int. Congr. Stereology, Chicago* (H. Elias, ed.). Springer-Verlag, New York.

Author Index

Numbers in *italics* indicate pages where references are given in full

J

K

L

T

U

Z

Subject Index

C

F

G

H

I